Feature Papers for Celebrating the Fifth Anniversary of the Founding of *Processes*

Feature Papers for Celebrating the Fifth Anniversary of the Founding of *Processes*

Special Issue Editor

Michael A. Henson

MDPI • Basel • Beijing • Wuhan • Barcelona • Belgrade

Special Issue Editor
Michael A. Henson
University of Massachusetts Amherst
USA

Editorial Office
MDPI
St. Alban-Anlage 66
4052 Basel, Switzerland

This is a reprint of articles from the Special Issue published online in the open access journal *Processes* (ISSN 2227-9717) from 2017 to 2019 (available at: https://www.mdpi.com/journal/processes/special_issues/Feature_Papers)

For citation purposes, cite each article independently as indicated on the article page online and as indicated below:

LastName, A.A.; LastName, B.B.; LastName, C.C. Article Title. *Journal Name* **Year**, *Article Number*, Page Range.

ISBN 978-3-03897-525-0 (Pbk)
ISBN 978-3-03897-526-7 (PDF)

Contents

About the Special Issue Editor

Michael A. Henson is a Professor of Chemical Engineering at the University of Massachusetts in Amherst, MA. His research focuses on the systems-level modeling and analysis of complex biological systems and chemical processes with applications in microbial communities, renewable fuels and chemicals, circadian timekeeping, and downstream pharmaceutical manufacturing. His research has produced 120 refereed journal publications and 70 invited presentations and seminars. Among his accomplishments are the NSF Career Award, the Alexander von Humboldt Fellowship, the UMass College of Engineering Outstanding Senior Faculty Award, and AIChE Fellow. He is the Founding Editor-in-Chief of the open access journal Processes, Co-Editor of *Engineering in Life Sciences*, Associate Editor for *IET Systems Biology*, and Executive Director of Computer Aids for Chemical Engineering (CACHE).

Preface to "Feature Papers for Celebrating the Fifth Anniversary of the Founding of *Processes*"

Processes is an open access journal focused on the experimental characterization and mathematical modeling of complex chemical, biological, and materials systems. To celebrate the fifth anniversary of this successful journal, a Special Issue providing broad coverage of these areas was developed. The Special Issue contains 18 papers from leading researchers in these fields and shows the diversity of applications and methods being explored to advance the design and optimization of these systems. In addition to advancing the current state-of-the-art, this collection of papers demonstrates that the broadly-defined process engineering field remains vibrant and is becoming increasingly diverse. *Processes* will continue to chart the progress of this field using the open access model.

Michael A. Henson
Special Issue Editor

Editorial

Special Issue on Feature Papers for Celebrating the Fifth Anniversary of the Founding of *Processes*

Michael A. Henson

Department of Chemical Engineering and the Institute for Applied Life Sciences, University of Massachusetts Amherst, N527 Life Sciences Laboratories, 240 Thatcher Way, Amherst, MA 01003, USA; mhenson@umass.edu; Tel.: +01-413-545-3481

Received: 5 November 2018; Accepted: 5 November 2018; Published: 1 January 2019

The Special Issue "Feature Papers for Celebrating the Fifth Anniversary of the Founding of *Processes*" represents a landmark for this open access journal covering chemical, biological, materials, pharmaceutical, and environmental systems as well as general computational methods for process and systems engineering. The Special Issue is available online at: https://www.mdpi.com/journal/processes/special_issues/Feature_Papers.

Chemical Processes

A major focus of *Processes* is chemical process engineering with applications to both traditional industries and emerging industries for renewable chemicals and energy production. The Special Issue contains four papers in this area. The first paper describes the development and simulation of a model prediction controller for achieving the desired viscosity curve in the continuous production of a non-Newtonian fluid [1]. The second paper addresses the problem of minimizing fuel consumption in fuel gas networks through a superstructure-based method and mixed-integer nonlinear programming [2]. The interesting possibility of extending the favorable characteristics of cuboid packed-bed devices originally designed for chromatographic separations to packed-bed chemical reactors is explored in the third paper [3]. The fourth paper addresses the optimal design of ammonia production plants based on adsorption for unreacted nitrogen and hydrogen recovery and wind energy for electricity generation [4].

Biological Systems

The development of system models and process technology for biological applications is a major focus area of *Processes*. The Special Issue contains five papers in this area. The first paper shows the development of a systems model for investigating patient outcomes from liver surgery and determining the processes responsible for liver failure [5]. The extension of the cybernetic modeling approach to lipid metabolism in mouse bone marrow-derived macrophage cells is the focus of the second paper [6]. The third paper addresses the metabolic modeling of cell-free protein expression through adaptation of a genome-scale metabolic model and direct incorporation of metabolite measurements [7]. The engineering of cuboid packed-bed devices is revisited in the fourth paper by investigating the effect of the length-to-width aspect ratio on chromatographic separation efficiency [8]. The fifth paper reports on the computational development of a novel method for constraint-based metabolic modeling of intracellular glycosylation reactions [9].

Materials Processes

Processes covers a wide range of materials related topics including fundamental modeling, process technology, and applications. This diversity is reflected in the Special Issue through six contributions. The first paper provides a literature-based comparison of rotor-stator mixing devices operated in batch and continuous modes [10]. Two contributions focus on challenges in pharmaceutical manufacturing. The first paper describes the development of a database of multiphase reactions for synthesizing small-molecule pharmaceutical to facilitate investigation of reaction-separation schemes [11], while the second paper provides a complementary study on developing an ontological information infrastructure for integrating data from pharmaceutical manufacturing plants and analytical laboratories [12]. The Special Issue also contains two contributions focusing on polymerization processes. One paper investigates the experimental validation of a kinetic model for thermal spontaneous polymerization of n-butyl acrylate) [13], while the other paper presents a MATLAB-based tool for copolymerization reactivity ratio estimation based on error-in-variables models [14]. The sixth paper in this area addresses the problem of materials processing through the development of a classification method for identifying a promising experimental design region from a literature database [15].

Computational Methods

The development of novel computational methods with applications to systems modeling and engineering is an important area of *Processes*. The Special Issue has two papers in this area. The first paper addresses the problem of model-based experimental design based on global parameter sensitivity measures and demonstrates the application of the methods through a case study on the synthesis of a precursor for protein kinase inhibitors [16]. The second contribution describes the development of a maximum entropy-based simulation method for analysis of metabolic pathways with application to the central metabolism of the mold *Neurospora crassa* [17].

Other Topics

Processes contains an "Other Topics" section to compile papers that do not fit well in the other four sections. The Special Issue contains two papers in this section, both focused on water systems. The first contribution presents a combined feedforward/feedback model predictive control strategy for minimizing disturbance effects in military water networks [18]. The second contribution investigates the problem of optimal capacity planning in seawater desalination systems from a multiscale perspective [19].

The Future of *Processes*

The Special Issue covers a broad range of topics consistent with the mission of *Processes* to become a highly visible outlet for publishing novel studies in systems modeling, process engineering, and associated applications. The journal will continue to solicit high-quality contributions in these domains.

Michael A. Henson
Founding Editor-in-Chief of Processes.

Funding: There is no funding supports.

Conflicts of Interest: The author declares no conflicts of interest.

References

1. Mei, R.; Grosso, M.; Corominas, F.; Baratti, R.; Tronci, S. Multivariable Real-Time Control of Viscosity Curve for a Continuous Production Process of a Non-Newtonian Fluid. *Processes* **2018**, *6*, 12. [CrossRef]
2. Li, J.; Demirel, S.; Hasan, M. Fuel Gas Network Synthesis Using Block Superstructure. *Processes* **2018**, *6*, 23. [CrossRef]
3. Ghosh, R. Cuboid Packed-Beds as Chemical Reactors? *Processes* **2018**, *6*, 44. [CrossRef]
4. Palys, M.; McCormick, A.; Cussler, E.; Daoutidis, P. Modeling and Optimal Design of Absorbent Enhanced Ammonia Synthesis. *Processes* **2018**, *6*, 91. [CrossRef]
5. Verma, B.; Subramaniam, P.; Vadigepalli, R. Modeling the Dynamics of Human Liver Failure Post Liver Resection. *Processes* **2018**, *6*, 115. [CrossRef]

6. Aboulmouna, L.; Gupta, S.; Maurya, M.; DeVilbiss, F.; Subramaniam, S.; Ramkrishna, D. A Cybernetic Approach to Modeling Lipid Metabolism in Mammalian Cells. *Processes* **2018**, *6*, 126. [CrossRef]
7. Dai, D.; Horvath, N.; Varner, J. Dynamic Sequence Specific Constraint-Based Modeling of Cell-Free Protein Synthesis. *Processes* **2018**, *6*, 132. [CrossRef]
8. Chen, G.; Ghosh, R. Effect of the Length-to-Width Aspect Ratio of a Cuboid Packed-Bed Device on Efficiency of Chromatographic Separation. *Processes* **2018**, *6*, 160. [CrossRef]
9. Hutter, S.; Wolf, M.; Papili Gao, N.; Lepori, D.; Schweigler, T.; Morbidelli, M.; Gunawan, R. Glycosylation Flux Analysis of Immunoglobulin G in Chinese Hamster Ovary Perfusion Cell Culture. *Processes* **2018**, *6*, 176. [CrossRef]
10. Håkansson, A. Rotor-Stator Mixers: From Batch to Continuous Mode of Operation—A Review. *Processes* **2018**, *6*, 32. [CrossRef]
11. Papadakis, E.; Anantpinijwatna, A.; Woodley, J.; Gani, R. A Reaction Database for Small Molecule Pharmaceutical Processes Integrated with Process Information. *Processes* **2017**, *5*, 58. [CrossRef]
12. Cao, H.; Mushnoori, S.; Higgins, B.; Kollipara, C.; Fermier, A.; Hausner, D.; Jha, S.; Singh, R.; Ierapetritou, M.; Ramachandran, R. A Systematic Framework for Data Management and Integration in a Continuous Pharmaceutical Manufacturing Processing Line. *Processes* **2018**, *6*, 53. [CrossRef]
13. Riazi, H.; Shamsabadi, A.; Corcoran, P.; Grady, M.; Rappe, A.; Soroush, M. On the Thermal Self-Initiation Reaction of n-Butyl Acrylate in Free-Radical Polymerization. *Processes* **2018**, *6*, 3. [CrossRef]
14. Scott, A.; Penlidis, A. Computational Package for Copolymerization Reactivity Ratio Estimation: Improved Access to the Error-in-Variables-Model. *Processes* **2018**, *6*, 8. [CrossRef]
15. McBride, M.; Persson, N.; Reichmanis, E.; Grover, M. Solving Materials' Small Data Problem with Dynamic Experimental Databases. *Processes* **2018**, *6*, 79. [CrossRef]
16. Schenkendorf, R.; Xie, X.; Rehbein, M.; Scholl, S.; Krewer, U. The Impact of Global Sensitivities and Design Measures in Model-Based Optimal Experimental Design. *Processes* **2018**, *6*, 27. [CrossRef]
17. Cannon, W.; Zucker, J.; Baxter, D.; Kumar, N.; Baker, S.; Hurley, J.; Dunlap, J. Prediction of Metabolite Concentrations, Rate Constants and Post-Translational Regulation Using Maximum Entropy-Based Simulations with Application to Central Metabolism of *Neurospora crassa*. *Processes* **2018**, *6*, 63. [CrossRef]
18. James, C.; Webber, M.; Edgar, T. Minimizing the Effect of Substantial Perturbations in Military Water Systems for Increased Resilience and Efficiency. *Processes* **2017**, *5*, 60. [CrossRef]
19. Baaqeel, H.; El-Halwagi, M.M. Optimal Multiscale Capacity Planning in Seawater Desalination Systems. *Processes* **2018**, *6*, 68. [CrossRef]

Article

Multivariable Real-Time Control of Viscosity Curve for a Continuous Production Process of a Non-Newtonian Fluid

Roberto Mei [1], Massimiliano Grosso [1], Francesc Corominas [2], Roberto Baratti [1] and Stefania Tronci [1,*]

[1] Dipartimento di Ingegneria Meccanica, Chimica e dei Materiali, Università degli Studi di Cagliari, Via Marengo 3, 09123 Cagliari, Italy; r.mei@dimcm.unica.it (R.M.); massimiliano.grosso@dimcm.unica.it (M.G.); roberto.baratti@dimcm.unica.it (R.B.)

[2] Procter & Gamble Eurocor N.V., Temselaan 100, 1853 Strombeek-Bever, Belgium; corominas.f@pg.com

* Correspondence: stefania.tronci@dimcm.unica.it; Tel.: +39-070-675-5050

Received: 22 December 2017; Accepted: 27 January 2018; Published: 30 January 2018

Abstract: The application of a multivariable predictive controller to the mixing process for the production of a non-Newtonian fluid is discussed in this work. A data-driven model has been developed to describe the dynamic behaviour of the rheological properties of the fluid as a function of the operating conditions using experimental data collected in a pilot plant. The developed model provides a realistic process representation and it is used to test and verify the multivariable controller, which has been designed to maintain viscosity curves of the non-Newtonian fluid within a given region of the viscosity-vs-shear rate plane in presence of process disturbances occurring in the mixing process.

Keywords: non-Newtonian fluid; multivariable control system; viscosity curve

1. Introduction

The industrial continuous production of complex fluids still encounters to date issues in terms of quality control of products. Parameters like shear rate dependent viscosity, which affects product quality, could easily go out of specifications if the ingredient characteristics or other variables of the process move away from the original formulation. The main issue is represented by the difficulty of controlling such processes because there aren't commercial sensors capable of providing real-time information about the rheological properties of the fluid. Available in-line sensors can measure the viscosity only for a punctual shear rate value, therefore their use for controlling the production process of non-Newtonian fluid is not possible. Furthermore, very few in-line rheometers have been tested and few studies have been presented for non-Newtonian and opaque industrial fluids [1].

Because of the importance of the continuous monitoring of rheological parameters of industrial fluids during production, in the last years many steps have been accomplished for designing and developing sensors capable of providing rheological proprieties of complex fluids in real-time. Many recent works focus their efforts on non-invasive techniques which seem the most promising for this type of problems. Kotzé et al. [2] studied a technique based on ultrasonic velocity profiling (UVP) which measures an instantaneous one-dimensional velocity profile in a fluid containing particles across the ultrasonic beam axis or measurement line. The method combines the UVP technique with pressure difference (PD) measurements, it is non-invasive, it can be used to measure opaque and concentrated suspensions. Preliminary results obtained in concentrated cement pastes showed that UVP is a feasible and promising technique for flow characterisation in viscous fluids. Meacci et al. [1] presented an efficient, fully programmable and integrated system for in-line fluids characterisation of a wide range of non-Newtonian and opaque fluids. This system, named Flow-Viz exploits ultrasounds to detect the velocity profile of the flow moving in the pipe, and it is designed for industrial use. Recently, Yoshida et al. [3] proposed a sensor based on ultrasonic spinning rheometry (USR), which is expected to provide details of various rheological properties. The proposed USR capabilities were assessed for three test fluids chosen as examples of thixotropic fluids, shear-thinning fluids, and multiphase fluids.

The recent results on innovative sensors encourage the efforts for developing control strategies which provide the target rheological characteristics in the production of non-Newtonian fluids. The present study is aimed to design a controller for the production of a detergent obtained by mixing different ingredients, which should lead to specific rheological and physical properties. The product is characterised by a complex rheological behavior and, in a previous work [4] the authors developed a one-point control of viscosity curve for the continuous production process, showing how the choice of the point on the viscosity-vs-shear rate curve was crucial for the quality of the product. Results also evidenced that it is possible to partly reject disturbances using only one point as controlled variable, but this configuration was not adequate for the severe specifics required in the plant under investigation. In this situation, a model predictive control (MPC) can improve the performance of the system. MPC can handle non-square system, therefore the two available manipulated inputs can be used to control more than two points on the viscosity-vs-shear rate curve in an efficient way. It is worth noting that the plant under consideration is characterised by high time delay, if compared to the characteristic time of the mixing process, and again the MPC algorithm is effective in this case [5]. The multivariable control performances have been assessed by simulating the process at different operating conditions, including time delay and measurement noise.

2. Process Description

The production of a detergent on pilot plant scale (facilities made available by the Brussels Innovation Centre in Belgium, BIC) is considered in this work. The product is a compound which contains several ingredients with different rheological properties, obtaining a final mixture with a highly complex rheological behavior. It is important to underline that blends rheology is affected by factors such rheology of single components but also temperature and polydispersity [6,7]. Details on the ingredients cannot be reported for confidentiality reasons.

The pilot plant (Figure 1) is designed as a main pipe connected to a series of tanks, each of them containing one ingredient, through secondary pipes. Ingredients are pumped in the main line and a series of static mixer between inputs ensure a good blending of the mixture. At the end of the line the product is collected in a tank for off-line rheological measurements (for further details see [8]).

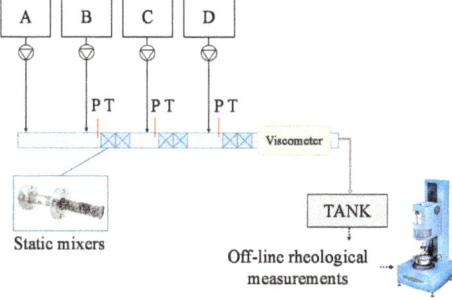

Figure 1. Scheme of the pilot plant at Brussels Innovation Centre (BIC) (Patent N. US 2013/0225468 A1).

An in-line Endress-Hauser Proline Promass 83I Coriolis flowmeter, located in the main pipe after the 4 inputs of the ingredients, is used to gain information about system dynamics, along with pressure and temperature sensors. Such a flowmeter can provide only a point measurement of viscosity.

Such measure is not exhaustive when dealing with non-Newtonian fluids but can provide a rough estimation about dynamics.

Off-line rheological measurements were carried out on the samples collected in the final tank with a rheometer AR-2000 TA stress controlled equipped with a 40 mm cone and plate fixture, used to obtain the viscosity dependence on shear rate.

3. Problem Statement

The quality of the product under study is determined through its rheological properties, which can be summarised by the viscosity-vs-shear rate curve. The rheological properties may deviate from the nominal ones, because of process disturbances, and they can be adjusted by varying the ingredients flow rates which have a major impact without affecting too much detergent characteristics.

As representative example, Figure 2 reports a typical viscosity curve for the detergent produced in the plant (schematised in Figure 1), with a target reported with the green continuous line and the region (grey shaded area) of accepted viscosity values limited by the upper and lower curves. Such region was obtained by considering a maximum error of 10% at low shear rate (lower than 10 s^{-1}), while 20% has been set at high shear rate. The red dashed lines are two examples of undesired output values, when the process is out-of-control. It is worth noting that the target of the control problem is not a point value but infinite points lying on a one-dimensional manifold (the viscosity curve) and the entire curve should stay within the grey region of allowable viscosity. The main issue is to individuate the points on the curve that guarantee the respect of the limits at every condition in the plant. Based on the previous investigation of the authors [4], the viscosities (η) of the studied compound at shear-rate ($\dot{\gamma}$) values equal to 0.1, 1, 10 and 1100 s^{-1} have been selected. Such points are the most representative for the rheological behavior of the considered system.

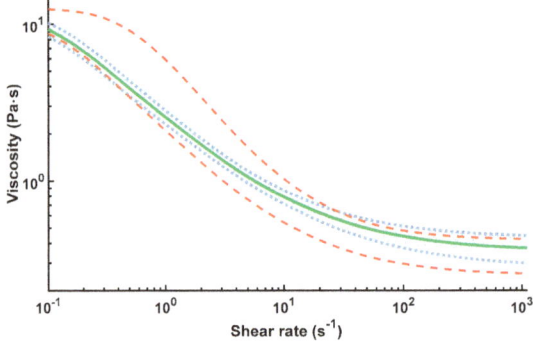

Figure 2. Rheological curves for the system under study. The shaded region indicates the viscosity values for in-control product. The red dotted lines are examples of off-control products.

4. Process Simulator

A process simulator has been developed for designing and evaluating a rheological controller for the continuous production of a detergent. The role of this simulator is to give time responses for the viscosities of the produced compound taking as inputs the ingredient amounts. Because the rheological behavior of the system is very complex, a first principle description of the process is not possible, and a data-driven model has been preferred. The experimental apparatus used in this work cannot give on-line information about the quality of the ingredients which are fed in the plant, meaning that it is only possible to describe the input-output relationship between ingredient flow rates and product's rheological properties. For model development, several step tests were carried out at different inlet flow rates and plant setup (distance between the last mixer and the sample collection), and the products were analysed analyzed off-line with the rheometer described in the previous section. The rheological information obtained with such tests can be related to steady state conditions. The transient response was on the other hand observed using the data collected with the in-line viscometer (Promass).

A simple description of the data collected in the pilot plant can be attained by means of a continuous-time Hammerstein model [9], where the nonlinear no-memory gain is calculated with a Neural Network (NN) model [10,11]. The time invariant state space representation of a nonlinear continuous-time Hammerstein system (Figure 3) that cascades a static nonlinearity followed by a linear dynamic system may be described by Equation (1)

$$\dot{x}(t) = Ax(t) + f_{NN}(u), \qquad (1)$$

where x is the n-dimensional state vector, u is the m-dimensional input vector, A is a constant matrix and f_{NN} indicates the memoryless NN model.

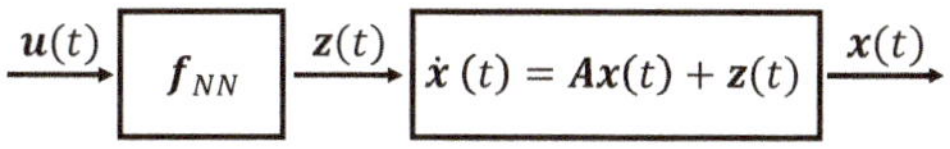

$$\xrightarrow{\;u(t)\;} \boxed{f_{NN}} \xrightarrow{\;z(t)\;} \boxed{\dot{x}\,(t) = Ax(t) + z(t)} \xrightarrow{\;x(t)\;}$$

Figure 3. Nonlinear continuous time system of the Hammerstein Model.

The neural network between the inputs u and the variable z on Figure 3 has been designed and tested to relate viscosities measured off-line at different shear rate values with the inputs of the pilot plant, in terms of ingredient mass fraction in the feed. Mass fractions were chosen instead of mass flows to avoid problems concerning the use of different order of mass flows magnitude. The outputs (z) are the off-line viscosities (η) of the studied compound at shear-rate ($\dot{\gamma}$) values equal to 0.1, 1, 10 and 1100 s^{-1}, as reported in the previous section. The NN structure is represented in Figure 4, where it is shown that the input vector consists of four units, the output layer has four neurons and the hidden layer has six neurons. For both the hidden and output layers a sigmoidal activation function is used. The experimental data used for model calibration were obtained varying the percentage of the ingredient flow rates within the interval reported in Table 1. In more details, 27 different operating conditions were used in the pilot plant and with the repeated experiments allowed the collection of 171 viscosity curves. The model parameter estimation was carried out using 70% of points for the training step, 15% for the validation and 15% for the test.

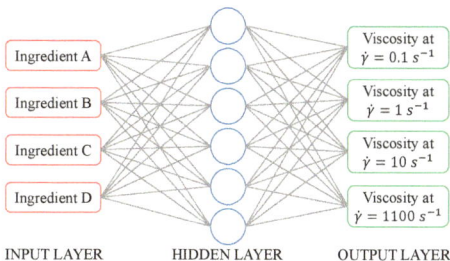

Figure 4. Structure of the neural network used for the function f_{NN} in the continuous Hammerstein model.

Table 1. Range of the ingredients' mass fraction.

	A	B	C	D
MIN	0.730	0.072	0.000	0.012
MAX	0.880	0.250	0.039	0.066

Model's performance is evaluated by considering the coefficient R^2 and the mean absolute deviation (MAD) defined in Equation (2), where y_i is the measured viscosity, f_i is the calculated viscosity, is the mean viscosity value and N is the number of experimental points.

$$R^2 = 1 - \frac{\sum_{i=1}^{N}(y_i - f_i)^2}{\sum_{i=1}^{N}(y_i - \bar{y})^2}, \ MAD = \frac{1}{N}\sum_{i=1}^{N}\left|\frac{y_i - f_i}{y_i}\right| \qquad (2)$$

The developed NN is capable to predict viscosities of the compound with good results, as reported in Table 2, where the performance indexes are shown for the four output variables. Measured viscosities against predicted NN viscosities are reported, by way of example, in Figure 5 for shear-rate equal to $0.1\ s^{-1}$. The test dataset is evidenced in the figure to better evaluate the performance of the model, because in this case the neural model uses data which have not been used during the training/validation step. The obtained results are in the main quite good, but at = $1100\ s^{-1}$, R^2 is relatively low (about 0.93) showing some predictability limits of the model. It should be considered that the viscosity variations at high shear-rate value are however rather small (from 0.1 to 0.5 Pa·s). Standard residuals (Figure 6) appear without a deterministic structure, indicating that the model captures the essential features of the data and they validate the reliability of the NN to predict the steady state values of off-line viscosities for different amounts of ingredients, at least for the investigated conditions.

The second block in the Hammerstein model (Figure 3) describes the dynamic behaviour and it has been obtained using the in-line measurements. Because only a point viscosity is available with Promass, a unique characteristic time is adopted for the four output variables, in the understanding that the rheological changes in the mixture happens at the same moment for each shear rate. By the same token, the time delay is assumed equal for each calculated output. A comparison between the measured viscosity and the one obtained using a first-order plus time delay system is reported in Figure 7, with the purpose of showing the dynamics assessment. It is worth noticing that it is not possible to compare the viscosity predicted by the NN, calculated offline, and the one of the flowing detergents in the plant, because it is not possible to relate the in-line viscosity given by Promass with the off-line values. This means that a comparison between the model and plant behaviour is possible only when the in-line sensor is available. In this work, the impact of process conditions on the characteristic time is neglected and it is set equal to 20 s. Time delay has been calculated considering one configuration of the plant and it is set equal to 8 s.

Table 2. Performance of the neural network.

$\dot{\gamma}\ (s^{-1})$	R^2	MAD (%)
0.1	0.9757	8.6
1	0.9666	7.0
10	0.9808	4.0
1100	0.9315	3.7

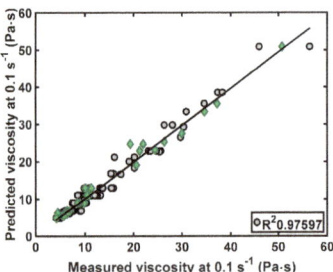

Figure 5. Predicted viscosities against measured viscosities at $0.1\ s^{-1}$. Training and validation: grey circle; test: green diamond.

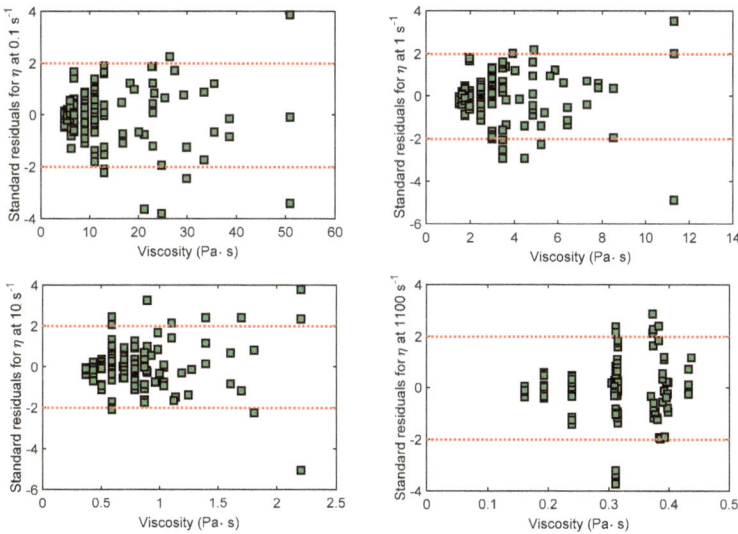

Figure 6. Standardised residuals with respect to predicted viscosities at: $0.1\ \mathrm{s}^{-1}$ (left upper panel), $1\ \mathrm{s}^{-1}$ (right upper panel), $10\ \mathrm{s}^{-1}$ (left lower panel) and $1100\ \mathrm{s}^{-1}$ (right lower panel).

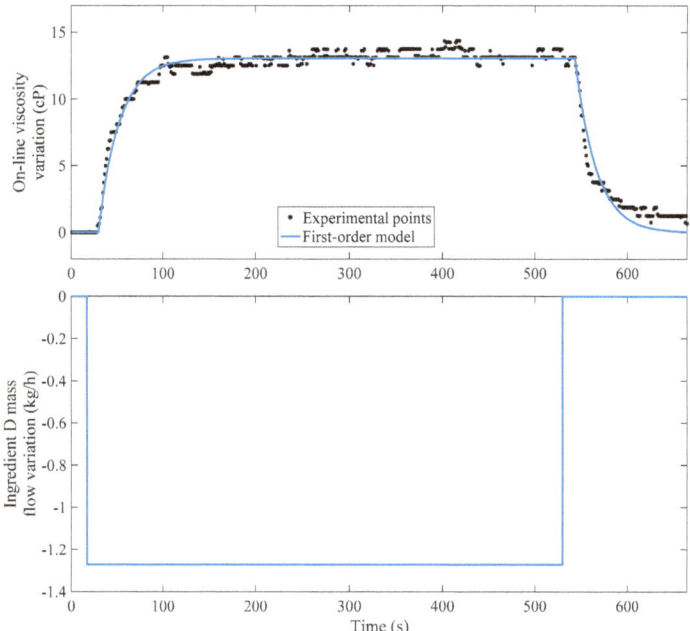

Figure 7. Viscosity measured on line (black dots) and values calculated (blue continuous line) with a first-order plus time delay system (upper panel) due to the step change of the ingredient D flow rate (lower panel).

5. Control Algorithm

The control problem is addressed using a model predictive control (MPC) in its basic formulation of dynamic matrix control (DMC), where the controlled outputs are selected among the ones given by the simulator. In more details, because the small variations at $\dot{\gamma} = 1100$ s^{-1} can be challenging for the robustness of the controller [12] only the viscosities at $\dot{\gamma} = 0.1$, 1 and 10 s^{-1} have been considered as controlled variable, while the ingredient B and D flow rates are the manipulated inputs. The evaluation of the control configuration has been carried out using the simulator reported in Section 4 to act as a virtual plant and a schematic representation of the control loop is reported in Figure 8.

The objective function J used to calculate the manipulated action in the DMC is reported in Equation (3)

$$J = [(e(k+1) - F\Delta u)^T W(e(k+1) - F\Delta u)] + [\Delta u]^T K[\Delta u] \tag{3}$$

where k denotes the time index, $e(k + 1)$ is the 3xH_p dimensional vector representing the difference between the desired outputs and the current output prediction without further control action, F is the dynamic matrix [5], H_p is the prediction horizon, u is 2xH_u dimensional vector of the future control moves, H_u is the control horizon, K is a block diagonal matrix used to penalise changes of manipulated inputs, W is a weighting matrix used as a tuning parameter. The prediction error is corrected by the measured outputs available at the sampling instant. More in details, W is a positive (3xH_p)x(3xH_p) block diagonal matrix which in turn is composed of three diagonal matrixes, one for each output (Equation (4)).

$$W = \begin{bmatrix} diag(w_1) & 0 & 0 \\ 0 & diag(w_2) & 0 \\ 0 & 0 & diag(w_3) \end{bmatrix} \tag{4}$$

The matrices $diag(w_i)$, with $i = 1, 2, 3$, have dimensions H_pxH_p and the element on the diagonal is the positive weight for a specific output. The matrix K is a positive (2xH_u)x(2xH_u) block diagonal matrix which in turn is composed of two diagonal matrixes, one for each input.

$$K = \begin{bmatrix} diag(k_1) & 0 \\ 0 & diag(k_2) \end{bmatrix} \tag{5}$$

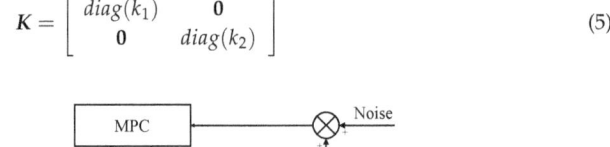

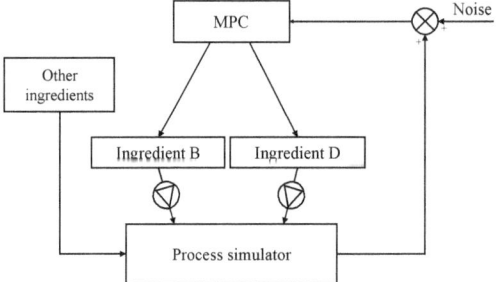

Figure 8. Schematic representation of the control loop for the model predictive control (MPC).

The matrices $diag(k_i)$, with $i = 1, 2$, have dimensions H_uxH_u and the element on the diagonal is positive and it penalises a specific input. The process simulator (Equation (1)) provides the four points of the rheological curve corresponding to the actual situation of the process. Random noise is then added to the outputs in order to simulate measurement noise.

A linear predictive model is required to implement the DMC in its traditional form. This task is addressed using the simulator to carry on step response tests from which the dynamic matrix F is obtained [5]. The simulator model is excited by varying the input corresponding to the manipulated variables, starting from the reference condition and considering step changes of different sizes [13,14]. It is worth noting that the simulator does not model the variations of ingredients quality (composition and rheological behavior). The coefficients of the dynamic matrix F are obtained by averaging the responses of the different step changes.

6. Results

Various tests have been carried out with the intention of assessing the performance of the control strategy. The main purpose of the simulations is to understand how the quality of the on-line measurements (delay and noise) can affect the rheological control behavior, in view of a future implementation of continuous monitoring of rheological parameters in industrial plants. Ultrasound sensors are indeed candidate to monitor and control the production of materials with complex characteristic behavior (e.g., [1]), but information on their in-line response such as time required for calculation of the viscosity curve, measurement noise, reliability, repeatability is not available yet for the fluid under investigation. In particular, it could be useful to understand what happens if the time interval for the ultrasound sensor to collect measurements and calculate the velocity profile is much greater than the characteristic time of the process.

6.1. MPC for Set-Point Tracking

The parameters related to the MPC development, such as prediction and control horizon, sampling time and weights, are found by analysing the dynamic response of the process and by tuning. Considering a measurement delay equal to 30 s, the following parameters have been selected in order to achieve an acceptable dynamic matrix conditioning while maintaining good controller performances: (i) the control action is applied every 10 s, and (ii) the dimension of the prediction horizon H_p is set equal to 16. The control horizon H_u is set equal to 4 for the entire control configuration, as suggested by [5], and only the first control move is applied at each sampling time. Table 3 summarises the process conditions and MPC parameters for a reference case.

Table 3. Specifications for the reference case.

Parameter	Value
Time interval for control action	10 s
Measurement delay	30 s
Time constants for simulator (for each controlled viscosity)	20 s, 20 s, 20 s
Time delays for simulator (for each controlled viscosity)	8 s, 8 s, 8 s
Noise in the measurements about $\pm 2\%$: -0.2 Pa·s < random noise for $\eta_{0.1s^{-1}}$ < $+0.2$ Pa·s -0.05 Pa·s < random noise for $\eta_{1s^{-1}}$ < $+0.05$ Pa·s -0.02 Pa·s < random noise for $\eta_{10s^{-1}}$ < $+0.02$ Pa·s	
Weights of the controller (for each controlled viscosity)	$w_1 = 1$, $w_2 = 12$, $w_3 = 40$
Parameters for the penalisation of the controller (Ingredients D and B respectively)	$k_1 = 3.5 \times 10^4$ $k_2 = 3.0 \times 10^2$

Figure 9 shows the system responses to step variation of the set points. The controller works well and it is capable to bring the three controlled viscosities (viscosity at shear-rates equal to 0.1, 1 and $10\ \mathrm{s}^{-1}$) to the desired values in a relatively low time. Actions on the two manipulated variables have been penalised to avoid excessive overshoots.

The effects of a greater measurement delay on the controlled process is represented in Figure 10. In this case measurements are available every 60 s, and H_p was set equal to 19. The controller is capable to bring the viscosities to the desired values but the response is slower than the reference case.

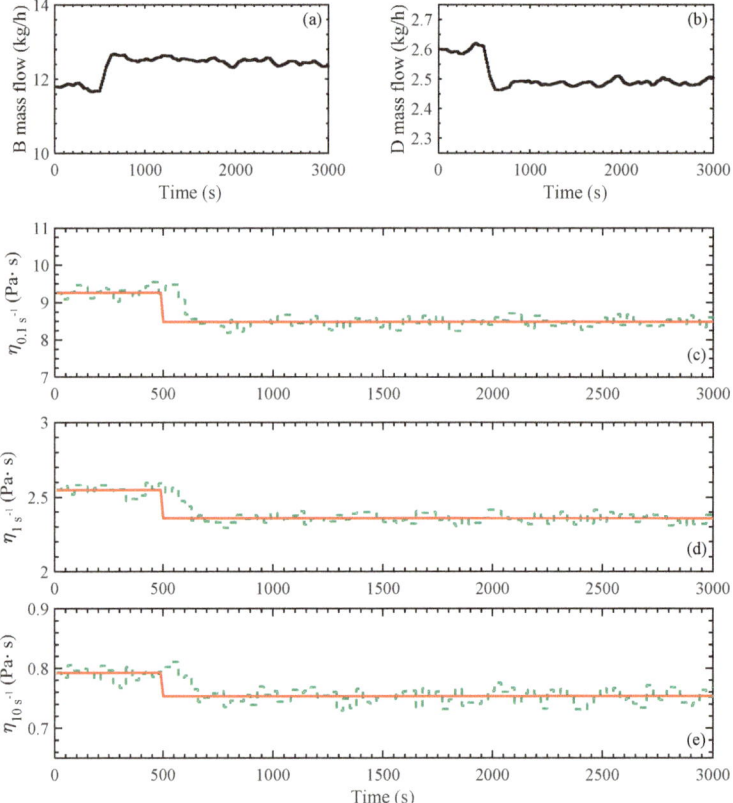

Figure 9. Performance of controller for the reference case: (**a**) mass flow of ingredient B; (**b**) mass flow of ingredient D; (**c**) response of the viscosity at shear rate equal to 0.1 s^{-1}; (**d**) response of the viscosity at shear rate equal to 1 s^{-1}; (**e**) response of the viscosity at shear rate equal to 10 s^{-1}. For (**c–e**) the following data are reported: target curve (red solid line), measured value (green dashed line).

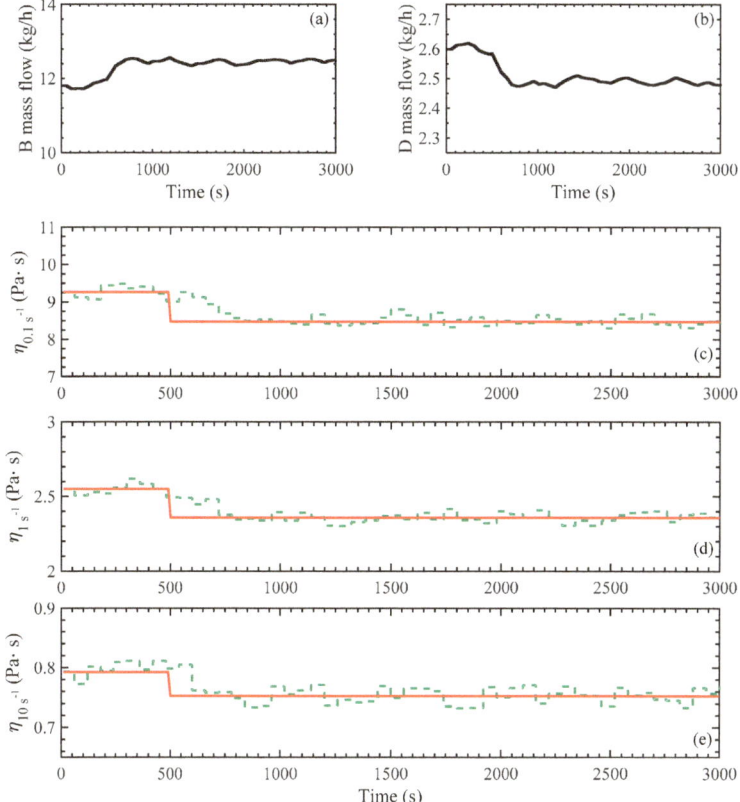

Figure 10. Performance of controller for the case with time sampling equal to 60 s: (**a**) mass flow of ingredient B; (**b**) mass flow of ingredient D; (**c**) response of the viscosity at shear rate equal to 0.1 s^{-1}; (**d**) response of the viscosity at shear rate equal to 1 s^{-1}; (**e**) response of the viscosity at shear rate equal to 10 s^{-1}. For (**c–e**) the following data are reported: target curve (red solid line), measured value (green dashed line).

6.2. MPC for Disturbance Rejection

To study the performance of the MPC controller when different batches of ingredients are used, another test has been designed and developed. More in details, a different type of ingredient B has been used in the process.

Another neural network (hereafter indicated as NN_d) with the same structure shown in Figure 3 has been trained using data coming from tests where another type of ingredient B (type 2) has been used. The second linear block of the Hammerstein model, which is only related to the dynamic behavior, is the same of the previous case. The new model NN_d has been used to simulate a disturbance entering the process, as explained in the following. Time intervals for process simulation, control action and sampling are the same reported in Table 2. The magnitude of noise in the measurements is respectively ± 0.2 Pa·s for $\eta_{0.1 \text{ s}^{-1}}$, ± 0.05 Pa·s for $\eta_{1 \text{ s}^{-1}}$ and ± 0.02 Pa·s for $\eta_{10 \text{ s}^{-1}}$.

To simulate the disturbance, at a certain point in the timeline of the simulation (when relative time is equal to 500 s), the neural network used for the process simulator (NN) has been replaced with the other one (NN_d). It's important to highlight that only the simulator is affected by this change, while the dynamic matrix **F** is the same. This action has the goal to simulate what happens to the real process when a different batch of ingredient B, which has a different rheological behavior with respect to the previous one, is suddenly used (maybe to replace a finished batch).

The response of the system to this change of ingredient is reported in Figure 11. As shown in the figure, the controller is capable to maintain the controlled viscosities to target. The dotted orange line represents the response of the process to the change of ingredient B in an open loop configuration, while the green dotted line represents the controlled process. In this case, also the behavior of the viscosity at high shear rate is reported in the figure, in order to evaluate if controllability for the four states is satisfied. Indeed, the viscosity at $1100\,\mathrm{s}^{-1}$ is not controlled, nonetheless the difference between target and actual value is reduced indicating a correct behavior of the proposed controller.

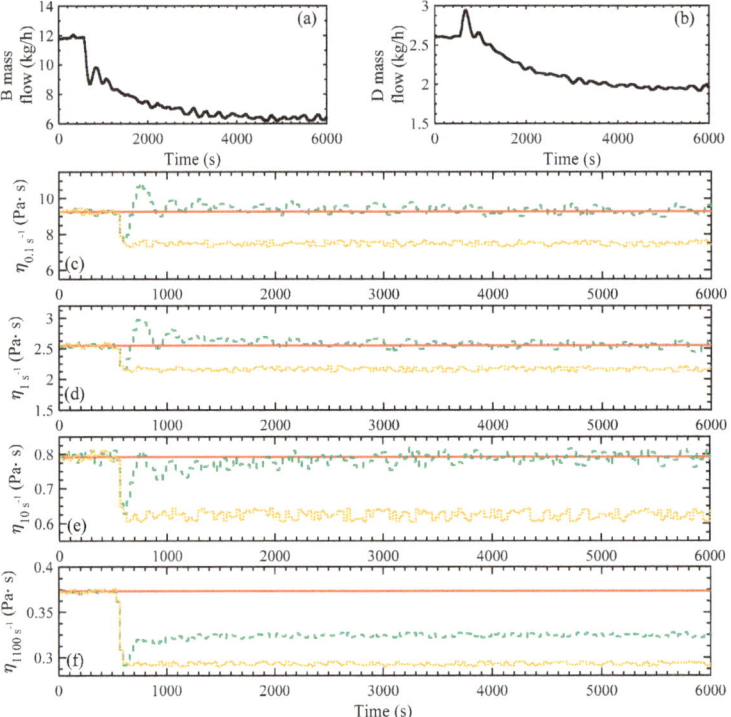

Figure 11. Performance of the controller responding to a disturbance (change of ingredient B) and comparison with the open loop response: (**a**) mass flow of ingredient B; (**b**) mass flow of ingredient D; (**c**) response of the viscosity at shear rate equal to $0.1\,\mathrm{s}^{-1}$; (**d**) response of the viscosity at shear rate equal to $1\,\mathrm{s}^{-1}$; (**e**) response of the viscosity at shear rate equal to $10\,\mathrm{s}^{-1}$; (**f**) response of the viscosity at shear rate equal to $1100\,\mathrm{s}^{-1}$. For (**c–f**) the following data are reported: target curve (red solid line), measured value (green dashed line), open loop evolution of the process (orange dotted line).

7. Conclusions

The present paper was focused on the control of the rheological properties of a detergent which is produced by means of a continuous process. The control target was to maintain the product's viscosity curve within a given region of the viscosity-vs-shear rate plane, and this issue was addressed selecting three points on the viscosity curve that were controlled manipulating two ingredient flow rates according to MPC algorithm. The paper is preparatory to the introduction of on-line innovative rheological sensors in the complex fluid production and showed that automatic control can effectively

Processes **2018**, *6*, 12

improve the quality of the product but it is important to reduce the time required by the sensor to compute the rheological curve.

Acknowledgments: This work has received funding from the European Union's Horizon 2020 research and innovation programme under grant agreement No. 636942.

Author Contributions: Massimiliano Grosso conceived and designed the experiments; Francesc Corominas performed the experiments and contributed with reagents/materials/analysis tools; Stefania Tronci and Roberto Baratti analysed the data for modeling and control development purposes; Roberto Mei developed the simulator and controller. Stefania Tronci and Roberto Mei wrote the paper.

Conflicts of Interest: The authors declare no conflict of interest.

References

1. Meacci, V.; Ricci, S.; Wiklund, J.; Birkhofer, B.; Kotz, R. Flow-Viz-An integrated digital in-line fluid characterization system for industrial applications. *IEEE Sens. Appl. Symp. (SAS)* **2016**, 1–6. [CrossRef]
2. Kotzé, R.; Wiklund, J.; Haldenwang, R. Application of ultrasound Doppler technique for in-line rheological characterization and flow visualization of concentrated suspensions. *Can. J. Chem. Eng.* **2016**, *94*, 1066–1075. [CrossRef]
3. Yoshida, T.; Tasaka, Y.; Murai, Y. Rheological evaluation of complex fluids using ultrasonic spinning rheometry in an open container. *J. Rheol.* **2017**, *61*, 537–549. [CrossRef]
4. Mei, R.; Grosso, M.; Tronci, S.; Baratti, R.; Corominas, F. Real-Time Control of Viscosity Curve for a Continuous Production Process of a Non-Newtonian Fluid. *Chem. Eng. Trans.* **2017**, *57*, 1099–1104. [CrossRef]
5. Ogunnaike, B.A.; Ray, H.W. *Process Dynamics, Modeling, and Control*; Oxford University Press: New York, NY, USA, 1995; ISBN 978-0-19-509119-9.
6. Mewis, J.; Wagner, N. *Colloidal Suspension Rheology*; Cambridge University Press: New York, NY, USA, 2012; ISBN 978-0-521-51599-3.
7. Reinheimer, K.; Grosso, M.; Hetzel, F.; Kübel, J.; Wilhelm, M. Fourier Transform Rheology as an innovative morphological characterization technique for the emulsion volume average radius and its distribution. *J. Colloid Interface Sci.* **2012**, *380*, 201–212. [CrossRef] [PubMed]
8. Corominas, F.; Beelen, L.; Akalay, M. Methods for Producing Liquid Detergent Products. U.S. Patent 2013/0225468 A1, 29 August 2013.
9. Daniel-Berhe, S.; Unbehauen, H. Identification of nonlinear continuous-time Hammerstein model via HMF-method. In Proceedings of the 36th IEEE Conference on Decision and Control, San Diego, CA, USA, 12 December 1997; Volume 3, pp. 2990–2995.
10. Tronci, S.; Coppola, S.; Bacchelli, F.; Grosso, M. Flow instabilities in rheotens experiments: Analysis of the impacts of the process conditions through neural network modelling. *Polym. Eng. Sci.* **2013**, *53*, 1241–1252. [CrossRef]
11. Tronci, S.; Baratti, R. A Gain-Scheduling PI Control Based on Neural Networks. *Complexity* **2017**, 9241254. [CrossRef]
12. Cogoni, G.; Tronci, S.; Baratti, R.; Romagnoli, J.A. Controllability of semibatch nonisothermal antisolvent crystallization processes. *Ind. Eng. Chem. Res.* **2014**, *53*, 7056–7065. [CrossRef]
13. Foscoliano, C.; Del Vigo, S.; Mulas, M.; Tronci, S. Predictive control of an activated sludge process for long term operation. *Chem. Eng. J.* **2016**, *304*, 1031–1044. [CrossRef]
14. Mulas, M.; Tronci, S.; Corona, F.; Haimi, H.; Lindell, P.; Heinonen, M.; Vahala, R.; Baratti, R. Predictive control of an activated sludge process: An application to the Viikinmäki wastewater treatment plant. *J. Process Control* **2015**, *35*, 89–100. [CrossRef]

Article

Fuel Gas Network Synthesis Using Block Superstructure

Jianping Li, Salih Emre Demirel and M. M. Faruque Hasan *

Artie McFerrin Department of Chemical Engineering, Texas A&M University,
College Station, TX 77843-3122, USA; ljptamu@tamu.edu (J.L.); emredemirel@tamu.edu (S.E.D.)
* Correspondence: hasan@tamu.edu; Tel.: +1-979-862-1449

Received: 2 February 2018; Accepted: 26 February 2018; Published: 1 March 2018

Abstract: Fuel gas network (FGN) synthesis is a systematic method for reducing fresh fuel consumption in a chemical plant. In this work, we address FGN synthesis problems using a block superstructure representation that was originally proposed for process design and intensification. The blocks interact with each other through direct flows that connect a block with its adjacent blocks and through jump flows that connect a block with all nonadjacent blocks. The blocks with external feed streams are viewed as fuel sources and the blocks with product streams are regarded as fuel sinks. An additional layer of blocks are added as pools when there exists intermediate operations among source and sink blocks. These blocks can be arranged in a $I \times J$ two-dimensional grid with $I = 1$ for problems without pools, or $I = 2$ for problems with pools. J is determined by the maximum number of pools/sinks. With this representation, we formulate FGN synthesis problem as a mixed-integer nonlinear (MINLP) formulation to optimally design a fuel gas network with minimal total annual cost. We revisit a literature case study on LNG plants to demonstrate the capability of the proposed approach.

Keywords: process integration; fuel gas network synthesis; block superstructure; optimization; MINLP

---

1. Introduction

Over 40% of the operating cost of a petrochemical plant is attributed to energy consumption [1]. Energy is needed for raw material preprocessing (preheating, purification), separation of products from intermediates or impurities (product refining), and material transportation. There are multiple energy sources that can be exploited in a refinery, such as liquefied petroleum gas, fuel gas, off-gas, etc. [2,3]. These energy sources either come from external process raw materials and purchased fuels or from internal process products, and byproducts. Depending on where these fuel sources originate from, they can be classified as fuel from feed (FFF, e.g., natural gas) or fuel from product (FFP, e.g., products, byproducts) [4]. In 2016, external fuels supplied to refinery industry in the United States mainly consisted of natural gas (31%), electricity (5%), purchased steam and coal (1%) [5]. About 63% of the energy consumed by the refining industry comes from byproducts of the refining process for heat and power. These energy sources can be converted to each other. For example, fuel gas, produced internally from the distillation columns, crackers and reformers [6], can be converted to steam, electricity or heat. Fuel gas accounts for 46% of all energy sources for the refining industry in the United States [5,7–9]. Fuel gas is often composed of hydrocarbons (methane, ethane, propane and butane), hydrogen, and carbon monoxide, which have large heating values [10]. In most cases, these fuels are flared to the atmosphere, leading to detrimental effects on the environment and loss of heating values [11,12].

Due to the importance of fuel gas and the environment concern of fuel gas emission, many efforts have been made on improving the equipment efficiency [13] or exploiting new energy sources to decrease fuel gas generation and pollution emission [14]. Although these works give insights and directions on improving design of equipment and operating conditions, a generic and systematic strategy for elucidating the effective utilization of fuel gas is crucial. For example, in a typical fuel gas system, multiple fuel gas sources with different qualities are available for various equipment (sinks). As a result, effective and systematic management of fuel gas flow among fuel gas sources and fuel gas sinks can provide economic benefits for process design by fully utilizing the heating value embedded in the fuel gas [15,16].

Optimization-based methods enable the user to address fuel gas network (FGN) synthesis problems, which are aimed at redistributing the fuel gas at the system level [1,4,17]. To this end, Hasan et al. formalized the FGN synthesis problem as a nonlinear programming problem (NLP) considering the integration of fuel gases appropriately though auxiliary equipment (valves, pipelines, compressors, heaters/coolers, etc.) to achieve best utilization of them [4]. They posed the FGN problem as a special class of pooling problem which leads a superstructure involving many practical features such as nonisobaric and nonisothermal operation, nonisothermal mixing, nonlinear fuel-quality specifications, and emission standards. Here the superstructure is defined as a superset of postulated process alternatives [18]. The proposed FGN superstructure in the work of Hasan et al. (shown in Figure 1) includes a set of fuel gas sources with temperature specification T_f, pressure specification P_f and feed availability specification F_f^{feed} for each source f, and a set of fuel gas sinks with demand range $\left[D_p^L, D_p^U\right]$, temperature range $\left[T_p^{min}, T_p^{max}\right]$, and pressure range $\left[P_p^{min}, P_p^{max}\right]$ for each sink stream p. To achieve the sink requirements, the intermediate operations such as cooling, compression, heating and expansion are considered in addition to mixing and splitting. Jagannath et al. [17] extended this work to include the multi-period FGN operation. This FGN design makes dynamic plant operation more robust and helps to reduce capital costs. Nassim et al. [1,19] modified the FGN model introduced by Hasan et al. [4] to include more constraints on addressing environmental issues and developed a novel methodology for grass-root and retrofit design of FGNs.

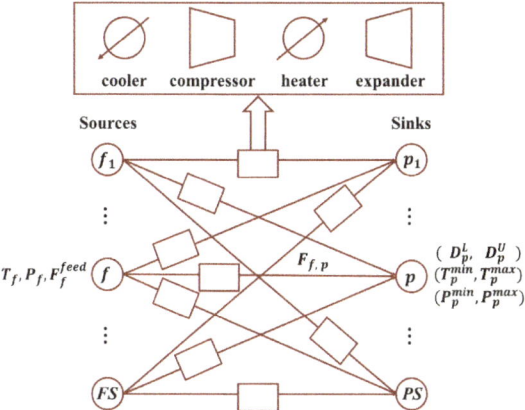

Figure 1. A classic superstructure for the fuel gas network.

The first step for many optimization-based methods is the construction of a superstructure. Hence the appropriate selection of superstructure representation method is critical. There are many representations such as state-task-network [20,21], state-equipment-network [21], P-graph [22,23], state-space [24,25] , HEN and MEN building blocks [26,27], phenomena building blocks [28–30], process-group contribution method [31], and unit-port-conditioning-stream (UPCS) approach [32,33]. We recently proposed a new superstructure representation method using building blocks for systematic process intensification [34–36]. The block superstructure has been constructed based on the dissection of various unit operations into fundamental building blocks. Later on, the proposed block-based approach is applied to address process synthesis problems [37].

In this work, we address the optimal synthesis of fuel gas networks using a block superstructure, originally proposed in our previous work for process synthesis and intensification [34–37]. Since the fuel gas network by its definition is a special class of pooling problem, our block representation method can be extended to general pooling problems as well. In this representation, each block allows multiple fuel gas inlet flows and single product outlet flow (unique composition for different product streams). The blocks with external feeds and external products serve as sources and sinks for fuel gas respectively. The material and energy flow among different blocks are achieved via jump flow streams connecting all nonadjacent blocks with each other and direct connecting streams connecting only adjacent blocks. The involvement of jump flows avoids the utilization of unnecessary intermediate blocks for inter-block connections. Each stream connecting two adjacent blocks are placed with compressors/expanders to adjust the pressure for achieving the sink requirements. Options for supplying extra hot/cold utility are provided to each block for allowing nonisothermal operation. When there is no direct connecting stream, the block boundary between adjacent blocks is regarded as completely restricted boundary. These blocks are collected in a two-dimensional grid to form a superstructure of blocks. We formulate the fuel gas synthesis problem as a mixed-integer nonlinear optimization (MINLP) problem. The model constraints involve mass and energy balance, flow directions, work calculation and logic constraints. The nonlinear terms of the proposed model arise from splitting, energy balances and work-related calculations.

The remaining of the article is structured as follows. First, we elaborate the representation of fuel gas network using block-based approach. Next, we present the MINLP formulation for fuel gas network synthesis problem. Finally, we demonstrate the applicability of our approach with one case study on FGN synthesis in an LNG plant.

2. Block-Based Representation of Fuel Gas Network

In this section, we describe how the classic fuel gas network superstructure such as the one proposed by Hasan et al. [4] can be represented using block-based approach [34,37] as a generic tool for designing fuel gas utilization system. First, we illustrate the classic FGN superstructure and analyze the operation involved in synthesizing a FGN. Next, we construct a block superstructure that can also include the same features. We provide block superstructures for fuel gas network with or without intermediate pools which bring additional mixing operations for more economic benefits.

In a classic FGN superstructure (Hasan et al. [4]), shown in Figure 1, there are FS number of fuel gas sources and PS number of fuel gas sinks. The source stream f has the temperature as T_f and the pressure as P_f. The sink stream p is obtained with temperature range as $\left[T_p^{min}, T_p^{max}\right]$, pressure range as $\left[P_p^{min}, P_p^{max}\right]$ and demand range as $\left[D_p^L, D_p^U\right]$. Each stream $F_{f,p}$ connecting a source f and a sink p passes through two utility exchangers (heater and/or cooler) and one mover (compressor or expander). The sources completely or partially come from different fuel gas sources and are mixed at different fuel gas sinks with different temperature, pressure and quality requirements. The operations in a FGN problem typically include mixing, cooling, heating, pressurizing and depressurizing.

Most FGN synthesis problems involve multiple sources and multiple sinks. In addition, there are similar equipment assignment that are assigned between sources and sinks. This allows us to develop a general block representation for FGN synthesis as shown in Figure 2. It involves I number of rows and J number of columns, where each row or column is a collection of blocks. Let $B_{i,j}$ represent the block at row i and column j. Each block allows multiple external feed streams $M_{i,j,k,f}$ to enter block $B_{i,j}$. The available amount of feed f can be partially or completely fed into a block $B_{i,j}$ with $z_{i,j,f}^{feedfrac}$ fraction of available amount F_f^{feed}. Similarly, product stream p can be withdrawn from each block with the component flowrate of $H_{i,j,k,p}$.

As shown in Figure 2b, the mass and energy transfer within the block superstructure is achieved through the direct connecting streams between adjacent blocks and jump connecting streams among all nonadjacent blocks. Direct connecting streams are achieved via inter-block flow $F_{i,j,k,d}$, which is the flowrate of component k between block $B_{i,j}$ and $B_{i,j+1}$ when the flow alignment $d = 1$ (the connecting flow between adjacent blocks is in horizontal direction) or the flowrate of component k between block $B_{i,j}$ and $B_{i+1,j}$ when the flow alignment $d = 2$ (the connecting flow between adjacent blocks is in vertical direction) . These direct connecting streams can be either positive when the flow is from block $B_{i,j}$ to $B_{i,j+1}$ for $d = 1$ (from block $B_{i,j}$ to $B_{i+1,j}$ for $d = 2$) or negative when the flow is from block $B_{i,j+1}$ to $B_{i,j}$ for $d = 1$ (from block $B_{i+1,j}$ to $B_{i,j}$ for $d = 2$). Also, these direct connecting stream flow across the block boundary between adjacent blocks. When there is no direct connecting stream ($F_{i,j,k,d} = 0$), the block boundary between $B_{i,j}$ and $B_{i,j+1}$ ($d = 1$) or between $B_{i,j}$ and $B_{i+1,j}$ ($d = 2$) is identified as completely restricted boundary. The jump connecting streams are depicted by $J_{i,j,i',j',k}$, which is the flowrate of component k from block $B_{i,j}$ to $B_{i',j'}$, where i' and j' designate the row number and column number of a different block. Because of this unidirectional feature, $J_{i,j,i',j',k}$ is a jump flow withdrawn from $B_{i,j}$ and supplied to $B_{i',j'}$.

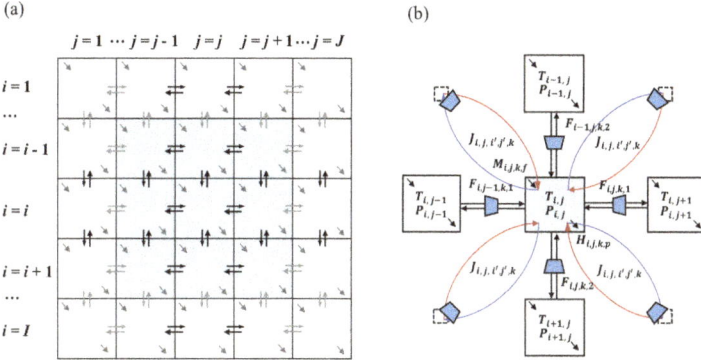

Figure 2. Construction of superstructure for fuel gas synthesis problems: (**a**) Block superstructure illustration; (**b**) Block interaction via connecting streams (blue line: jump connecting flow from the block $B_{i,j}$; red line: jump connecting flow into the block $B_{i,j}$; blocks at diagonal positions are ignored for simplicity).

With these direct and jump connecting streams, blocks with multiple inlets and multiple outlets can serve as stream mixers and splitters, respectively. Source block is identified when multiple external feed streams enter into a block and get mixed, while blocks with external product stream are sinks. Note that splitting of the source streams is not regarded as a splitting operation because it can be achieved through the splitting fraction $z_{i,j,f}^{feedfrac}$ of source stream f into block $B_{i,j}$ and thus can be regarded as supplies of multiple source streams with the same specification.

The operation equipment (heaters/coolers, compressors/expanders) is embedded in the block superstructure through auxiliary units. To represent the pressurizing/depressurizing operation, both direct connecting streams and jump connecting streams are assigned with compressor or expander (only one of them would be selected). Each stream leaving block $B_{i,j}$ has a pressure designated as $P_{i,j}$, which is also the inlet pressure for the compressors or expanders on these streams. The inlet temperature for these compressors/expanders arranged at outlet streams ($F_{i,j,k,d}$ and $J_{i,j,i',j',k}$) of $B_{i,j}$ is the block temperature $T_{i,j}$, which is also the common temperature of outlet streams from $B_{i,j}$. The heating and cooling operations are achieved through the heat duty $Q_{i,j}^h$ and cold duty $Q_{i,j}^c$, which are obtained from the energy balance around block $B_{i,j}$.

The general block superstructure for FGN synthesis problem developed in Figure 2 can be reduced to block superstructure with smaller size if the number of intermediate pools is known beforehand. As an illustrative example, we first consider the case without intermediate pools. Knowing certain number of sources and sinks together with their specification and requirement, the classic superstructure is built by connecting each source and sink and shown in Figure 3a. Here all stream heaters/coolers and expanders/compressors are ignored for representation simplicity. As is shown in Figure 3b, we use a $1 \times N$ block superstructure to incorporate the classic superstructure. In this case, the column number is directly equal to number of sinks ($J = PS$). Since there are no intermediate pools, row number $I = 1$. Each block serves as sink block, from which product streams are withdrawn. Meanwhile, each block could also function as feed block, where multiple types of source streams are fed. Specifically, taking the first sink block $B_{1,1}$ as an example, there could be at most FS number of source streams entering this block. The activation of connectivity between sources and sinks could be reflected by the feed fraction $z_{i,j,f}^{feedfrac}$ of different sources f. If the feed fraction $z_{i,j,f}^{feedfrac}$ of source stream at the sink block $B_{i,j}$ is zero, then there is no connectivity between the source f and the sink p in block $B_{i,j}$; source-sink connectivity exists as long as the feed fraction of source stream $z_{i,j,f}^{feedfrac}$ is nonzero. Besides, the horizontal connecting streams between adjacent blocks in Figure 3b are also allowed. This additional feature physically indicates the material flowing between two fuel gas sinks.

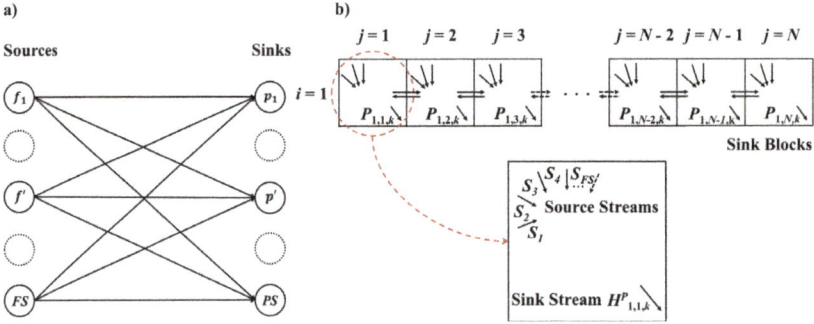

Figure 3. Block representation for fuel gas network problem: (**a**) Classical superstructure for fuel gas network; (**b**) Equivalent block superstructure for fuel gas network.

As for the more general case of the fuel gas network superstructure, between the sources and sinks layer, there can be another layer consisting of L number of intermediate pools, as is shown in Figure 4. Source streams first come into the intermediate pools, where certain operations such as mixing, purifying are executed according to different sink requirements. The outlet streams coming from the intermediate pools are further directed to the sinks or to the other different pools (shown as the blue line in Figure 4a). One way to incorporate the general superstructure is to utilize a block superstructure with larger size so that pools (involving mixing and splitting operations) can be included into the system. With this new feature of intermediate pools, the updated block superstructure is shown in Figure 4b. The first row consists of L number of pool blocks (grey blocks) and the second row consists of PS number of sink blocks (blue blocks). In this case, the number of columns can be taken as $J = max\{L, PS\}$. The existence of intermediate pools make the row number as $I = 2$, one row to accommodate pools and another row for sinks. The distribution of source streams into each pool blocks is achieved through splitting operation of source streams. In the first row, the jump flows are withdrawn from each block as outlet streams of intermediate pools. Specifically, taking the first column of block superstructure in Figure 4b as an example, the jump flow $J_{1,1,k}^P$ (the summation of all the jump connecting streams to other blocks from block $B_{1,1}$) is withdrawn and directed to other blocks as inlet flows. The inlet jump flows, $J_{1,j,2,1,k}$, from nonadjacent blocks $B_{1,j}$ ($j \in [2, J]$) are mixed in the second row at sink block $B_{2,1}$ and then taken as the final product $H_{2,1,k}^p$ (the overall component flowrate for all product stream p).

As is discussed above, the block superstructure can be converted from the classic superstructure of fuel gas network. When there is no prior information provided on flow connectivity among sources, pools and sinks, the block superstructure can be constructed by simply setting the row number I and column number J (i.e., $J = max\{L, PS\}$), which then involves as many process alternatives as possible. The benefit for block representation method is on its generic feature that each block follows the same pattern with multiple inlet streams and outlet streams. Based on the representation method, we now develop the MINLP formulation for the FGN synthesis problem.

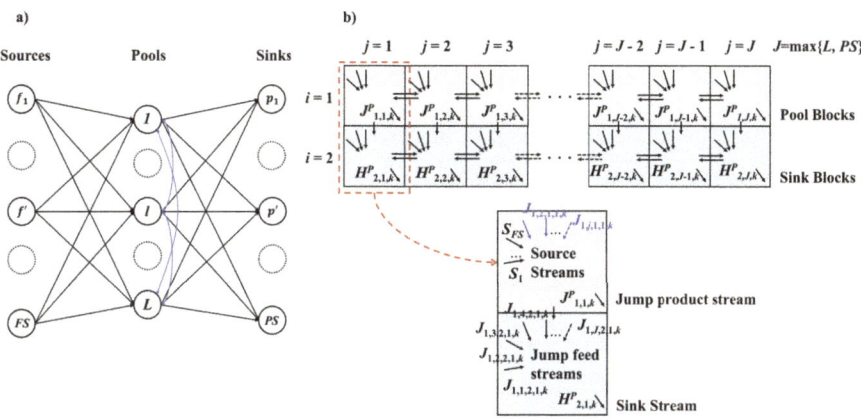

Figure 4. General superstructure for fuel gas network synthesis problem with intermediate pools: (a) General superstructure for fuel gas network with intermediate pools; (b) equivalent block superstructure.

3. FGN Synthesis Problem Statement

This section gives the formal problem description for FGN synthesis problem using block superstructure. The sets given for this problem are the set $K = \{k | k = 1, ..., |K|\}$ of components, the set $FS = \{f | f = 1, ..., |FS|\}$ of fuel gas sources with component specification $y_{k,f}^{feed}$, a set $PS = \{p | p = 1, ..., |PS|\}$ of fuel gas sinks with the material demand range as $[D_p^L, D_p^U]$, energy demand range as $[De_p^L, De_p^U]$, purity range as $[y_{k,p}^{min,prod}, y_{k,p}^{max,prod}]$ for species k as well as ranges for other specifications $[q_{s,p}^{min,prod}, q_{s,p}^{max,prod}]$. The objective is to synthesize a fuel gas network that systematically utilizes the arrangement of fuel gas resources and minimizes the total annual cost. The set $D = \{d | d = 1, 2\}$ designates the flow alignment. The flow alignment $d = 1$ when the stream is flowing in the horizontal direction, i.e., from block $B_{i,j}$ to $B_{i,j+1}$; $d = 2$ when the stream is flowing in the vertical direction, i.e., from block $B_{i,j}$ to $B_{i+1,j}$. The temperature range and flowrate range for all connecting flows including direct connecting flow and jump connecting flow are set as $[T^{min}, T^{max}]$ and $[FL, FU]$ respectively.

We consider the assumptions for this work as constant properties (heat capacity, lower heating value, etc.), continuous steady-state operation, ideal gas condition, adiabatic expansion/compression, and ideal mixing. With this, we now describe the MINLP model for fuel gas network synthesis based on block superstructure.

4. MINLP Model Formulation for Block-Based Fuel Gas Network Synthesis

The main constraints for the MINLP model involve block material balance, flow directions, block energy balance, work calculation and task assignment/logic constraints. The objective of the FGN synthesis is to minimize the total annual cost.

4.1. Block Material Balance

The general material balance for each block $B_{i,j}$ considers the material flows of component k including horizontal inlet flow $F_{i,j-1,k,1}$, the horizontal outlet flow $F_{i,j,k,1}$, vertical inlet flow $F_{i-1,j,k,2}$, vertical outlet flow $F_{i,j,k,2}$, external feed stream $M^f_{i,j,k}$, external product stream $H^p_{i,j,k}$, jump flow into the block $J^f_{i,j,k}$, and jump flow from the block $J^p_{i,j,k}$. Specifically, the material balance relation is presented as follows.

$$F_{i,j-1,k,1} - F_{i,j,k,1} + F_{i-1,j,k,2} - F_{i,j,k,2} + M^f_{i,j,k} - H^p_{i,j,k} + J^f_{i,j,k} - J^p_{i,j,k} = 0, \quad i \in I, j \in J, k \in K \quad (1)$$

The last four terms in the above relation are obtained though the following constraints.

$$M^f_{i,j,k} = \sum_{f \in FS} M_{i,j,k,f} \quad i \in I, j \in J, k \in K \quad (2)$$

$$H^p_{i,j,k} = \sum_{p \in PS} H_{i,j,k,p} \quad i \in I, j \in J, k \in K \quad (3)$$

$$J^f_{i,j,k} = \sum_{(i',j') \in LN} J_{i',j',i,j,k} \quad i \in I, j \in J, k \in K \quad (4)$$

$$J^p_{i,j,k} = \sum_{(i',j') \in LN} J_{i,j,i',j',k} \quad i \in I, j \in J, k \in K \quad (5)$$

All variables including $M^f_{i,j,k}$, $H^p_{i,j,k}$, $J^f_{i,j,k}$ and $J^p_{i,j,k}$ are obtained by summing multiple inlet streams or outlet streams within single block $B_{i,j}$. The positive continuous variable $M_{i,j,k,f}$ indicates the amount of component flowrate k into block $B_{i,j}$ carried by feed stream f. The amount of component k taken from block $B_{i,j}$ through product stream p is designated by positive continuous variable $H_{i,j,k,p}$. The material flowrate for component k from block $B_{i,j}$ to $B_{i',j'}$ is $J_{i,j,i',j',k}$. The index i' and j' indicate row position and column position of a block $B_{i',j'}$ that is different from $B_{i,j}$. The subset $LN(i,j,i',j')$ designates the connection between block $B_{i,j}$ and block $B_{i',j'}$. It should be noted that for jump connecting flow $J_{i,j,i',j',k}$, $i \neq i'$ and $j \neq j'$ so as to avoid remixing in block $B_{i,j}$. The stream connectivities at the outer boundary of block superstructure are set as $F_{i=I,j,k,1} = F_{i,j=J,k,2} = 0$ to ensure that the interaction between the superstructure and the environment is only achieved through external feeds and products.

The flowrate $M_{i,j,k,f}$ for each feed f into block $B_{i,j}$ is completely or partially from the overall available amount F^{feed}_f. The distribution of feed stream f is achieved by the feed fraction $z^{feedfrac}_{i,j,f} \geq 0$ in block $B_{i,j}$. Hence $M_{i,j,k,f}$ can be determined as follows:

$$M_{i,j,k,f} = F^{feed}_f y^{feed}_{k,f} z^{feedfrac}_{i,j,f}, \quad i \in I, j \in J, k \in K, f \in FS \quad (6)$$

$$0 \leq \sum_{i \in I} \sum_{j \in J} z^{feedfrac}_{i,j,f} \leq 1, \quad f \in FS \quad (7)$$

Typically, headers receiving fuel gas have purity requirement for inlet streams to ensure the required operating conditions of the corresponding equipment. This is achieved through the following inequality constraints:

$$y^{min,prod}_{k,p} \sum_{k' \in K} H_{i,j,k',p} \leq H_{i,j,k,p} \leq y^{max,prod}_{k,p} \sum_{k' \in K} H_{i,j,k',p}, \quad i \in I, j \in J, (k,p) \in kp \quad (8)$$

Here, the purity range for component k in product stream p is given by $[y_{k,p}^{min,prod}, y_{k,p}^{max,prod}]$. The set kp relates the key component k with product stream p with purity specifications. The product stream p have no purity restrictions when it does not appear in set kp.

On top of purity requirement of key component k in product stream p, possible requirement on ratio of different component k in product stream p is also considered.

$$P_{i,j,k=k',p} \geq \sum_{k'' \in K} \pi_{k',k'',p}^{prod} P_{i,j,k'',p} \quad i \in I, j \in J, p \in Ps \tag{9}$$

where $\pi_{k',k'',p}^{prod}$ is the minimum product ratio requirement between component k' and component k'' for product p.

We also impose the demand constraint for product p supplied to different headers:

$$D_p^L \leq \sum_{i \in I} \sum_{j \in J} \sum_{k \in K} H_{i,j,k,p} \leq D_p^U, \quad p \in PS \tag{10}$$

Here, D_p^L and D_p^U are minimum and maximum allowed amount for product stream p respectively. Hence if there is no specification existing for D_p^L, D_p^U or both, we set $D_p^L = 0$ and $D_p^U = \max\limits_{f \in FS} F_f^{feed}$.

Besides, energy demands De_p for each product stream p should be satisfied based on the following constraint:

$$\sum_{i \in I} \sum_{j \in J} \sum_{k \in K} H_{i,j,k,p} LHV_k \geq De_p, \quad p \in PS \tag{11}$$

where LHV_k refers to lower heating value for each component k, which measures energy content per unit mass or volume of pure combustible component.

Furthermore, each product stream should have acceptable limits on other certain specifications including lower heating value (LHV), reverse specific gravity ($1/SG$), etc. Assuming that all the considered specifications are linearly related with mixture compositions, the following constraint is supplied below for each product stream p [4].

$$q_{s,p}^{min,prod} \sum_{i \in I} \sum_{j \in J} \sum_{k \in K} H_{i,j,k,p} \leq \sum_{i \in I} \sum_{j \in J} \sum_{k \in K} H_{i,j,k,p} q_{s,k} \leq q_{s,p}^{max,prod} \sum_{i \in I} \sum_{j \in J} \sum_{k \in K} H_{i,j,k,p}, \quad p \in PS \tag{12}$$

Here the parameter $q_{s,k}$ denote the value of specification s for component k, and $[q_{s,p}^{min,prod}, q_{s,p}^{max,prod}]$ is the acceptable range of specification s for product stream p. Note that the quality specification $q_{s,k}$ is component flowrate-based instead of total flowrate-based, which is considered in the work of Hasan et al. [4].

To obtain the total flowrate for all streams associated with the block $B_{i,j}$, we sum all components in each stream. Specifically, we obtain the total flowrate $FP_{i,j,d}^T$, $FN_{i,j,d}^T$, $J_{i,j,i',j'}^T$, $M_{i,j,f}^T$, and $H_{i,j,p}^T$ from the component flowrate for $FP_{i,j,k,d}$, $FN_{i,j,k,d}$, $J_{i,j,i',j',k}$, $M_{i,j,k,f}$, and $H_{i,j,k,p}$ through the following relations.

$$FP_{i,j,d}^T = \sum_{k \in K} FP_{i,j,k,d}, \quad i \in I, j \in J, d \in D \tag{13}$$

$$FN_{i,j,d}^T = \sum_{k \in K} FN_{i,j,k,d}, \quad i \in I, j \in J, d \in D \tag{14}$$

$$J_{i,j,i',j'}^T = \sum_{k \in K} J_{i,j,i',j',k}, \quad (i,j,i',j') \in LN(i,j,i',j') \tag{15}$$

$$M_{i,j,f}^T = \sum_{k \in K} M_{i,j,k,f}, \quad i \in I, j \in J, f \in FS \tag{16}$$

$$H_{i,j,p}^T = \sum_{k \in K} H_{i,j,k,p}, \quad i \in I, j \in J, s \in PS \tag{17}$$

With the total flowrate information, we model the splitting operation for achieving identical composition for all outlet streams as follows:

$$FP_{i,j,k,d} = y^b_{i,j,k} FP^T_{i,j,d} \quad i \in I, j \in J, d \in D \tag{18}$$

$$FN_{i,j-1,k,1} = y^b_{i,j,k} FN^T_{i,j-1,1} \quad i \in I, j \in J \tag{19}$$

$$FN_{i-1,j,k,2} = y^b_{i,j,k} FN^T_{i-1,j,2} \quad i \in I, j \in J \tag{20}$$

$$J_{i,j,i',j',k} = y^b_{i,j,k} J^T_{i,j,i',j'} \quad (i,j,i',j') \in LN(i,j,i',j'), k \in K \tag{21}$$

$$H_{i,j,k,p} = y^b_{i,j,k} H^T_{i,j,p} \quad i \in I, j \in J, k \in K, s \in PS \tag{22}$$

Here the positive continuous variable $y^b_{i,j,k}$ refers to the block composition of component k. This block composition refers to the composition of component k in all the outlet streams from block $B_{i,j}$.

4.2. Flow Directions

The direct connectivity $F_{i,j,k,d}$ among adjacent blocks is a bidirectional flow with its positive component $FP_{i,j,k,d}$ and negative component $FN_{i,j,k,d}$. Only one of them is active when the connecting flow $F_{i,j,k,d}$ is chosen to be nonzero. The selection of flow direction is a decision variable, which is achieved through the following binary variable:

$$z^{Plus}_{i,j,d} = \begin{cases} 1 & \text{if } F_{i,j,k,d} \text{ is from block } B_{i,j} \text{ to } B_{i,j+1} \ (d=1) \text{ or from block } B_{i,j} \text{ to } B_{i+1,j} \ (d=2) \\ 0 & \text{otherwise} \end{cases}$$

As a result, the flow direction determination is achieved though the following constraints:

$$F_{i,j,k,d} = FP_{i,j,k,d} - FN_{i,j,k,d} \quad i \in I, j \in J, k \in K, d \in D \tag{23}$$

$$FP_{i,j,k,d} \leq FU z^{Plus}_{i,j,d}, \quad i \in I, j \in J, k \in K, d \in D \tag{24}$$

$$FN_{i,j,k,d} \leq FU(1 - z^{Plus}_{i,j,d}), \quad i \in I, j \in J, k \in K, d \in D \tag{25}$$

4.3. Block Energy Balance

The enthalpy terms for block energy balance includes inlet and outlet inter-block stream enthalpy, feed enthalpy, product enthalpy, external heating/cooling, work energy associated with expansion/compression. Then the steady-state energy balance for block $B_{i,j}$ is formulated as follows:

$$EF_{i,j-1,1} - EF_{i,j,1} + EF_{i-1,j,2} - EF_{i,j,2} + EM_{i,j} - EP_{i,j} + EJ^f_{i,j} - EJ^p_{i,j} + Q_{i,j} + W_{i,j} = 0, \ i \in I, j \in J \tag{26}$$

where, $EF_{i,j,d}$ represents the stream enthalpy carried by the material flow $F_{i,j,k,d}$ in flow direction d, $EM_{i,j}$ is the overall enthalpy brought into block $B_{i,j}$ along with external feed streams , $EP_{i,j}$ is overall enthalpy taken away by external product streams, $EJ^f_{i,j}$ is overall enthalpy carried into block $B_{i,j}$ through jump flows, $EJ^p_{i,j}$ is overall enthalpy taken out from block $B_{i,j}$ through jump flows, $Q_{i,j}$ represents amount of heat added into or removed from the block $B_{i,j}$, $W_{i,j}$ indicates the amount of work energy added into or withdrawn from block $B_{i,j}$. These energy flow variables are shown in Figure 5.

The stream enthalpy is determined as follows with the information provided on flowrate, component heat capacities and the block temperature. Depending on the flow direction, in flow alignment $d = 1$, the inlet temperature for block $B_{i,j}$ is either $T_{i,j}$ from block $B_{i,j}$ to $B_{i,j+1}$ or $T_{i,j+1}$ from block $B_{i,j+1}$ to $B_{i,j}$; in flow alignment $d = 2$, the inlet temperature for block $B_{i,j}$ is either $T_{i,j}$ from block $B_{i,j}$ to $B_{i+1,j}$ or $T_{i+1,j}$ from block $B_{i+1,j}$ to $B_{i,j}$.

$$EF_{i,j,1} = \sum_{k \in K} FP_{i,j,k,1} Cp_k T_{i,j} - \sum_{k \in K} FN_{i,j,k,1} Cp_k T_{i,j+1} \tag{27}$$

$$EF_{i,j,2} = \sum_{k \in K} FP_{i,j,k,2} Cp_k T_{i,j} - \sum_{k \in K} FN_{i,j,k,2} Cp_k T_{i+1,j} \tag{28}$$

where Cp_k is the heat capacity of component k.

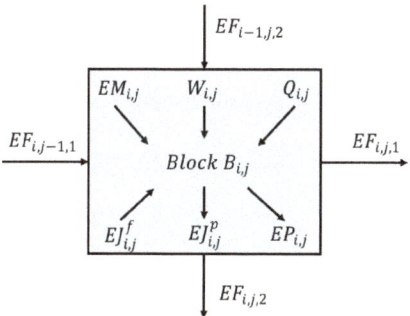

Figure 5. Illustration of energy balance on block $B_{i,j}$.

The enthalpy amount brought into or withdrawn from block $B_{i,j}$ through jump flows are determined as follows:

$$EJ^f_{i,j} = \sum_{k \in K} \sum_{(i',j') \in LN} J_{i',j',i,j,k} Cp_k T_{i',j'} \quad i \in I, j \in J \tag{29}$$

$$EJ^p_{i,j} = \sum_{k \in K} \sum_{(i',j') \in LN} J_{i,j,i',j',k} Cp_k T_{i,j} \quad i \in I, j \in J \tag{30}$$

It should be noted that the inlet temperature of jump flow is always the temperature of the source block, $T_{i,j}$. Likewise, the feed enthalpy and product enthalpy are determined with the following constraints:

$$EM_{i,j} = \sum_{k \in K} \sum_{f \in F} M_{i,j,k,f} Cp_k T^f \quad i \in I, j \in J \tag{31}$$

$$EP_{i,j} = \sum_{k \in K} \sum_{p \in P} P_{i,j,k,p} Cp_k T_{i,j} \quad i \in I, j \in J \tag{32}$$

The amount of hot/cold utility consumed in block $B_{i,j}$ can be evaluated through the amount of heat introduced into ($Q^h_{i,j}$) or withdrawn from ($Q^c_{i,j}$) block $B_{i,j}$.

$$Q_{i,j} = Q^h_{i,j} - Q^c_{i,j} \tag{33}$$

The work energy can be also determined by the amount of work added into or taken out of block $B_{i,j}$, which are denoted as $W^{com}_{i,j}$ for compression and $W^{exp}_{i,j}$ for expansion respectively. The calculation of $W^{com}_{i,j}$ and $W^{exp}_{i,j}$ is explained later in this Section 4.6.

$$W_{i,j} = W^{com}_{i,j} - W^{exp}_{i,j} \tag{34}$$

Finally, to prevent condensation in the FGN and ensure sufficient superheating, the following constraints are supplied for product stream p in block $B_{i,j}$ [4].

$$\sum_{k \in K} H_{i,j,k,p} C p_k T_{i,j} \geq (MDP_p + \frac{5}{9}(5.15\frac{P_{i,j}}{100} - 312) \sum_{k \in K} H_{i,j,k,p} C p_k \quad i \in I, j \in J, p \in PS \quad (35)$$

$$\sum_{k \in K} H_{i,j,k,p} C p_k T_{i,j} \leq (HDP_p + \frac{5}{9}(2.33(\frac{P_{i,j}}{100})^2 - 2.8\frac{P_{i,j}}{100} - 305) \sum_{k \in K} H_{i,j,k,p} C p_k \quad i \in I, j \in J, p \in PS \quad (36)$$

where parameter MDP_p is moisture dew-point temperature and parameter HDP_p is the hydrocarbon dew-point temperature for the product p.

4.4. Product Stream Assignments and Logical Constraints

We define binary variables for each product stream p at block $B_{i,j}$ to determine whether they are active in $B_{i,j}$ or not:

$$z_{i,j,p}^{product} = \begin{cases} 1 & \text{if product stream } p \text{ is withdrawn from block } B_{i,j} \\ 0 & \text{otherwise} \end{cases}$$

The identification of block as product block is achieved through the following logical relation, which involves product binary variable:

$$\sum_{k \in K} P_{i,j,k,p} \leq D_p^U z_{i,j,p}^{product} \quad i \in I, j \in J, p \in PS \quad (37)$$

For each block, there are at most one type of product stream present in block $B_{i,j}$. The logic proposition is illustrated as follows:

$$\sum_{p \in PS} z_{i,j,p}^{product} \leq 1 \quad i \in I, j \in J \quad (38)$$

Each product stream p appears in the block superstructure for at least once so as to ensure the supply of fuel gas header.

$$\sum_{i \in I} \sum_{j \in J} z_{i,j,p}^{product} \geq 1 \quad p \in PS \quad (39)$$

The temperature range for block with product stream p is from T_p^{min} to T_p^{max}.

$$T_p^{min} z_{i,j,p}^{product} + T^{min}(1 - z_{i,j,p}^{product}) \leq T_{i,j} \leq T_p^{max} z_{i,j,p}^{product} + T^{max}(1 - z_{i,j,p}^{product}) \quad i \in I, j \in J, p \in PS \quad (40)$$

Likewise, the pressure range for product block is $\left[P_p^{min} \text{ to } P_p^{max}\right]$.

$$P_p^{min} z_{i,j,p}^{product} + P^{min}(1 - z_{i,j,p}^{product}) \leq T_{i,j} \leq P_p^{max} z_{i,j,p}^{product} + P^{max}(1 - z_{i,j,p}^{product}) \quad i \in I, j \in J, p \in PS \quad (41)$$

4.5. Boundary Assignment

The boundary type between adjacent blocks can be either completely restricted or not. If there is no direct connecting stream between adjacent blocks, then the inter-block boundary is identified as completely restricted boundary. The decision of boundary type is achieved through the following binary variable $z_{i,j,d}^{cr}$.

$$z_{i,j,d}^{cr} = \begin{cases} 1 & \textit{If boundary between } B_{i,j} \textit{ and } B_{i,j+1} \textit{ for } d = 1 \textit{ (between } B_{i,j} \textit{ and } B_{i+1,j} \textit{ for } d = 2) \\ & \textit{is completetly restricted} \\ 0 & \textit{Otherwise} \end{cases}$$

According to the definition of completely restricted boundary, the following constraints are supplied to relate flowrate $F_{i,j,k,d}$ with boundary type.

$$F_{i,j,k,d} \leq FU(1 - z_{i,j,d}^{cr}), \quad i \in I, j \in J, d \in D \tag{42}$$

4.6. Work Calculation

The work term $W_{i,j}$ consists of compression work term $W_{i,j}^{com}$ and expansion work term $W_{i,j}^{exp}$. Both $W_{i,j}^{com}$ and $W_{i,j}^{exp}$ consist of work components for direct connecting streams ($W_{i,j,d}^{comp,FP}$ and $W_{i,j,d}^{exp,FP}$ for compression and expansion work of positive component, $W_{i,j,d}^{comp,FN}$ and $W_{i,j,d}^{exp,FN}$ for compression and expansion work of negative component), feed streams ($W_{i,j,f}^{comp,FS}$ and $W_{i,j,f}^{exp,FS}$ for compression and expansion work respectively), and jump connecting streams ($W_{i',j',i,j}^{comp,JF}$ and $W_{i',j',i,j}^{exp,JF}$ for compression and expansion work respectively). Accordingly,

$$W_{i,j}^{com} = \sum_{d \in D}(W_{i,j,d}^{comp,FP} + W_{i,j,d}^{comp,FN}) + \sum_{f \in FS}W_{i,j,f}^{comp,FS} + \sum_{(i',j') \in LN(i,j,i',j')}W_{i',j',i,j}^{comp,JF}, \quad i \in I, j \in J \tag{43}$$

$$W_{i,j}^{exp} = \sum_{d \in D}(W_{i,j,d}^{exp,FP} + W_{i,j,d}^{exp,FN}) + \sum_{f \in FS}W_{i,j,f}^{exp,FS} + \sum_{(i',j') \in LN(i,j,i',j')}W_{i',j',i,j}^{exp,JF}, \quad i \in I, j \in J \tag{44}$$

We define the positive variable $PR_{i,j,d}^{F}$ to designate the pressure ratio between the block $B_{i,j+1}$ and $B_{i,j}$ for flow alignment $d = 1$ or between the block $B_{i+1,j}$ and $B_{i,j}$ for flow alignment $d = 2$. In horizontal direction, the pressure ratio is determined as follows:

$$PR_{i,j,1}^{F} = \frac{P_{i,j+1}}{P_{i,j}} \quad i \in I, j \in J \tag{45}$$

Similarly, in vertical direction, the pressure ratio is determined as follows:

$$PR_{i,j,2}^{F} = \frac{P_{i+1,j}}{P_{i,j}} \quad i \in I, j \in J \tag{46}$$

For feed stream f, the pressure ratio is taken as the ratio between block pressure $P_{i,j}$ and parameter P_{f}^{feed} for feed pressure .

$$PR_{i,j,f}^{feed} = \frac{P_{i,j}}{P_{f}^{feed}} \quad i \in I, j \in J, f \in FS \tag{47}$$

From these pressure ratio definitions, we calculate the isentropic work on direct connecting streams, feed streams and jump connecting streams. In the horizontal direction, the inlet isentropic work is determined as follows:

$$\eta W_{i,j,1}^{comp,FP} - W_{i,j,1}^{exp,FP}/\eta = \sum_{k \in K}FP_{i,j-1,k,1}T_{i,j-1,1}^{s}R_{gas}\frac{\gamma}{\gamma-1}\{(PR_{i,j-1,1}^{F})^{\frac{\gamma-1}{\gamma}} - 1\} \quad i \in I, j \in J \tag{48}$$

$$\eta W_{i,j,1}^{comp,FN} - W_{i,j,1}^{exp,FN}/\eta = \sum_{k \in K}FN_{i,j,k,1}T_{i,j,1}^{s}R_{gas}\frac{\gamma}{\gamma-1}\{(\frac{1}{PR_{i,j,1}^{F}})^{\frac{\gamma-1}{\gamma}} - 1\} \quad i \in I, j \in J \tag{49}$$

Here R_{gas} is the gas constant and γ is the adiabatic compression coefficient for process streams. γ is taken as heat capacity ratio. η is the adiabatic compression efficiency. Similarly, the isentropic work for a vertical entering stream is calculated as follows:

$$\eta W_{i,j,2}^{comp,FP} - W_{i,j,2}^{exp,FP}/\eta = \sum_{k \in K} FP_{i-1,j,k,2} T_{i-1,j,2}^s R_{gas} \frac{\gamma}{\gamma-1} \{(PR_{i-1,j,2}^F)^{\frac{\gamma-1}{\gamma}} - 1\} \quad i \in I, j \in J \quad (50)$$

$$\eta W_{i,j,2}^{comp,FN} - W_{i,j,2}^{exp,FN}/\eta = \sum_{k \in K} FN_{i,j,k,2} T_{i,j,2}^s R_{gas} \frac{\gamma}{\gamma-1} \{(\frac{1}{PR_{i,j,2}^F})^{\frac{\gamma-1}{\gamma}} - 1\} \quad i \in I, j \in J \quad (51)$$

The work terms related to feed streams and jump connecting streams are calculated in a similar way:

$$\eta W_{i,j,f}^{comp,FS} - W_{i,j,f}^{exp,FS}/\eta = \sum_{k \in K} M_{i,j,k,f} T_f^{feed} R_{gas} \frac{1}{n_{fs}} \{(PR_{i,j,f}^{feed})^{n_{fs}} - 1\} \quad i \in I, j \in J, f \in FS \quad (52)$$

$$\eta W_{i,j,i',j'}^{comp,JF} - W_{i,j,i',j'}^{exp,JF}/\eta = J_{i,j,i',j'}^T T_{i,j} R_{gas} \frac{\gamma}{\gamma-1} \{(\frac{P_{i',j'}}{P_{i,j}})^{\frac{\gamma-1}{\gamma}} - 1\} \quad (i,j,i',j) \in LN(i,j,i',j') \quad (53)$$

Here n_{fs} is the adiabatic compression coefficient for source stream f.

4.7. Objective Function

We consider the components of economic objective in the work of Hasan et al. [4] and derive the objective function for the FGN synthesis as follows.

$$\begin{aligned} \min \quad TAC = &\sum_{f \in FS}(\sum_{i \in I}\sum_{j \in J} UFC_f F_f^{feed} z_{i,j,f}^{feedfrac} + Di_f(F_f^{feed} - \sum_{i \in I}\sum_{j \in J} F_f^{feed} z_{i,j,f}^{feedfrac})) \\ &- \sum_{p \in PS} Rev_p(\sum_{k \in K} LHV_k(\sum_{i \in I}\sum_{j \in J} H_{i,j,k,ps}) - De_p) + \sum_{i \in I}\sum_{j \in J}\sum_{f \in FS} \pi_f F_f^{feed} z_{i,j,f}^{feedfrac} \\ &+ CC^{HU}\sum_{i \in I}\sum_{j \in J} Qh_{i,j} + CC^{CU}\sum_{i \in I}\sum_{j \in J} Qc_{i,j} + CC^{exp}\sum_{i \in I}\sum_{j \in J} W_{i,j}^{exp} + CC^{com}\sum_{i \in I}\sum_{j \in J} W_{i,j}^{com} \end{aligned} \quad (54)$$

This objective function aims at minimizing total annual cost (TAC). Here parameter UFC_f is the unit cost of different source streams, Di_f is the unit cost of treatment cost for unused source streams, Rev_p is the unit profit from excess energy in product stream p. Besides, the parameter π_f denotes the unit transportation cost for source stream f. Parameters CC^{HU}, CC^{CU}, CC^{exp} and CC^{com} denote the unit cost of heaters, coolers, expansion operations and compression operations, respectively. The first term in the objective function consists of source stream purchase cost and disposal cost. The second term corresponds to the profit gained from the released excess amount of energy in product stream p. The third term indicates the transporting cost of source streams. The last four terms refer to overall cost (both capital cost and operating cost) for heaters, coolers, expansion operations and compression operations.

This completes the MINLP model for block-based FGN synthesis. It should be noted that commercial solvers can handle the proposed FGN design problems with small number of sinks or pools. However, when a large-scale problem is considered, further simplification can be made by fixing the streams associated with unused blocks to zero when the number of pools and sinks do not match each other. This fixing ensures that the number of blocks in the first row is only equal to the number of pools assigned in the system and the number of blocks in the second row is equal to the number of sinks.

5. Case Study

In this section, the FGN synthesis problem in an LNG plant is presented to demonstrate the application of block superstructure in synthesis of FGN. We consider two cases for the FGN synthesis problem: case 1 for representation without intermediate pools; case 2 for representation with intermediate pools. The case study is from the work of Hasan et al. [4] and all problem instances are solved using ANTIGONE 1.1. [38] in GAMS 24.4 on a Dell Optiplex 9020 computer (Intel 8 Core i7-4770 CPU 3.4 GHz, 15.5 GB memory) running Springdale Linux.

5.1. Case Study Description

Natural gas (NG) utilization has expanded from residential utilization to industrial productions due to its lower waste emission compared with fossil fuel [39]. NG is delivered to destination by transporting through pipelines or transporting as liquefied natural gas (LNG) [40]. For long-distance transportation, LNG is preferred for economical, technical, safety-related, and political considerations [41]. A conventional LNG plant flowsheet is found in Figure 6. Typically, an LNG process train contains acid-gas removal, dehydration and mercury removal, liquefaction, nitrogen rejection, and sulfur recovery systems, etc. [41,42]. The fuel gas system in the LNG plant normally takes the natural gas as a feed (FFF) to generate steam, provide power and supply electricity to the LNG process, while large amount of energy is lost through flares, turbine exhausts, flash gas, etc. if they are not integrated to fuel sinks in the process. Hence, identification of other fuel gas sources (FFP) in the LNG plant can help to effectively exploit their heating value and reduce the consumption of FFF.

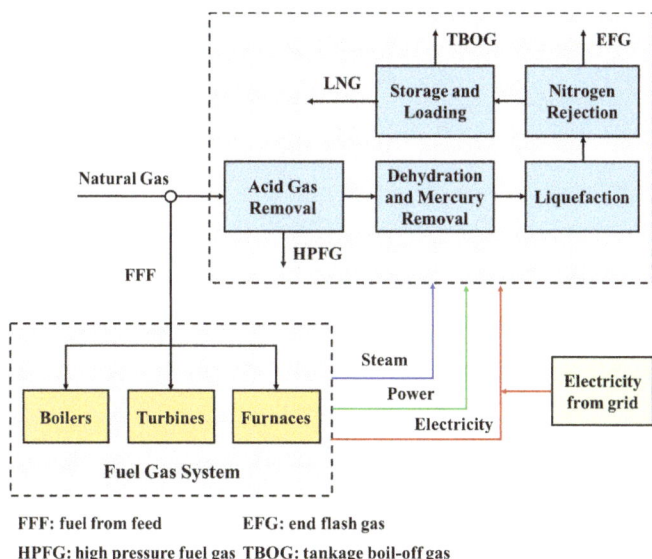

Figure 6. Process diagram for a conventional LNG plant.

As can be seen from Figure 6, there are three fuel gas sources from byproducts (FFP) for the fuel gas system: high-pressure fuel gas (HPFG) from acid gas removal, end flash gas (EFG) from nitrogen rejection, and tankage boil-off gas (TBOG) from storage process. These fuel gas sources EFG, HPFG, TBOG, FFF are represented as S_1, S_2, S_3, and S_4 respectively. The main components in each source streams are methane, ethane, propane, C_{3+}, CO and N_2. Although the definition of fuel gas network is taken from the literature (Hasan et al. [4]), the model we utilized in this work is not based on the total flowrate but the component flowrate. Because of the model discrepancy, we keep part of the fuel gas source data from the literature in Table 1 and update other required component parameters in Table 2. These component parameters include lower heating value (LHV) and reverse specific gravity ($1/SG$) corresponding to the sink requirements on these specifications as well as component heat capacity (Cp_k) required for block energy balance.

Within the LNG process, liquefaction is the most energy-intensive process section and the majority of energy is required to run refrigeration compressors which are driven by frame-type gas turbine drivers (GTD) and gas turbine generators (GTG). Besides, boilers consume certain amount of fuel gas to generate steam for the LNG process. According to similarity of specifications among fourteen units [four gas turbine generators (GTG) for power generation, two gas turbine drivers (GTD) for the propane cycle, three GTDs for the mixed refrigerant (MR) cycle, and five boilers] that consume fuel in the LNG plant, five process sinks are identified (C_1, C_2, C_3, C_4, and C_5). The sink data is directly taken from the literature without any changes and listed in Table 3. The sink specifications include energy demand, material demand, temperature and pressure specification, moisture dew-point temperature (MDP_p) and hydrocarbon dew-point temperature (HDP_p), lower heating value (LHV) and reverse specific gravity ($1/SG$). We do not consider the profit from excess energy in sink streams C_1, C_2, C_3 and C_4. It should be noted that all the data have been converted to standard units.

Table 1. Sources streams specifications.

Specification/Parameter	EFG	HPFG	TBOG	FFF
Adiabatic compression coefficient, n_{fs}	0.254	0.2	0.18	0.2
Availability, F_f^{feed} (kmol/s)	0.92938	0.05310	0.18255	<7.30229
Temperature, T_f^{feed} (K)	240	325	113	298
Pressure, p_f^{feed} (bar)	1.72369	7.58423	1.72369	26.20007
Methane, CH_4 (%)	60.0	81.0	92.0	85.0
Ethane, C_2H_6 (%)	0.0	6.0	0.0	5.0
Propane, C_3H_8 (%)	0.0	5.0	0.0	4.0
C_{3+} (%)	0.0	2.5	0.0	2.0
CO (%)	0.0	0.0	0.0	0.05
N2 (%)	40.0	5.5	8.0	3.95
Source unit cost, UFC_f ($/kmol)	0.0	0.0	0.0	4.184
Source disposal cost, Di_f ($/kmol)	0.209	0.292	0.209	0
Feed transporting cost, π_f ($/kmol)	0.0008	0.0008	0.0008	0.0008

EFG: end flash gas; **HPFG**: high-pressure fuel gas; **TBOG**: tankage boil-off gas; **FFF**: fuel from feed.

Table 2. Component quality parameters.

Parameter	Methane	Ethane	Propane	C_{3+}	CO	N2
LHV(MJ/kmol)	800.234	1425.580	2041.113	2654.134	282.637	0
1/SG (28.96/mol wt)	1.8060	0.9636	0.6571	0.4985	1.0344	1.0342
Cp [KJ/(kmol K)]	37.16	57.40	80.30	114.93	29.20	29.15

Table 3. Specification for product streams (sinks).

Specification/Parameter	C1	C2	C3	C4	C5
Energy demand, De_p (MJ/s)	152.309	149.378	120.305	149.378	87.921
Material demand, $[D_p^L, D_p^U]$ (kmol/s)	0.159–0.172	0.156–0.169	0.159–0.172	0.149–0.169	0.132–0.199
Temperature range, $[T_p^{min}, T_p^{max}]$ (K)	113–1000	113–1000	113–1000	113–1000	113–1000
Pressure range, $[P_p^{min}, P_p^{max}]$ (bar)	1.72–24.82	1.72–24.82	1.72–24.82	1.72–24.82	1.72–24.82
MDP_p (K)	277	277	277	277	277
HDP_p (K)	277	277	277	277	277
LHV(MJ/kmol)	264.885–8829.500	264.885–8829.500	264.885–8829.500	264.885–8829.500	264.885–8829.500
$1/SG$ (28.96/mol wt)	1.0–2.4	1.0–2.4	1.0–2.4	1.0–2.4	1.0–2.4
Methane, CH_4 (%)	>85.0	>85.0	>85.0	>85.0	>65.0
Ethane, C_2H_6 (%)	<15.0	<15.0	<15.0	<15.0	<15.0
Propane, C_3H_8 (%)	<15.0	<15.0	<15.0	<15.0	<15.0
C_{3+} (%)	<5.0	<5.0	<5.0	<5.0	<5.0
CO (%)	<10.0	<10.0	<10.0	<10.0	<10.0
N2 (%)	<15.0	<15.0	<15.0	<15.0	<15.0
Treatment factor, ψ_{sp}	1.0	1.0	1.0	1.0	1.0
Unit profit, Rev_p ($/KJ)	0	0	0	0	6.6347×10^{-6}

Table 4 lists the cost parameters including capital expenditure (CAPEX) and operating expenditure (OPEX) for various FGN units (heaters/coolers, and compressors/expanders). Finally, we assign temperature lower bound as T^{min} = 113 K, temperature upper bound as T^{max} = 1000 K. The transporting cost for each source stream f is $\pi_f = 8.37 \times 10^{-4}$ $/kmol. The adiabatic compression coefficient for process streams is $\gamma = 0.286$. The operating time per year is 365 days.

Table 4. CAPEX and OPEX Coefficients for Various Equipment Units.

Unit	CAPEX ($/KWh)	OPEX ($/KWh)	Total ($/KWh)
Compressor	10	0.01	$CC^{com} = 10.01$
Expander	1	0.05	$CC^{exp} = 1.05$
Heater	5	0.01	$CC^{HU} = 5.01$
Cooler	5	0.02	$CC^{CU} = 5.02$

5.2. Case 1: FGN Synthesis Without Pools

In this case, the block representation of FGN shown in Figure 3 is used. To avoid part of product stream recycled as feed into adjacent blocks through direct connecting flow, all the horizontal and vertical material flow, namely $F_{i,j,k,d=1}$ and $F_{i,j,k,d=2}$, are ignored for each product block. Accordingly, horizontal ($d = 1$) and vertical ($d = 2$) energy flow, $EF_{i,j,d}$, as well as their associated work terms are removed from energy balance. Also jump connecting streams from product blocks are fixed to be zero since they make the product blocks as intermediate pools.

The model for FGN without intermediate pools has 397 continuous variables, 45 binary variables, 849 bilinear terms, 243 signomial terms. The solution is obtained within 565 CPU seconds with optimal total annual cost as 70,136,064 $/year and optimality gap as 0.1%. The optimal solution reported in the literature [4] is 79,943,071 $/year. This 12.27 % reduction in TAC could be attributed to the facts that: (1) we do not consider the nonlinear quality in this work, i.e., wobbe index, which brings less strict requirement on network design; and (2) we assume ideal gas instead of real gas for expansion operation. The optimal block configuration for FGN and its corresponding optimal network are shown in Figure 7a,b respectively.

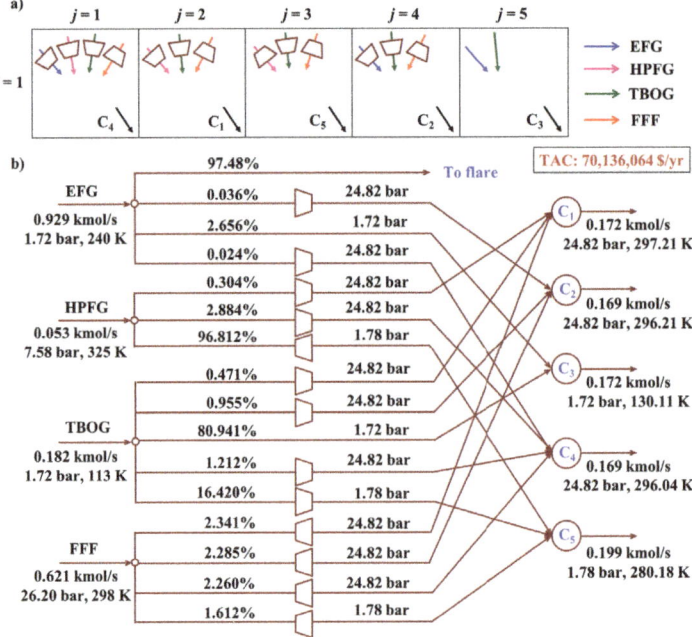

Figure 7. Block representation and process flowsheet for the optimal solution of FGN without intermediate pools: (**a**) Block representation for the optimal solution of FGN; (**b**) process flowsheet for the optimal solution of FGN.

In the block representation of the optimal result (Figure 7a), the block $B_{1,1}$ takes compressed streams from source EFG, HPFG, and TBOG and expanded stream from source FFF while supplying sink stream to header C_4. Both the block $B_{1,2}$ and block $B_{1,3}$ collect part of compressed streams from source HBFG and TBOG and expanded stream from source FFF to generate sink stream for header C_1 and C_5 respectively. In block $B_{1,4}$, partial compressed streams from source EFG and TBOG mix with expanded stream from source FFF. This block yields the sink stream for header C_2. The block $B_{1,5}$ blend streams from source EFG and TBOG to yield a product stream for header C_3.

This obtained block representation is converted into FGN network shown in Figure 7b. It utilizes both HPFG and TBOG fully. Among all sink streams, only C_4 uses all source streams EFG, HPFG, TBOG, FFF while C_1, C_5 only use HPFG, TBPG, and FFF as source streams. Sink C_2 blends streams from EFG, HPFG, and TBOG. Sink C_3 takes source streams from EFG, and TBOG. It should be noted that both C_2 and C_3 accept part of EFG. The whole FGN network could only utilize 2.716% of EFG and the rest of it goes to flare. The reason is that EFG contains low methane (60%) and high inert content (40%). To utilize EFG as much as possible, it should be mixed with other source streams; however, such mixing could bring unacceptable large flows to sinks so EFG is only partially utilized in the system. Regarding the FFF, none of sinks are taking it alone and sink C3 does not use FFF at all.

The optimal header pressures are 24.82, 24.82, 1.72, 24.82, and 1.78 bar for header C_1–C_5 respectively. The flow rate of sink streams at headers at C_1–C_5 are 0.172, 0.169, 0.172, 0.169, 0.199 kmol/s respectively. The optimal header temperatures are found as 297.21, 296.21, 130.11, 296.04, 280.18 K for header C_1–C_5 respectively. HPFG needs expanders before mixing with TBOG (1.72 bar) and FFF (1.72 bar) in C_5 because of its high pressure (7.58 bar). Similarly, all the FFF (26.20 bar) needs expanders so as to mix with other flows in C_1, C_2, C_4 and C_5. However, EFG and TBOG do not need any compressors or expanders before entering C_3, which are already at 1.72 bar. No heating and cooling operations are required in the optimal FGN.

5.3. Case 2: FGN Synthesis With Pools

To investigate the influence of existence of pools on improving the economic performance of FGN, we use the representation shown in Figure 4. Note that the problem specifications are the same for case 1 and case 2. The only difference is that the row number $I = 1$ for case 1 and it is $I = 2$ for case 2. The material balance involving jump connecting streams is utilized to build connection between pool blocks and product blocks. For the jump flow $J^p_{i,j,k}$ withdrawn from the pool block, it is distributed back into other pool blocks or product block. To avoid self-recycle of the jump connecting stream, the inlet streams coming via jump flows from the same block is fixed to be zero, $J_{i,j,i',j',k} = 0$, where $i = i'$ and $j = j'$. External feed streams are only allowed to enter into the first row, i.e., pool blocks while external product streams are only withdrawn from the second row, i.e., sink blocks.

The model contains 1741 continuous variables, 58 binary variables, 9530 bilinear terms and 1321 signomial terms. The comparison of model statistics for these two cases are summarized in Table 5.

Table 5. Summary of model statistics for case 1 and case 2.

	Case 1	Case 2
Continuous variable	397	1741
Binary variable	45	58
Bilinear terms	849	9530
Signomial terms	243	1321
CPU time (second)	565	935
Solution (MM$/year)	70.1	69.3

The solution is obtained within 935 CPU seconds with optimal TAC as 69,259,363 $/year and optimality gap as 0.1%. The involvement of intermediate pools results in a reduction of total annual cost by 1.25%, compared to the one reported as 70,136,064$/year for the fuel gas network without intermediate pools. The detailed TAC comparison is shown in Table 6. Through this additional row for intermediate pools, less source cost (specifically FFF) for purchase and disposal is needed although the revenue obtained from excess energy in product stream C_5 is also decreased. Less cost is spent on expansion and compression operation in case 2. No heating and cooling operations are required in the optimal FGN for both cases. Figure 8 shows the optimal fuel gas network configuration.

Table 6. Total annual cost comparison for case 1 and case 2.

TAC Component ($/Year)	Case 1	Case 2
Source cost	87,852,764	68,530,679
Revenue from excess energy in sinks	17,736,117	16,329,735
Source transportation cost	23,246	23,246
Heaters cost	0	0
Coolers cost	0	0
Expansion operation cost	5850	3016
Compression operation cost	3168	2744

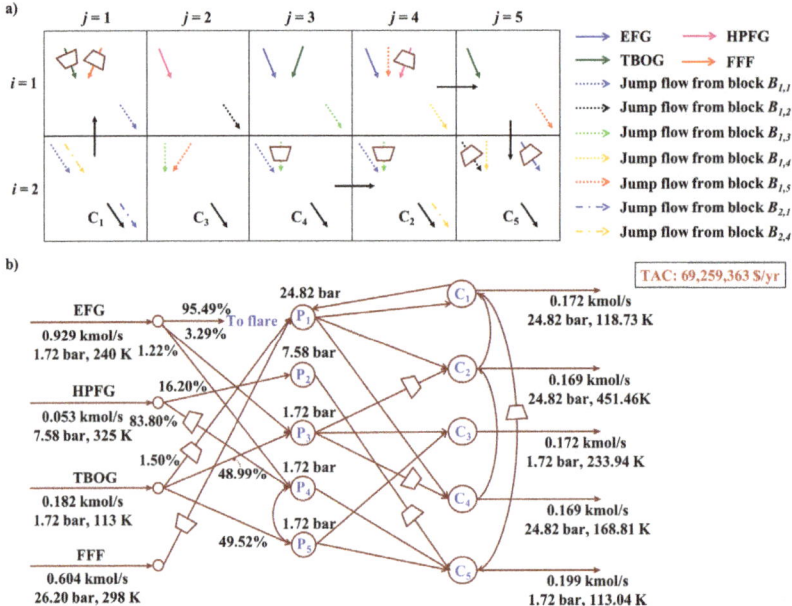

Figure 8. Block representation and process flowsheet for the optimal solution of FGN with intermediate pools: (**a**) Block representation for the optimal solution of FGN; (**b**) process flowsheet for the optimal solution of FGN.

The obtained block representation for the FGN is given in Figure 8a. Feed stream EFG is distributed into block $B_{1,3}$ and $B_{1,4}$. Part of feed stream HPFG is expanded and then enters into block $B_{1,4}$ while extra amount distributes into block $B_{1,2}$. In addition, feed stream TBOG is partially supplied to block $B_{1,3}$ and block $B_{1,5}$. Some other amount of TBOG is compressed and then enter block $B_{1,1}$. The feed stream FFF only enters the block $B_{1,1}$ after expanding operation. The blocks $B_{1,j}$ in the first row (column number j ranges from 1 to 5) collect the mixed stream and yield the outlet jump flow which are supplied into the second row. Hence, these blocks are identified as intermediate pools, i.e., P_1, P_2, P_3, P_4, and P_5. At the second row of the block representation, the jump flow withdrawn from block $B_{1,1}$ and block $B_{2,4}$ mix at block $B_{2,1}$, which supplies product stream to sink C_1 and generates outlet jump flows entering block $B_{2,5}$. The block $B_{2,2}$ blends the outlet jump flows from block $B_{1,3}$ and $B_{1,5}$ to obtain sink stream C_3. The outlet jump flows from block $B_{1,3}$ split into three parts. One part compresses first and mixes with outlet jump flow from block $B_{1,1}$ at block $B_{2,3}$, where the sink stream C_4 is generated. Another part is compressed and mixes with outlet jump flow from block $B_{1,1}$ at block $B_{2,4}$, where the sink stream C_2 is obtained. The last part mixes with outlet jump flow from block $B_{1,5}$ at block $B_{2,2}$, where the sink stream C_3 is obtained. Besides, an outlet jump flow is withdrawn from block $B_{2,4}$ and fed into block $B_{2,1}$. The outlet jump flows from block $B_{2,1}$, $B_{1,2}$ are compressed and mixed with other outlet jump flows from block $B_{1,4}$ and $B_{1,5}$ to supply the sink stream C_5.

The corresponding network structure is shown in Figure 8b. The optimal network consumes both HPFG and TBOG fully. Since all the blocks in the first row embed the inlet flow for mixing, five pools can be identified. Pool P_1 accepts source stream from TBOG and FFF, which only supply feed to P_1. P_2 takes part of source stream HPFG. P_3 blends streams from EFG and TBOG. Part of external stream from EFG and HPFG enter pool P_4 while P_5 only takes stream from source TBOG. The outlet flow from pool P_1 is distributed into sink C_1, C_2 and C_4. The outlet flows from pool P_2 and P_4 are directly transported to sink C_5. Sink C_2, C_3 and C_4 accept inlet flow withdrawn from pool P_3. Part of the outlet flow from pool P_5 is recycled back to pool P_4 and another is transported into sink C_3. Part of product streams from C_2 and C_1 are recycled back to C_1 and C_5. The utilization of EFG in the whole FGN network is only 4.51% and the rest of it goes to flare.

The header pressures are 24.82, 24.82, 1.72, 24.82, and 1.72 bar for C_1–C_5 respectively. Expanders are arranged on inlet stream to P_4 from HPFG and inlet stream to P_1 from FFF. TBOG needs compressor before mixing with FFF in P_1 because of its low pressure (7.58 bar). Compressors are placed on the outlet streams of P_3 to sink C_2 and C_4 respectively. Expanders are arranged on the stream from P_2 to sink C_5 as well as on the stream from sink C_1 to sink C_5 so as to meet the pressure requirement. The temperature for sink streams C_1–C_5 are 118.73, 451.46, 233.94, 168.81 and 113.04 K. In addition, headers C_1–C_5 collect the flow rate of sink streams as 0.172, 0.169, 0.172, 0.169 and 0.199 kmol/s respectively.

To summarize for the case study section, the block-based representation method can effectively handle the FGN synthesis problem and the involvement of intermediate pools helps to improve the management of FGN network, which decreases the total annual cost.

6. Conclusions

We present an abstract superstructure representation for FGN synthesis, which is based on a block-based arrangement of sources and sinks. Each block allows multiple external fuel gas source streams and single fuel gas sink streams. The direct connecting streams between adjacent blocks and jump connecting streams among all nonadjacent blocks enable many alternative ways for the mass and energy flow from sources to sinks. The blocks with multiple inlet streams serve as mixers and the blocks with multiple outlet streams are splitters. These blocks form a superstructure when arranged in a two-dimensional grid. The row number is determined by the number of intermediate pool layers and the number of sink layers. The column number is determined by the number of intermediate pools and the number of sinks. With the representation method, an MINLP model for fuel gas network synthesis problem was proposed with constraints on material balance, energy balance, flow directions, logical relations, and work calculation. A case study from LNG plant was presented for two instances: one without intermediate pools and another with intermediate pools. The FGN with pools reduces the total annual cost by 1.25% to 69.3 MM\$/year, compared to TAC (70.1 MM\$/year) of the FGN without intermediate pools. This case study revealed that the block-based representation method enables the synthesis of fuel gas network and helps to find novel network design. Note that the block-based representation method is initially proposed for systematic process intensification, and then applied to process synthesis. The application of block-based approach for FGN integration suggests a general framework towards process intensification, integration and synthesis.

Acknowledgments: The authors gratefully acknowledge financial support from the U.S. National Science Foundation (NSF CBET-1606027).

Author Contributions: J.L., S.E.D. and M.M.F.H. conceived the model and prepared the manuscript.

Conflicts of Interest: The authors declare no conflict of interest.

Nomenclature

Sets and Indices

I	Set of row numbers indexed by i
J	Set of column numbers indexed by j
D	Set of flow alignments indexed by d
K	Set of components indexed by k
FS	Set of feed streams indexed by f
PS	Set of product streams indexed by p

Subsets

$LN(i,j,i',j')$	Set designating the connection between block $B_{i,j}$ and block $B_{i',j'}$
$kp(k,p)$	Set relating the key component k with product stream p with purity specifications

Variables

$F_{i,j,k,d}$	Flowrate of component k between block $B_{i,j}$ and $B_{i,j+1}$ in the flow alignment d
$Q_{i,j}$	Amount of heat/cold utility consumed in block $B_{i,j}$
$W_{i,j}$	Amount of work energy added into or withdrawn from block $B_{i,j}$
TAC	Total annual cost

Positive Continuous Variables

$F_{f,p}$	Stream connecting source f and sink p in the classic superstructure
$M_{i,j,k,f}$	Component flowrate for k in external feed stream f into block $B_{i,j}$
$M_{i,j,k}^f$	Component flowrate k of external feed stream into block $B_{i,j}$
$z_{i,j,f}^{feedfrac}$	Distribution of feed f into block $B_{i,j}$
$H_{i,j,k,p}$	Amount of component k in external product stream p withdrawn from block $B_{i,j}$
$H_{i,j,k}^p$	Component flowrate k of external product stream withdrawn from block $B_{i,j}$
$J_{i,j,i',j',k}$	Flowrate of component k from block $B_{i,j}$ to another block $B_{i',j'}$
$J_{i,j,k}^f$	Overall component flowrate k of jump connecting flow into block $B_{i,j}$
$J_{i,j,k}^p$	Overall component flowrate k of jump connecting flow withdrawn from block $B_{i,j}$
$FP_{i,j,k,d}$	Positive component of flow $F_{i,j,k,d}$
$FN_{i,j,k,d}$	Negative component of flow $F_{i,j,k,d}$
$FP_{i,j,d}^T$	Total flowrate for flow $FP_{i,j,k,d}$
$FN_{i,j,d}^T$	Total flowrate for flow $FN_{i,j,k,d}$
$J_{i,j,i',j'}^T$	Total flowrate for flow $J_{i,j,i',j',k}$
$M_{i,j,f}^T$	Total flowrate for flow $M_{i,j,k,f}$
$H_{i,j,p}^T$	Total flowrate for flow $H_{i,j,k,p}$
$y_{i,j,k}^b$	Block composition of component k in block $B_{i,j}$
$P_{i,j}$	Pressure designation in block $B_{i,j}$
$T_{i,j}$	Temperature designation in block $B_{i,j}$
$Q_{i,j}^h$	Heat amount supplied into block $B_{i,j}$
$Q_{i,j}^c$	Heat amount withdrawn from block $B_{i,j}$
$EF_{i,j,d}$	Stream enthalpy carried by the material flow $F_{i,j,k,d}$ in flow direction d
$EM_{i,j}$	Overall enthalpy brought into block $B_{i,j}$ along with feed streams
$EP_{i,j}$	Overall enthalpy taken away by product streams at block $B_{i,j}$
$EJ_{i,j}^f$	Overall enthalpy carried into block $B_{i,j}$ through jump flow
$EJ_{i,j}^p$	Overall enthalpy taken out from block $B_{i,j}$ through jump flow

$W_{i,j}^{com}$	Work energy associated with compression operation
$W_{i,j}^{exp}$	Work energy associated with expansion operation
$W_{i,j,d}^{comp,FP}$	Compression work for positive component of flow $F_{i,j,k,d}$
$W_{i,j,d}^{comp,FN}$	Compression work for negative component of flow $F_{i,j,k,d}$
$W_{i,j,f}^{comp,FS}$	Compression work for feed stream f
$W_{i',j',i,j}^{comp,JF}$	Compression work for jump flow $J_{i,j,i',j',k}$
$W_{i,j,d}^{exp,FP}$	Expansion work for positive component of flow $F_{i,j,k,d}$
$W_{i,j,d}^{exp,FN}$	Expansion work for negative component of flow $F_{i,j,k,d}$
$W_{i,j,f}^{exp,FS}$	Expansion work for feed stream f
$W_{i',j',i,j}^{exp,JF}$	Expansion work for jump flow $J_{i,j,i',j',k}$
$PR_{i,j,d}^{F}$	Pressure ratio between the block $B_{i,j+1}$ and $B_{i,j}$ for flow alignment $d = 1$ or between the block $B_{i+1,j}$ and $B_{i,j}$ for flow alignment $d = 2$

Binary Variables

$z_{i,j,d}^{Plus}$	1 if $F_{i,j,k,d}$ is from block $B_{i,j}$ to $B_{i,j+1}$ ($d = 1$) or from block $B_{i,j}$ to $B_{i+1,j}$ ($d = 2$)
$z_{i,j,p}^{product}$	1 if product stream p is withdrawn from block $B_{i,j}$
$z_{i,j,d}^{cr}$	1 if boundary between $B_{i,j}$ and $B_{i,j+1}$ for $d = 1$ (between $B_{i,j}$ and $B_{i+1,j}$ for $d = 2$) is completely restricted

Parameters

T_f	Temperature of feed stream f
P_f	Pressure of feed stream f
T^{min}	Minimum temperature in the process
T^{max}	Maximum temperature in the process
FL	Flowrate lower bound in the process
FU	Flowrate upper bound in the process
T_p^{min}	Minimum temperature of product stream p
T_p^{max}	Maximum temperature of product stream p
P_p^{min}	Minimum pressure of product stream p
P_p^{max}	Maximum pressure of product stream p
F_f^{feed}	Available amount of feed stream f
$y_{k,f}^{feed}$	Specification of component k in feed stream f
D_p^{L}	Minimum amount requirement for product p
D_p^{U}	Maximum amount requirement for product p
De_p^{L}	Minimum energy demand for product p
De_p^{U}	Maximum energy demand for product p
$y_{k,p}^{min,prod}$	Minimum purity requirement of component k in product stream p
$y_{k,p}^{max,prod}$	Maximum purity requirement of component k in product stream p
$q_{s,p}^{min,prod}$	Minimum requirement of specification s in product stream p
$q_{s,p}^{max,prod}$	Maximum requirement of specification s in product stream p
$\pi_{k',k'',p}^{prod}$	Minimum product ratio requirement between component k' and component k'' for product p
LHV_k	Lower heating value for component k
$q_{s,k}$	Specification s for component k
Cp_k	Heat capacity of component k
MDP_p	Moisture dew-point temperature for the product p
HDP_p	Hydrocarbon dew-point temperature for the product p
R_{gas}	Gas constant
γ	Adiabatic compression coefficient for process streams
η	Adiabatic compression efficiency
n_{fs}	Adiabatic compression coefficient for feed f
UFC_f	Unit cost of different source streams

Di_f	Unit cost of treatment cost for the remaining source stream
Rev_p	Unit profit from excess energy in product stream p
π_f	Unit transportation cost for source stream f
CC^{HU}	Unit cost of heaters
CC^{CU}	Unit cost of coolers
CC^{exp}	Unit cost of expansion operations
CC^{com}	Unit cost of compression operations

Abbreviations

FGN	Fuel Gas Network
MINLP	Mixed-integer nonlinear programming problem
NLP	Nonlinear programming problem
HEN	Heat exchange network
MEN	Mass exchange network
UPCS	unit-port-conditioning-stream approach
L	Lower bound
U	Upper bound
min	Minimum
max	Maximum
prod	Product
feed	Feed stream
T	Total
$B_{i,j}$	The block at row i and column j
NG	Natural gas
EFG	End flash gas
HPFG	High-pressure fuel gas
TBOG	Tankage boil-off gas
FFF	Fuel from feed
FFP	Fuel from products or byproducts
GTG	Gas turbine generators
GTD	Gas turbine drivers
MR	Mixed refrigerant

References

1. Tahouni, N.; Gholami, M.; Panjeshahi, M.H. Integration of flare gas with fuel gas network in refineries. *Energy* **2016**, *111*, 82–91.
2. Zhang, J.; Zhu, X.; Towler, G. A simultaneous optimization strategy for overall integration in refinery planning. *Ind. Eng. Chem. Res.* **2001**, *40*, 2640–2653.
3. Pellegrino, J.; Brueske, S.; Carole, T.; Andres, H. *Energy and Environmental Profile of the US Petroleum Refining Industry*; Technical Report; EERE Publication and Product Library: Washington, DC, USA, 2007.
4. Hasan, M.M.F.; Karimi, I.A.; Avison, C.M. Preliminary synthesis of fuel gas networks to conserve energy and preserve the environment. *Ind. Eng. Chem. Res.* **2011**, *50*, 7414–7427.
5. *U.S. Department of Energy (DOE): Refinery Capacity 2017*; Number Energy Information Administration: Washington, DC, USA, 2017.
6. De Carli, A.; Falzini, S.; Liberatore, R.; Tomei, D. Intelligent management and control of fuel gas network. In Proceedings of the IECON 02 IEEE 2002 28th Annual Conference of the Industrial Electronics Society, Sevilla, Spain, 5–8 November 2002; Volume 4, pp. 2921–2926.
7. Zhou, L.; Liao, Z.; Wang, J.; Jiang, B.; Yang, Y.; Du, W. Energy configuration and operation optimization of refinery fuel gas networks. *Appl. Energy* **2015**, *139*, 365–375.
8. Zhang, J.; Rong, G. An MILP model for multi-period optimization of fuel gas system scheduling in refinery and its marginal value analysis. *Chem. Eng. Res. Des.* **2008**, *86*, 141–151.
9. Zhang, J.; Rong, G.; Hou, W.; Huang, C. Simulation based approach for optimal scheduling of fuel gas system in refinery. *Chem. Eng. Res. Des.* **2010**, *88*, 87–99.
10. White, D.C. Advanced automation technology reduces refinery energy costs. *Oil Gas J.* **2005**, *103*, 45–53.

11. Ismail, O.S.; Umukoro, G.E. Global impact of gas flaring. *Energy Power Eng.* **2012**, *4*, 290–302.
12. Fawole, O.G.; Cai, X.M.; MacKenzie, A. Gas flaring and resultant air pollution: A review focusing on black carbon. *Environ. Pollut.* **2016**, *216*, 182–197.
13. Quan, C.; Gao, N.; Wu, C. Utilization of NiO/porous ceramic monolithic catalyst for upgrading biomass fuel gas. *J. Energy Inst.* **2017**, doi:10.1016/j.joei.2017.02.008.
14. Mokheimer, E.M.; Dabwan, Y.N.; Habib, M.A. Optimal integration of solar energy with fossil fuel gas turbine cogeneration plants using three different CSP technologies in Saudi Arabia. *Appl. Energy* **2017**, *185*, 1268–1280.
15. Friedler, F. Process integration, modelling and optimisation for energy saving and pollution reduction. *Appl. Therm. Eng.* **2010**, *30*, 2270–2280.
16. El-Halwagi, M.M. Pollution prevention through process integration. *Clean Prod. Process.* **1998**, *1*, 5–19.
17. Jagannath, A.; Hasan, M.M.F.; Al-Fadhli, F.M.; Karimi, I.A.; Allen, D.T. Minimize flaring through integration with fuel gas networks. *Ind. Eng. Chem. Res.* **2012**, *51*, 12630–12641.
18. Chen, Q.; Grossmann, I. Recent developments and challenges in optimization-based process synthesis. *Annu. Rev. Chem. Biomol. Eng.* **2017**, *8*, 249–283.
19. Tahouni, N.; Gholami, M.; Panjeshahi, M. Reducing energy consumption and GHG emission by integration of flare gas with fuel gas network in refinery. *Int. J. Chem. Nucl. Mater. Metall. Eng.* **2014**, *8*, 900–904.
20. Kondili, E.; Pantelides, C.; Sargent, R. A general algorithm for short-term scheduling of batch operations—I. MILP formulation. *Comput. Chem. Eng.* **1993**, *17*, 211–227.
21. Yeomans, H.; Grossmann, I.E. A systematic modeling framework of superstructure optimization in process synthesis. *Comput. Chem. Eng.* **1999**, *23*, 709–731.
22. Friedler, F.; Tarjan, K.; Huang, Y.; Fan, L. Graph-theoretic approach to process synthesis: Axioms and theorems. *Chem. Eng. Sci.* **1992**, *47*, 1973–1988.
23. Friedler, F.; Tarjan, K.; Huang, Y.; Fan, L. Graph-theoretic approach to process synthesis: Polynomial algorithm for maximal structure generation. *Comput. Chem. Eng.* **1993**, *17*, 929–942.
24. Bagajewicz, M.J.; Manousiouthakis, V. Mass/heat-exchange network representation of distillation networks. *AIChE J.* **1992**, *38*, 1769–1800.
25. Bagajewicz, M.J.; Pham, R.; Manousiouthakis, V. On the state space approach to mass/heat exchanger network design. *Chem. Eng. Sci.* **1998**, *53*, 2595–2621.
26. Papalexandri, K.P.; Pistikopoulos, E.N. Generalized modular representation framework for process synthesis. *AIChE J.* **1996**, *42*, 1010–1032.
27. Proios, P.; Goula, N.F.; Pistikopoulos, E.N. Generalized modular framework for the synthesis of heat integrated distillation column sequences. *Chem. Eng. Sci.* **2005**, *60*, 4678–4701.
28. Lutze, P.; Gani, R.; Woodley, J.M. Process intensification: A perspective on process synthesis. *Chem. Eng. Process. Process Intensif.* **2010**, *49*, 547–558.
29. Lutze, P.; Babi, D.K.; Woodley, J.M.; Gani, R. Phenomena based methodology for process synthesis incorporating process intensification. *Ind. Eng. Chem. Res.* **2013**, *52*, 7127–7144.
30. Babi, D.K.; Holtbruegge, J.; Lutze, P.; Górak, A.; Woodley, J.M.; Gani, R. Sustainable process synthesis—Intensification. *Comput. Chem. Eng.* **2015**, *81*, 218–244.
31. Tula, A.K.; Eden, M.R.; Gani, R. Process synthesis, design and analysis using a process-group contribution method. *Comput. Chem. Eng.* **2015**, *81*, 245–259.
32. Wu, W.; Henao, C.A.; Maravelias, C.T. A superstructure representation, generation, and modeling framework for chemical process synthesis. *AIChE J.* **2016**, *62*, 3199–3214.
33. Wu, W.; Yenkie, K.; Maravelias, C.T. A superstructure-based framework for bio-separation network synthesis. *Comput. Chem. Eng.* **2017**, *96*, 1–17.
34. Demirel, S.E.; Li, J.; Hasan, M.M.F. Systematic process intensification using building blocks. *Comput. Chem. Eng.* **2017**, *105*, 2–38.
35. Li, J.; Demirel, S.E.; Hasan, M.M.F. Simultaneous Process Synthesis and Process Intensification using Building Blocks. *Comput. Aided Chem. Eng.* **2017**, *40*, 1171–1176.
36. Demirel, S.E.; Li, J.; Hasan, M.M.F. A General Framework for Process Synthesis, Integration and Intensification. In Proceedings of the 13th International Symposium on Process System Engineering, San Diego, CA, USA, 1–5 July 2018. (accepted)

37. Li, J.; Demirel, S.E.; Hasan, M.M.F. Process Synthesis using Block Superstructure with Automated Flowsheet Generation and Optimization. *AIChE J.* **2018**, under review.

38. Misener, R.; Floudas, C.A. ANTIGONE: Algorithms for continuous/integer global optimization of nonlinear equations. *J. Glob. Optim.* **2014**, *59*, 503–526.

39. Aslambakhsh, A.H.; Moosavian, M.A.; Amidpour, M.; Hosseini, M.; AmirAfshar, S. Global cost optimization of a mini-scale liquefied natural gas plant. *Energy* **2018**, doi:10.1016/j.energy.2018.01.127.

40. Alabdulkarem, A.; Mortazavi, A.; Hwang, Y.; Radermacher, R.; Rogers, P. Optimization of propane pre-cooled mixed refrigerant LNG plant. *Appl. Therm. Eng.* **2011**, *31*, 1091–1098.

41. Lim, W.; Choi, K.; Moon, I. Current status and perspectives of liquefied natural gas (LNG) plant design. *Ind. Eng. Chem. Res.* **2013**, *52*, 3065–3088.

42. Hasan, M.M.F. Modeling and Optimization of Liquefied Natural Gas Process. Ph.D. Thesis, National University of Singapore, Singapore, 13 August 2009.

Article

Cuboid Packed-Beds as Chemical Reactors?

Raja Ghosh

Department of Chemical Engineering, McMaster University, 1280 Main Street West,
Hamilton, ON L8S 4L7, Canada; rghosh@mcmaster.ca; Tel.: +1-905-525-9140 (ext. 27415)

Received: 12 April 2018; Accepted: 23 April 2018; Published: 1 May 2018

Abstract: Columns are widely used as packed-bed or fixed-bed reactors in the chemical process industry. Packed columns are also used for carrying out chemical separation techniques such as adsorption, distillation, extraction and chromatography. A combination of the variability in flow path lengths, and the variability of velocity along these flow paths results in significant broadening in solute residence time distribution within columns, particularly in those having low bed height to diameter ratios. Therefore, wide packed-column reactors operate at low efficiencies. Also, for a column of a particular bed height, the ratio of heat transfer surface area to reactor volume varies inversely as the radius. Therefore, with wide columns, the available heat transfer area could become a limiting factor. In recent papers, box-shaped or cuboid packed-bed devices have been proposed as efficient alternatives to packed columns for carrying out chromatographic separations. In this paper, the use of cuboid packed-beds as reactors for carrying out chemical and biochemical reactions has been proposed. This proposition is primarily supported in terms of advantages resulting from superior system hydraulics and narrower residence time distributions. Other potential advantages, such as better heat transfer attributes, are speculated based on geometric considerations.

Keywords: packed-bed reactor; packed column; cuboid packed-bed; chemical reactor; residence time distribution; heat transfer

1. Introduction

Packed-bed or fixed-bed reactors are ubiquitous in the chemical process industry [1–3]. With the increasing popularity of flow chemistry-based synthesis, their use is likely to increase [4–6]. A column is the most common configuration for a packed-bed reactor. It is a cylindrical vessel within which a bed of particles is tightly held in place under compression (see Figure 1). In certain specialized forms of packed-beds, such as monoliths [7,8], the particles are fused together. Packed columns are also used for chemical separation techniques such as adsorption, distillation, extraction and chromatography [9–12].

A packed-column reactor generally facilitates contact between the surface of catalyst particles and reactant/s present in the fluid flowing through it. The most common example of the use of such catalytic packed-column reactors is ammonia synthesis. Columns packed with inert material could also be used to bring about intimate contact between reactants in a simple and continuous flow-through mode. In a separation process, the solid material comprising the packed column is generally the separation media, e.g., an adsorbent, being used to separate a solute of interest. Packed columns are also used to increase the interfacial area between immiscible fluids in processes such as distillation and extraction.

The residence time distribution or RTD is an important attribute in packed-bed reactors [13–15]. By the rule of thumb, the wider the column, and the poorer (i.e., the broader) is the RTD. Therefore, wider columns are less efficient than longer columns. Also, with a column of a particular bed height, the heat transfer surface area to reactor volume ratio (i.e., specific surface area, excluding the headers) varies inversely as the radius. Therefore, when columns with large cross-sectional areas are used, the available heat transfer area could become a limiting factor [16–18]. Where this is the case, multi-tube packed-column reactors are commonly used [19–22]. While this multi-tube configuration increases heat-transfer, there are penalties in the form of design complexity, and higher equipment cost.

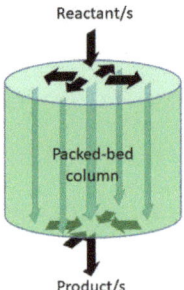

Figure 1. Idealized flow paths in a packed column.

Chromatography is traditionally carried out using packed columns. In recent papers from my group, we have proposed the use of box-shaped or cuboid packed-bed devices as alternative to columns, for carrying out chromatographic separation of proteins [23–26]. The design of these cuboid packed-bed devices was inspired by laterally fed membrane chromatography (or LFMC) devices [27–31], also developed in recent years in my research group. A diagrams of a cuboid packed-bed device is shown in Figure 2. A set of lateral flow channels similar to those utilized in LFMC devices [27–31] are used to distribute the feed liquid into the cuboid packed-bed and collect the effluent liquid emerging from it on the other side. Such flow arrangement ensures uniformity in solute flow path lengths, which in turn contributes towards the narrowing of the solute RTD within the device [24,29]. Figures 1 and 2 show images of a packed column and a cuboid packed-bed respectively, in each case, showing the idealized flow-paths within these devices. As mentioned earlier, a combination of variability in the flow path lengths and variability of velocity along these flow paths results in a significant broadening in solute RTD within a column [23]. This contributes towards peak broadening and poor peak-resolution in multi-component separation. By contrast, the flow path lengths within a cuboid packed-bed are fairly uniform and the velocities along these do not vary that significantly [23]. The cuboid packed-bed devices described in our papers outperformed their corresponding equivalent chromatography columns (i.e., packed with same media, and having the same bed height and area of cross-section) in terms of multiple separation metrics such as the number of theoretical plates, peak shape, peak width, peak asymmetry, and resolution in multi-component protein separation [23–26]. The dispersion effects in these cuboid packed-bed devices were significantly lower than those in their equivalent columns as evident from their greater numbers of theoretical plates per unit bed height. Based on the above considerations, could it be assumed that box-shaped or cuboid packed-bed reactors would outperform packed-column reactors, at least in terms of their hydraulic properties and RTD attributes?

This paper proposes the use of cuboid packed-beds as reactors for carrying out chemical and biochemical reactions. The rationale this is provided in terms of advantages resulting from superior system hydraulics and the resultant narrower RTD. The residence time distribution predictions are based on mathematical models discussed in our earlier papers [24,29]. Other potential advantages, such as superior heat transfer attributes are speculated based on geometric considerations.

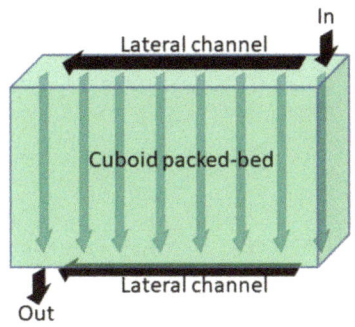

Figure 2. Idealized flow paths in a cuboid packed-bed.

2. Materials and Methods

Sodium chloride was purchased from Sigma-Aldrich (St. Louis, MO, USA). Sodium chloride solution was prepared using water obtained from a SIMPLICITY 185 water purification unit Millipore (Molsheim, France). All solutions were micro-filtered and degassed prior to use using PVDF micro-filter (VVLP04700, 0.1 µm pore size, Millipore, Billerica, MA, USA). HiTrap Capto S (5 mL, product number 17-5441-2) columns, and Capto S (product number 17-5441-01) chromatographic media were purchased from GE Healthcare Biosciences, QC, Canada. The HiTrap Capto S column was used as the surrogate packed-column reactor while a custom designed cuboid packed-bed device filled with Capto S resin particles served as the surrogate cuboid packed-bed reactor. The basic design of the cuboid packed-bed devices has been described in our previous papers [23–26]. The central housing for the cuboid packed-bed was made of polyvinyl chloride (PVC). The upper and lower plates were made of acrylic. The lateral channels engraved in the plates were pillared for efficient flow distribution and collection respectively. The dimensions of the cuboid packed-bed were respectively 25 mm (length) × 8 mm (width) × 25 mm (height), the effective bed volume being 5 mL. The packed-bed was separated from the lateral channels using nylon meshes (0.002 inch opening, product number 9318T48, McMaster Carr, Chicago, IL, USA) for retaining and holding the chromatographic media in place.

An AKTA prime liquid chromatography system (GE Healthcare Biosciences, Quebec, QC, Canada) was used for salt tracer experiments. E-curve type profiles for the packed column and the cuboid packed-bed device were obtained using 0.8 M sodium chloride as tracer and 0.4 M sodium chloride as base solution. A 0.1 mL sample loop was used for injecting the tracer pulse for characterizing the hydraulics of the column and the cuboid packed-bed.

3. Results and Discussion

In a previous paper [23], the performance of a chromatographic column having 9 mL bed volume (30 mm diameter and 12.7 mm bed height) and packed with Capto S media was compared with an equivalent cuboid packed-bed (58.9 mm length × 12 mm width × 12.7 mm height). At a flow rate of 5 mL/min which corresponded to a superficial velocity of 42.4 cm/h, the number of theoretical plates per meter were found to be 2628 and 4651 respectively, for the column and the cuboid packed-bed. To verify whether such difference was due to flow maldistribution within the column, the RTD within the column and the cuboid packed-bed device were simulated using the using the methods described in our previous papers [24,29]. Figure 3 shows the RTD within the column as function of dimensionless radius (r/R) and dimensionless column volume (V/V_T) at a superficial velocity of 42.4 cm/h. The radial distance step used in the simulation was 0.001 cm. The shortest residence time corresponded to $r = 0$, i.e., the center of the column, and the value increased significantly towards the periphery. This was consistent with the idealized flow paths picturized in Figure 1. In the top header, the liquid flowed in a radially outward direction while in the lower header, it flowed in a radially inward direction. This flow pattern resulted in significant variability in the path lengths, i.e., a flow path closer to the center was significantly shorter than one closer to the periphery. Also, as discussed in a previous paper [23], the radial velocity in the top header decreased very significantly in an outward direction due to an increase in available cross-sectional area, and the loss of liquid due to entry in to the packed-bed.

$$\frac{dv_r}{dr} = -\left(\frac{v_r}{r} + \frac{v_s}{h}\right) \tag{1}$$

In the lower header, the radial velocity increased in an inward direction due to a decrease in available cross-sectional area and cumulative collection of liquid from the packed-bed. Quite clearly, the increased transit time in the column headers for fluid elements traversing the peripheral regions of the column contributed significantly to the increase in residence time. Both profiles shown in Figure 3 indicated the same trend, i.e., the residence time increased very significantly closer to the column periphery. However, this increase was a bit more pronounced when the residence time was plotted versus (V/V_T). This reflected the fact that in a column, more media was located towards the periphery than near the center. Figure 4 shows a plot of (V/V_T) versus (r/R) which indicated that 50% of the media existed in the outer 29% radial locations of the column. Hence, the increased residence time closer to the periphery of the column affected a very significant volume fraction of media packed within the column.

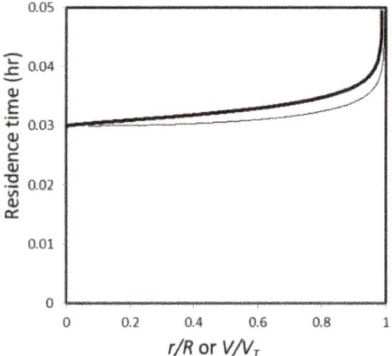

Figure 3. Residence time within a column as function of dimensionless radius (r/R) and dimensionless volume (V/V_T) (media: Capto Q, dimensions: 30 mm diameter and 12.7 mm bed height, bed volume: 9 mL, flow rate: 5 mL/min, superficial velocity = 42 cm/h).

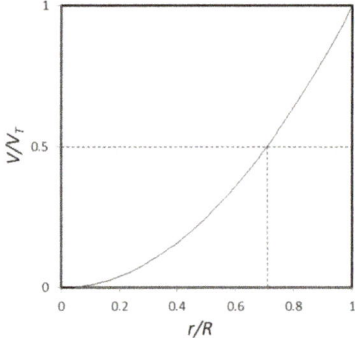

Figure 4. Plot of dimensionless volume (V/V_T) versus dimensionless radius (r/R).

Figure 5 shows the RTD within the cuboid packed-bed device as function of dimensionless length (z/l). The lateral distance step used in the simulation was 0.001 cm. In a recent paper (24) it has been shown that the residence time as function of length within a cuboid packed-bed could be obtained by:

$$\tau = \left[\frac{h_B}{v_S}\right] + \left[\frac{2z}{2v_0 - z\left|\frac{dv_U}{dz}\right|}\right] + \left[\frac{2(l-z)}{2v_0 - (l-z)\left|\frac{dv_U}{dz}\right|}\right] \tag{2}$$

When $\left|\frac{dv_U}{dz}\right|$ is small:

$$\tau \cong \frac{h_B}{v_S} + \frac{l}{v_0} \tag{3}$$

Equation (3) implies that when the change in velocity within the lateral channels of the cuboid packed-bed is small, the residence time is largely independent of location. Figure 5 shows that for the cuboid packed-bed device under consideration [23], the residence time was slightly lower at the mid-point of the device. However, the difference in residence time between the mid-point and the ends was negligible compared to that observed with the packed column. From Equation (2), it could be shown that the minimum residence time corresponding to the mid-point of the cuboid packed-bed device could be obtained by [24]:

$$\tau_{\min} = \left[\frac{h_B}{v_S}\right] + \left[\frac{2l}{2v_0 - \frac{l}{2}\left|\frac{dv_U}{dz}\right|}\right] \tag{4}$$

Likewise, the maximum residence time corresponding to the ends could be obtained by [24]:

$$\tau_{\max} = \left[\frac{h_B}{v_S}\right] + \left[\frac{2l}{2v_0 - l\left|\frac{dv_U}{dz}\right|}\right] \tag{5}$$

For the optimum functioning of a cuboid packed-bed device, the difference between the maximum and minimum residence time should be as small as possible [24]:

$$|\tau_{\max} - \tau_{\min}| = \left[\frac{2l}{2v_0 - l\left|\frac{dv_U}{dz}\right|}\right] - \left[\frac{2l}{2v_0 - \frac{l}{2}\left|\frac{dv_U}{dz}\right|}\right] \tag{6}$$

Tracer experiments using sodium chloride solution were carried out with the 5 mL HiTrap Capto S column (i.e., the surrogate packed-column reactor) and the 5 mL custom designed cuboid packed-bed device filled with Capto S resin particles (i.e., the surrogate cuboid packed-bed reactor). These experiments were performed in triplicate. Figure 6 shows representative salt tracer peaks obtained from these experiments which were carried at 4 mL/min flow rate (150 cm/h) using 0.4 M sodium chloride as mobile phase and 100 µL of 0.8 M sodium chloride solution as the tracer pulse (i.e., 2% of the packed-bed volume). The peak obtained with the cuboid packed-bed device was higher, sharper and more symmetric than that obtained with the column. The calculated theoretical plate data (N) obtained from these are summarized in Table 1. The number of theoretical plates was calculated using the following equation [32]:

$$N = 5.545(t_R/w_{0.5})^2 \tag{7}$$

The results shown in Figure 6 and Table 1 clearly demonstrate that the cuboid packed-bed device had vastly superior hydraulic attributes than the equivalent packed column. For the same bed height, the cuboid packed-bed had 3.8 times the number of theoretical plates. These new experimental results are consistent with the RTD results shown in Figures 3 and 5. Overall, these results suggest that a reactor based on the cuboid packed-bed design would perform better than its equivalent packed column, at least in terms of the hydraulic properties and RTD. Future studies will involve detailed analysis of E-curves and F-curves obtained from such reactors. Also, the experiments discussed above were carried out using small reactors (i.e., 5 mL bed volume). Experimental work based on 200 mL scale reactors are currently being carried out in my laboratory. Computational fluid dynamics (CFD) studies which allow the study of larger reactors through simulations are also being carried out in parallel.

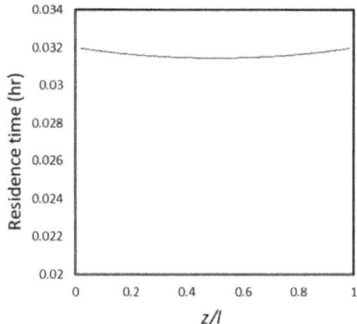

Figure 5. Residence time within a cuboid packed-bed as function of dimensionless length (z/l) (media: Capto Q, dimensions: 58.9 mm length × 12 mm width × 12.7 mm bed height, bed volume: 9 mL, Flow rate: 5 mL/min, superficial velocity = 42 cm/h).

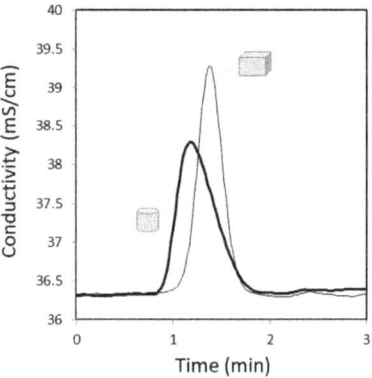

Figure 6. Salt tracer peaks obtained using the surrogate packed-column reactor and the surrogate cuboid packed-bed reactor (bed volume: 5 mL, bed height 25 mm, flow rate: 4 mL/min, superficial velocity: 150 cm/h, media: Capto S, mobile phase: 0.4 M sodium chloride, tracer: 0.8 M sodium chloride, tracer volume: 1 microliters).

Table 1. Number of theoretical plates in the surrogate packed-column reactor and the surrogate cuboid packed-bed reactor (bed volume: 5 mL, bed height 25 mm, flow rate: 4 mL/min, superficial velocity: 150 cm/h, media: Capto S, mobile phase: 0.4 M sodium chloride, tracer: 0.8 M sodium chloride, tracer volume: 1 microliters).

Device	N
Packed column	35
Cuboid packed-bed	133
Error: ±5%	

The design features of the cuboid packed-bed device also provide certain potential advantages from a heat transfer perspective. With a packed column of a given bed height, the heat transfer area per unit reactor volume (excluding the headers) varies inversely with the diameter, i.e., the larger the reactor diameter, the lower is the specific heat transfer area (HTA). This implies that removal or addition of heat to a packed column with a large diameter for endothermic and exothermic reactions respectively would be challenging. By contrast, the HTA of a cuboid packed-bed device could be changed in a very flexible manner by changing the length to width (L/W) ratio, to increase or decrease the values of HTA as required. To make an objective comparison, the HTA of a cuboid packed-bed device was compared with that of a packed column having the same bed height and bed-volume. For a packed column, the HTA is fixed based on the diameter. Figure 7 shows a plot of the ($HTA_{cuboid}/HTA_{column}$) ratio for the two reactors as function of the (L/W) ratio of the cuboid packed-bed device. The minimum ($HTA_{cuboid}/HTA_{column}$) ratio corresponded to an (L/W) ratio of 1, i.e., when the cuboid packed-bed device had a square cross-sectional area. As the (L/W) ratio was increased, the ($HTA_{cuboid}/HTA_{column}$) ratio could be increased quite significantly. Therefore, a cuboid packed-bed reactor would have a significant advantage over a packed column for conducting endothermic and exothermic reactions.

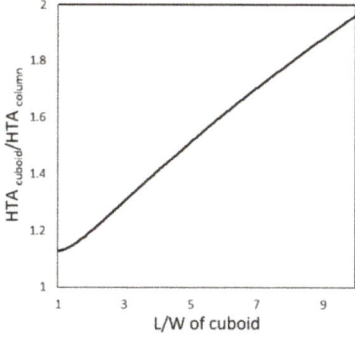

Figure 7. ($HTA_{cuboid}/HTA_{column}$) ratio as function of the (L/W) ratio of a cuboid packed-bed reactor.

The heat transfer advantage of the cuboid packed-bed device over a packed column could also be hypothesized from a heat transfer distance perspective. Heat transfer distance is a vital factor determining thermal gradients within packed-beds, i.e., the greater the distance, the greater is the likelihood of temperature variation. As shown in Figure 8, the maximum heat transfer distance (MHTD) for a packed column corresponded to its radius. On the other hand, the average heat transfer distance (AHTD) corresponded to 29% of the radius, since 50% of the packed material existed within the zone extending up to 71% of the radius from the center. Therefore, both MHTD and AHTD for a packed column varied directly with the diameter, i.e., the greater the diameter, the greater these heat transfer distances. Therefore, for a packed column having a particular bed height and bed volume, both MHTD and AHTD are fixed. Once again, these heat transfer distances for a cuboid packed-bed could be adjusted in a flexible manner by changing the (L/W) ratio. To make an objective comparison, the MHTD and AHTD of a cuboid packed-bed device were compared with those of a packed column having the same bed height and the same bed volume. With the cuboid packed-bed device, the MHTD was equal to half the width while the AHTD was equal to one fourth the width. Figure 8 shows a plot of the MHTD and AHTD ratios for the two reactors (cuboid/column) as function of the (L/W) ratio of the cuboid packed-bed device. The MHTD of the cuboid packed-bed device was consistently lower at all values of (L/W). The AHTD was greater with cuboid packed-bed device for (L/W) values lower than 2.4 but was lower are greater values of (L/W). Once again, these results suggest that a cuboid packed-bed reactor would have a significant advantage over a packed column for conducting endothermic and exothermic reactions.

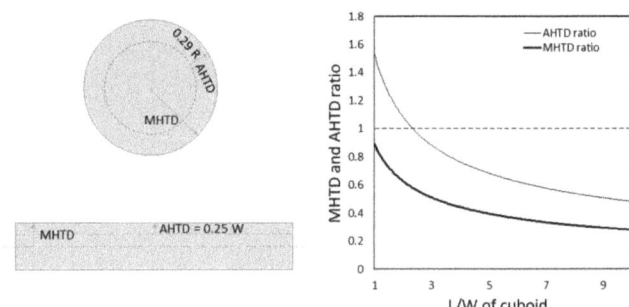

Figure 8. ($\text{MHTD}_{\text{cuboid}} / \text{MHTD}_{\text{column}}$) and ($\text{AHTD}_{\text{cuboid}} / \text{AHTD}_{\text{column}}$) ratios as function of the (L/W) ratio of a cuboid packed-bed reactor.

4. Conclusions

The above discussion supports the basic proposition of this paper, i.e., cuboid packed-beds could potentially be used as reactors for carrying out chemical and biochemical reactions. The principal rationale behind this proposition is that the narrower RTD of the cuboid packed-bed makes it better than a packed column for carrying out chemical reaction. The superior RTD attributes of the cuboid packed-bed were demonstrated using mathematical models and new experimental data obtained using surrogate packed column and cuboid packed-bed reactors. In addition to superior RTD attributes, the cuboid packed-bed could potentially also have superior heat transfer attributes, relevant for conducting exothermic and endothermic reactions. These potential advantages, were speculated-based purely on geometric considerations. For packed columns of a given bed height, the heat transfer area per unit reactor volume varies inversely with the diameter. Therefore, if a packed column with a large diameter is to be used for conducting endothermic or exothermic reactions, removal or addition of heat would be challenging. By contrast, the specific heat transfer area of a cuboid packed-bed device could be changed in a very flexible manner by changing the length to width ratio. The cuboid packed-bed also compares favorably in terms of lower heat transfer distances. Therefore, a cuboid packed-bed reactor could potentially have significant advantages over a packed column, also from a heat transfer perspective.

5. Patents

CHROMATOGRAPHY DEVICE AND METHOD FOR FILTERING A SOLUTE FROM A FLUID R Ghosh—US Patent App. 15/616, 333, 2017.

Funding: This research was funded by the Natural Sciences and Engineering Research Council (NSERC) of Canada, and the Ontario Research Fund—Research Excellence Programme.

Acknowledgments: This work is supported by the natural Science and Engineering Research Council (NSERC) of Canada and the Ontario Research Fund—Research Excellence (ORF-RE) Program. The Author thanks Paul Gatt (department of Chemical Engineering, McMaster University) for fabricating the cuboid packed-bed devices used in this study based on design provided by the author.

Conflicts of Interest: The author declares no conflict of interest.

Notation

v_r	Radial velocity (m/s)
r	Dimension in radial direction (m)
v_s	Superficial (or linear) velocity (m/s)
h	Header height (m)
V	Volume (m^3)
V_T	Total volume of packed column (m^3)

R	Total radius of packed column (m)
τ	Residence time (s)
h_B	Bed height (m)
z	Dimension along length of cuboid packed-bed (m)
v_0	Inlet velocity in upper lateral channel (m/s)
v_U	Velocity in upper lateral channel (m/s)
l	Length of cuboid packed-bed (m)
τ_{min}	Minimum residence time (s)
τ_{max}	Maximum residence time (s)
N	Number of theoretical plates (-)
t_R	Retention time (s)
$w_{0.5}$	Peak width at half height (s)

References

1. Andrigo, P.; Bagatin, R.; Pagani, G. Fixed bed reactors. *Catal. Today* **1999**, *52*, 197–221. [CrossRef]
2. Chica, A.; Corma, A.; Dómine, M.E. Catalytic oxidative desulfurization (ODS) of diesel fuel on a continuous fixed-bed reactor. *J. Catal.* **2006**, *242*, 299–308. [CrossRef]
3. Yuan, T.; Tahmasebi, A.; Yu, J. Comparative study on pyrolysis of lignocellulosic and algal biomass using a thermogravimetric and a fixed-bed reactor. *Bioresour. Technol.* **2015**, *175*, 333–341. [CrossRef] [PubMed]
4. Szloszár, A.; Mándity, I.M.; Fülöp, F. Sustainable synthesis of *N*-methylated peptides in a continuous-flow fixed bed reactor. *J. Flow Chem.* **2018**, *8*, 21–27.
5. Brzozowski, M.; O'Brien, M.; Ley, S.V.; Polyzos, P. Flow Chemistry: Intelligent processing of gas–liquid transformations using a tube-in-tube reactor. *Acc. Chem. Res.* **2015**, *48*, 349–362. [CrossRef] [PubMed]
6. Adamo, A.; Beingessner, R.L.; Behnam, M.; Chen, J.; Jamison, T.F.; Jensen, K.F.; Monbaliu, J.C.M.; Myerson, A.S.; Revalor, E.M.; Snead, D.R.; et al. On-demand continuous-flow production of pharmaceuticals in a compact, reconfigurable system. *Science* **2016**, *352*, 61–67. [CrossRef] [PubMed]
7. Sauer, M.L.; Ollis, D.F. Acetone oxidation in a photocatalytic monolith reactor. *J. Catal.* **1994**, *149*, 81–91. [CrossRef]
8. Wang, Y.; Sun, G.; Dai, J.; Chen, G.; Morgenstern, J.; Wang, Y.; Kang, S.; Zhu, M.; Das, S.; Cui, L.; et al. A high-performance, low-tortuosity wood-carbon monolith reactor. *Adv. Mater.* **2017**, *29*, 1604257. [CrossRef] [PubMed]
9. Lin, C.-C.; Liu, W.-T.; Tan, C.-S. Removal of carbon dioxide by absorption in a rotating packed-bed. *Ind. Eng. Chem. Res.* **2003**, *42*, 2381–2386. [CrossRef]
10. Bader, A.J.; Afacan, A.; Sharp, D.; Chuang, K.T. Effect of liquid-phase properties on separation efficiency in a randomly packed distillation column. *Can. J. Chem. Eng.* **2015**, *93*, 1119–1125. [CrossRef]
11. Lesellier, E.; West, C. The many faces of packed column supercritical fluid chromatography—A critical review. *J. Chromatogr. A* **2015**, *1382*, 2–46. [CrossRef] [PubMed]
12. West, C.; Khalikova, M.A.; Lesellier, E.; Héberger, K. Sum of ranking differences to rank stationary phases used in packed column supercritical fluid chromatography. *J. Chromatogr. A* **2015**, *1409*, 241–250. [CrossRef] [PubMed]
13. Guo, K.; Guo, F.; Feng, Y.; Chen, J.; Zheng, C.; Gardner, N.C. Synchronous visual and RTD study on liquid flow in rotating packed-bed contactor. *Chem. Eng. Sci.* **2000**, *55*, 1699–1706. [CrossRef]
14. Yang, C.; Teixeira, A.R.; Si, Y.; Born, S.; Lin, H.; Song, Y.L.; Peer, M.; Martin, B.; Schenkel, B.; Jensen, K.F. Catalytic hydrogenation of N-4-nitrophenyl nicotinamide in a micro-packed-bed reactor. *Green Chem.* **2018**, *20*, 886–893. [CrossRef]
15. Dadgar, F.; Venvik, H.J.; Pfeifer, P. Application of hot-wire anemometry for experimental investigation of flow distribution in micro-packed-bed reactors for synthesis gas conversion. *Chem. Eng. Sci.* **2018**, *177*, 110–121. [CrossRef]
16. Dixon, A.G.; Cresswell, D.L. Theoretical prediction of effective heat transfer parameters in packed-beds. *AIChE J.* **1979**, *25*, 663–676. [CrossRef]

17. Mears, D.E. Tests for transport limitations in experimental catalytic reactors. *Ind. Eng. Chem. Proc. Des. Dev.* **1971**, *10*, 541–547. [CrossRef]
18. Nijemeisland, M.; Dixon, A.G. Comparison of CFD simulations to experiment for convective heat transfer in a gas–solid fixed bed. *Chem. Eng. J.* **2001**, *82*, 231–246. [CrossRef]
19. Olbert, G.; Corr, F.; Reuter, P.; Wambach, L.; Hammon, U. Multi-Tube Fixed-Bed Reactor, Especially for Catalytic Gas Phase Reactions. U.S. Patent 7,226,567, 5 June 2007.
20. Sugiyama, M.; Ando, Y.; Taniguchi, Y. Fixed Bed Multitube Reactor. U.S. Patent 7,588,739, 15 September 2009.
21. Westerman, D.; Schrauwen, F.J.M. Multitube Reactor. U.S. Patent 4,894,205, 16 January 1990.
22. Jiang, B.; Hao, L.; Zhang, L.; Sun, Y.; Xiao, X. Numerical investigation of flow and heat transfer in a novel configuration multi-tubular fixed bed reactor for propylene to acrolein process. *Heat Mass Transf.* **2015**, *51*, 67–84. [CrossRef]
23. Ghosh, R. Using a box instead of a column for process chromatography. *J. Chromatogr. A* **2016**, *1468*, 164–172. [CrossRef] [PubMed]
24. Ghosh, R.; Chen, G. Mathematical modelling and evaluation of performance of cuboid packed-bed devices for chromatographic separations. *J. Chromatogr. A* **2017**, *1515*, 138–145. [CrossRef] [PubMed]
25. Chen, G.; Ghosh, R. Effects of process parameters on the efficiency of chromatographic separations using a cuboid packed-bed device. *J. Chromatogr. B* **2018**, in press. [CrossRef] [PubMed]
26. Chen, G.; Gerrior, A.; Ghosh, R. Feasibility study for high-resolution multi-component separation of protein mixture using a cation-exchange cuboid packed-bed device. *J. Chromatogr. A* **2018**, *1549*, 25–30. [CrossRef] [PubMed]
27. Madadkar, P.; Wu, Q.; Ghosh, R. A laterally-fed membrane chromatography module. *J. Membr. Sci.* **2015**, *487*, 173–179. [CrossRef]
28. Madadkar, P.; Nino, S.L.; Ghosh, R. High-resolution, preparative purification of PEGylated protein using a laterally-fed membrane chromatography device. *J. Chromatogr. B* **2016**, *1035*, 1–7. [CrossRef] [PubMed]
29. Ghosh, R.; Madadkar, P.; Wu, Q. On the workings of laterally-fed membrane chromatography. *J. Membr. Sci.* **2016**, *516*, 26–32. [CrossRef]
30. Madadkar, P.; Umatheva, U.; Hale, G.; Durocher, Y.; Ghosh, R. Ultrafast separation and analysis of monoclonal antibody aggregates using membrane chromatography. *Anal. Chem.* **2017**, *89*, 4716–4720. [CrossRef] [PubMed]
31. Madadkar, P.; Sadavarte, R.; Butler, M.; Durocher, Y.; Ghosh, R. Preparative separation of monoclonal antibody aggregates by cation-exchange laterally-fed membrane chromatography. *J. Chromatogr. B* **2017**, *1055*, 158–164. [CrossRef] [PubMed]
32. Moldoveanu, S.C.; David, V. *Essentials in Modern HPLC Separations*; Elsevier: Amsterdam, The Netherlands, 2013; p. 63.

Article

Modeling and Optimal Design of Absorbent Enhanced Ammonia Synthesis

Matthew J. Palys, Alon McCormick, E. L. Cussler and Prodromos Daoutidis *

Department of Chemical Engineering and Materials Science, University of Minnesota, Minneapolis, MN 5405, USA; palys003@umn.edu (M.J.P.); mccormic@umn.edu (A.M.); cussler@umn.edu (E.L.C.)
* Correspondence: daout001@umn.edu; Tel.: +1-612-625-8818

Received: 18 June 2018; Accepted: 12 July 2018; Published: 18 July 2018

Abstract: Synthetic ammonia produced from fossil fuels is essential for agriculture. However, the emissions-intensive nature of the Haber–Bosch process, as well as a depleting supply of these fossil fuels have motivated the production of ammonia using renewable sources of energy. Small-scale, distributed processes may better enable the use of renewables, but also result in a loss of economies of scale, so the high capital cost of the Haber–Bosch process may inhibit this paradigm shift. A process that operates at lower pressure and uses absorption rather than condensation to remove ammonia from unreacted nitrogen and hydrogen has been proposed as an alternative. In this work, a dynamic model of this absorbent-enhanced process is proposed and implemented in gPROMS ModelBuilder. This dynamic model is used to determine optimal designs of this process that minimize the 20-year net present cost at small scales of 100 kg/h to 10,000 kg/h when powered by wind energy. The capital cost of this process scales with a 0.77 capacity exponent, and at production scales below 6075 kg/h, it is less expensive than the conventional Haber–Bosch process.

Keywords: ammonia synthesis; dynamic modeling; design optimization

---

1. Introduction

Synthetic ammonia is an important commodity in present day society. In 2015, 160 million tonnes of ammonia were produced globally, the majority of which was used either directly or as a building block for nitrogen fertilizer, and the global demand for ammonia is expected to grow steadily at an annual rate of 1.5% [1]. However, the hydrogen required for ammonia production is conventionally obtained from fossil fuels such as natural gas or coal [2]. Furthermore, conventional ammonia production is energy intensive; in fact, ammonia synthesis for nitrogen-based fertilizers is responsible for 1% of global energy consumption [3]. A finite and depleting supply of fossil resources, as well as a desire for increased sustainability of fertilizer production have motivated the idea of producing ammonia using renewable energy. Additionally, ammonia has the potential as an energy-dense, carbon neutral liquid fuel, which, when made using renewables, could be used in various applications such as long-term energy storage or transportation to further reduce carbon intensity [4,5]. A proposed production pathway for renewable ammonia is to use electricity generated from renewable sources such as wind or solar to obtain hydrogen from electrolysis, nitrogen from air and to power the ammonia synthesis process itself. Small-scale, distributed production of ammonia better enables the use of this renewable energy. Pursuant to this notion of small-scale renewable-powered ammonia synthesis, a 65-kg/day (2.71 kg/h) wind-powered Haber–Bosch process has been constructed in Morris, MN [6].

The economics of renewable-powered ammonia synthesis using the Haber–Bosch process have also been investigated. In [7], wind-powered ammonia production for the purpose of fuel on remote islands was considered. Small-scale ammonia production was shown to be economically viable only if diesel costs more than \$10/gallon. This result may be viable for isolated communities, but does not lend itself to more widespread adoption. In [8], the economic feasibility of a grid-connected, offshore wind-powered ammonia facility was examined, and the synthesis loop was found to account for over 20% of the ammonia system capital cost. An isolated (not grid-connected) ammonia energy storage system was proposed in [9], and it was determined that the ammonia synthesis loop accounted for up to 25% of the total system capital cost. Evidently, reducing the capital cost of ammonia synthesis would aid in improving the economics of these systems. Taking a more expansive view, optimal ammonia fertilizer supply chains for Iowa and Minnesota were determined [10]. In addition to purchasing from conventional producers, the option to install small-scale wind-powered Haber–Bosch processes was available, but the optimal supply chains only included renewable ammonia synthesis when a carbon tax was imposed. This result provides further support for efforts to reduce the capital cost of ammonia synthesis.

The main driver of the capital cost in the Haber–Bosch process is high pressure, which leads to expensive compressors, as well as high wall thickness requirements for piping and vessels. However, significantly reducing the pressure in the Haber–Bosch process is difficult because doing so reduces reactor single pass conversion and also requires even lower temperatures for condensation of ammonia from unreacted hydrogen and nitrogen, thus increasing refrigeration demand. An alternative approach is absorbent-enhanced ammonia synthesis in which a bed of supported alkali metal salt replaces the conventional condenser [11,12]. This absorbent allows for more complete ammonia separation as compared to condensation, and subsequently, high ammonia production rates can be achieved while operating at lower pressures [13]. Additionally, ammonia absorption can occur around 200 °C, meaning that cooling water can be used, rather than refrigeration, which is required in the Haber–Bosch process.

The promise of this absorbent-enhanced process has motivated the present work. First, we develop a model of this process (which is dynamic because absorption is inherently transient). We then use the model for design optimization to determine process design and operating conditions that minimize its net present cost (NPC) under the condition that the process is powered using only stranded wind energy at multiple small scales. The rest of the paper is structured as follows. Section 2 provides a description of the absorbent-enhanced process. Section 3 outlines the dynamic model of this process. Section 4 provides the formulation of the optimal design problem. The results of the optimal design problem at scales of 100 kg/h, 500 kg/h, 1000 kg/h, 5000 kg/h and 10,000 kg/h are presented in Section 5. The implications of the optimal design results as they pertain to the economics of the Haber–Bosch process are discussed in Section 6.

2. Process Description

A flow diagram of the absorbent-enhanced ammonia synthesis process is given in Figure 1. Absorption is inherently transient, but the use of two beds allows for operation at a cyclic steady state. The process can naturally be partitioned into two distinct parts: reaction-absorption and desorption (regeneration). The feed to the reaction-absorption loop is a stoichiometric mixture of nitrogen at 1.013 bar (atmospheric pressure) and hydrogen at 10 bar (the outlet pressures of pressure-swing adsorption and electrolysis, respectively). These gases are compressed to the process operating pressure, which is between 15 and 30 bar (one order of magnitude lower than industrial conditions in the Haber–Bosch process), before being sent to a heat exchanger, which uses the reactor effluent to heat the feed gas to reaction temperature, which is around 400 °C. The nitrogen and hydrogen then react over a fixed bed of wustite-based iron catalyst, which is the industrial standard, to form ammonia. The reactor effluent is partially cooled using the previously-mentioned heat exchanger and further cooled with cooling water to the absorption temperature, which is in the range of 100–200 °C. The absorber is a fixed bed of magnesium chloride supported on silica [12], which serves to remove ammonia selectively through its absorption into the solid. The absorber effluent, which is primarily nitrogen and hydrogen with some unabsorbed ammonia, is then cooled to ensure safe operation of the recycle compressor, for its discharge temperature cannot exceed 150 °C. These gases are mixed with the fresh feed.

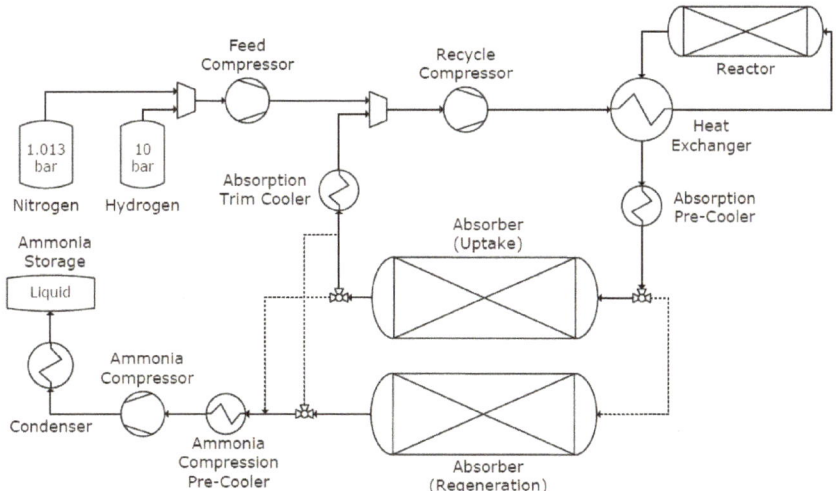

Figure 1. Absorbent-enhanced ammonia synthesis process flow diagram.

At the end of an absorption cycle, once the absorber is deemed to be saturated, it is regenerated through a combination of heating to between 400 and 500 °C and decreasing pressure by opening one end of the bed. This causes a reduction in the ammonia storage capacity of the absorbent, and subsequently, ammonia is released to the gas phase. This gas leaves the bed through the open end due to the imposed pressure gradient. The absorber effluent is cooled (to not damage the subsequent compressor), compressed and condensed so that it can be stored as a liquid. The release of ammonia occurs until the absorbent has been emptied and the bed pressure is equal to the pressure outside the vessel. Once this occurs, one bed volume of stoichiometric nitrogen and hydrogen is fed to the bed to displace the gaseous ammonia. The emptied bed is re-pressurized to the reaction-absorption pressure with a stoichiometric mixture of nitrogen and hydrogen and then reconnected to the reaction-absorption loop.

3. Mathematical Model

A dynamic model of the absorbent-enhanced process was created using gPROMS ModelBuilder 5.1 [14]. The individual unit models are described in this section, though they are largely standard to the PML Library in gPROMS. Physical properties such as mass-specific enthalpy $\hat{H}$, density ρ and viscosity μ are calculated in Multiflash 6.1 [15] using temperature T, pressure P and mass composition w_i as inputs. These calculations are based on the Redlich–Kwong–Soave equation of state for thermodynamic properties and the SuperTRAPP model [16] for transport properties.

3.1. Reactor

The ammonia synthesis reactor is modeled as an adiabatic pseudo-homogeneous plug flow reactor, with spatial variation in the axial direction z. Reactor specifications, which are taken as constant in the model, are given in Table 1. The mass balance for species "i" is:

$$\varepsilon_{bed}\frac{\partial \tilde{m}_i}{\partial t} = \frac{-1}{A_t}\frac{\partial \dot{m}_i}{\partial z} + \nu_i MW_i(1 - \varepsilon_{bed})r_{NH_3} \tag{1}$$

where $\tilde{m}_i$ is the species mass concentration in kg/m^3, $\dot{m}_i$ is the species mass flow rate in kg/s, r_{NH3} is the rate of ammonia formation in mol/(m$^3_{cat}$ s), ν_i is the species stoichiometric coefficient, MW_i

is the species molecular weight in kg/mol, ε_{bed} is the bed void fraction and A_t is the reactor tube cross-sectional area in m². Writing the species balance in terms of mass, rather than moles, as is often the case, aids with the numerical solution in gPROMS. The energy balance is:

$$\frac{\partial \tilde{U}}{\partial t} = -\frac{1}{A_t}\frac{\partial(\dot{m}\hat{H})}{\partial z} + (-\Delta H_{rxn})(1 - \varepsilon_{bed})r_{NH_3} \tag{2}$$

where $\tilde{U}$ is the volume-specific internal energy in kJ/m³, $\dot{m}$ is the total mass flow rate in kg/s, $\hat{H}$ is the mass specific enthalpy in kJ/kg and ΔH_{rxn} is the enthalpy of the ammonia synthesis reaction in kJ/mol. The volume specific internal energy is given by:

$$\tilde{U} = \rho\hat{H} - P \tag{3}$$

where ρ is the gas density in kg/m³ and P is pressure in bar. The pressure drop through the bed is given by the Ergun equation:

$$-\frac{\partial P}{\partial z} = v_s\left[\frac{150(1 - \varepsilon_{bed})^2}{\varepsilon_{bed}^3 d_p^2}\mu + \frac{1.75(1 - \varepsilon_{bed})}{\varepsilon_{bed}^3 d_p}\rho v_s\right] \tag{4}$$

where v_s is the superficial velocity in m/s, d_p is the particle diameter in m and μ is the gas viscosity in kg/(m s). In the considered range of reactor operating conditions, the reactor particle Reynolds number is in the transitional region.

Table 1. Ammonia synthesis reactor specifications.

Parameter	Symbol	Specification
Bed Void Fraction [1]	ε_{bed}	0.4
Catalyst Density, kg/m³	ρ_{cat}	3000
Catalyst Particle Diameter, m	d_p	2×10^{-3}

The bed void fraction is determined from the experimental setup in [13].

The rate of ammonia formation is given by the expression proposed in [17]:

$$r_{NH_3} = \frac{k(p_{N_2}K_a^2 - p_{NH_3}^2/p_{H_2}^3)}{(1 + K_{NH_3}p_{NH_3}/p_{H_2}^\omega)^{2\alpha}} \tag{5}$$

where p_i is the partial pressure of species "i" in atm and ω and α are fixed constants, the values of which are given in Table 2. The rate constant k, ammonia adsorption constant K_{NH3} and equilibrium constant K_a are given by:

$$k = k_o\exp(-E_A/RT) \tag{6}$$

$$K_{NH_3} = K_{NH_{3}o}\exp(E_{NH_3}/RT) \tag{7}$$

$$\log_{10}(K_a) = \log_{10}(K_{a1}) + K_{a2}T + K_{a3}T^2 + K_{a4}/T + K_{a5} \tag{8}$$

where T is the temperature in the bed in K, and values for all constants appearing in these expressions, such as E_A, are given in Table 2. This rate expression was chosen because it is non-infinite for zero ammonia partial pressure. This is not a concern when modeling the Haber–Bosch process because condensation never results in complete removal of ammonia from unreacted gases, but the condition of low ammonia partial pressure in the reactor inlet could potentially be achieved by the absorbent used in this process. It is noted that this kinetic expression is assumed to be valid for the lower total pressures investigated in this work and has not been validated experimentally at those conditions.

Table 2. Ammonia synthesis reaction rate expression parameters.

Parameter	Value
k_o, mol/(m$^3_{cat}$ s atm)	1.096×10^{10}
E_A, J/mol	46,737
K_{NH_3o}, atm$^{(1-\omega)}$	2.94×10^{-4}
E_{NH_3}, J/mol	100,628
ω	1.564
α	0.64
K_{a1}	-2.691122
K_{a2}, 1/K	-5.519265×10^{-5}
K_{a3}, 1/K^2	1.848863×10^{-7}
K_{a4}, K	2001.6
K_{a5}	2.6899

Low ammonia partial pressures also cause internal diffusion limitations in the ammonia synthesis catalyst [18]. Rigorously, this limitation would be accounted for by considering a heterogeneous reactor model with the simultaneous solution of fluid and particle phase mass balances. In order to reduce the computational complexity of the model while still accounting for mass transfer limitations, we generated an empirical expression for the ammonia partial pressure dependence of the catalyst effectiveness factor (η):

$$\eta(p_{NH_3}) = \frac{a_2 p_{NH_3}^2 + a_1 p_{NH_3} + a_o}{b_2 p_{NH_3}^2 + b_1 p_{NH_3} + 1} \tag{9}$$

where values for the fitted constants are given in Table 3. This expression was obtained following the approach described in [19]. Effectiveness factor data were generated by the repeated "offline" solution of the particle mass balance for varying ammonia partial pressures. The empirical expression above was subsequently fit to the data.

Table 3. Ammonia reaction effectiveness factor parameters.

Parameter	a_o	a_1	a_2	b_1	b_2
Value	0.03582	8.366	35.94	7.705	36.11

3.2. Absorber

The model for the bulk fluid phase of the absorber considers axial dispersion and convection of mass and energy (the axial coordinate is z). The specifications, which are taken as constant in the absorber model, are given in Table 4. The mass balance for species "i" in an absorber tube is:

$$\varepsilon_{total}\frac{\partial \tilde{m}_i}{\partial t} = -\frac{1}{A_t}\frac{\partial \dot{m}_i}{\partial z} + \varepsilon_{bed}D_i\frac{\partial^2 \tilde{m}_i}{\partial z^2} + MW_i(1 - \varepsilon_{total})\rho_{abs}(r_{des,i} - r_{abs,i}) \tag{10}$$

where D_i is the species effective dispersion coefficient in m^2/s, which is calculated using the Wakao correlation for gas phase dispersion in packed beds [20], $r_{des,i}$ and $r_{abs,i}$ are species desorption and absorption rates in mol/(kg$_{abs}$ s), ε_{total} is the total void fraction (accounting for the bed void fraction and absorbent particle porosity), ρ_{abs} is the absorbent density in kg/m^3 and A_t is the absorber tube cross-sectional area in m^2. It is noted that for the range of conditions in which absorption operates, the magnitude of dispersion coefficients is quite low such that dispersion could be neglected if desired. The absorber is non-adiabatic to accommodate the temperature difference required between absorption and desorption modes. The energy balance for an absorber tube is:

$$\frac{\partial \tilde{U}}{\partial t} = -\frac{1}{A_t}\frac{\partial(\dot{m}\hat{H})}{\partial z} + \varepsilon_{bed}\lambda_{bed}\frac{\partial^2 T}{\partial z^2} + \sum_{i=1}^{N_s}\Delta H_{abs}(1 - \varepsilon_{total})\rho_{abs}(r_{des,i} - r_{abs,i}) + \frac{4q_{ext}}{d_T} \tag{11}$$

where T is the temperature of the fluid and absorbent in K (assumed to be identical), λ_{bed} is the bed effective thermal conductivity in kW/(m K), which is calculated using the Specchia correlation [21], ΔH_{abs} is the enthalpy of absorption of species "i" in kJ/mol, q_{ext} is the external heat addition or removal in kW and d_T is the absorber tube diameter in m. The volumetric internal energy holdup is given by:

$$\tilde{U} = \varepsilon_{total}(\rho\hat{H} - P) + (1 - \varepsilon_{total})\rho_{abs}\hat{c}_{p,abs}T \tag{12}$$

where $\hat{c}_{p,abs}$ is the absorbent heat capacity in kJ/(kg$_{abs}$ K), the value of which is given in Table 4. Under this definition of volume-specific internal energy, sufficiently fast heat transfer between the solid and fluid phases is considered, such that their temperatures can be assumed to be identical [22]. As with the reactor, the pressure drop through the bed is given by the Ergun equation (Equation (4)), and the particle Reynolds number is again in the transitional region.

Table 4. Absorber specifications.

Parameter	Symbol	Specification
Bed Void Fraction [1]	ε_{bed}	0.32
Absorbent Void Fraction [1]	ε_{abs}	0.60
Total Void Fraction [1]	ε_{total}	0.728
Absorbent Density, kg/m^3	ρ_{cat}	2507
Absorbent Particle Diameter, m	d_p	2×10^{-4}
Absorbent Heat Capacity [2], kJ/(kg$_{abs}$ K)	$\hat{c}_{p,abs}$	1.21

[1] The bed and absorbent void fractions are determined from the experimental setup in [23]. [2] The absorbent heat capacity is assumed to be constant and is from [24].

The solid phase (the absorbent) is volume-averaged; its mass balance has no explicit spatial dependence:

$$\frac{dq_i}{dt} = r_{abs,i} - r_{des,i} \tag{13}$$

where q_i is the absorbed concentration of species "i" in mol/kg$_{abs}$. The absorbent used in this system is selective to ammonia, and thus, the rates of absorption (and desorption) for hydrogen and nitrogen are zero. The rates of ammonia absorption and desorption were determined using data in [23]. The rate of ammonia absorption is given by:

$$r_{abs} = \begin{cases} \frac{k_{abs}[p_{NH_3} - p_{eq}(T)]^7}{K_{abs} + [p_{NH_3} - p_{eq}(T)]^6}, & \text{for } p_{NH_3} > p_{eq}, \; q_{NH_3} < q_{NH_3}^{max} \\ 0, & \text{for } p_{NH_3} < p_{eq}, \; q_{NH_3} = q_{NH_3}^{max} \end{cases} \tag{14}$$

It is noted that the exponents of 7 and 6 in the numerator and denominator, respectively, are chosen for fitting purposes, but are generally representative of a higher order (non-linear) rate. The rate of ammonia desorption is given by:

$$r_{des} = \begin{cases} k_{des}[p_{eq}(T) - p_{NH_3}], & \text{for } p_{NH_3} < p_{eq}, \; q_{NH_3} > 0 \\ 0, & \text{for } p_{NH_3} < p_{eq}, \; q_{NH_3} = 0 \end{cases} \tag{15}$$

where p_{NH_3} is ammonia partial pressure in bar, $q_{NH_3}^{max}$ is the maximum capacity of the absorbent based on stoichiometry in mol/kg$_{abs}$ and $p_{eq}(T)$ is the absorbent equilibrium pressure, in bar, at a given temperature T. Values for constants k_{abs}, K_{abs} and k_{des} are given in Table 5. For this work, the maximum stoichiometric capacity of the MgCl$_2$ absorbent is assumed (conservatively) to be that given by a 1:1 molar ratio of ammonia to salt, which corresponds to a maximum absorbed concentration of 10.5 mol$_{NH_3}$/kg$_{MgCl_2}$. Given that the solid phase contains only 40% salt by weight, the effective maximum absorbed concentration is 4.2 mol$_{NH_3}$/kg$_{abs}$.

Table 5. Ammonia absorption and desorption rate parameters.

Parameter	Value
k_a, $mol_N H_3/(kg_{abs}$ bar s)	0.4668
K, bar^6	5×10^{-24}
k_d, $mol_N H_3/(kg_{abs}$ bar s)	7.002×10^{-3}

The temperature dependence of the equilibrium pressure is given by:

$$\ln\left(\frac{p_{crit}(T)}{p_{crit,ref}}\right) = -\frac{\Delta H_{abs}}{R}\left(\frac{1}{T} - \frac{1}{T_{ref}}\right) \tag{16}$$

where the reference equilibrium pressure-temperature pair ($p_{crit,ref}$ and T_{ref}) and the heat of absorption ΔH_{abs} are from [25] and are given in Table 6.

Table 6. Absorbent equilibrium pressure temperature dependence parameters.

Parameter	Value
$p_{crit,ref}$, bar	1
ΔH_{abs}, J/mol	−87,000
T_{ref}, K	648.05

3.3. Compressor

Compression is assumed to by polytropic, so the compressor outlet temperature T_{out} and power requirement $\dot{W}_{comp}$ are given by the following equations:

$$T_{out} = T_{in}\left(\frac{P_{out}}{P_{in}}\right)^{\frac{n-1}{n}} \tag{17}$$

$$\dot{W}_{comp} = \dot{m}\frac{n}{(n-1)\eta_{poly}\eta_{elec}}\frac{z_{in}RT_{in}}{MW_{gas}}\left[\left(\frac{P_{out}}{P_{in}}\right)^{\frac{n-1}{n}} - 1\right] \tag{18}$$

where T_{in} is the inlet temperature in K, P_{in} and P_{out} are the inlet and outlet pressures in bar, $\dot{m}$ is the mass flow rate in kg/s and z_{in} is the compressibility factor at inlet conditions. The values of polytropic efficiency η_{poly} and electric efficiency η_{elec} are taken to be 70% and 75%, respectively. The polytropic index n is given by:

$$\frac{n-1}{n} = \frac{1}{\eta_{poly}}\frac{\gamma - 1}{\gamma} \tag{19}$$

where γ is the heat capacity ratio at the inlet conditions.

3.4. Heat Exchanger

The heat exchanger has a counter-current flow configuration. It is governed by a steady-state energy balance for simplicity [26]. This energy balance is:

$$\dot{m}_{hot}(\hat{H}_{in,hot} - \hat{H}_{out,hot}) = \dot{m}_{cold}(\hat{H}_{out,cold} - \hat{H}_{in,cold}) = UA\Delta T_{\ell m} \tag{20}$$

where $\dot{m}$ is the mass flow rate in kg/s of hot and cold streams (denoted by subscripts), $\hat{H}_{in}$ and $\hat{H}_{out}$ are the mass-specific enthalpies in kJ/kg at the inlet and outlet conditions (hot and cold streams denoted by subscripts), A is heat transfer area in m², U is the overall heat transfer coefficient, which is taken to be 0.015 kW/(m² K), and $\Delta T_{\ell m}$ is the log mean temperature difference in K, which is given by:

$$\Delta T_{\ell m} = \frac{(T_{in,hot} - T_{out,cold}) - (T_{out,hot} - T_{in,cold})}{\ln\left(\frac{T_{in,hot} - T_{out,cold}}{T_{out,hot} - T_{in,cold}}\right)} \tag{21}$$

where T is the temperature in K of each of the four streams (denoted by subscripts).

3.5. Cooler with External Heat Removal

As with the heat exchanger, a steady-state energy balance governs this unit:

$$Q = \dot{m}(\hat{H}_{out} - \hat{H}_{in}) \tag{22}$$

where Q is the heat input rate in kW (negative in the case of the cooler) and $\hat{H}_{in}$ and $\hat{H}_{out}$ are the mass specific enthalpies in kJ/kg at the inlet and outlet conditions. It is also noted that this model is used for a condenser, where the difference between outlet and inlet enthalpies accounts for the enthalpy of condensation.

In the range of temperatures observed or potentially observed in this process, 40 °C to 500 °C, cooling water can be used. Thus, the external duty cooler is modeled as a heat exchanger with cooling water in the shell side. The area A of such a heat exchanger is:

$$A = \frac{Q}{U\Delta T_{\ell m}} \tag{23}$$

where U is the overall heat transfer coefficient, the value of which is given in Table 7, and $\Delta T_{\ell m}$ is the log mean temperature difference, given by Equation (21) where water is used in the cold stream. The inlet and outlet water temperatures are given in Table 7. The required flow rate of cooling water $\dot{m}_{cw}$ is calculated as:

$$\dot{m}_{cw} = \frac{Q}{\hat{c}_{p,water}(T_{out,cw} - T_{in,cw})} \tag{24}$$

where water heat capacity $\hat{c}_{p,water}$ is given in Table 7.

Table 7. Cooler specifications.

Parameter	Symbol	Specification
Overall Heat Transfer Coefficient, kW/(m² K)	U	0.03
Cooling Water Inlet Temperature, °C	$T_{in,cw}$	20
Cooling Water Outlet Temperature, °C	$T_{out,cw}$	40
Cooling Water Heat Capacity, kJ/(kg K)	$\hat{c}_{p,water}$	4.16

4. Design Optimization Formulation

4.1. Objective Function

The objective of the optimization is to minimize the 20-year net present cost of the absorbent-enhanced process under the condition that the process is powered entirely using stranded wind energy. The net present cost of the process is the sum of capital C^{cap} and operating C^{op} costs:

$$NPC = C^{cap} + C^{op} \tag{25}$$

The process units under consideration are the feed compressor (*comp, feed*), recycle compressor (*comp, rcy*), reactor (*reactor*), heat exchanger (*HEx*), absorber precooler (*cooler, pre*), two absorbers (*abs*), absorber trim cooler (*cooler, trm*), ammonia compression precooler (*cooler, NH3comp*), ammonia compressor (*comp, NH3*) and ammonia condenser (*condenser*). It is noted that the capital cost of cooling units (all coolers and the condenser) consists of costs both for the heat exchange infrastructure, as well as the cooling water recirculation pump.

4.1.1. Capital Costs

The capital cost of each unit is calculated using a power law relationship based on a reference size parameter. The capital cost of unit j is given by:

$$C_j^{cap} = C_{j,fixed}^{cap} + C_{j,ref}^{cap} \left(\frac{X_j}{X_{j,ref}}\right)^{\beta_j} \tag{26}$$

The capital cost parameters for each unit are given in Table 8. The cost correlations are from [27]. The fixed cost parameter represents the cost of unit control systems, while the reference cost parameter takes labor and maintenance into account. Both fixed and reference cost parameters have been scaled with the chemical engineering plant cost index (CEPCI) from 2007 dollars (CEPCI = 525.4 [27]) to 2015 dollars (CEPCI = 556.8 [28]).

Table 8. Capital cost correlations.

Unit	Basis	$C_{j,fixed}$	$C_{j,ref}$	$X_{j,ref}$	β_j
Reactor	Volume, m^3	66,800	268,000	20	0.52
Absorber	Area of Tubes, m^2	66,800	1,039,000	100	0.68
Compressor [1]	Rated Power, kW	7400	7,690,000	1000	0.9
Heat Exchanger	Area, m^2	28,600	208,000	100	0.71
Pump	Rated Power	7400	15,000	23	0.29

Material of construction for feed, recycle and ammonia compressors assumed to be stainless steel [29].

A pressure multiplier f_P that accounts for the need to increase vessel wall thicknesses at higher pressures applies to the reactor and absorbers. Its value is given by [27]:

$$f_P = 0.125\left(\frac{P}{10}\right) + 0.875 \tag{27}$$

where pressure P has units of bar. Ammonia synthesis catalyst is assumed to cost \$15.50/kg [9], and so, the overall cost of the reactor is:

$$C_{reactor}^{cap} = C_{fixed,reactor}^{cap} + C_{ref,reactor}^{cap} \left(\frac{V_{reactor}}{20 \text{ m}^3}\right)^{0.52} f_P + 15.50 W_{cat} \tag{28}$$

The absorbent is a mixture of 40 wt% MgCl$_2$, which is assumed to cost \$0.35/kg [30], and 60 wt% silica gel, which has a cost that scales with weight as follows [31]:

$$C_{silica} = \$61.33(W_{silica})^{0.563} \tag{29}$$

Thus, the overall cost of each absorber is:

$$C_{abs}^{cap} = C_{fixed,abs}^{cap} + C_{ref,abs}^{cap} \left(\frac{A_{abs}}{100 \text{ m}^2}\right)^{0.68} f_P + 0.35(0.4 W_{abs}) + 61.33(0.6 W_{abs})^{0.563} \tag{30}$$

It is assumed that compressors are available in any required size, which allows all pressure decisions to be unconstrained and compressor capital cost to be continuous. In the event that only certain sizes of compressors would be available from manufacturers, additional constraints requiring that optimal rated power be equal to that of one of the available compressors would be needed, as would binary decisions for selecting a certain compressor size for each application.

As noted above, the pump cost correlation is included in Table 8 because it is needed to pump cooling water. The overall capital cost of an external duty cooler, which uses cooling water, is thus:

$$C_{cooler}^{cap} = C_{HEx}^{cap}(A_{Cooler}) + C_{Pump}^{cap}(\hat{W}_{cw}\dot{m}_{cw}) \tag{31}$$

where cooler area and cooling water flow rate are calculated using Equations (23) and (24), respectively. The parameter $\hat{W}_{cw}$ is the specific energy required for cooling water recirculation and is taken to be 1.898 kJ/kg [32].

In the optimization, the possibility exists for the installation of multiple feed compressors with inter-stage cooling. This could potentially be required to achieve some range of feed pressures while ensuring that the feed compressor outlet temperatures do not exceed 150 °C. In all subsequent equations, the variable $C^{cap}_{comp,feed}$ denotes the capital cost for all feed compressor units and their respective inter-coolers (heat exchanger and cooling water pump as described by Equation (31)) and is the sum of the cost of each of those units.

The total installed cost of the absorbent-enhanced ammonia synthesis process is thus given by the sum of installed costs of each unit in the process:

$$
\begin{aligned}
C^{cap} =& C^{cap}_{comp,feed} + C^{cap}_{comp,rcy} + C^{cap}_{reactor} + C^{cap}_{HEx} + C^{cap}_{cooler,pre} \\
& + 2C^{cap}_{absorber} + C^{cap}_{cooler,trim} + C^{cap}_{cooler,NH3comp} + C^{cap}_{comp,NH3} + C^{cap}_{condenser}
\end{aligned}
\tag{32}
$$

4.1.2. Operating Costs

The operating costs considered in this work are those resulting from purchasing wind energy to power. These power requirements can be divided into compression, cooling water pumping (for coolers) and desorption. The power required for compression is given by Equation (18). The power required for pumping cooling water is given by:

$$
\dot{W}^{cw} = \hat{W}_{cw}\dot{m}_{cw}
\tag{33}
$$

It is noted that the variable $\dot{W}_{comp,feed}$ denotes the energy requirements for all compressors and cooling water pumps in the feed compressor train. Since this process is powered entirely using electricity, desorption heating is achieved with an electric heater. Assuming electric heating has an efficiency of 100%, the power required for desorption is simply external heat addition during desorption q_{ext}. Thus, the overall electric energy consumption during a single 30-min absorption-desorption cycle is given by:

$$
\begin{aligned}
E_{cycle} = \int_0^{1800} & [\dot{W}_{comp,feed} + \dot{W}_{comp,rcy} + +\dot{W}^{cw}_{cooler,pre} + \dot{W}^{cw}_{cooler,trim} \\
& + q_{ext} + \dot{W}^{cw}_{cooler,NH3comp} + \dot{W}_{comp,NH3} + \dot{W}^{cw}_{condenser}]dt
\end{aligned}
\tag{34}
$$

Over each year of operation, the power requirements for each cycle are assumed to be identical, and therefore, the annual energy requirement of the process is:

$$
E_{year} = 48\frac{\text{cycles}}{\text{day}} \times 365\frac{\text{days}}{\text{year}}E_{cycle}
\tag{35}
$$

Finally, the 20-year net present operating cost is given by:

$$
C^{op} = \sum_{y=1}^{20} \frac{1}{(1+d)^{y-1}}c_{wind}E_{year}
\tag{36}
$$

where d is the discount rate and c_{wind} is the unit cost of wind energy. For this work, the discount rate is taken to be 8.3%, and wind energy is assumed to cost \$0.03/kWh [33].

4.2. Decision Variables

The decision variables of the capital cost minimization problem are: the recycle mass flow rate, the reaction-absorption loop pressure (defined as the pressure to which the feed gases are compressed), the reaction temperature (defined at the inlet of the adiabatic reactor), the length and diameter of the reactor, the area of the heat exchanger that uses the reactor effluent to heat the reactor feed, the absorption temperature (defined as the outlet temperature of the absorption precooler), the number of absorber tubes, as well as the length and diameter of those tubes, the outlet temperature of the trim cooler, the desorption pressure and temperature, the outlet temperature of the ammonia compression precooler and the ammonia storage pressure and temperature.

In general, these variables must be positive, but are otherwise unconstrained. Two exceptions to this are the reaction temperature, which is constrained to a lower bound of 370 °C due to the lack of reliable kinetic expressions at lower temperatures [17] and the desorption temperature, which is constrained to an upper bound of 500 °C to remain safely below the autoignition temperature of hydrogen [34].

4.3. Constraints

In addition to decision variable bounds and the underlying dynamic model described in Section 3, there are additional operational constraints to which the optimal solution must adhere. The amount of ammonia produced in a 30-min cycle, defined as ammonia leaving the bed during desorption, must meet the required demand:

$$\int_0^{1800} \dot{m}_{NH3,des}dt \geq m_{NH3,target} \tag{37}$$

This ammonia must be stored at a temperature and pressure such that it is liquid:

$$P_{storage} \geq P_{NH_3}^{vap}(T_{storage}) \tag{38}$$

The reactor and absorbers are constrained to have length-to-diameter ratios of at least two:

$$\frac{L_{reactor}}{D_{reactor}} \geq 2 \tag{39}$$

$$\frac{L_{absorber}}{D_{absorber}} \geq 2 \tag{40}$$

The reactor and absorber outlet temperatures are not allowed to exceed 500 °C in order to remain safely below the autoignition temperature of hydrogen [34]:

$$T_{reactor}(t, z = L) \leq 773.15 \qquad \forall t \tag{41}$$

$$T_{absorber}(t, z = L) \leq 773.15 \qquad \forall t \tag{42}$$

Additionally, the outlet temperatures of all compressors are constrained to be less than 150 °C, so as to not damage the components of the compressor [27]:

$$T_{comp}^{discharge}(t) \leq 423.15 \qquad \forall t \tag{43}$$

4.4. Problem Summary and Computation

The optimal design problem is to minimize Equation (25) subject to the underlying process model, the lower bound on reactor temperature, the upper bound on desorption temperature and performance and safety constraints (Equations (37) through (43)). This is a mixed integer non-linear problem (MINLP) with 15 continuous decisions and one integer decision, the number of absorber tubes. The model consists of 14,954 equations after spatial discretization. The reactor and absorber are discretized into 50 and 250 points, respectively. The mixed integer non-linear optimization was solved using the outer approximation/equality relaxation/augmented penalty (OAERAP) solver. At each iteration in the OAERAP algorithm, an underlying non-linear problem with fixed integer variables is solved using the non-linear problem sequential quadratic programming (NLPSQP) solver. The optimal design problem was solved at scales of 100 kg/h, 500 kg/h, 1000 kg/h, 5000 kg/h and 10,000 kg/h.

5. Results

The optimal 20-year NPC and capital cost in millions of UDS (MM$) and specific energy consumption for each scale are given in Table 9. In the chemical process industries, power law correlations are commonly used to relate the capital cost of entire processes at different scales [32]. A correlation of this type for the absorbent-enhanced process is fit to the five optimal capital cost data in Figure 2.

Table 9. Absorbent-enhanced ammonia synthesis optimal design results.

Scale, kgNH$_3$/h	100	500	1000	5000	10,000
Net Present Cost, MM$	3.2	11.2	20.2	85.8	164.3
Capital Cost, MM$	2.3	7.0	11.8	43.9	80.2
Specific Energy Consumption, kWh/kgNH$_3$	3.12	3.07	3.06	3.07	3.08

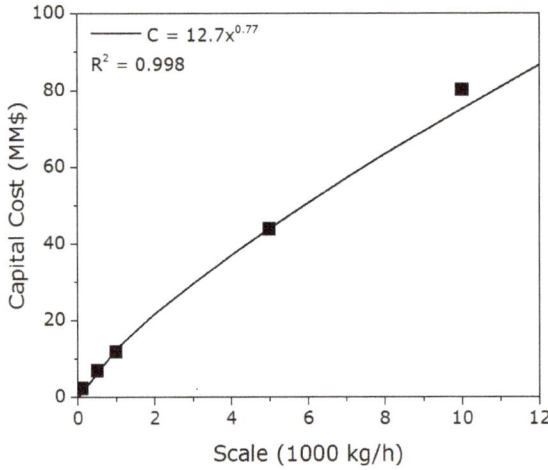

Figure 2. Absorbent-enhanced ammonia synthesis capital cost power law.

5.1. Analysis of Optimal Design

The following discussion provides a plausible physical explanation and justification for key aspects of the optimal design decisions, which are given in Table 10. Certain reaction-absorption loop characteristics, which are helpful in interpreting these results, are given in Table 11.

Table 10. Absorbent-enhanced ammonia synthesis optimal design decisions.

Scale, kgNH$_3$/h	100	500	1000	5000	10,000
Recycle Ratio	12.5	12.7	12.7	12.8	12.8
Reaction-Absorption Pressure, bar	20.3	20.3	20.3	20.3	20.3
Reactor Temperature, °C	370	370	370	370	370
Reactor Length, m	1.48	2.68	3.52	7.68	10.68
Reactor Diameter, m	0.74	1.33	1.76	3.84	5.34
Heat Exchanger Area, m^2	247	1250	2520	12,700	30,200
Absorber Temperature, °C	167	174	176	178	176
Number of Absorber Tubes	4	29	68	512	1183
Absorber Tube Length, m	1.14	1.00	0.96	0.84	0.80
Absorber Tube Diameter, m	0.57	0.50	0.48	0.42	0.40
Trim Cooler Temperature, °C	141	142	142	141	97
Desorption Pressure, bar	13.4	13.4	13.4	13.4	13.4
Desorption Temperature, °C	500	500	500	500	500
Ammonia Compressor Precooler Temperature, °C	57	50	48	43	42
Ammonia Storage Pressure, bar	23.4	21.2	20.5	19.1	18.6
Ammonia Storage Temperature, °C	56	52	50	48	47

Table 11. Reaction-absorption loop characteristics.

Scale, kgNH$_3$/h	100	500	1000	5000	10,000
Reactor Volume, m^3	0.64	3.72	8.58	88.6	239
Distance from Equilibrium, %	0.07	0.04	0.02	0.001	0.0001
Reactor Pressure Drop, bar	0.04	0.14	0.24	0.57	0.85
Total Pressure Drop, bar	1.23	1.11	1.12	1.28	1.50

The reaction-absorption pressure was 20.29 bar at all scales. This is the limiting pressure, which results in a feed compressor outlet temperature of 150 °C. A higher reaction-absorption pressure would require multi-stage compression with inter-cooling for the feed, and the additional incurred cost would not be worth the increased reactor productivity resulting from the pressure increase.

The reactor inlet temperature was 370 °C at all scales. This lowest allowable temperature shifts the reaction equilibrium towards ammonia, which, from a thermodynamic perspective, allows for the highest possible outlet mole fraction of ammonia at the given pressure. The combination of the recycle ratio and reactor dimensions at each scale takes advantage of the favorable equilibrium shift; it resulted in ammonia mole fractions at the reactor outlet of 4.5% to 4.6%, which is very close to equilibrium at reactor outlet conditions. It is clear that the optimal design maximized single-pass reactor productivity under the constraint of using only one feed compressor. The largest possible single-pass conversion corresponds to the smallest possible recycle rate, which serves to minimize recycle compression capital cost and energy consumption, both because power is a function of flow rate and specific work and because it minimizes the reaction-absorption loop pressure drop. Additionally, the lowest possible reaction temperature and recycle rates minimized the amount of cooling required after the reactor/before the absorber, which is the product of enthalpy (temperature) difference and flow rate, resulting in capital (smaller heat exchanger and cooler) and operating (less pumping of cooling water) cost savings.

The absorption temperature had a complex effect on the capital and operating cost of the process. A low absorption temperature decreased the equilibrium pressure, resulting in less ammonia leaving the absorber and therefore a lower recycle rate. It also reduced the demand on the trim cooler, which is needed after the absorber due to the temperature rise caused by the exothermic absorption. On the other hand, a high absorption temperature resulted in a lessened precooling demand, and more importantly, less heating was needed to move from the absorption temperature to the desorption temperature. From scales of 100 kg/h to 5000 kg/h, absorption temperature increased from 167 °C to

178 °C. This reflects the fact that a high absorption temperature primarily reduces the operating cost (less heating for desorption), which increases linearly with production rate, whereas a low absorption temperature primarily lowers capital cost (smaller recycle compressor), which scales up well in comparison. At these scales, the trim cooler outlet temperature was such that the recycle compressor reached 150 °C. The optimal design at 10,000 kg/h does not follow the trend of the smaller scales. The absorption temperature in this case was 176 °C, and the trim cooler temperature was 97 °C, much lower than in the designs at other scales. This can be explained by the fact that the reaction-absorption loop pressure drop is higher at this scale (see Table 11), primarily due to the reactor pressure drop, so these lower temperatures were chosen to reduce the recycle compressor power: a lower absorption temperature gives a lower recycle rate, and compression power was linear in the inlet temperature (trim cooler temperature in this case).

As scale increases, the absorber was designed such that its pressure drop decreases; the factor by which the number of absorber tubes increases is greater than or equal to the factor by which the ammonia production scale increases. This occurs primarily because compressor cost (capacity exponent of 0.9) does not scale as well as absorber cost (capacity exponent of 0.68) and also to counteract the increase in reactor pressure drop with scale-up.

The desorption temperature was 500 °C in all cases. This high temperature was chosen to give a high absorbent critical pressure. For example, the critical pressure of the absorbent at this temperature was 13.64 bar, whereas at 400 °, the critical pressure was only 1.83 bar. This allows ammonia to leave the bed at a higher pressure and subsequently minimizes the size of the compressor required to bring the ammonia to storage pressure. At all scales, the ammonia compression capital and energy cost reduction that resulted outweighed the significant energy costs that came from this high desorption temperature. As the scale of ammonia production was increased, a decrease in ammonia storage pressure and temperature was observed; this serves to reduce the size and energy consumption of the ammonia compressor. This occurs because the cost of coolers and the condenser (capacity exponent of 0.71) scales better than compressor cost (capacity exponent of 0.9).

5.2. Analysis of Energy Consumption

The relative contribution of different unit types of the overall energy consumption is given in Table 12. These fractions are the same at each scale under consideration.

Table 12. Operation type contribution to optimal energy consumption.

Operation	Value
Compression	0.16
Pumping cooling water	0.02
Desorption	0.82

It is evident that desorption was the dominant form of energy consumption. This is expected, given that conventional chemical processes often rely on combustion of fossil fuels to achieve high temperatures, but that is not an option in the wind-powered absorbent-enhanced process. On the other hand, it is promising that cooling used such a small fraction of the energy in this process; this supports the hypothesis that the higher separation temperatures afforded by using absorption instead of condensation are beneficial.

6. Discussion

The absorbent-enhanced ammonia synthesis capital cost correlation determined in this work is:

$$C^{cap} = 12.7 \left(\frac{X_{NH_3}}{1000 \text{ kg/h}} \right)^{0.77} \tag{44}$$

where C^{cap} is in MM$ and χ_{NH_3} is the ammonia production capacity in kg/h. This result compares well to those in the literature for small-scale Haber–Bosch processes. The ammonia supply chain optimization work in [10] uses the following capital cost correlation for a small-scale Haber–Bosch process [35]:

$$C^{cap} = 1.482\left(\frac{\chi_{NH_3}}{3 \text{ kg/h}}\right)^{0.67} \tag{45}$$

where C^{cap} is in MM$ and χ_{NH_3} is the ammonia production capacity in kg/h. Comparing the correlation for the absorbent-enhanced process to this one, capital cost reductions of 86% at 100 kg/h to 78% at 10,000 kg/h are observed.

However, the Haber–Bosch cost correlation in [35] may be overly conservative based on the small size of the initial installation (only 3 kg/h). For another comparison at a more relevant reference scale, a mini Haber–Bosch unit with a capacity of 3 tons/day (125 kg/h) was quoted at 3.8 MM$ in [9]. In comparison, the absorbent-enhanced process would only cost 2.55 MM$ at that scale, a cost reduction of 33%. The 0.77 scaling exponent for the absorbent-enhanced process is higher than the conventional chemical industry exponent of 0.67, which suggests that this process is more well suited to implementation at a small scale. Conversely, this higher-than-average capacity exponent indicates that the absorbent-enhanced process may not scale up well. For example, if the 0.67 exponent is applied to the Haber–Bosch process using the process from [9] as the reference scale, this Haber–Bosch becomes less expensive than the absorbent enhanced process at a scale of 6075 kg/h (145 ton/day) and larger.

Although the capital cost of the absorbent enhanced process is lower than the Haber–Bosch one at a small scale, this is not true for energy consumption. The current optimal design of the absorbent-enhanced process uses between 3.06 and 3.12 kWh/kg NH$_3$, whereas the process in [35] uses 2.12 kWh/kg NH$_3$. At a larger scale, the 100 tons/day (4167 kg/h) Haber–Bosch process in [8] used only 0.64 kWh/kg NH$_3$. This result is not entirely surprising, as the heating needed for desorption is very electricity-intensive. However, given that in the overall wind-to-ammonia pathway using electrolysis and PSA, electrolysis accounts for the majority of energy consumption (for example, 93% in [8]), the increased ammonia synthesis energy consumption may be worth the decrease in capital cost in the economic outlook of the overall system.

Additionally, the Haber–Bosch process has been optimized over a century, whereas this absorbent-enhanced process is novel, so there are avenues for further capital and operating cost reduction via process improvement. The key novelty in the process is the use of absorption rather than condensation to remove the ammonia from unreacted hydrogen and nitrogen, so naturally, the absorbent is the part of the process to which the most significant improvements can be made. In this work, absorbent capacity was estimated to be a ratio of 1 mol NH$_3$ to 1 mol MgCl$_2$ (5.6 kg of salt are needed to absorb 1 kg NH$_3$) based on the work in [12]. Coupled with the fact that the salt only makes up 40% of the absorbent mixture, every absorbed kg of NH$_3$ requires 14 kg of total absorbent. This results in large absorbent beds as compared to what is theoretically possible, which is absorbing 6 mol NH$_3$ on 1 mol MgCl$_2$ [25]. Efforts to increase the working capacity of these absorbents or to discover new ammonia absorbents with higher capacities will be beneficial in reducing process cost by reducing the size of absorbent beds and subsequently the pressure drop through the bed. Furthermore, the diameter of the currently-used absorbent pellets is on the order of 200 μm [23], an order of magnitude smaller than what is conventionally used in industrial packed beds. As a result, many tubes of absorbent are required, especially at the larger scales under investigation, to prevent prohibitively high pressure drop. The design of larger absorbent pellets would almost certainly be beneficial in that some compromise between reduction in absorber size and recycle compression power could be achieved. An absorbent that can be regenerated at lower temperatures and/or with less temperature variation between absorption and desorption but still at pressures in the range of 15 bar would be beneficial given that desorption requires over 80% of the energy consumed in this process (see Table 12). Regeneration of this absorbent would require less energy consumption and, hence, reduce operating cost, without increasing the size (cost) of the compressor required to liquefy the ammonia, as is the case when the currently-used absorbent is regenerated at lower temperatures. Evidently, efforts to improve ammonia absorbent performance will significantly ameliorate the economics of this process.

Additional efforts to expand the understanding of ammonia synthesis kinetics would also be beneficial. It is clear that running the ammonia synthesis reaction at lower temperature is helpful at the low pressure of this process; the lower bound of 370 °C was chosen as the reaction temperature at all scales. It is possible that an even lower reaction temperature would be beneficial, but a reliable low temperature ammonia synthesis kinetic expression does not currently exist in the literature. The study of ammonia synthesis at lower temperatures would aid in elucidating a true optimal reaction temperature and would perhaps lead to further capital cost and/or operating cost reduction.

Author Contributions: This paper is a joint collaboration between the authors. A.M., E.L.C. and P.D. devised the process concept and contributed to the conception of the overall scope of the work. All authors contributed to the modeling and optimization framework. M.J.P. carried out the computational aspects of the work under the supervision of P.D. All authors contributed to writing and editing of the manuscript.

Funding: The information, data, or work presented herein was funded in part by the Advanced Research Projects Agency-Energy (ARPA-E), U.S. Department of Energy, under Award Number DE-AR0000804. The views and opinions of authors expressed herein do not necessarily state or reflect those of the United States Government or any agency thereof.

Conflicts of Interest: The authors declare no conflict of interest.

References

1. FAO. *World Fertilizer Trends and Outlook to 2018*; Annual Report 14; Food and Agriculture Organization of the United Nations (FAO): Rome, Italy, 2015; ISBN 978-92-5-108692-6.
2. Smil, V. *Enriching the Earth: Fritz Haber, Carl Bosch, and the Transformation of World Food Production*; MIT Press: Massachusetts, MA, USA, 2004.
3. Swaminathan, B.; Sukalac, K. Technology transfer and mitigation of climate change: The fertilizer industry perspective. In Proceedings of the IPCC Expert Meeting on Industrial Technology Development, Transfer and Diffusion, Tokyo, Japan, 21–23 September 2004.
4. Rees, N.V.; Compton, R.G. Carbon-free energy: A review of ammonia-and hydrazine-based electrochemical fuel cells. *Energy Environ. Sci.* **2011**, *4*, 1255–1260. [CrossRef]
5. Wang, G.; Mitsos, A.; Marquardt, W. Conceptual design of ammonia-based energy storage system: System design and time-invariant performance. *AIChE J.* **2017**, *63*, 1620–1637. [CrossRef]
6. Reese, M.; Marquart, C.; Malmali, M.; Wagner, K.; Buchanan, E.; McCormick, A.; Cussler, E.L. Performance of a small-scale Haber process. *Ind. Eng. Chem. Res.* **2016**, *55*, 3742–3750. [CrossRef]
7. Morgan, E.; Manwell, J.; McGowan, J. Wind-powered ammonia fuel production for remote islands: A case study. *Renew. Energy* **2014**, *72*, 51–61. [CrossRef]
8. Morgan, E.R.; Manwell, J.F.; McGowan, J.G. Sustainable ammonia production from US offshore wind farms: A techno-economic review. *ACS Sustain. Chem. Eng.* **2017**, *5*, 9554–9567. [CrossRef]
9. Bañares-Alcántara, R.; Dericks, G.; Fiaschetti, M.; Grünewald, P.; Lopez, J.M.; Tsang, E.; Yang, A.; Ye, L.; Zhao, S. *Analysis of Islanded Ammonia-Based Energy Storage Systems*; University of Oxford: Oxford, UK, 2015.
10. Allman, A.; Daoutidis, P.; Tiffany, D.; Kelley, S. A framework for ammonia supply chain optimization incorporating conventional and renewable generation. *AIChE J.* **2017**, *63*, 4390–4402. [CrossRef]
11. Wagner, K.; Malmali, M.; Smith, C.; McCormick, A.; Cussler, E.; Zhu, M.; Seaton, N.C. Column absorption for reproducible cyclic separation in small scale ammonia synthesis. *AIChE J.* **2017**, *63*, 3058–3068. [CrossRef]
12. Malmali, M.; Le, G.; Hendrickson, J.; Prince, J.; McCormick, A.V.; Cussler, E.L. Better Absorbents for Ammonia Separation. *ACS Sustain. Chem. Eng.* **2018**, *6*, 6536–6546. [CrossRef]
13. Malmali, M.; Wei, Y.; McCormick, A.; Cussler, E.L. Ammonia synthesis at reduced pressure via reactive separation. *Ind. Eng. Chem. Res.* **2016**, *55*, 8922–8932. [CrossRef]
14. *gPROMS ModelBuilder*, version 5.1.1; Software for Advanced Process Modelling Platform; Process Systems Enterprise: London, UK, 2018.

15. *Multiflash*, version 6.1; Software for PVT Modelling and Flow Assurance; KBC: Surrey, UK, 2017.

16. Huber, M. (Ed.) *NIST Thermophysical Properties of Hydrocarbon Mixtures Database (SUPERTRAPP)*, version 3.2; National Institute of Standards and Technology: Gaithersburg, MD, USA, 2007.

17. Nielsen, A.; Kjaer, J.; Hansen, B. Rate equation and mechanism of ammonia synthesis at industrial conditions. *J. Catal.* **1964**, *3*, 68–79. [CrossRef]

18. Liu, H. *Ammonia Synthesis Catalysts: Innovation and Practice*, 1st ed.; World Scientific: Singapore; Chemical Industry Press: Beijing, China, 2013.

19. Dyson, D.; Simon, J. Kinetic expression with diffusion correction for ammonia synthesis on industrial catalyst. *Ind. Eng. Chem. Fundam.* **1968**, *7*, 605–610. [CrossRef]

20. Wakao, N.; Funazkri, T. Effect of fluid dispersion coefficients on particle-to-fluid mass transfer coefficients in packed beds: Correlation of Sherwood numbers. *Chem. Eng. Sci.* **1978**, *33*, 1375–1384. [CrossRef]

21. Specchia, V.; Baldi, G.; Sicardi, S. Heat transfer in packed bed reactors with one phase flow. *Chem. Eng. Commun.* **1980**, *4*, 361–380. [CrossRef]

22. Shafeeyan, M.S.; Daud, W.M.A.W.; Shamiri, A. A review of mathematical modeling of fixed-bed columns for carbon dioxide adsorption. *Chem. Eng. Res. Des.* **2014**, *92*, 961–988. [CrossRef]

23. Smith, C.; Malmali, M.M.; Liu, C.Y.; McCormick, A.; Cussler, E.L. Rates of Ammonia Absorption and Desorption in Calcium Chloride. *ACS Sustain. Chem.* **2018**, submitted.

24. Mofidi, S.A.H.; Udell, K.S. Study of heat and mass transfer in mgcl2/nh3 thermochemical batteries. *J. Energy Resour. Technol.* **2017**, *139*, 032005. [CrossRef]

25. Sørensen, R.Z.; Hummelshøj, J.S.; Klerke, A.; Reves, J.B.; Vegge, T.; Nørskov, J.K.; Christensen, C.H. Indirect, reversible high-density hydrogen storage in compact metal ammine salts. *J. Am. Chem. Soc.* **2008**, *130*, 8660–8668. [CrossRef] [PubMed]

26. Jinasena, A.; Lie, B.; Glemmestad, B. Dynamic Model of an Ammonia Synthesis Reactor Based on Open Information. In Proceedings of the 9th EUROSIM Congress on Modelling and Simulation, Oulu, Finland, 12–16 September 2016; pp. 943–948.

27. Woods, D.R. *Rules of Thumb in Engineering Practice*, 1st ed.; Wiley-VCH: Weinheim, Germany, 2007.

28. Chemical Engineering. *Chemical Engineering Economic Indicators*; Technical Report for Chemical Engineering; Chemical Engineering: Rockville, MD, USA, 2016.

29. Committee of Stainless Steel Producers. *Stainless Steel in Ammonia Production*; Technical Report; American Iron and Steel Institute: Washington, DC, USA, 1978.

30. *Indicative Chemical Prices A-Z*; 2006 Price for Magnesium Chloride; Independent Chemical Information Service: Sutton, UK, 2008.

31. *717177 ALDRICH Silica Gel*; 2018 Price for Technical Grade, Pore Size 60 A, 70-230 mesh, 63-200 μm Silica Gel; Millapore Sigma: Darmstadt, Germany, 2018.

32. Towler, G.; Sinnott, R.K. *Chemical Engineering Design: Principles, Practice and Economics of Plant and Process Design*, 2nd ed.; Elsevier: New York, NY, USA, 2012.

33. Ilas, A.; Ralon, P.; Rodriguez, A.; Taylor, M. *Renewable Power Generation Costs in 2017*; International Renewable Energy Agency: Masdar City, UAE, 2018.

34. Lee, D.; Hochgreb, S. Hydrogen autoignition at pressures above the second explosion limit (0.6–4.0 MPa). *Intern. J. Chem. Kinet.* **1998**, *30*, 385–406. [CrossRef]

35. Tiffany, D.; Reese, M.; Marquart, C. *Economic Evaluation of Deploying Small to Moderate Scale Ammonia Production Plants in Minnesota Using Wind and Grid-Based Electrical Energy Sources*; Technical Report; Minnesota Corn Research and Promotion Council: Shakopee, MN, USA, 2015.

Article

Modeling the Dynamics of Human Liver Failure Post Liver Resection

Babita K. Verma [1,2], Pushpavanam Subramaniam [2] and Rajanikanth Vadigepalli [1,*]

[1]   Daniel Baugh Institute for Functional Genomics/Computational Biology, Department of Pathology,
      Anatomy and Cell Biology, Thomas Jefferson University, Philadelphia, PA 19107, USA;
      ch14d207@smail.iitm.ac.in
[2]   Department of Chemical Engineering, Indian Institute of Technology-Madras, Chennai 600036, India;
      spush@iitm.ac.in
*   Correspondence: rajanikanth.vadigepalli@jefferson.edu; Tel.: +1-215-955-0576

Received: 29 June 2018; Accepted: 1 August 2018; Published: 4 August 2018

Abstract: Liver resection is an important clinical intervention to treat liver disease. Following liver resection, patients exhibit a wide range of outcomes including normal recovery, suppressed recovery, or liver failure, depending on the regenerative capacity of the remnant liver. The objective of this work is to study the distinct patient outcomes post hepatectomy and determine the processes that are accountable for liver failure. Our model based approach shows that cell death is one of the important processes but not the sole controlling process responsible for liver failure. Additionally, our simulations showed wide variation in the timescale of liver failure that is consistent with the clinically observed timescales of post hepatectomy liver failure scenarios. Liver failure can take place either instantaneously or after a certain delay. We analyzed a virtual patient cohort and concluded that remnant liver fraction is a key regulator of the timescale of liver failure, with higher remnant liver fraction leading to longer time delay prior to failure. Our results suggest that, for a given remnant liver fraction, modulating a combination of cell death controlling parameters and metabolic load may help shift the clinical outcome away from post hepatectomy liver failure towards normal recovery.

Keywords: liver regeneration; liver failure; liver resection; virtual patient; dynamic modeling; cell death

1. Introduction

Liver regeneration is a unique repair mechanism underlying physiological recovery following hepatic injury and enables surgical treatment as a viable clinical intervention into liver disease, small for size liver transplant and live donor liver transplantation. The process of liver regeneration takes place via hyperplasia and hypertrophy of hepatocytes as well as other liver resident cell types to reconstitute the liver morphology and function. In this process, the liver parenchymal cells, that is, differentiated post-mitotic hepatocytes, respond to signals from non-parenchymal cells that are stimulated by the hepatic injury and re-enter cell cycle, leading to multiple rounds of cell proliferation, thus compensating for the lost tissue mass [1–3].

Liver resection is widely performed for hepatocellular carcinoma (HCC), metastatic colorectal cancer and benign liver disease [4]. Liver response to resection is also relevant in live donors for liver transplantation, where the remnant liver in the donor needs to regenerate after removal of a portion of the liver for transplantation in the recipient. Advances in surgical techniques and expertise, careful patient selection and post-operative patient care have resulted in better clinical outcomes in the recent years [5]. However, Post hepatectomy liver failure (PHLF) is not uncommon due to a variety of risk factors, including patient-based factors such as age, weight, diabetes [6,7]. Preexisting liver disease such as cirrhosis, cholestasis, steatosis result in impaired liver regeneration [8].

Surgery related factors such as excessive blood loss, portal vein hypertension, ischemia-reperfusion injury and sepsis can result in liver failure. One of the most important factor that determines PHLF is how much liver can be resected which depends on the functional capacity of the remnant liver fraction. The percentage of liver failure post hepatectomy lies in the range of 0.7 to 9.1% [9]. To improve the outcome post-surgery, it is important to be able to predict the liver regeneration response based on patients' pre- and post-operative assessment. Mathematical modeling of liver regeneration can play a significant role in predicting the potential for PHLF prior to the surgical intervention, thus enabling consideration of alternative interventions to prevent or reduce chances of PHLF.

In this study, we started with our previously developed computational model of liver regeneration [10] and fine-tuned the parameters based on liver volumetric data from patients that underwent liver resection [11]. The main objective of this work is to model the dynamics of the liver failure scenario following a surgical resection. We focused on the model parameters that control the cell death process and generated a virtual patient cohort by sampling across the parameter space involving metabolic load and cell death sensitivity. We employed these virtual patients to analyze different modes of potential response; normal recovery, suppressed recovery and liver failure. Our simulations revealed wide variation in the timescale of liver failure consistent with the range of observed timescales in human PHLF scenarios. Simulations indicate that liver failure can either happen instantaneously or after a varying delay. We analyzed the distribution of this delay in a virtual patient cohort and found that the remnant liver fraction plays a significant role in controlling the time of delay in liver failure cases. Specifically, our simulations suggest that lower remnant liver fraction can lead to faster timescale of failure, depending on a balance of a subset of intrinsic parameters. We analyzed the effect of these controlling parameters corresponding to the metabolic load and cell death sensitivities on the overall cell death post resection. Our results suggest that specific combinations of these parameters can lead to a reasonable match to the clinically observed timescales of PHLF.

2. Materials and Methods

2.1. Mathematical Model

The mechanism of liver regeneration involves increase in portal pressure and production of cytokines from the Kupffer cells (resident macrophages in the liver). The cytokines activate the hepatocytes resulting in the priming and initiation of hypertrophy of the cells. Once the hepatocytes are primed they become responsive to the growth factor released from extracellular matrix via the action of matrix metalloproteinases (MMPs), as a consequence of which the parenchymal cells advance into the cell cycle and replicate leading to hyperplasia. The mathematical model (Cook et al. [10]) considered for this work incorporates the above-mentioned processes using eleven ordinary differential equations representing the hepatocytes in the three phases of the cell cycle quiescent (*Q*), priming (*P*) and replicating (*R*) states. The model comprises of seven equations for the molecular regulation, three equations for the cell state balances and the last equation accounts for relative cell mass growth, that is, hypertrophy. The system of equations is considered lumped since we assume the process of regeneration to be homogeneous in the remaining liver mass. The model equations are:

$$\frac{dQ}{dt} = -k_{QP}([IE] - [IE_0])Q + k_{RQ}[ECM]R + k_{req}\sigma_{req}P - k_{cd}\sigma_{cd}Q \tag{1}$$

$$\frac{dP}{dt} = k_{QP}([IE] - [IE_0])Q - k_{PR}([GF] - [GF_0])P - k_{req}\sigma_{req}P - k_{cd}\sigma_{cd}P \tag{2}$$

$$\frac{dR}{dt} = k_{PR}([GF] - [GF_0])P - k_{RQ}[ECM]R + k_{prol}R - k_{cd}\sigma_{cd}R \tag{3}$$

$$\frac{d[IL6]}{dt} = k_{IL6}\frac{M}{N+\varepsilon} - \frac{V_{JAK}[IL6]}{[IL6] + K_M^{JAK}} - \kappa_{IL6}[IL6] + k_1 \tag{4}$$

$$\frac{d[JAK]}{dt} = \frac{V_{JAK}[IL6]}{[IL6] + K_M^{JAK}} - \kappa_{JAK}[JAK] + k_2 \tag{5}$$

$$\frac{d[STAT3]}{dt} = \frac{V_{ST3}[JAK][proSTAT3]^2}{[proSTAT3]^2 + K_M^{ST3}\left(1 + [SOCS3]/K_I^{SOCS3}\right)}$$
$$- \frac{V_{IE}[STAT3]}{[STAT3] + K_M^{IE}} - \frac{V_{SOCS3}[STAT3]}{[STAT3] + K_M^{SOCS3}} - \kappa_{ST3}[STAT3] + k_3 \tag{6}$$

$$\frac{d[SOCS3]}{dt} = \frac{V_{SOCS3}[STAT3]}{[STAT3] + K_M^{SOCS3}} - \kappa_{SOCS3}[SOCS3] + k_4 \tag{7}$$

$$\frac{d[IE]}{dt} = \frac{V_{IE}[STAT3]}{[STAT3] + K_M^{IE}} - \kappa_{IE}[IE] + k_5 \tag{8}$$

$$\frac{d[GF]}{dt} = k_{GF}\frac{M}{N + \varepsilon} - k_{up}[GF][ECM] - \kappa_{GF}[GF] + k_7 \tag{9}$$

$$\frac{d[ECM]}{dt} = -k_{\deg}[IL6][ECM] - \kappa_{ECM}[ECM] + k_6 \tag{10}$$

$$\frac{dG}{dt} = k_G\left(\frac{M}{N + \varepsilon}\right) - k_G M \tag{11}$$

where,

$$\varepsilon = 0.01 \tag{12}$$

$$\sigma_{cd} = 0.5\left(1 + \tanh\left(\frac{\theta_{cd} - (N + \varepsilon)/M}{\beta_{cd}}\right)\right) \tag{13}$$

$$\sigma_{req} = 0.5\left(1 + \tanh\left(\frac{\theta_{req} - [GF]}{\beta_{req}}\right)\right) \tag{14}$$

$$N = Q + G(P + R) \tag{15}$$

$$k_1 = \frac{V_{JAK}}{1 + K_M^{JAK}} - k_{IL6}\frac{M}{N_{ss} + \varepsilon} + \kappa_{IL6} \tag{16}$$

$$k_2 = \kappa_{JAK} - \frac{V_{JAK}}{1 + K_M^{JAK}} \tag{17}$$

$$k_3 = -\frac{V_{ST3}[proSTAT3]^2}{[proSTAT3]^2 + K_M^{ST3}\left(1 + 1/K_I^{SOCS3}\right)} + \frac{V_{IE}}{1 + K_M^{IE}} +$$
$$\frac{V_{SOCS3}}{1 + K_M^{SOCS3}} + \kappa_{ST3} \tag{18}$$

$$k_4 = -\frac{V_{SOCS3}}{1 + K_M^{SOCS3}} + \kappa_{SOCS3} \tag{19}$$

$$k_5 = -\frac{V_{IE}}{1 + K_M^{IE}} + \kappa_{IE} \tag{20}$$

$$k_6 = k_{\deg} + \kappa_{ECM} \tag{21}$$

$$k_7 = -k_{GF}\frac{M}{N_{SS} + \varepsilon} + k_{up} + \kappa_{GF} \tag{22}$$

$$N_{SS} = 0.99 \tag{23}$$

Here, $k_1 \ldots k_7$ are constants defined such that the molecular species are at steady state under the normal functioning of the liver. The initial conditions for solving the above-mentioned system of equations are:

$$Q_0 = \text{remnant liver fraction}; P_0 = 0; R_0 = 0$$
$$[\text{IL6}_0] = 1; [\text{JAK}_0] = 1; [\text{STAT3}_0] = 1; [\text{SOCS3}_0] = 1;$$
$$[\text{IE}_0] = 1; [\text{GF}_0] = 1; [\text{ECM}_0] = 1 \tag{24}$$
$$G_0 = 1$$
$$N_0 = Q_0 + G_0(P_0 + R_0) = Q_0$$

The system of equations for this model are stiff and were solved in Matlab using *ode15s*. The initial conditions for the quiescent state and molecular species signify the steady state under normal functioning of liver in which both primed and replicating cells are at zero levels.

2.2. Defining Virtual Patient Cohort

We employed the emerging approach of simulating a cohort of virtual patients [12,13] to assess the distribution of potential responses, in an attempt to account for the clinically observed wide variation of outcomes. In the present study, the virtual patients were defined based on sampling key parameters within a range of 2X, 3.125X and 7X, respectively, from the nominal values for the three controlling parameters (M, β_{cd}, θ_{cd}). The sampling was performed in an unbiased approach to span the full range of the three-dimensional volume using Sobol sampling [14,15] to generate a cohort of 9000 virtual patients. The range of the controlling parameters were so chosen to capture the different response modes in the cohort of virtual patients under study.

2.3. Identification of Distinct Response Modes

The simulated results were classified into distinct response modes based on the level of liver mass fraction at the end of 2.5 years post liver resection—Normal recovery: 0.9–1.1, Suppressed recovery: <0.8. Patients are classified as exhibiting a liver failure scenario if the liver mass fraction dropped below 0.1 by 2.5 years post liver resection. The liver failure cases were further classified as exhibiting delayed failure or immediate failure based on the time taken to undergo failure. Immediate liver failure was deemed as occurring if the liver mass fraction is below 0.1 within the 5th day post resection and the remaining liver failure cases were classified as corresponding to the delayed failure scenarios.

2.4. Parameter Optimization

All the 33 parameters of the model (Table A1) were optimized using *fmincon* in Matlab using the elastic net approach with the regularization weightage of 0.001 for both Ridge and Lasso [16,17]. We explored the effect of using different regularization weights over a wide range from 0.0001 to 0.1. However, some of the regularization weights resulted in an integration error or terminated with a premature solution for a subset of the patients. Following this approach resulted in patient-specific regularization weights and thus likely yielded an overfitted model in some cases. We found that the regularization weight of 0.001 worked for all the patients and thus avoided over-fitting of the model.

The liver fraction at any given time point at and post resection is calculated as the ratio of remnant liver volume to the total liver volume prior to resection, based on the volumetric data from Yamamoto et al. [11]. However, note that the present ODE model considers fractional liver mass and not volume, as we incorporated the tissue growth dynamics in the model. For the purpose of the study, we considered the liver volume fraction information from the clinical data set to be equivalent to the liver mass fraction calculated in the simulations, under a reasonable assumption that the density of the tissue does not alter significantly after resection. This also permits us to compare the response profiles across individual patients that may have wide variation in the liver size. The model dynamics are sensitive to the three controlling parameters; metabolic load (M) and cell death sensitivities (β_{cd}, θ_{cd}). The initial guess for the parameter optimization was considered from the parameter values of Cook et al. [10] for humans, except for the values of the three controlling parameters (M, θ_{cd}, β_{cd}). These values were fixed from a representative case of a virtual patient undergoing delayed liver failure (Table A2).

2.5. Clinical Dataset

The human liver volumetry dataset used for the present work was obtained from the published literature [11], which contains information about the liver volume post resection in 196 patients. The dataset was analyzed for volume changes over time and a subset of 7 patients that showed liver failure were considered for further analysis and matching to model dynamics.

2.6. Model Repeatability and Reproducibility

The model equations and parameters for various simulations presented in the manuscript are available in the main text and Appendix, respectively. The Matlab code for virtual patient cohort simulation and response mode classification is available in the supplemental file S1. The Matlab code for parameter optimization is available as supplemental file S2. The network model in Equations (1)−(24) was independently implemented in the Systems Biology Markup Language (SBML) by a laboratory colleague not involved in the original study, based on the equations and information provided in the manuscript. This model was set up based on the parameter values in the "Nominal Value" column of the Table A2 in the Appendix. An SBML code corresponding to this model implementation is included as supplemental file S3. Select figures for exploring the results from virtual patient cohort analysis are provided in the supplement in the Matlab FIG format. A brief discussion of the biological context and assumptions motivating the model is included in the Methods and some of the limitations were detailed in the Discussion.

3. Results

3.1. Modeling the Range of Response to Liver Resection in a Virtual Patient Cohort

We started with a computational model of liver regeneration, previously developed by our group [10]. The mathematical model of Cook et al. [10] considers hypertrophy and hyperplasia in liver following resection [18], by accounting for hepatocytes in different phases of the cell cycle. The molecular aspects of this network model of liver regeneration are largely based on the dynamics of cytokine and growth factor pathways stimulated by the liver resection injury. Immediately post resection, the hepatocytes undergo priming in response to the cytokines released by Kupffer cells, activating the Janus Kinase (JAK)—Signal Transducer and Activator of Transcription (STAT) signaling pathway in the hepatocytes. Primed hepatocytes become responsive to the growth factors linking the priming phase to the cell-cycle progression. In the present computational model, response to liver regeneration following partial hepatectomy is triggered by metabolic load per unit liver mass. The network model is shown in Figure 1A and the corresponding systems biology graphic notation (SBGN) [19] diagram for the liver regeneration model is provided as supplementary information (Figure S1). This model can exhibit distinct modes of response ranging from complete recovery of liver mass to liver failure. As an illustrative example, Figure 1B shows the regeneration profile of two virtual patients that underwent 1/3rd liver resection but showed opposite responses, that is, recovery versus failure. Phase portrait [20,21] for the same virtual patients is shown in Figure 1C, depicting the evolution of hepatocytes in different phase of the cell cycle over time. Depending on the model parameters which corresponds to the two different virtual patients, for the same initial remnant liver mass, there exist two different stable steady states corresponding to recovery and failure.

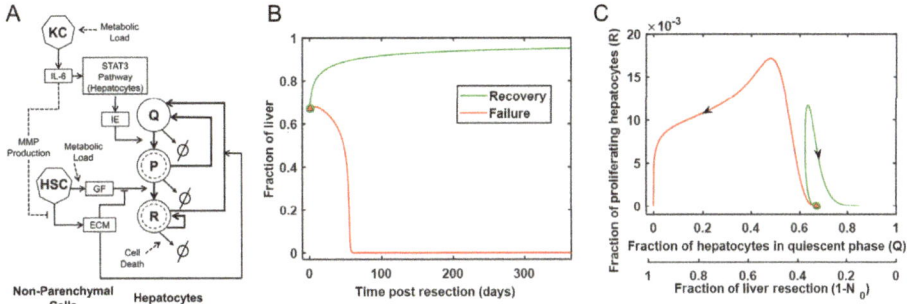

Figure 1. (**A**) Network diagram of the liver regeneration model. (**B**) Regeneration profile of two virtual patients with liver recovery and failure respectively for 1/3rd resection, their model parameters are same as Cook et al. but with different values of the controlling parameters; Recovery: M = 11.4645, β_{cd} = 0.0064, θ_{cd} = 0.0262; Failure: M = 10.7312, β_{cd} = 0.0219, θ_{cd} = 0.0224. (**C**) Phase plane of the same virtual patients showing recovery and failure with the evolution of hepatocytes in quiescent and replicating phase of the cell cycle.

We simulated the dynamics of liver mass in response to resection in a virtual patient cohort generated by varying key model parameters (see Methods). We analyzed the distribution of responses in the virtual patient cohort and classified the virtual patients into multiple response categories based on the liver mass outcome at the end of 2.5 years following resection. We analyzed the responses of 9000 virtual patients to unravel the mechanisms that lead to different response modes. The virtual patient cohort was generated by varying the metabolic load and the cell death sensitivities via Sobol sampling, while the remaining 30 parameters (out of 33) were held at the same values as that of Cook et al. [10] for the case of human. We chose to vary metabolic load (M) and cell death sensitivities (β_{cd}, θ_{cd}) from the 33-dimensional parameter space since these 3 parameters control the cell death process as incorporated in the network model (see Equations (1)–(3) and (13)). In this formulation, liver failure is considered as likely to occur in two ways: (i) either the liver regeneration does not commence at the rate necessary to meet the increased functional demand post resection, or (ii) even as the regeneration process is initiated, the cell death rate is sufficiently high as to result in liver failure [22].

Our simulation-based analysis suggests that the virtual patients can be categorized into four distinct classes based on the response to the resection: normal recovery, suppressed recovery, delayed failure and immediate failure. The regeneration profiles of different classes of response for 10% and 33.3% are shown in Figure 2A,B,D,E. We note that the suppressed patients did not show fast recovery immediately after surgery. In Figure 2C we show a representative patient data demonstrating recovery post-surgery from Yamamoto et al. [11]. From Figure 2F we observe that a subset of the virtual patients exhibited instantaneous liver failure, whereas others showed a delayed liver failure), demonstrating a good match in the timescale of the liver failure based on volumetric data from patients in Yamamoto et al. [11]. Simulated regeneration profiles of liver failure response show a sudden drop in fraction of liver either immediately post resection or after a delay of few days. However, such a sudden drop was not readily apparent in the clinical data. It is unlikely that patients undergoing liver failure would be subjected to liver volumetric analysis in the clinic. Also, in reality, the liver mass does not become zero as is seen in the numerical simulations, since the patient's body might succumb to the non-functioning liver even before such condition of extremely low levels of liver mass is reached. However, the drop in the liver mass in simulations is instructive on the dynamics of the process, informing the potential trajectories taken by a range of patients, consistent with the wide range of time delays observed in patients exhibiting liver failure [11].

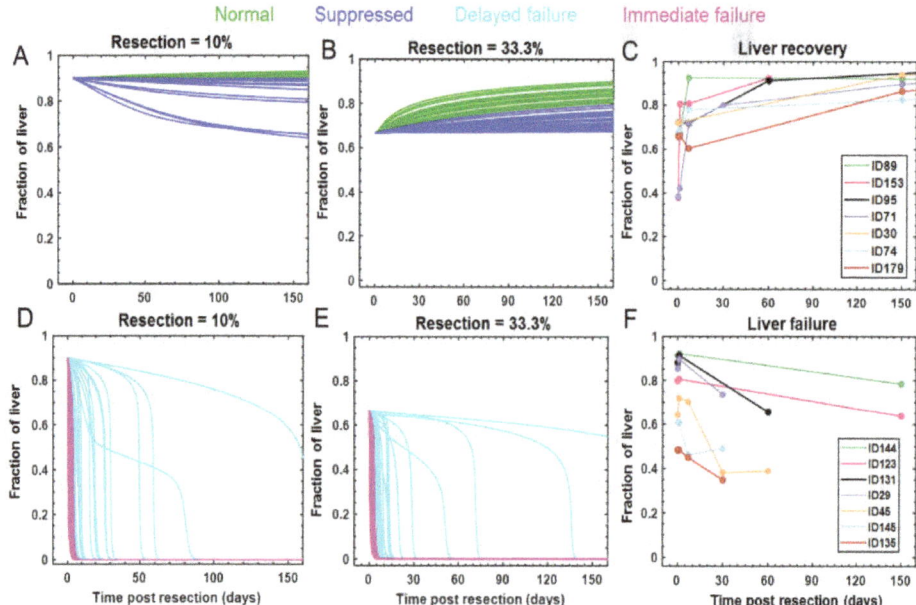

Figure 2. Regeneration profiles of virtual patients depicting two types of liver recovery (Normal and suppressed) for different levels of resection (**A**) 10% resection (**B**) 33.3% resection (**C**) Volumetric liver data from Yamamoto et al. [11] for patients that exhibited liver recovery. Response modes of virtual patients that underwent delayed and immediate liver failure for (**D**) 10% resection (**E**) 33.3% resection (**F**) Volumetric liver data for patients showing liver failure from Yamamoto et al. [11].

We visualized the distribution of different classes of response modes—Normal, suppressed, delayed failure and immediate failure—as a function of the three controlling parameters (Figure 3). These classes were represented, albeit in different proportions, for varying levels of resection (10% and 33.3%). Figure 3A–F show the projection of classes of virtual patients for increasing levels of cell death sensitivity (θ_{cd}) at 10% and 33.3% resection. For low cell death sensitivity (θ_{cd}), we observe that liver failure occurs at high levels of metabolic load (M) and cell death sensitivity (β_{cd}) and corresponds to only delayed liver failure (Figure 3A,D). For higher level of resection, we observe that the normal recovery space is enveloped by suppressed recovery space. As the cell death sensitivity θ_{cd} increases, it becomes apparent that the parameter regions where normal recovery and liver failure occur are separated by a narrow region corresponding to the suppressed recovery (Figure 3A–F). For either level of resection, low metabolic load and cell death sensitivities led to normal recovery. This appears to be governed primarily by low cell death (Figure 4). By contrast, high values of these controlling parameters result in liver failure. Figure 3G,H show the pattern of the parameter region corresponding to the virtual patients that exhibited delayed failure for remnant liver fraction of 0.9 and 0.667 (Supplementary information Figures S2 and S3, Video S1, S2). The proportion of cases that exhibited delayed liver failure varied from 8.73% to 40.22% of the total number of virtual patients for remnant liver fraction of 0.9 and 0.667, respectively (i.e., 10% and 33.3% resection). At the same time, the proportion of cases that exhibited suppressed recovery changed from 1.63% to 37.25% and the proportion of normal recovery cases reduced from 37.23% to 8.17%, as the remnant liver fraction decreased from 0.9 to 0.667 (i.e., increase in the level resection from 10% to 33%). This shift in the response of different virtual patients with increasing resection implies that the remnant liver mass significantly regulates the qualitative outcome of liver resection surgery. In the next section, we focus on the changes in the cell death process as a function of the three key controlling factors involving metabolic load and cell death sensitivities.

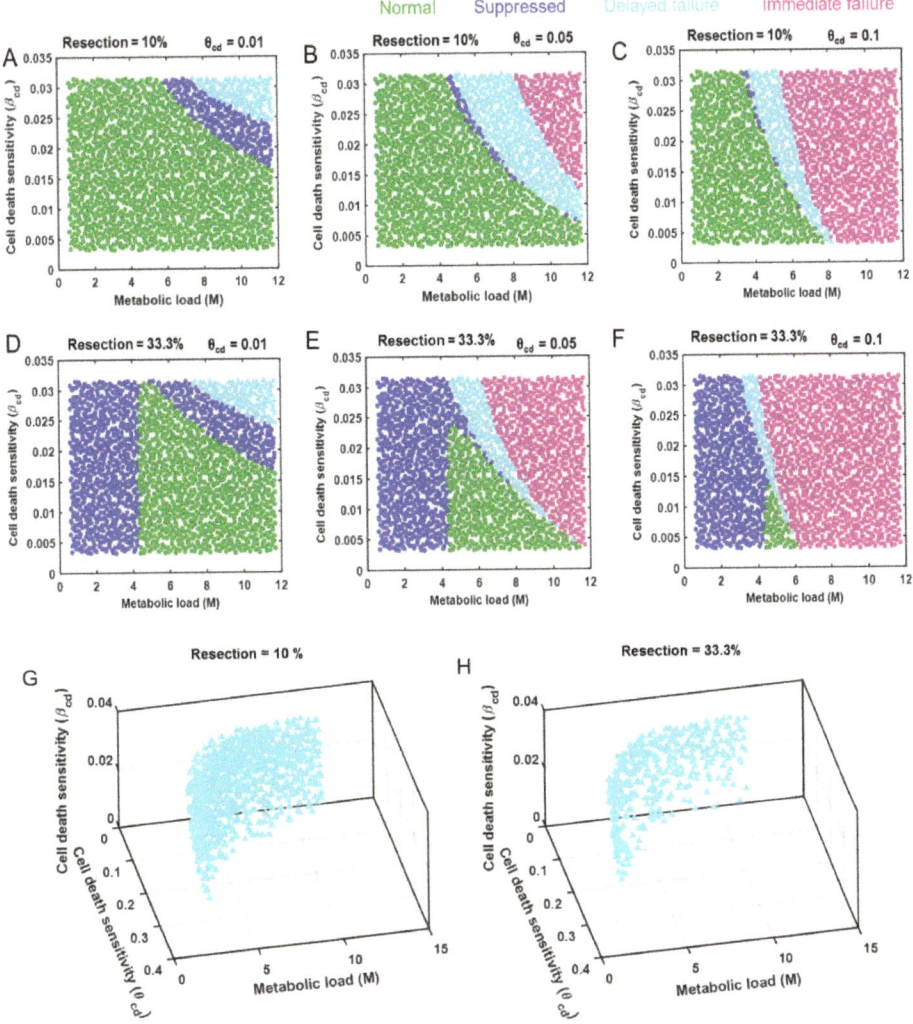

Figure 3. (**A–F**) Two-dimensional distribution of response modes of virtual patients sampled over a range of metabolic load (M) and cell death sensitivity (β_{cd}), for increasing levels of cell death sensitivity (θ_{cd}) at (**A–C**) 10% and (**D–F**) 33.3% resection. (**G,H**) Parameter space of the three controlling parameters showing the distribution of virtual patients with delayed liver failure post hepatectomy for 10% and 33.3% resection, respectively. Each marker represents a virtual patient.

3.2. Cell Death as a Function of the Controlling Parameters in Different Classes of Patients

Following partial hepatectomy, both positive and negative stimulus are triggered resulting in liver regeneration or failure based on a balance of these stimulus. Cell death is likely an important regulatory process in liver regeneration, levels of which determine whether the organ can recover following injury. Analysis of the cellular equations (Equations (1)–(3)) of the model along with the cell death sigmoidal function σ_{cd} (Equation (13)) reveals that there are three controlling parameters: metabolic demand (M) and the two cell death sensitivity parameters (θ_{cd}) and (β_{cd}) that influence decay of cells post resection. In this section, we analyze the cell death profile as a function (Equation (13)) of these three parameters (M, θ_{cd}, β_{cd}), individually and in combination. A change in the cell death sensitivity β_{cd} results in a shift of the angle or slope of the cell death function (Figure 4A). At a high level of resection (i.e., low fraction of remnant liver), increasing β_{cd} results in a downward shift in the cell death function, lowering the cell death (Figure 4A). However, at low level of resection (i.e., high remnant liver mass fraction) the cell death increases with increasing β_{cd} (Figure 4A). The cell death profile as a function of increasing cell death sensitivity (θ_{cd}) or metabolic load (M) results in a horizontal shift towards higher levels of remnant liver fraction (Figure 4B,C). Consequently, for a given level of remnant liver fraction, increasing metabolic load (M) or cell death sensitivity (θ_{cd}) lead to higher levels of cell death. The cell death profile when both the cell death sensitivity (β_{cd}) and metabolic load are changed simultaneously show a wide variation in both slope and the switching threshold of the sigmoidal cell death function (Figure 4D).

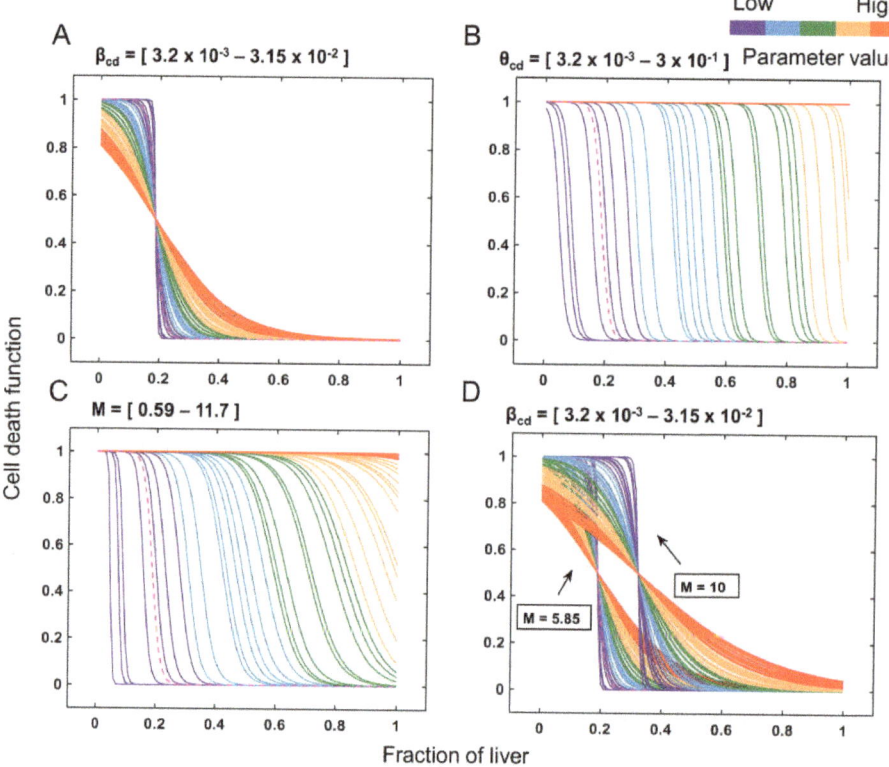

Figure 4. Changes in the cell death function upon varying the parameters that control the cell death process. (**A**) $\beta_{cd} = [3.2 \times 10^{-3} - 3.15 \times 10^{-2}]$; M = 5.8507, θ_{cd} = 0.0320. (**B**) $\theta_{cd} = [3.2 \times 10^{-3} - 3 \times 10^{-1}]$; M = 5.8507, β_{cd} = 0.0045. (**C**) M = [0.59 − 11.7]; θ_{cd} = 0.0320, β_{cd} = 0.0045. (**D**) $\beta_{cd} = [3.2 \times 10^{-3} - 3.15 \times 10^{-2}]$ and M = 5.85 or 10, θ_{cd} = 0.0320.

We analyzed the cell death function of the 9000 virtual patients that were categorized for response modes in Section 3.1. The cell death functions corresponding to the four response mode classes of patients are shown in Figure 5. The cell death functions of both recovery classes, normal and suppressed growth, are steep and shifted towards lower remnant liver fraction (Figure 5A,B,E,F), which resulted in almost negligible cell death at low levels of resection (i.e., high remnant liver fraction). The cell death functions for the liver failure classes show wide variation and are more inclined as compared to normal and suppressed recovery classes, resulting in higher cell death for any given level of resection (Figure 5C,D,G,H). The ranges of cell death functions are overlapping across the response modes, suggesting that cell death function is not the sole determining factor for classifying the fate of the patient post resection surgery. Specifically, the level of resection also plays a crucial role in determining the dynamics of liver recovery and potential failure.

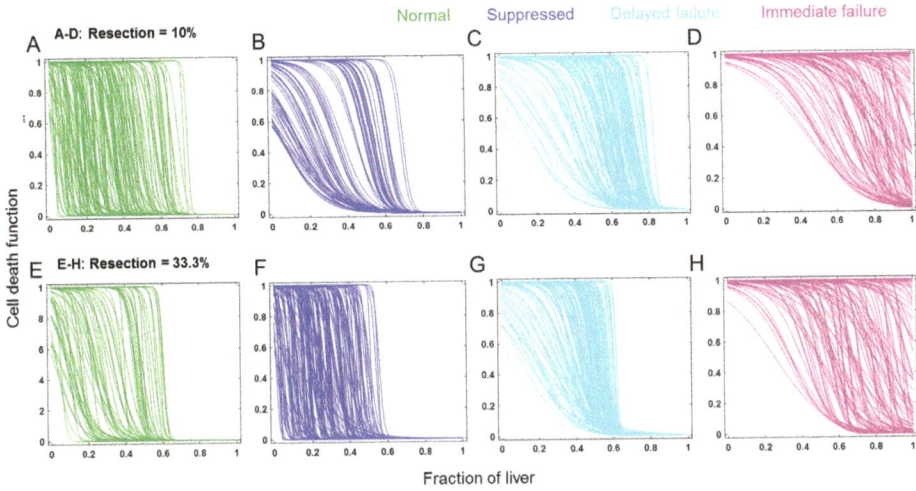

Figure 5. The distribution of profiles of cell death function for the four classes of patients categorized by the response mode to a given level of resection. (**A,E**) Normal growth. (**B,F**) Suppressed recovery. (**C,G**) Delayed liver failure. (**D,H**) Immediate liver failure. (**A–D**) 10% resection and (**E–H**) 33.3% resection.

3.3. Comparison of Cell Death between Recovery versus Failure Scenarios for Varying Level of Resection

In this section, we compare the responses of two virtual patients showing liver recovery and liver failure to different levels of resection to examine the distribution of responses. Figure 6A,D shows the regeneration profiles of a representative normal recovery and a liver failure cases from the virtual patient cohort for 1/3rd resection. The response profiles were simulated for different levels of resection. The virtual patient shows recovery for varying levels of resection, until a threshold value of 83%, beyond which liver failure occurred due to lack of a robust regenerative response (Figure 6A). By contrast, the response of the patient with controlling parameters leading to liver failure was insensitive to the level of remnant liver mass (Figure 6D). Simulations suggest that at higher levels of resection (i.e., lower remnant liver fraction), the patient can exhibit an inverse response in which the liver mass enters an initial recovery phase while undergoing a continuous loss of liver tissue that overcomes the recovery, resulting in liver failure. The failure is reflected in the response profile as a fast decline in the remnant liver fraction and the time delay of failure was longer with higher levels of remnant liver mass (Figure 6D).

For the above two virtual patient cases, we examined the evolution of hepatocytes in quiescent versus proliferating phase over time for different levels of resection. The corresponding projections of the phase spaces shown in Figure 6B,E suggest the existence of two attractors, one corresponding to failure and the other to recovery. For the first virtual patient case, a threshold of 83% resection separates the two attractors, such that increasing level of resection results in a shift in the stability of the system from one attractor to another, that is, exhibiting a transition from recovery to failure (Figure 6B).

By contrast, the threshold of failure for the second virtual patient was low, likely corresponding to the existence of only one attractor for this case that is associated to liver failure (Figure 6E). We compared the cell death function for the two cases for uncovering potential differences over time. The surface plots shown in Figure 6C,F correspond to the two virtual patient cases and the trajectories on the surface depict the cell death process for a specific level of resection. For the first virtual patient case, the cell death is negligible for low levels of resection as seen by the trajectories largely spanning the lower part of the cell death function (Figure 6C). At a resection level beyond the threshold of failure, there is a drastic change in the cell death profile with the trajectories distributed in the upper portion of the cell death function, resulting in high cell death. In the second virtual patient case, the cell death trajectories show initial fluctuations at low levels of the cell death function, before shifting upward and saturating to the maximum cell death which leads to a fast decline in the remnant liver fraction (Figure 6F).

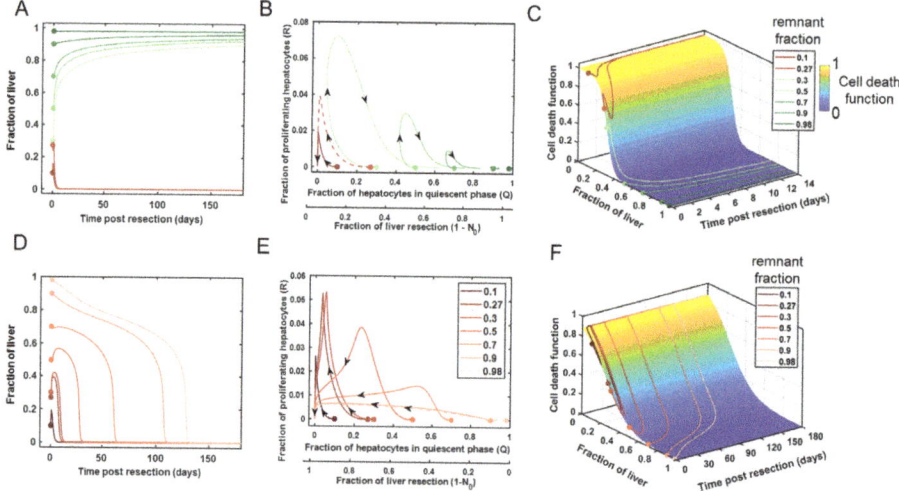

Figure 6. (**A,D**) Temporal profile of remnant liver fraction for two virtual patients for varying levels of resection. These cases were chosen based on the recovery and failure outcomes respectively for 1/3rd resection. (Recovery: M = 11.4645, β_{cd} = 0.0064, θ_{cd} = 0.0262; Failure: M = 10.7312, β_{cd} = 0.0219, θ_{cd} = 0.0224) The corresponding (**B,E**) phase plane and (**C,F**) cell death function for the two virtual patients are shown.

3.4. Tuning the Model for Different Liver Failure Patients

The analysis so far has provided insights into the effect of key biochemical and biophysical parameters in the liver regeneration process. We used this knowledge to tune the model parameters to capture the timescale of response to resection as reflected in the volumetric liver data for the patients that exhibited liver failure available in Yamamoto et al. [11]. Figure 7 shows the regeneration profiles predicted and optimized parameters for 7 real patients who exhibited liver failure. The model parameters were optimized for each patient-specific data set using elastic net (see Methods). The initial values for parameter optimization were chosen such that the values of the controlling parameters (M, θ_{cd}, β_{cd}) were set from a representative case of a virtual patient belonging to the class of delayed liver failure and the remaining 30 parameters were set from Cook et al. [10]. The simulated liver fraction profiles from models with patient-specific optimized parameters showed a good match with the volumetric data and captured both the slow (ID114, 123, 131, 29) and fast (ID45, 245, 135) timescales of liver failure (Figure 7A). Comparing the optimized parameters to the initial values show that the priming of hepatocytes and JAK-STAT pathway contribute to the difference in the timescale of liver failure (Figure 7B). The table of optimized parameter values for the individual patients is given in Appendix A.

The profiles of remnant liver fraction for the slow and fast timescale of liver failure cases are shown in Figure 8A,F. The corresponding phase space portrait projection and the cell death functions of patients undergoing liver failure at the slow and fast timescales are shown in Figure 8A–E and Figure 8F–I, respectively. Even as the regeneration profiles of the patients with slow timescale of response appear to match a suppressed recovery case, the phase plane depicts that the patients likely undergo liver failure, as the fraction of remnant liver in each case is progressing towards the attractor of zero levels. Analysis of the patient-specific cell death functions and trajectories suggest that the cell death trajectories of patients showing slow timescale of failure were largely restricted to the lower levels of cell death function. By contrast, the cell death trajectories of patients with relatively faster timescale of liver failure continually progressed from lower levels with a steep ascent to reach maximal levels of cell death function, corresponding to a rapid decline in the predicted liver fraction.

We note that the optimized values of the controlling parameters of all the patients were similar (Appendix A) and yet the cell death trajectories were rather divergent across the patients (Figure 8B–E,G–I). This is because cell death is not only dependent on the identified controlling parameters but also on the remnant liver fraction, which depends on the regenerative processes including priming pathways and proliferation. The parameters for the regenerative capacity are different between the patients (Figure 7B) and these differences manifest in the net amount of recovery, or lack thereof, in the fractional liver, indirectly affecting the cell death trajectory. Our simulations suggest that patients with slow timescale of liver failure sustain low cell death for a longer duration compared to the patients with a relatively faster timescale of failure that experience high cell death with a rapid decline in the remnant liver fraction leading to failure and patient death. These results from analysis of patient-specific model parameterization and simulation provide insights into how the cell death trajectories likely control the timescale of response to resection, particularly for the liver failure cases.

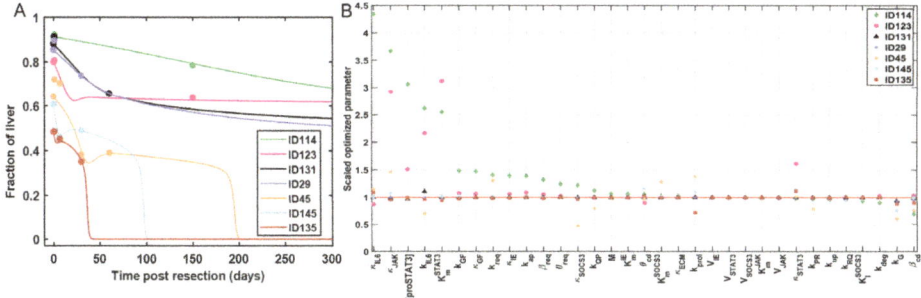

Figure 7. (**A**) Temporal profile of fractional liver of different patients that exhibited liver failure. The model fits are based on patient-specific optimized parameters obtained using elastic net. (**B**) Optimized parameters scaled to the initial parameter values for the individual patients.

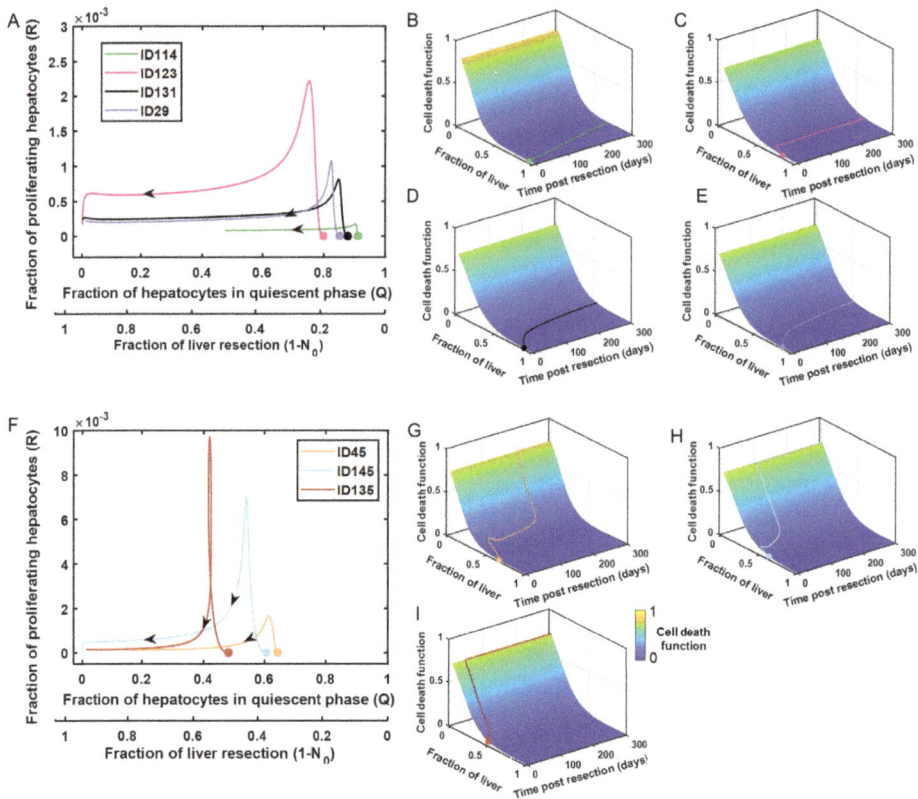

Figure 8. (**A,F**) Phase plane of a select set of patients exhibiting liver failure. (**B–E**) Patient-specific cell death functions and trajectories for patients with slow timescale of liver failure. (**G–I**) Patient-specific cell death functions and trajectories for patients with fast timescale of liver failure.

4. Discussion

The focus of this study is to understand the interplay between the different processes that are activated after a liver resection and can lead to post hepatectomy liver failure. We started with a computational model of liver regeneration process based on the published cellular network model of Cook et al. [10]. We simulated the model over a wide range of parameters and found that the model captures the wide variability in patient outcomes of liver regeneration and accounts for the variation in the observed timescales of liver failure. The variability in response to liver resection may arise due to inter-individual differences in the sensitivity to injury, which is likely due to disease etiology as well as intrinsic genetic factors [23]. For instance, hepatic steatosis alters the potential for cell survival and proliferation after liver resection as compared to the normal liver [10,24].

We identified the controlling parameters in the model which govern the timescale of the post hepatectomy liver failure in humans. These controlling parameters correspond to the metabolic load and cell death sensitivities. We generated a cohort of virtual patients by varying the controlling parameters and classified the virtual patients based on their model-predicted outcomes post liver resection. Our analysis suggests an inverse relationship between the levels of resection on the proportion of cases showing normal growth, suppressed recovery, with concomitant increase in the liver failure cases. These results are consistent with expectations of increased potential for failure with increasing level of resection in humans and animals [6]. Our model-based virtual patient analysis revealed that the parameter space corresponding to the transition from recovery to failure outcomes corresponds to delayed liver failure and the proportion of the delayed liver failure is dependent on the level of resection. These results show that the remnant liver fraction is an important decision making variable that not only discriminates between liver failure and recovery but also governs the timescale of failure especially when a patient experiences delayed liver failure. These results provide new insights in interpreting the wide range of variation seen in clinical data sets [11].

Our model-based results on the critical role of a combination of cell death process and remnant liver fraction in controlling the liver regenerative response is consistent with the findings from animal studies showing that beyond a certain threshold of resection, cell death increases substantially, resulting in high likelihood of failure [6,25]. Importantly, reducing the cell death rescues the animals from undergoing failure after partial hepatectomy [26]. Similarly, preventing the hepatocyte death by blocking mitochondrial permeability transition can improve liver regeneration even for low remnant liver fraction in animals [27,28]. Our model-based results place these experimental findings in the context of cellular networks that underlie the regeneration process, as well as point out additional parameters that can be potentially manipulated to help shift the trajectory of the response modes to the recovery zone. The key controlling parameters were determined such that their combination decided the level of cell death that occurred post resection. With our analysis, we concluded that though cell death is an important process which influences the liver regeneration it is not the sole factor which determines liver failure. Liver failure is conditional upon the load and other perioperative conditions. This is likely to be patient specific and needs to be correlated with epidemiological, clinical measures and disease etiology [29]. Modulation of these additional factors likely improve the probability of liver recovery and hence have the potential to yield additional clinical options to reduce the chances of liver failure due to lack of regenerative response.

Various parameters of this model were derived based on biochemical data and animal/clinical observations (Furchtgott et al. [30]; Cook et al. [10]). The validation relevant to our present study is regarding the dynamics of liver failure. Model simulations predict a wide range of time to failure, which is consistent with the variability of the time scale of liver failure, seen in human data (Yamamoto et al. [11]). We note that the virtual patient analysis approach differs from patient-specific modeling that identifies parameters for individual patients and then performs comparative analysis across patients. By contrast, the virtual patient simulations are evaluated against the types of behavior (e.g., recovery versus failure), distribution of outcomes (e.g., range of time scales of failure) and so forth., that are exhibited in a cohort (An et al. [12,13]). In the present study, information from the virtual patient analysis on delayed liver failure response was used for optimization of the model parameters for patient-specific data, which were not used to build the original model. Our analysis identifies cell death relevant parameters and metabolic load as key factors controlling the time scale of liver failure post-surgical resection. These results are consistent with Yamamoto et al. [11] findings from analysis of patient data that suggest the extent of blood loss during surgery as a key factor discriminating between failure versus recovery scenarios. It is likely that perioperative factors such as blood loss manifest their effects on liver at least partly through increased cell death and higher load on the remaining functional organ.

Our study was focused on the regulatory networks at the cellular and pathway scales in the liver during the response to resection. However, failure or success of the resection surgery also depends on the recovery of the metabolic function in addition to the liver mass. The present computational framework can be extended to incorporate metabolic components to account for relationship between functional liver fraction and key metabolic functions, for example, glucose homeostasis [31], ammonia metabolism and urea cycle [32], or genome-scale metabolic models [33]. Liver failure can also occur due to extra-hepatic factors including immune response, as well as inter-organ metabolic and physiological relationships. For example, hepatic encephalopathy is an important condition that is guarded against by monitoring ammonia homeostasis, which is considered as a clinically-relevant indicator of liver metabolic function after liver surgery [34]. These processes are not explicitly accounted for in

Processes **2018**, *6*, 115

the present network model. One can consider that metabolic load parameter serves as a lumped, phenomenological factor that accounts for the stress on liver, with metabolic load per unit of liver mass (M/N in Equations (4), (9) and (11)) serving as a stimulus for regeneration as well as cell death [10,30,35]. Opportunities exist for integrating the liver regeneration model with multi-organ physiology models to expand the utility of the model into a whole body context. The model- based research discussed in this work on liver regeneration has been based on a lumped model considering the hepatocyte functional states. This approach can be extended to incorporate functional states of other liver cell types including hepatic stellate cells, sinusoidal endothelial cells and Kupffer cells (resident macrophages of the liver) [36], as well as potential emerging rescue approaches such as stem cell transplant [37]. Taken together, the above detailed extensions to the network model will permit relating the key controlling parameters to inter-cellular, tissue-scale and whole body physiological scale parameters and will likely provide additional insights into novel venues for clinical management and intervention.

Our use of a virtual patient cohort approach helped delineate the key parameter intervals and combinatorial dependencies that correspond to a wide range of predicted outcomes post resection. Fruitful next steps could be to correlate these model-predicted parameter subspaces to patient demographical data as well as preoperative clinical information and disease etiology. Development of such a correspondence between patient information and model parameters is likely to aid in generalized application of the dynamic modeling to a wide range of liver surgery scenarios. Such a model-based approach informed by patient data to constrain the parameters can assist in clinical decision making by predicting the categorical outcome prior to the clinical intervention. Specifically, our model-based approach can aid in estimating the safe level of resection a patient can undergo along with predicting the need for necessary manipulations of cell death sensitivities and metabolic load to maximize the chances of recovery.

Supplementary Materials: The following are available online at http://www.mdpi.com/2227-9717/6/8/115/s1, File S1: Matlab code for classification and analysis of virtual patient cohort response modes. (.m 16 KB), File S2: Matlab code for parameter optimization for liver failure scenario using elastic net technique. (.m 11 KB), File S3: SBML model containing an independent implementation of the network model and model equations in the Methods (.xml 102 KB), Figure S1: SBGN diagram for the liver regeneration model (.pdf 522 KB), Figure S2: Parameter space of the three controlling parameters showing the distribution of virtual patients across different response modes for a 10% resection scenario (.fig 455 KB), Figure S3: Parameter space of the three controlling parameters showing the distribution of virtual patients across different response modes for a 33.3% resection scenario (.fig 454 KB), Video S1: Parameter space of the three controlling parameters showing the distribution of virtual patients across different response modes for a 10% resection (.mp4 9.98 MB), Video S2: Parameter space of the three controlling parameters showing the distribution of virtual patients across different response modes for a 33.3% resection (.mp4 10 MB)

Author Contributions: R.V. and B.K.V. conceived and designed the in silico experiments; B.K.V. performed the in silico experiments; B.K.V., P.S. and R.V. analyzed the simulation data; B.K.V., P.S. and R.V. wrote the paper.

Funding: This research was funded by National Institute on Alcohol Abuse and Alcoholism grant number R01 AA018873, National Institute of Biomedical Imaging and Bioengineering grant number U01 EB023224 and National Science Foundation grant number EAGER 1747917.

Acknowledgments: We thank Timothy Josephson for independently generating the SBML file and the SBGN diagram for the liver regeneration network model under study, which have been included in the supplementary material.

Conflicts of Interest: The authors declare no conflict of interest. The funding sponsors had no role in the design of the study; in the collection, analyses, or interpretation of data; in the writing of the manuscript and in the decision to publish the results.

Appendix A

Table A1. Nomenclature of the model parameters and values of upper and lower bounds used in the optimization.

Parameter	Range	Description
M	1.00411–100.41126	Mechanical stress or increased nutrient and detoxification demand
k_{IL6}	0.15–15	Rate of production of IL6 from non-parenchymal cells
κ_{IL6}	0.09–9	Rate of IL6 degradation
V_{JAK}	2000–200,000	Maximum rate of activation of JAK
K_M^{JAK}	1000–100,000	Concentration of JAK, when the rate of activation of JAK is half of the maximum rate
κ_{JAK}	0.04–4	Rate of degradation of JAK
[proSTAT3]	0.01–20	Relative concentration of monomeric STAT3
V_{ST3}	75–7500	Maximum rate of STAT3 phosphorylation
K_M^{ST3}	0.04–40	Michaelis-Menten concentration of proSTAT3
κ_{ST3}	0.05–1	Rate of dephosphorylation of proSTAT3
V_{SOCS3}	14,000–34,000	Maximum rate of SOCS3 activation
K_M^{SOCS3}	0.0004–0.001	Concentration of SOCS3, when it's rate of activation is half the maximum rate
κ_{SOCS3}	0.1–0.7	Rate of degradation of SOCS3
K_I^{SOCS3}	0.005–0.025	SOCS3 inhibition constant on STAT3
V_{IE}	150–350	Maximum rate of activation of IE gene
K_M^{IE}	15–21	Concentration of IE gene, when it's rate of activation is half the maximum rate
κ_{IE}	3–7	Rate of degradation of IE gene
k_{deg}	5–9	Rate of degradation of ECM by MMPs
κ_{ECM}	30–36	Rate of degradation of ECM
k_{GF}	0.05–0.2	Rate of production of growth factor from non-parenchymal cells
κ_{GF}	0.1–0.35	Rate of degradation of growth factor
k_{up}	0.04–0.08	Rate of binding of growth factor to ECM
k_{QP}	0.005–0.009	Rate of hepatocytes transition from quiescence to primed state
k_{PR}	0.003–0.006	Rate of hepatocytes transition from primed to replicating state
k_{RQ}	0.04–0.065	Rate of hepatocytes transition from replicating to quiescence state
k_{prol}	0.01–0.03	Rate of proliferation of hepatocytes
k_{req}	0.05–0.15	Rate of requiescence of primed hepatocytes
θ_{req}	6–10	Requiescence parameter in the sigmoidal function (σ_{req}) defining the threshold of requiescence
β_{req}	2–4	Requiescence parameter in the sigmoidal function (σ_{req}) defining the threshold of requiescence
k_{cd}	0.05–0.15	Cell death rate of damaged hepatocytes
θ_{cd}	0.00110–0.11050	Cell death sensitivity parameter of the sigmoidal function (σ_{cd})
β_{cd}	0.00260–0.26046	Cell death sensitivity parameter of the sigmoidal function (σ_{cd}) defining the cell death threshold
k_G	0.0003–0.0007	Rate of growth of relative cell mass

Table A2. Table of the 33 model parameters optimized using the initial value as given in the table for the 7 patients (studied in section 3.4) from Yamamoto et al. [11] that showed liver failure.

Parameter	Nominal Value	Patient Identification Number						
		ID114	ID123	ID131	ID29	ID45	ID145	ID135
M	10.041128	10.7480	10.0211	10.0173	10.0396	9.9112	10.0552	10.0425
k_{IL6}	1.5	3.9280	3.2484	1.6583	1.5000	1.0530	1.4620	1.4464
κ_{IL6}	0.9	3.9040	0.7807	0.9043	0.9000	1.0276	0.9117	0.9800
V_{JAK}	20,000	20,000	20,000	20,000	20,000	20,000	20,000	20,000
K_M^{JAK}	10,000	10,000	10,000	10,000	10,000	10,000	10,000	10,000
κ_{JAK}	0.4	1.4695	1.1701	0.3941	0.4000	0.5844	0.4330	0.3804
[proSTAT3]	2	6.1276	3.0179	1.9367	1.9736	1.9422	1.9866	1.9897
V_{ST3}	750	750.0337	750.0022	750.0003	750.0001	750.0006	750.0000	749.9999
K_M^{ST3}	0.4	1.0194	1.2485	0.4001	0.4000	0.4115	0.4096	0.3752
κ_{ST3}	0.1	0.0987	0.1614	0.1000	0.1000	0.1017	0.1152	0.1116
V_{SOCS3}	24000	24000	24000	24000	24000	24000	24000	24000
K_M^{SOCS3}	0.0007	0.0007	0.0007	0.0007	0.0007	0.0009	0.0007	0.0007
κ_{SOCS3}	0.4	0.4871	0.3970	0.4000	0.4000	0.1913	0.3855	0.3847
K_I^{SOCS3}	0.015	0.0139	0.0149	0.0150	0.0150	0.0150	0.0147	0.0150
V_{IE}	250	250.12	250.01	250.00	249.99	249.99	249.99	249.99
K_M^{IE}	18	19.1762	18.1366	18.0041	18.0055	18.0474	18.0014	17.9778
κ_{IE}	5	6.9573	5.3009	5.0022	4.9925	4.9951	5.0061	4.9493
k_{deg}	7	6.3119	7.2057	7.0017	6.9977	7.1109	6.9982	7.0171
κ_{ECM}	33	33.9672	33.0645	32.9999	33.0004	32.9871	33.0004	32.9939
k_{GF}	0.113	0.1677	0.1214	0.1130	0.1130	0.1128	0.1166	0.1103
κ_{GF}	0.23	0.3384	0.2453	0.2300	0.2300	0.2291	0.2298	0.2302
k_{up}	0.06	0.0577	0.0590	0.0600	0.0600	0.0598	0.0600	0.0600
k_{QP}	0.007	0.0079	0.0069	0.0070	0.0070	0.0056	0.0070	0.0070
k_{PR}	0.0044	0.0043	0.0044	0.0044	0.0044	0.0034	0.0044	0.0044
k_{RQ}	0.054	0.0517	0.0536	0.0540	0.0540	0.0509	0.0507	0.0547
k_{prol}	0.02	0.0202	0.0202	0.0200	0.0200	0.0277	0.0220	0.0144
k_{req}	0.1	0.1407	0.0999	0.1000	0.1000	0.1308	0.1013	0.0987
θ_{req}	8	9.9514	8.2151	7.9998	7.9994	7.9845	8.0009	8.0182
β_{req}	3	3.9615	3.1587	3.0021	3.0000	2.9657	2.9990	2.9563
k_{cd}	0.1	0.1390	0.1089	0.1000	0.1000	0.1001	0.1006	0.1003
θ_{cd}	0.011050	0.0117	0.0099	0.0111	0.0111	0.0108	0.0128	0.0114
β_{cd}	0.026046	0.0180	0.0270	0.0257	0.0249	0.0198	0.0260	0.0234
k_G	0.000657	0.000588	0.000601	0.000613	0.000614	0.000400	0.000497	0.000574

References

1. Fausto, N.; Campbell, J.S. The role of hepatocytes and oval cells in liver regeneration and repopulation. *Mech. Dev.* **2003**, *120*, 117–130. [CrossRef]
2. Taub, R. Liver regeneration: From myth to mechanism. *Nat. Rev. Mol. Cell Biol.* **2004**, *5*, 836–847. [CrossRef] [PubMed]
3. Michalopoulos, G.K. Hepatostat: Liver regeneration and normal liver tissue maintenance. *Hepatology* **2017**, *65*, 1384–1392. [CrossRef] [PubMed]
4. Lorenzo, C.S.F.; Limm, W.M.L.; Lurie, F.; Wong, L.L. Factors affecting outcome in liver resection. *HPB* **2005**, *7*, 226–230. [CrossRef] [PubMed]
5. Dimick, J.B. Hepatic Resection in the United States. *Arch. Surg.* **2003**, *138*, 185. [CrossRef] [PubMed]
6. Guglielmi, A.; Ruzzenente, A.; Conci, S.; Valdegamberi, A.; Iacono, C. How much remnant is enough in liver resection? *Dig. Surg.* **2012**, *29*, 6–17. [CrossRef] [PubMed]
7. Garcea, G.; Maddern, G.J. Liver failure after major hepatic resection. *J. Hepatobiliary Pancreat. Surg.* **2009**, *16*, 145–155. [CrossRef] [PubMed]
8. Mann, D.V.; Lam, W.W.M.; Hjelm, N.M.; So, N.M.C.; Yeung, D.K.W.; Metreweli, C.; Lau, W.Y. Human liver regeneration: Hepatic energy economy is less efficient when the organ is diseased. *Hepatology* **2001**, *34*, 557–565. [CrossRef] [PubMed]

9.   Van Den Broek, M.A.J.; Olde Damink, S.W.M.; Dejong, C.H.C.; Lang, H.; Malagó, M.; Jalan, R.; Saner, F.H. Liver failure after partial hepatic resection: Definition, pathophysiology, risk factors and treatment. *Liver Int.* **2008**, *28*, 767–780. [CrossRef] [PubMed]

10. Cook, D.; Ogunnaike, B.A.; Vadigepalli, R. Systems analysis of non-parenchymal cell modulation of liver repair across multiple regeneration modes. *BMC Syst. Biol.* **2015**, *9*, 1–24. [CrossRef] [PubMed]

11. Yamamoto, K.N.; Ishii, M.; Inoue, Y.; Hirokawa, F.; MacArthur, B.D.; Nakamura, A.; Haeno, H.; Uchiyama, K. Prediction of postoperative liver regeneration from clinical information using a data-led mathematical model. *Sci. Rep.* **2016**, *6*, 34214. [CrossRef] [PubMed]

12. An, G.; Mi, Q.; Dutta-moscato, J.; Vodovotz, Y. Agent-based models in translational systems biology. *Syst. Biol. Med.* **2009**, *1*, 159–171. [CrossRef] [PubMed]

13. An, G.; Mi, Q.; Dutta-moscato, J.; Vodovotz, Y. Mathematical models of the acute inflammatory response. *Curr. Opin. Crit. Care* **2004**, *10*, 383–390.

14. Sobol, I.M. Global sensitivity indices for nonlinear mathematical models and their Monte Carlo estimates. *Math. Comput. Simul.* **2001**, *55*, 271–280. [CrossRef]

15. Miller, G.M.; Ogunnaike, B.A.; Schwaber, J.S.; Vadigepalli, R. Robust dynamic balance of AP-1 transcription factors in a neuronal gene regulatory network. *BMC Syst. Biol.* **2010**, *4*, 1–17. [CrossRef] [PubMed]

16. Zou, H.; Hastie, T. Regularization and variable selection via the elastic net. *J. R. Stat. Soc. Ser. B Stat. Methodol.* **2005**, *67*, 301–320. [CrossRef]

17. Howsmon, D.P.; Hahn, J. Regularization techniques to overcome over-parameterization of complex biochemical reaction networks. *IEEE Life Sci. Lett.* **2017**, *3*, 31–34. [CrossRef]

18. Marongiu, F.; Marongiu, M.; Contini, A.; Serra, M.; Cadoni, E.; Murgia, R.; Laconi, E. Hyperplasia vs hypertrophy in tissue regeneration after extensive liver resection. *World J. Gastroenterol.* **2017**, *23*, 1764–1770. [CrossRef] [PubMed]

19. Le Novère, N.; Hucka, M.; Mi, H.; Moodie, S.; Schreiber, F.; Sorokin, A.; Demir, E.; Wegner, K.; Aladjem, M.I.; Wimalaratne, S.M.; et al. The Systems Biology Graphical Notation. *Nat. Biotechnol.* **2009**, *27*, 735–742. [CrossRef] [PubMed]

20. Strogatz, S.H. *Nonlinear Dynamics and Chaos with Applications to Physics, Biology, Chemistry and Engineering*, 1st ed.; Westview Press: Boulder, CO, USA, 2000.

21. Wiggins, S. *Introduction to Applied Nonlinear Dynamical Systems and Chaos*, 2nd ed.; Springer: Berlin, Germany, 2006; ISBN 0387217495.

22. Takeda, K.; Togo, S.; Kunihiro, O.; Fujii, Y.; Kurosawa, H.; Tanaka, K.; Endo, I.; Takimoto, A.; Sekido, H.; Hara, M.; et al. Clinicohistological features of liver failure after excessive hepatectomy. *Hepatogastroenterology* **2002**, *49*, 354–358. [PubMed]

23. Nagasue, N.; Yukaya, H.; Ogawa, Y.; Kohno, H.; Nakamura, T. Human liver regeneration after major hepatic resection. A study of normal liver and livers with chronic hepatitis and cirrhosis. *Ann. Surg.* **1987**, *206*, 30–39. [CrossRef] [PubMed]

24. Tanoue, S.; Uto, H.; Kumamoto, R.; Arima, S.; Hashimoto, S.; Nasu, Y.; Takami, Y.; Moriuchi, A.; Sakiyama, T.; Oketani, M.; et al. Liver regeneration after partial hepatectomy in rat is more impaired in a steatotic liver induced by dietary fructose compared to dietary fat. *Biochem. Biophys. Res. Commun.* **2011**, *407*, 163–168. [CrossRef] [PubMed]

25. Meier, M.; Andersen, K.J.; Knudsen, A.R.; Nyengaard, J.R.; Hamilton-Dutoit, S.; Mortensen, F.V. Liver regeneration is dependent on the extent of hepatectomy. *J. Surg. Res.* **2016**, *205*, 76–84. [CrossRef] [PubMed]

26. Van Poll, D.; Parekkadan, B.; Cho, C.H.; Berthiaume, F.; Nahmias, Y.; Tilles, A.W.; Yarmush, M.L. Mesenchymal stem cell-derived molecules directly modulate hepatocellular death and regeneration in vitro and in vivo. *Hepatology* **2008**, *47*, 1634–1643. [CrossRef] [PubMed]

27. Kroemer, G.; Reed, J.C. Mitochondrial control of cell death. *Nat. Med.* **2000**, *6*, 513–519. [CrossRef] [PubMed]

28. Lemasters, J.J.; Nieminen, A.L.; Qian, T.; Trost, L.C.; Elmore, S.P.; Nishimura, Y.; Crowe, R.A.; Cascio, W.E.; Bradham, C.A.; Brenner, D.A.; et al. The mitochondrial permeability transition in cell death: a common mechanism in necrosis, apoptosis and autophagy. *Biochim. Biophys. Acta* **1998**, *1366*, 177–196. [CrossRef]

29. Lafaro, K.; Buettner, S.; Maqsood, H.; Wagner, D.; Bagante, F.; Spolverato, G.; Xu, L.; Kamel, I.; Pawlik, T.M. Defining Post Hepatectomy Liver Insufficiency: Where do We stand? *J. Gastrointest. Surg.* **2015**, *19*, 2079–2092. [CrossRef] [PubMed]

30. Furchtgott, L.A.; Chow, C.C.; Periwal, V. A model of liver regeneration. *Biophys. J.* **2009**, *96*, 3926–3935. [CrossRef] [PubMed]
31. Huang, J.; Rudnick, D.A. Elucidating the metabolic regulation of liver regeneration. *Am. J. Pathol.* **2014**, *184*, 309–321. [CrossRef] [PubMed]
32. Schenk, A.; Ghallab, A.; Hofmann, U.; Hassan, R.; Schwarz, M.; Schuppert, A.; Schwen, L.O.; Braeuning, A.; Teutonico, D.; Hengstler, J.G.; et al. Physiologically-based modelling in mice suggests an aggravated loss of clearance capacity after toxic liver damage. *Sci. Rep.* **2017**, *7*, 1–13. [CrossRef] [PubMed]
33. Gille, C.; Bölling, C.; Hoppe, A.; Bulik, S.; Hoffmann, S.; Hübner, K.; Karlstädt, A.; Ganeshan, R.; König, M.; Rother, K.; et al. HepatoNet1: A comprehensive metabolic reconstruction of the human hepatocyte for the analysis of liver physiology. *Mol. Syst. Biol.* **2010**, *6*. [CrossRef] [PubMed]
34. Wright, G.; Noiret, L.; Olde Damink, S.W.M.; Jalan, R. Interorgan ammonia metabolism in liver failure: The basis of current and future therapies. *Liver Int.* **2011**, *31*, 163–175. [CrossRef] [PubMed]
35. Cook, D.; Vadigepalli, R. Computational Modeling as an Approach to Study the Cellular and Molecular Regulatory Networks Driving Liver Regeneration. In *Liver Regeneration Basic Mechanisms, Relevant Models and Clinical Applications*; Academic Press: Boston, MA, USA, 2015; pp. 185–198, ISBN 9780124201286.
36. Michalopoulos, G.K. Liver regeneration after partial hepatectomy: Critical analysis of mechanistic dilemmas. *Am. J. Pathol.* **2010**, *176*, 2–13. [CrossRef] [PubMed]
37. Stutchfiel, B.M.; Forbes, S.J.; Wigmore, S.J. Prospects for Stem Cell Transplantation in the Treatment of Hepatic Disease. *Liver Transplant.* **2010**, *16*, 827–836. [CrossRef] [PubMed]

Article

A Cybernetic Approach to Modeling Lipid Metabolism in Mammalian Cells

Lina Aboulmouna [1], Shakti Gupta [2], Mano R. Maurya [2], Frank T. DeVilbiss [1], Shankar Subramaniam [2,3,*] and Doraiswami Ramkrishna [1,*]

[1] The Davidson School of Chemical Engineering, Purdue University, West Lafayette, IN 47907, USA; laboulmo@purdue.edu (L.A.); fdevilbis@gmail.com (F.T.D.)

[2] Department of Bioengineering and San Diego Supercomputer Center, University of California San Diego, La Jolla, CA 92093, USA; shgupta@ucsd.edu (S.G.); mano@sdsc.edu (M.R.M.)

[3] Departments of Computer Science and Engineering, Cellular and Molecular Medicine, and the Graduate Program in Bioinformatics, University of California San Diego, La Jolla, CA 92093, USA

* Correspondence: shankar@ucsd.edu (S.S.); ramkrish@ecn.purdue.edu (D.R.);
Tel.: +1-858-822-3228 (S.S.); +1-765-494-4066 (D.R.)

Received: 16 July 2018; Accepted: 7 August 2018; Published: 12 August 2018

Abstract: The goal-oriented control policies of cybernetic models have been used to predict metabolic phenomena such as the behavior of gene knockout strains, complex substrate uptake patterns, and dynamic metabolic flux distributions. Cybernetic theory builds on the principle that metabolic regulation is driven towards attaining goals that correspond to an organism's survival or displaying a specific phenotype in response to a stimulus. Here, we have modeled the prostaglandin (PG) metabolism in mouse bone marrow derived macrophage (BMDM) cells stimulated by Kdo2-Lipid A (KLA) and adenosine triphosphate (ATP), using cybernetic control variables. Prostaglandins are a well characterized set of inflammatory lipids derived from arachidonic acid. The transcriptomic and lipidomic data for prostaglandin biosynthesis and conversion were obtained from the LIPID MAPS database. The model parameters were estimated using a two-step hybrid optimization approach. A genetic algorithm was used to determine the population of near optimal parameter values, and a generalized constrained non-linear optimization employing a gradient search method was used to further refine the parameters. We validated our model by predicting an independent data set, the prostaglandin response of KLA primed ATP stimulated BMDM cells. We show that the cybernetic model captures the complex regulation of PG metabolism and provides a reliable description of PG formation.

Keywords: lipids; prostaglandin metabolism; omics data; cybernetic modeling; optimization; metabolic objective functions

1. Introduction

Engineering methodologies are critical for a quantitative understanding of the physiological mechanisms in normal and disease states. Systems biology relies upon the use of models to organize biological knowledge and make predictions of complex processes. A variety of multi-omic data and mathematical approaches are available for modeling with varying (simple to complex) degrees of resolution [1,2]. One approach, cybernetic modeling, has been used for over three decades to predict a variety of metabolic phenomena [3,4]. Cybernetic modeling of metabolism, at its core, embodies a framework of ordinary differential equations for kinetic modeling, which describes the time-dependent evolution of metabolite concentrations, enzyme concentrations, and cellular growth. In cells, these changes in concentrations, both inside and outside of the cell, are governed by the directed actions of complex biological processes.

The cybernetic modeling framework distinguishes itself from traditional kinetic modeling by indirectly accounting for the unknown regulatory processes in the cell. These regulations are a cooperative cascade of molecular mechanisms that enhance cellular function, such as growth or survival. In the absence of high resolution knowledge of all cellular signaling and metabolic events, cybernetic regulation offers a significant advantage and modulates the level of key enzymes through the introduction of cybernetic variables for induction (u_i) and activation (v_i). Cybernetic models of metabolism were first formulated to describe the growth behavior of cells in multi-substrate environments. These models build on the assumption that the synthesis and activity of the enzymatic machinery are regulated to maximize a return on investment (ROI), such as, biomass, carbon uptake, etc. [5–7]. For example, in the classic scenario of diauxic growth, *E. coli* regulates its transport enzymes and prioritizes the utilization of the substitutable substrates based on an optimal growth rate [5]. While cybernetic models have focused on bacterial systems in the past, we presently adapt this framework to model the dynamic behavior of prostaglandin (PG) formation as an inflammatory response of bone marrow derived macrophages (BMDM) in a mammalian system [7–10].

Inflammation is an active defense mechanism of multicellular organisms in response to various harmful stressors. The primary role of inflammation is to counter the effects of these stressors and to initiate cell and tissue repair. Multiple factors in the immune system respond to inflammation; for example, macrophages are a type of white blood cell of the immune system designed to target substances that lack surface proteins associated with healthy body cells [11]. Upon infection, macrophage cells are activated via induced metabolic changes associated with lipids [12,13]. Consequently, we focus our study on a sub-category of fatty acyls known to contribute to inflammation, the eicosanoids. Eicosanoids are derived from arachidonic acid (AA), a 20-carbon fatty acid and are further classified into prostaglandins, thromboxanes, leukotrienes, and other oxidized products [14]. PGs have been found to mediate pain, fever, and other symptoms associated with inflammation [15]. PGs are synthesized from AA via the enzyme Prostaglandin G/H synthase (EC 1.14.99.1; cyclooxygenase (COX)), which has been targeted for treating inflammation, musculoskeletal pain, and other conditions. COX inhibitors are common and found in daily-use drugs such as aspirin and ibuprofen [16].

Lipid biosynthesis often requires the active transport and the chemical transformation of several intermediates. Lipid biosynthesis is further regulated at the corresponding enzyme synthesis levels starting from enzyme transcription through RNA processing, translation, and posttranslational modifications. We previously developed an approach to model the flux of AA and its downstream metabolites and applied it to the data from the murine macrophage-like (RAW 264.7) cells [17]. We also extended this model to bone marrow derived macrophages (BMDM) primed with the lipopolysaccharide (LPS) analogue KDO_2-Lipid A (KLA) and activated with a purinergic P2X7 receptor agonist adenosine triphosphate (ATP) [18]. Despite their ability to effectively predict the metabolite levels, these models either do not incorporate biological regulatory mechanisms or only account for simple regulation, such as at the gene expression level [17–20]. Given that biological processes are regulated at many other stages, such as posttranslational protein modification and interaction with a protein or substrate molecule, we have used the cybernetic modeling framework to account for such regulations in the modeling of lipid metabolism in this work.

The work presented here serves to provide a predictive kinetic model incorporating cellular regulatory mechanisms for eicosanoid metabolism, and signaling using the cybernetic framework, with inflammation as the system objective. While there is no single entity that represents the totality of inflammation by itself, the cytokine tumor necrosis factor alpha (TNFα) is well-known for its role in the generation of systemic inflammation and is a product of the response of macrophages to ATP and LPS [21,22]. We hypothesize that PG metabolism is regulated to maximize inflammation, characterized by the amount of TNFα generated by the system. Using the lipid pathways derived from the Kyoto Encyclopedia of Genes and Genomes (KEGG) pathway database and the time-course data from LIPID MAPS, our cybernetic approach to model the macrophage system provides a quantitative model of eicosanoid metabolism initiated with changes in the levels of AA (input) and resulting in the inflammatory outcome represented by TNFα [23–27]. The present study is an exemplar that highlights the potential for cybernetic approaches.

2. Materials and Methods

To describe the time-dependent formation of PGs, a kinetic model is generated. This description approximates the conversion of AA into an intermediate product, prostaglandin H2 (PGH_2), and its subsequent conversion into downstream products, prostaglandin D2 (PGD_2), prostaglandin E2 (PGE_2), and prostaglandin F2α ($PGF_{2\alpha}$). In this simple network of PG formation, the main focus is on how the PGH_2 conversion into the three downstream PG products is regulated, which may represent a central decision point in the lipid metabolic system in the macrophage inflammatory response (Figure 1a). The behavior of this network is modeled in three separate conditions, a control, a treatment with ATP, and a combined treatment of ATP and KLA. Measurements were made at 0, 0.25, 0.5, 1, 2, 4, 8, and 20 h after ATP stimulation (Figure 1b). The data for all these conditions was taken from LIPID MAPS [28–31]. Details of the experimental procedure can be obtained from the LIPID MAPS website (protocol available at: www.lipidmaps.org/protocols/PP0000004702.pdf).

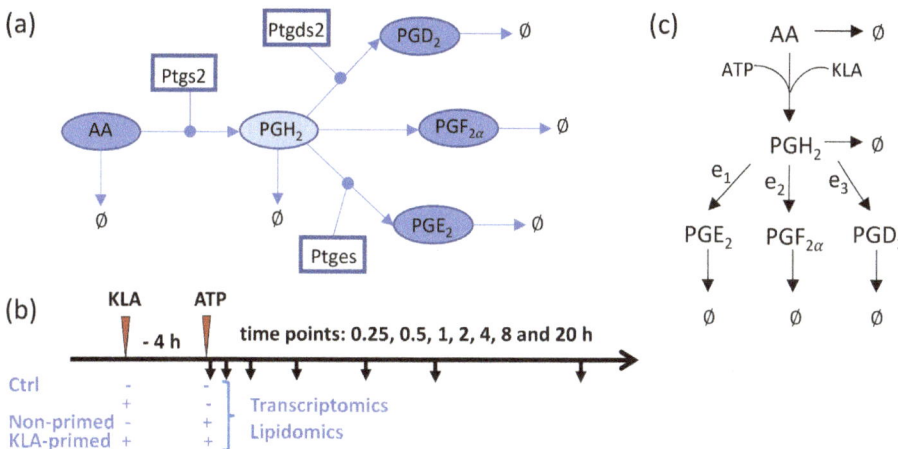

Figure 1. (**a**) The arachidonic acid metabolic pathway map for the breakdown of arachidonic acid into respective prostaglandin products via prostaglandin H2 (PGH_2) is shown: (rectangles) enzymes, (ellipses) lipid metabolites, (shaded) measured metabolites, (arrows) enzymatic and non-enzymatic reactions; (**b**) bone marrow derived macrophages (BMDM) were pretreated with or without KLA for 4 h and then stimulated with or without ATP. The media and cells were collected for lipidomic, tumor necrosis factor alpha (TNFα) and transcriptomic analysis at 0, 0.25, 0.5, 1, 2, 4, 8, and 20 h after ATP stimulation; (**c**) depiction of the simplified system network used for kinetic modeling illustrates PGH_2 as a control point and e_1, e_2, and e_3 as cybernetic enzymes regulated via cybernetic variables for the regulation of PGD_2, PGE_2, and $PGF_{2\alpha}$ fluxes.

2.1. Development of the Kinetic Model

The structure of the kinetics for this reaction network is arranged into two segments (Figure 1c). The first describes the conversion of AA into PGH_2 using simple linear kinetics. The kinetics of this reaction is modeled as three separate mechanisms, including a basal rate of synthesis, generation due to ATP stimulation, and KLA priming of cells.

$$r_{AA \rightarrow PGH_2} = k_{PGH_2}[AA](1 + k_{ATP}[ATP] + k_{KLA}[KLA])$$

To capture the effect of ATP, the treatments to the culture are modeled as a piecewise function, $f(t)$. This piecewise function ramps up to a maximum value of 1 at 0.5 h ($k_s = 2$ h^{-1}) and decreases exponentially ($k_d = 17.2$ h^{-1}) following the initial half hour of the experiment. For the KLA treatment case, the same $f(t)$ was used with a 4 h adjustment to account for the 4 h priming of KLA prior to the ATP stimulation.

$$f(t) = \begin{cases} k_s t & if \ t \leq 0.5 \\ e^{-k_d(t-0.5)} & if \ t > 0.5 \end{cases}$$

The primary difference in this function from previous work is in the second term, which includes exponential decay instead of a linear function to describe desensitization of cells to a given stimulus [32].

2.2. The Cybernetic Framework

The other segment of this model employs the cybernetic framework to capture the regulation between the different metabolic options [32]. In the cybernetic framework, there are two descriptions of the reaction kinetics. The first is the raw, enzyme-dependent rate of reaction, which we termed the kinetic rate of reaction, r^{kin}. This kinetic rate includes an enzyme quantity, e_i, which represents the amount of enzyme devoted to the conversion of PGH$_2$ to a PG product.

$$r^{kin}_{PGH_2 \rightarrow PG_i} = e_i k_{PG_i}[PGH_2]$$

The second description uses the cybernetic approach, which assumes a certain metabolic objective, namely, the optimal production of PG derivatives leading to maximum TNFα production. The framework views each pathway as a metabolic option to achieve such an objective and describes metabolic regulation in terms of their optimal combinations. Flux through the ith pathway is modeled as regulated by the control of the enzyme level and its activity, that is:

$$r^{reg}_{PGH_2 \rightarrow PG_i} = v_i r^{kin}_{PGH_2 \rightarrow PG_i}$$

where v_i is the cybernetic variable controlling enzyme activity. The resulting ordinary differential equations (ODEs) for each metabolite incorporated into the model (Figure 1c) can be written as a combination of regulated rates, r^{reg}, and degradation, where γ is the degradation rate constant, of the metabolites, as follows:

$$\frac{d[PGH_2]}{dt} = r_{AA \rightarrow PGH_2} - \sum_{i=1} r^{reg}_{PGH_2 \rightarrow PG_i} - \gamma_{PGH_2}[PGH_2]$$

$$\frac{d[PG_i]}{dt} = r^{reg}_{PGH_2 \rightarrow PG_i} - \gamma_{PG_i}[PG_i]$$

The enzyme level, e_i, is governed by the following dynamic equations:

$$r^{kin}_{e_i} = k_{e_i}[PGH_2]$$

$$r^{reg}_{e_i} = u_i r^{kin}_{e_i}$$

$$\frac{de_{PG_i}}{dt} = \alpha + r^{reg}_{e_i} - \beta e_{PG_i}$$

where u_i is the second cybernetic variable regulating the induction of the enzyme synthesis. The three terms on the right-hand side denote constitutive rate, α, and inducible rate of enzyme synthesis modulated by the cybernetic variable, u_i, and the decrease of the enzyme levels by degradation, where β is the degradation rate constant. The cybernetic control variables, u_i and v_i, are computed from the Matching and Proportional laws, respectively, as follows:

$$u_i = \frac{\rho_i}{\sum_k \rho_k}$$

$$v_i = \frac{\rho_i}{max_k(\rho_k)}$$

where the ROI, ρ_i, is defined by the flux through a particular pathway and is determined based on the designated system goal or objective [33].

2.3. Defining the Cybernetic Goal or Objective

PGs are well-characterized for their roles in the inflammatory response. Thus, in this paper, we focus on the regulation of PG synthesis as a function of TNFα, a marker of inflammation, for the selection of the model's objective function. To quantify the relationship between the PGs and TNFα, a simple, linear model of TNFα level is developed as a function of PG_i levels using their time-series data, as follows:

$$[TNF\alpha] = \sum_i c_i [PG_i]$$

We can also approximate the time derivative of TNFα concentration as a linear combination of time derivatives of PG_i concentrations over the time course. Additionally, due to the difference in magnitude of the different PG_i levels, a scaling was used to determine the contribution of each PG_i pathway leading to TNFα production. Thus, we define the weights, w_i, as follows:

$$w_i = \frac{c_i \overline{[PG_i]}}{\sum_j c_j \overline{[PG_j]}}$$

where w_i is the weight associated with the PG_i branch leading to TNFα production (w_i values of 0.2114, 0.2201, and 0.5685 for $i = 1$, 2, and 3 correspond to PGE_2, $PGF_{2\alpha}$, and PGD_2, respectively). Weights were obtained from c_i and $\overline{[PG_i]}$, the average concentration of PG_i across time, via regression of PG_i and TNFα data (Matlab® function 'fmincon', using the interior-point algorithm) using eight time points across ATP stimulated and control conditions; c_i does not change with time. Of the three pathways modeled, there is a varying degree of inflammation that results from the generation of each PG_i as described by the objective function. In this particular system, the ROI for each pathway is assumed to be the amount of TNFα that each unregulated pathway can yield at each instant in time, which is described by ρ_i.

$$\rho_i = w_i r^{kin}_{PGH_2 \rightarrow PG_i}$$

2.4. Estimation of the Kinetic Rate Parameters and Uncertainty Analysis

The model was parameterized using data from two of the three conditions, the control and the ATP treatment cases. Data was available for the AA, PGE_2, $PGF_{2\alpha}$, and PGD_2 metabolites as an eight-point time series over a 20 h time window. PGH_2 is an unstable intermediate metabolite we could not measure experimentally; in our model, we constrained the maximum concentration of PGH_2 to be ~10 pmol/µg DNA based on the total amount of PGs produced. The magnitudes of the different metabolites varied from 0.001 to 10 pmol/µg of DNA. To fit the model to the data, a least squared fit error was computed from the scaled profiles of the lipid, with respect to its maximum value, to ensure that the varying magnitude of each PG's level did not skew the parameters towards the sole fit of the PGs with higher magnitudes. The overall objective function for fitting the data was to minimize the fit-error between the experimental and the predicted metabolite concentrations [17], as follows:

$$\min_{K,X_o} \left(\sum_{i=1}^{nsp} \left(\sum_{j=1}^{ni} \left(y_{i,j,exp} - y_{i,j,pred}(K,X_0) \right)^2 \right) \right)$$

where K is the set of parameters or rate constants; X_0 is the set of initial conditions of the enzyme concentrations; ni is the number of time-points, 21, interpolated from 0 to 20 h (indexed as j) in order to provide equal weightage to later time points in the model fit; and nsp is the total number of species (indexed as i). The ODEs in the model were solved using ode15s for the stiff systems in MATLAB (version 9.4.0.813654, R2018a, Natick, MA, USA). The parameters (Table A1) were optimized using a two-step hybrid optimization procedure that started with a genetic algorithm seeded with random initial parameter values and evolved up to 100 generations in order to determine near optimal parameter values (Matlab® function 'ga'). The results from the application of the genetic algorithm-based optimization were then further refined using a generalized constrained non-linear optimization employing a gradient search method (Matlab® function 'fmincon', using the interior-point algorithm).

The quality of our model fit was assessed by comparing the variance for the fitted data to the variance in the experimental (replicate) data (treatment and control data combined) using the F-test, as follows [18]:

$$F = \frac{\frac{SSE_{fit}}{(2 \times nt)}}{\frac{SSE_{exp}}{(2 \times nt \times (nr-1))}}$$

$$F = \frac{\frac{\left(\sum_{j=1}^{nt} \left(Y_j^{trt} - \overline{X}_j^{trt} \right)^2 + \sum_{j=1}^{nt} \left(Y_j^{ctrl} - \overline{X}_j^{ctrl} \right)^2 \right)}{(2 \times nt)}}{\frac{\left(\sum_{j=1}^{nt} \sum_{i=1}^{nr} \left(X_{ij}^{trt} - \overline{X}_j^{trt} \right)^2 + \sum_{j=1}^{nt} \sum_{i=1}^{nr} \left(X_{ij}^{ctrl} - \overline{X}_j^{ctrl} \right)^2 \right)}{(2 \times nt \times (nr-1))}}$$

where Xj, $\overline{X}j$, and Yj denote the experimental data, mean experimental data, and simulated (fitted) data at time point j, respectively; nr is the number of replicates ($nr = 3$, indexed as i); nt is the number of experimental time points ($nt = 8$, indexed as j); and trt and ctrl are treatment and control groups, respectively. The degrees of freedom for determining the F distribution are $df_1 = (2 \times nt)$ and $df_2 = (2 \times nt \times (nr - 1))$. F statistic values smaller than $F_{0.95}(16, 32) = 1.97$ indicate statistically equal variance in simulated (fitted) and experimental data; whereas, F values smaller than $F_{0.05}(16, 32) = 0.4580$ indicate the fit-error is statistically smaller than the experimental error.

3. Results

3.1. Development of the Kinetic Model for the COX Pathway

Our cybernetic model describes the conversion of AA into the intermediate product, PGH_2, and its subsequent conversion into downstream prostaglandin products, PGE_2, $PGF_{2\alpha}$, and PGD_2. In this simple network of PG formation, the primary intent is on the regulation of PGH_2 conversion into the three downstream PG products. To address the latter, cybernetic regulation (implementation of u_i and v_i variables) was used at this branch point. The model for the COX pathway was described by 7 ODEs and 18 kinetic parameters (Table A1) in total; these 18 rate constants were estimated using a hybrid optimization approach (Materials and Methods). Using the optimized parameters, the eicosanoid profiles for the control and ATP stimulated cases were simulated (Figure 2). For most time points, the difference between the simulated and experimental data in both the treatment and control conditions fell within the standard error of the mean. The goodness of fit for the model was further examined by performing the F-test, indicating that the fit-error was less than the experimental measurement error (Table 1).

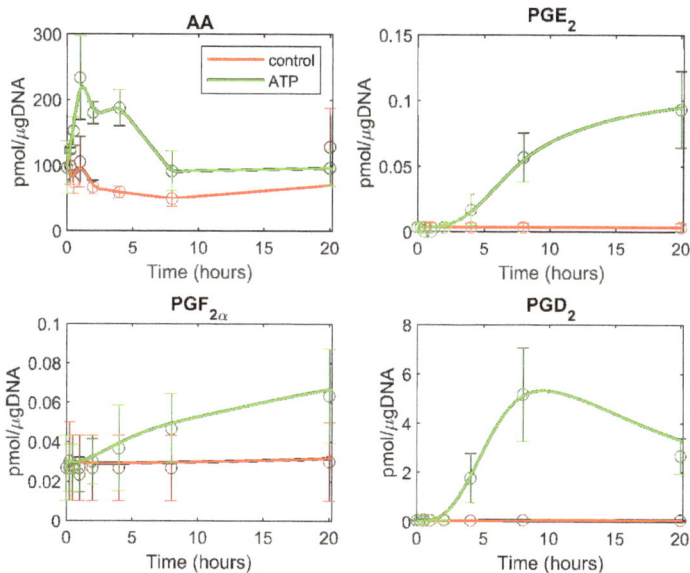

Figure 2. The computational simulation of the eicosanoid profile is generated using the cybernetic model in ATP stimulated BMDM. The mean experimental data (circles) with associated standard error of the mean (SEM) from three replicate experiments ($n = 3$) for the ATP stimulated (green) and control (red) cases are taken from the mass spectrometry measurements of lipids. The simulation results are shown for the treatment and control cases (solid green and red curves, respectively).

Table 1. Goodness of fit, *F*-test, for simulated/optimized (adenosine triphosphate (ATP) stimulated data) and predicted (Kdo2-Lipid A (KLA) primed and ATP stimulated) cases. *F* values smaller than $F_{0.05}(16, 32) = 0.4580$ indicate that the fit-error is statistically smaller than the experimental error; whereas the *F* values smaller than $F_{0.95}(16, 32) = 1.97$ indicate that the fit-error is statistically comparable to the experimental error. PGD_2—prostaglandin D2; PGE_2—prostaglandin E2; $PGF_{2\alpha}$—prostaglandin F2α.

Metabolite	Model Fit to ATP Data	Model Fit to KLA and ATP Data
PGE_2	0.0312	0.2421
$PGF_{2\alpha}$	0.0470	0.0342
PGD_2	0.2636	0.1192

The eicosanoid model robustness was evaluated by performing a parametric sensitivity analysis. Each parameter was varied individually by ± two-fold of the original optimized value and the maximum difference in the treatment and control case concentrations of each metabolite was plotted. The slopes of each metabolite sensitivity curve were calculated to evaluate the sensitivity for each parameter at the optimized value of that parameter. A heat map of the slopes was then generated (Figure 3). Small to moderate sensitivities in most of the parameters were observed. As expected, very little or no variation in the degradation parameters for PGD_2, PGE_2, and $PGF_{2\alpha}$ or in the KLA parameter is seen in response to metabolite changes. This is especially relevant to note given that the data set in which the parameter set was optimized for simulation was not treated with KLA and, consequently, would not have a dependence on this parameter. Based on these results, our model of eicosanoid metabolism is shown to be robust with respect to parametric perturbations.

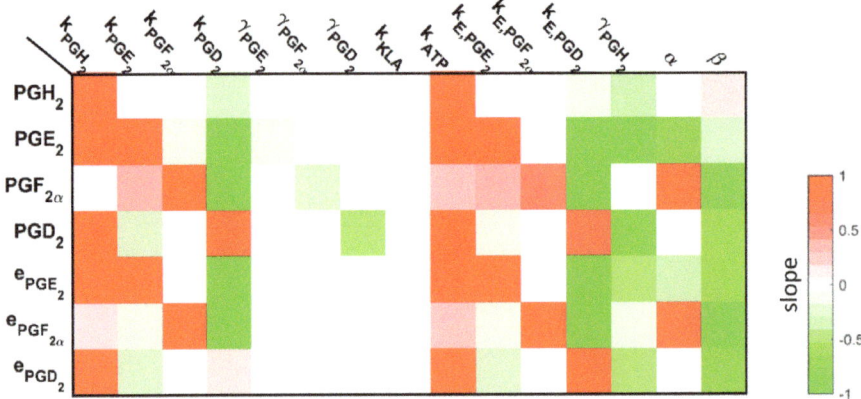

Figure 3. The slope of the sensitivity curves of the arachidonic acid (AA) metabolism are shown as a heat map. For example, the changes in the parameter associated with a conversion of AA into prostaglandin H2 (PGH_2) resulted in an increase in all of the metabolites; whereas, changes in the degradation of PGH_2 resulted in a decrease in all of the metabolites. This is expected, given that PGH_2 is in the upper part of the network, so the changes associated with these parameters will result in an impact on all of the corresponding downstream metabolites.

3.2. Prediction of the Eicosanoid Profile in KLA Primed ATP Stimulated Macrophages

To test the validity of the above obtained parameters, we used the parameter values to predict the new data set, the eicosanoid profile in KLA primed ATP stimulated BMDM cells. When the profiles were predicted with the optimized parameter values, the model prediction did not fit the experimental data well. We allowed up to 30% variability in the originally optimized parameter values for the re-estimation of parameters in the KLA primed ATP stimulated case. The range of 30% variability was chosen based on previous work by our group in determining the uncertainty of the calculated parameters in the ATP stimulated model [18]. The prediction with the relaxed bounds on the parameters yields a good fit to the experimental data (Figure 4 and see the results of *F*-test in Table 1). This prediction of an independent experimental dataset (KLA primed and ATP stimulated case), which was not used to fit the ATP stimulation data, further validated the model and parameter values. The mathematical model reflects the AA metabolic network dynamics in BMDM cells.

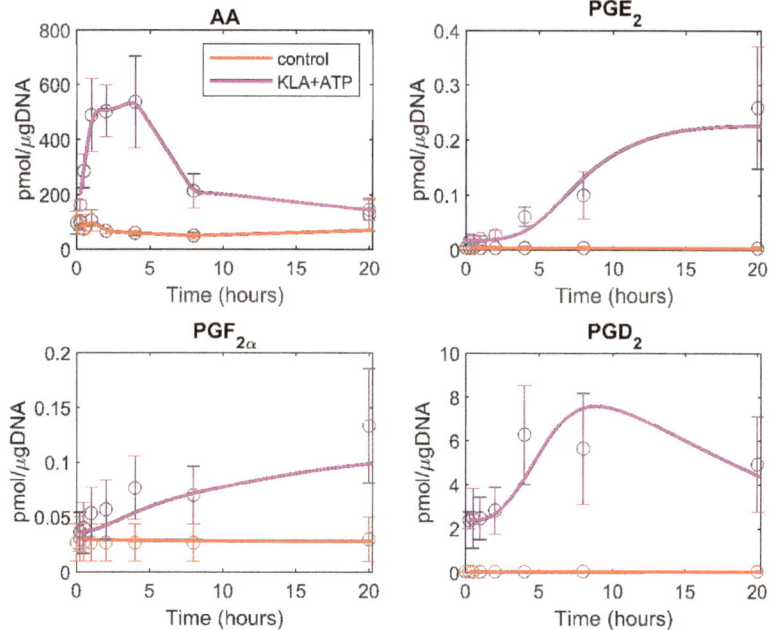

Figure 4. The computational prediction of the eicosanoid profile is generated using the cybernetic model in KLA primed and ATP stimulated BMDM. The mean experimental data (circles) with associated standard error of the mean (SEM) from three replicate experiments (*n* = 3) for KLA primed ATP-treated (magenta) and control (red) cases are taken from the mass spectrometry measurements of the lipids. The prediction results are shown for the treatment and control cases (solid magenta and red curves, respectively).

Table 2. Enzymes were identified from the Kyoto Encyclopedia of Genes and Genomes (KEGG) database and other selected resources for each pathway downstream of prostaglandin H2 (PGH$_2$) in prostaglandin synthesis. There is not a specific enzyme associated with the regulation of PGH$_2$ into PGF$_{2\alpha}$.

Entrez ID	Pathway	Gene Symbol	Name
64292	PGH$_2$ → PGE$_2$	Ptges	prostaglandin E synthase
54486	PGH$_2$ → PGD$_2$	Hpgds/Ptgds2	hematopoietic prostaglandin D2 synthase

3.3. Understanding the Role of Regulation in the Cybernetic Variables

In order to validate the cybernetic control mechanism that drives the modulation of the reaction rates in the model, the scaled gene expression data (representative of the enzymes synthesized) were compared to the scaled versions of the predicted enzyme levels. The qualitative trends among both the gene expression data and the predicted enzyme levels are expected to be similar [34]. Simply stated, if the gene expression level of the enzyme for one of the pathway branches is increasing over a certain time period, the cybernetic variable for the enzyme synthesis control and, therefore, the predicted enzyme levels should also be increasing.

For comparative purposes, we first identified the genes related to the respective branch in the eicosanoid metabolic pathways. These genes were selected using the KEGG database (Table 2) [24,27]. For two of the three branches in the pathway modeled, there are genes associated with enzymes for the catalysis of those pathways in the network. However, the $PGF_{2\alpha}$ branch is a non-enzymatically regulated process and does not have an associated gene for comparison with the corresponding cybernetic variable.

Gene expression data inform the relative levels of the enzymes present. To validate the cybernetic approach, we qualitatively compare the gene expression measurements with the corresponding cybernetic enzyme levels. Given that the gene expression data is represented as a fold change with respect to the control case, we have also taken fold changes of the enzyme levels from the cybernetic model in the treatment cases with respect to their corresponding value in the control cases. Both the gene expression and cybernetic enzyme level data were normalized by their corresponding maximum values ($e_i/e_{i,\max}$) to visualize a clear comparison of the dynamic trends. These comparisons are made for both the ATP and the combined KLA primed ATP stimulated treatment conditions. Overall, the scaled predicted enzyme profiles in solid green (ATP stimulated case) and magenta (KLA primed ATP stimulated case) match the general behavior of their corresponding genes, identified in Table 2, which are denoted by dashed black lines (Figure 5). There are some discrepancies such as in the Ptges profile for KLA primed and ATP stimulated BMDM, as well as, the Hpgds profile for ATP stimulated BMDM. This could be attributed to the different types of data used in the comparison plot of the cybernetic variables, which represent the modeled proteomic levels of the enzymes with the available transcriptomic data.

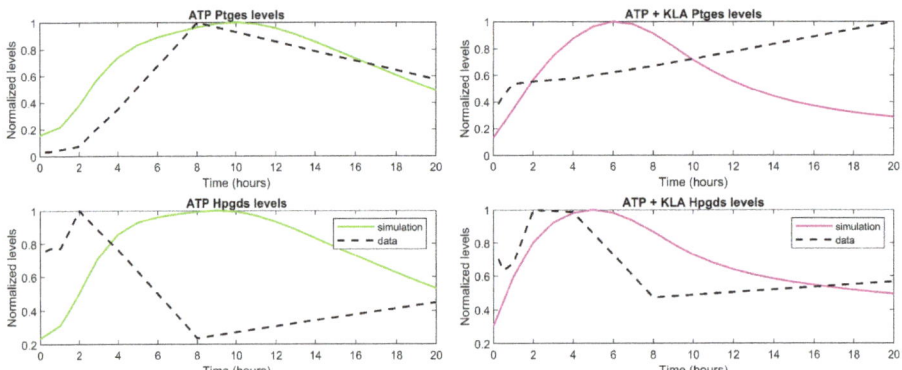

Figure 5. The behavior of the scaled cybernetic model enzyme level simulations (green in ATP stimulated case and magenta in KLA primed, followed by ATP stimulated case) generally match the trends of the scaled gene expression values (black dashed lines) for Ptges and Hpgds/Ptgds2 pathways in (left) ATP and (right) combined KLA primed ATP stimulated treatments.

4. Discussion

The cybernetic approach differs substantially from other modeling methods. For traditional kinetic modeling, detailed metabolic regulatory mechanisms are necessary [35,36]; however, the cybernetic approach models these regulatory actions as a collective process with an optimal system objective. Cybernetic enzymes and variables are used to describe a succinct mode of regulation related to the organism's goal. Cybernetic models have been useful in not only describing complex substrate uptake patterns [7] but have also yielded successful predictions of intracellular fluxes [37], gene-knockout behaviors [38], and multiplicity of steady states in chemostats [39]. While we show that cybernetic modeling predicts complex cellular phenomena, we also validate the assumption that the cybernetic control mechanisms mimic cellular regulation [34]. The cybernetic variables for enzyme synthesis and activity, u_i and v_i, are compared with experimental data representative of the regulatory mechanisms in cells. We show that the scaled predicted enzyme profiles generally match the behavior of their corresponding genes identified from literature (Section 3.3). These predicted enzyme levels are calculated from metabolomics data based on the assumption that the enzymes for different pathways are regulated in such a way as to optimize the objective function—in this case, the formation of TNFα. The predicted enzyme levels, as informed by the e_i control variables, are independent of the gene expression data. Qualitatively comparing the behavior of the dynamic gene expression profiles with the predicted enzyme levels further validates our cybernetic model (Figure 5) and serves to validate the idea that modeling macrophage cells from a goal-oriented perspective is useful.

The objective function used in the model, maximizing the rate of TNFα formation, is a central postulate of the cybernetic model presented here. While TNFα is well characterized as a signaling molecule generated in the macrophage response of LPS binding to the TLR4 receptor, other inflammatory cytokines, such as the interleukins (ILs) like IL-1, IL-6, and IL-12, can also be used to describe the goal of the system [40]. Control goals related to other functions of PGs, besides inflammation, are also of interest; however, given that the response of the macrophages to ATP and KLA is an inflammatory one, the objective function centered around TNFα is most relevant within the context of the system and conditions studied in this paper. To generate the cybernetic model, it is necessary to first make an assumption of the control goal. This assumption was further tested on the basis of whether or not the model is capable of making predictions of data beyond what the model is trained on. The fact that the model, with the TNFα objective function, is able to make predictions of the KLA primed and ATP stimulated treatment case, as well as of the gene expression trends, validates the use of TNFα as a control assumption central to the model.

The statistical analysis of multi-omics data coupled with the development of mathematical models aid in the unraveling of complex biological systems. We used a two-step, hybrid optimization approach in our study to estimate the rate constants of the AA metabolic network in BMDM using time-course lipidomic data. All of the kinetic parameters in our model were estimated through a nonlinear optimization approach based on the experimental data. Therefore, this study, using a multi-omics, data-driven, systems biology approach, is useful for understanding in vitro eicosanoid metabolism. Our model showed a good fit to the experimental data as seen from the goodness of fit performed by the *F*-test (Table 1), which suggests that the model captured the key characteristics of the lipid metabolic network in the BMDM cells. After fitting the parameters to two conditions (i.e., the control and ATP treatment conditions), the model provided the fits, which are shown in Figure 2. Effectively, this model is reliable as it is also evident that the model correctly explains the evolution of the metabolite concentrations for the different conditions involved in the fit. In the control condition, we see a relatively low rate of prostaglandin formation as we would expect. The ATP treatment case shows a good agreement with all of the prostaglandin products generated, and the kinetics of the model are cross-validated using an additional treatment condition, KLA primed ATP stimulated BMDM cells (Figures 2 and 3).

We then compared the effective rate constants associated with the enzymes of Hpgds/Ptgds2 (EC 5.3.99.2) and Ptges (EC 5.3.99.3) and their corresponding values reported in the literature to confirm the reliability of our optimized parameter values [41–47]. In order to compare the enzyme activities obtained from concentrations in the LIPID MAPS experimental data and literature values of enzyme-enriched protein, we used appropriate conversion factors, as discussed in Kihara et al., 2014 [18]. In summary, the calculated values from our simulation of the eicosanoid metabolism for Ptgds2 activity and Ptges were within the expected range of the values reported in the literature. The reported flux of PGD$_2$ in macrophages is not detectable and is determined to be less than 1 nmol/min/mg of total protein, which is consistent with our model value, 1×10^{-5} nmol/min/mg of total protein [46]. For the flux of PGE$_2$, the reported literature value of 0.4 pmol/min/mg of the total protein is of the same order as our computed value, 0.1 pmol/min/mg of protein [43]. Our computed values are consistent with those reported in the literature and further validate our computational model.

In conclusion, we have developed a quantitative model of the eicosanoid metabolic pathway by using cybernetic regulation in primary macrophages under control (basal) and ATP stimulated conditions. Additionally, we have been successful in predicting the metabolite levels of the eicosanoid profiles and capturing the relative changes in gene expression of relevant enzymes under a set of conditions different from that used for calculating the model rate constants. In particular, we successfully predicted the eicosanoid profiles for the KLA primed ATP stimulated case. We have demonstrated the use of the cybernetic approach to model the regulation of mammalian lipid metabolism. The cybernetic model provides a robust description of metabolite formation and can be used to predict perturbations to metabolism. Our computational model assists in understanding the complexity of eicosanoid metabolism and in examining complex regulatory phenomena.

Author Contributions: Conceptualization, D.R., S.S., L.A., S.G., M.R.M., and F.T.D.; methodology, L.A., S.G., M.R.M., and F.T.D.; validation, L.A., S.G., and M.R.M.; formal analysis, L.A. and S.G.; investigation, L.A., S.G., and M.R.M.; resources, D.R. and S.S.; data curation, S.G., L.A., and M.R.M.; writing (original draft preparation), L.A.; writing (review and editing), L.A., S.G., M.R.M., F.T.D., D.R., and S.S.; visualization, L.A. and S.G.; supervision, D.R. and S.S.; project administration, D.R. and S.S.; funding acquisition, D.R. and S.S.

Funding: This work is supported by the Center for Science of Information (CSoI), a National Science Foundation Science and Technology Center, under grant agreement CCF-0939370 (S.S. and D.R.) and NIH Research Grants R01HL106579 (to S.S.), HL108735 (to S.S.), U01CA198941 (to S.S.), R01LM012595 (to S.S. and M.R.M.), and U19AI090023 (to S.S.).

Acknowledgments: We acknowledge Raja Hemanth—an undergraduate summer research student from the Department of Chemical Engineering, Indian Institute of Technology, Bombay, Mumbai, India—for his help with the initial weight calculations.

Conflicts of Interest: The authors declare no conflict of interest.

Appendix A

Table A1. Reaction parameters were estimated for the eicosanoid metabolism model. The simulated and predicted columns refer to the parameters optimized for ATP stimulated BMDM cells and KLA primed ATP stimulated BMDM cells, respectively. The predicted parameters were further optimized from the simulated parameters within 30% variability.

PARAMETER	SIMULATED	PREDICTED
K_{PGH2}	0.0022	0.0016
K_{PGE2}	0.0044	0.0031
$K_{PGF2\alpha}$	0.0326	0.0339
K_{PGD2}	0.0533	0.0585
γ_{PGE2}	0.0062	0.0044
$\gamma_{PGF2\alpha}$	0.0205	0.0197
γ_{PGD2}	0.1275	0.0893
K_{KLA}	17.3923	0.0001
K_{ATP}	11.9112	8.3379
$K_{E,PGE2}$	8.0801	10.4215
$K_{E,PGF2\alpha}$	0.2078	0.1478
$K_{E,PGD2}$	0.2243	0.157
γ_{PGH2}	0.2603	0.3384
α	0.2244	0.2918
β	0.7757	1.0082
$E_{0,PGE2}$	0.3974	0.5094
$E_{0,PGF2\alpha}$	0.0133	0.0105
$E_{0,PGD2}$	0.2601	0.3379

References

1.   Maurya, M.R.; Subramaniam, S. *Computational Challenges in Systems Biology*; Systems Biomedicine: Concepts and Perspectives; Elsevier: Oxford, UK, 2010; pp. 177–223.

2. Gupta, S.; Ashok, R.D.; Merril, J.G.; Maurya, M.R.; Subramaniam, S. Omics Approaches to Macrophage Biology. In *Macrophages: Biology and Role in the Pathology of Diseases*; Springer: Berlin, Germany, 2014; pp. 587–615.
3. Ramkrishna, D. A Cybernetic Perspective of Microbial Growth. In *Foundations of Biochemical Engineering: Kinetics and thermodynamics in biological systems*; American Chemical Society: Washington, DC, USA, 1982; pp. 161–178.
4. Ramkrishna, D.; Song, H.S. *Cybernetic Modeling for Bioreaction Engineering*; Cambridge University Press: Cambridge, UK, 2018.
5. Dhurjati, P.; Flickinger, M.C.; Tsao, G.T. A cybernetic view of microbial growth: Modeling of cells as optimal strategists. *Biotechnol. Bioeng.* **1985**, *27*, 1–9. [PubMed]
6. Kompala, D.S.; Ramkrishna, D.; Jansen, N.B.; Tsao, G.T. Investigation of bacterial growth on mixed substrates: Experimental evaluation of cybernetic models. *Biotechnol. Bioeng.* **1986**, *28*, 1044–1055. [CrossRef] [PubMed]
7. Ramakrishna, R.; Ramkrishna, D.; Konopka, A.E. Cybernetic modeling of growth in mixed, substitutable substrate environments: Preferential and simultaneous utilization. *Biotechnol. Bioeng.* **1996**, *52*, 141–151. [PubMed]
8. Song, H.S.; Ramkrishna, D. Cybernetic models based on lumped elementary modes accurately predict strain-specific metabolic function. *Biotechnol. Bioeng.* **2011**, *108*, 27–40.
9. Turner, B.G.; Ramkrishna, D.; Jansen, N.B. Cybernetic modeling of bacterial cultures at low growth rates: Mixed-substrate systems. *Biotechnol. Bioeng.* **1988**, *32*, 46–54. [CrossRef] [PubMed]
10. Young, J.D.; Henne, K.L.; Morgan, J.A.; Konopka, A.E. Integrating cybernetic modeling with pathway analysis provides a dynamic, systems-level description of metabolic control. *Biotechnol. Bioeng.* **2008**, *100*, 542–559. [CrossRef] [PubMed]
11. Koh, T.J.; DiPietro, L.A. Inflammation and wound healing: The role of the macrophage. *Expert Rev. Mol. Med.* **2011**, *13*, e23. [CrossRef] [PubMed]
12. Remmerie, A.; Scott, C.L. Macrophages and lipid metabolism. *Cell Immunol.* **2018**. [CrossRef] [PubMed]
13. Hubler, M.J.; Kennedy, A.J. Role of lipids in the metabolism and activation of immune cells. *J. Nutr. Biochem.* **2016**, *34*, 1–7. [CrossRef] [PubMed]
14. Funk, C.D. Prostaglandins and leukotrienes: Advances in eicosanoid biology. *Science* **2001**, *294*, 1871–1875. [CrossRef] [PubMed]
15. Shimizu, T. Lipid mediators in health and disease: Enzymes and receptors as therapeutic targets for the regulation of immunity and inflammation. *Annu. Rev. Pharmacol. Toxicol.* **2009**, *49*, 123–150. [CrossRef] [PubMed]
16. Smith, W.L.; DeWitt, D.L.; Garavito, R.M. Cyclooxygenases: Structural, cellular, and molecular biology. *Annu. Rev. Biochem.* **2000**, *69*, 145–182. [CrossRef] [PubMed]
17. Gupta, S.; Maurya, M.R.; Stephens, D.L.; Dennis, E.A.; Subramaniam, S. An integrated model of eicosanoid metabolism and signaling based on lipidomics flux analysis. *Biophys. J.* **2009**, *96*, 4542–4551. [CrossRef] [PubMed]
18. Kihara, Y.; Gupta, S.; Maurya, M.R.; Armando, A.; Shah, I.; Quehenberger, O.; Glass, C.K.; Dennis, E.A.; Subramaniam, S. Modeling of eicosanoid fluxes reveals functional coupling between cyclooxygenases and terminal synthases. *Biophys. J.* **2014**, *106*, 966–975. [CrossRef] [PubMed]
19. Gupta, S.; Kihara, Y.; Maurya, M.R.; Norris, P.C.; Dennis, E.A.; Subramaniam, S. Computational modeling of competitive metabolism between omega3- and omega6-polyunsaturated fatty acids in inflammatory macrophages. *J. Phys. Chem. B* **2016**, *120*, 8346–8353. [CrossRef] [PubMed]
20. Gupta, S.; Maurya, M.R.; Merrill, A.H., Jr; Glass, C.K.; Subramaniam, S. Integration of lipidomics and transcriptomics data towards a systems biology model of sphingolipid metabolism. *BMC Syst. Biol.* **2011**, *5*, 26. [CrossRef] [PubMed]
21. Suzuki, T.; Hide, I.; Ido, K.; Kohsaka, S.; Inoue, K.; Nakata, Y. Production and release of neuroprotective tumor necrosis factor by P2X7 receptor-activated microglia. *J. Neurosci.* **2004**, *24*, 1–7. [CrossRef] [PubMed]
22. Van der Bruggen, T.; Nijenhuis, S.; van Raaij, E.; Verhoef, J.; van Asbeck, B.S. Lipopolysaccharide-induced tumor necrosis factor alpha production by human monocytes involves the raf-1/MEK1-MEK2/ERK1-ERK2 pathway. *Infect. Immun.* **1999**, *67*, 3824–3829. [PubMed]
23. Kanehisa, M.; Sato, Y.; Kawashima, M.; Furumichi, M.; Tanabe, M. KEGG as a reference resource for gene and protein annotation. *Nucleic Acids Res.* **2016**, *44*, D457–462. [CrossRef] [PubMed]

24. Tanabe, M.; Kanehisa, M. Using the KEGG database resource. *Curr. Protoc. Bioinf.* **2012**, *38*, 1.12.1–1.12.43. Available online: https://doi.org/10.1002/0471250953.bi0112s38 (accessed on 7 August 2018).

25. Fahy, E.; Cotter, D.; Sud, M.; Subramaniam, S. Update of the LIPID MAPS comprehensive classification system for lipids. *J. Lipid Res.* **2009**, *50*, S9–S14. [CrossRef] [PubMed]

26. Sud, M.; Fahy, E.; Cotter, D.; Brown, A.; Dennis, E.A.; Glass, C.K.; Merrill, A.H.Jr.; Murphy, R.C.; Raetz, C.R.; Russell, D.W.; et al. LMSD: LIPID MAPS structure database. *Nucleic Acids Res.* **2007**, *35*, D527–532. [CrossRef] [PubMed]

27. Ogata, H.; Goto, S.; Sato, K.; Fujibuchi, W.; Boho, H.; Kanehisa, M. KEGG: Kyoto encyclopedia of genes and genomes. *Nucleic Acids Res.* **1999**, *27*, 29–34. [CrossRef] [PubMed]

28. The LIPID MAPS Lipidomics Gateway. Available online: http://www.lipidmaps.org/ (accessed on 7 August 2018).

29. Dennis, E.A.; Deems, R.A.; Harkewicz, R.; Quehenberger, O.; Brown, H.A.; Milne, S.B.; Myers, D.S.; Glass, C.K.; Hardiman, G.; Reichart, D.; et al. A mouse macrophage lipidome. *J. Biol. Chem.* **2010**, *285*, 39976–39985. [CrossRef] [PubMed]

30. Dinasarapu, A.R.; Gupta, S.; Ram Maurya, M.; Fahy, E.; Min, J.; Sud, M.; Gersten, M.J.; Glass, C.K.; Subramaniam, S. A combined omics study on activated macrophages–enhanced role of STATs in apoptosis, immunity and lipid metabolism. *Bioinformatics* **2013**, *29*, 2735–2743. [CrossRef] [PubMed]

31. Subramaniam, S.; Fahy, E.; Gupta, S.; Sud, M.; Byrnes, R.W.; Cotter, D.; Dinasarapu, A.R.; Maurya, M.R. Bioinformatics and systems biology of the lipidome. *Chem. Rev.* **2011**, *111*, 6452–6490. [CrossRef] [PubMed]

32. DeVilbiss, F.T. Is Metabolism Goal-Directed? Investigating the Validaty of Modeling Biological Systems with Cybernetic Control via Omic Data. Ph.D Thesis, Purdue University, West Lafayette, IN, USA, 2016.

33. Young, J.D.; Ramkrishna, D. On the matching and proportional laws of cybernetic models. *Biotechnol. Prog.* **2007**, *23*, 83–99. [CrossRef] [PubMed]

34. DeVilbiss, F.; Mandli, A.; Ramkrishna, D. Consistency of cybernetic variables with gene expression profiles: A more rigorous test. *Biotechnol. Prog.* **2018**. In press. [CrossRef] [PubMed]

35. Jahan, N.; Maeda, K.; Matsuoka, Y.; Sugimoto, Y.; Kurata, H. Development of an accurate kinetic model for the central carbon metabolism of Escherichia coli. *Microb. Cell Fact* **2016**, *15*, 1–19. [CrossRef] [PubMed]

36. Kotte, O.; Zaugg, J.B.; Heinemann, M. Bacterial adaptation through distributed sensing of metabolic fluxes. *Mol. Syst. Biol.* **2010**, *6*, 355–364. [CrossRef] [PubMed]

37. Song, H.S.; Ramkrishna, D.; Pinchu, G.E.; Belaiev, A.S.; Konopka, A.E.; Frederickson, J.K. Dynamic modeling of aerobic growth of Shewanella oneidensis. Predicting triauxic growth, flux distributions, and energy requirement for growth. *Metab. Eng.* **2013**, *15*, 25–33. [CrossRef] [PubMed]

38. Song, H.S.; Ramkrishna, D. Prediction of dynamic behavior of mutant strains from limited wild-type data. *Metab. Eng.* **2012**, *14*, 69–80. [CrossRef] [PubMed]

39. Kim, J.I.; Song, H.S.; Sunkara, S.R.; Lali, A.; Ramkrishna, D. Exacting predictions by cybernetic model confirmed experimentally: Steady state multiplicity in the chemostat. *Biotechnol. Prog.* **2012**, *28*, 1160–1166. [CrossRef] [PubMed]

40. Balkwill, F.R.; Burke, F. The cytokine network. *Immunol. Today* **1989**, *10*, 299–304. [CrossRef]

41. Boulet, L.; Ouellet, M.; Bateman, K.P.; Ethier, D.; Percival, M.D.; Riendeau, D.; Mancini, J.A.; Méthot, N. Deletion of microsomal prostaglandin E2 (PGE2) synthase-1 reduces inducible and basal PGE2 production and alters the gastric prostanoid profile. *J. Biol. Chem.* **2004**, *279*, 23229–23237. [CrossRef] [PubMed]

42. Schomburg, I.; Chang, A.; Ebeling, C.; Gremse, M.; Heldt, C.; Huhn, G.; Schomburg, D. BRENDA, the enzyme database: Updates and major new developments. *Nucleic Acids Res.* **2004**, *32*, D431–433. [CrossRef] [PubMed]

43. Lazarus, M.; Kubata, B.K.; Eguchi, N.; Fujitani, Y.; Urade, Y.; Hayaishi, O. Biochemical characterization of mouse microsomal prostaglandin E synthase-1 and its colocalization with cyclooxygenase-2 in peritoneal macrophages. *Arch. Biochem. Biophys.* **2002**, *397*, 336–341. [CrossRef] [PubMed]

44. Tanioka, T.; Nakatani, Y.; Semmyo, N.; Murakami, M.; Kudo, I. Molecular identification of cytosolic prostaglandin E2 synthase that is functionally coupled with cyclooxygenase-1 in immediate prostaglandin E2 biosynthesis. *J. Biol. Chem.* **2000**, *275*, 32775–32782. [CrossRef] [PubMed]

45. Chan, G.; Boyle, J.O.; Yang, E.K.; Zhang, F.; Sacks, P.G.; Shah, J.P.; Edelstein, D.; Soslow, R.A.; Koki, A.T.; Woerner, B.M. Cyclooxygenase-2 expression is up-regulated in squamous cell carcinoma of the head and neck. *Cancer Res.* **1999**, *59*, 991–994. [PubMed]

46. Urade, Y.; Ujihara, M.; Horiguchi, Y.; Igarashi, M.; Nagata, A.; Ikai, K.; Hayaishi, O. Mast cells contain spleen-type prostaglandin D synthetase. *J. Biol. Chem.* **1990**, *265*, 371–375. [PubMed]

47. Shimizu, T.; Yamamoto, S.; Hayaishi, O. Purification and properties of prostaglandin D synthetase from rat brain. *J. Biol. Chem.* **1979**, *254*, 5222–5228. [PubMed]

Article

Dynamic Sequence Specific Constraint-Based Modeling of Cell-Free Protein Synthesis

David Dai †, Nicholas Horvath † and Jeffrey Varner *

Robert Frederick Smith School of Chemical and Biomolecular Engineering, Cornell University, Ithaca, NY 14853, USA; wd226@cornell.edu (D.D.); ngh36@cornell.edu (N.H.)
* Correspondence: jdv27@cornell.edu; Tel.: +1-607-255-4258
† These authors contributed equally to this work.

Received: 8 June 2018; Accepted: 9 August 2018; Published: 17 August 2018

Abstract: Cell-free protein expression has emerged as an important approach in systems and synthetic biology, and a promising technology for personalized point of care medicine. Cell-free systems derived from crude whole cell extracts have shown remarkable utility as a protein synthesis technology. However, if cell-free platforms for on-demand biomanufacturing are to become a reality, the performance limits of these systems must be defined and optimized. Toward this goal, we modeled *E. coli* cell-free protein expression using a sequence specific dynamic constraint-based approach in which metabolite measurements were directly incorporated into the flux estimation problem. A cell-free metabolic network was constructed by removing growth associated reactions from the *i*AF1260 reconstruction of K-12 MG1655 *E. coli*. Sequence specific descriptions of transcription and translation processes were then added to this metabolic network to describe protein production. A linear programming problem was then solved over short time intervals to estimate metabolic fluxes through the augmented cell-free network, subject to material balances, time rate of change and metabolite measurement constraints. The approach captured the biphasic cell-free production of a model protein, chloramphenicol acetyltransferase. Flux variability analysis suggested that cell-free metabolism was potentially robust; for example, the rate of protein production could be met by flux through the glycolytic, pentose phosphate, or the Entner-Doudoroff pathways. Variation of the metabolite constraints revealed central carbon metabolites, specifically upper glycolysis, tricarboxylic acid (TCA) cycle, and pentose phosphate, to be the most effective at training a predictive model, while energy and amino acid measurements were less effective. Irrespective of the measurement set, the metabolic fluxes (for the most part) remained unidentifiable. These findings suggested dynamic constraint-based modeling could aid in the design of cell-free protein expression experiments for metabolite prediction, but the flux estimation problem remains challenging. Furthermore, while we modeled the cell-free production of only a single protein in this study, the sequence specific dynamic constraint-based modeling approach presented here could be extended to multi-protein synthetic circuits, RNA circuits or even small molecule production.

Keywords: dynamic constraint-based modeling; cell-free protein synthesis; systems biology

1. Introduction

Cell-free protein expression has become a widely used research tool in systems and synthetic biology, and a promising technology for personalized point of use biotechnology [1]. Cell-free systems offer many advantages for the study, manipulation and modeling of metabolism compared to in vivo processes. Central amongst these is direct access to metabolites and the biosynthetic machinery, without the interference of a cell wall or the complications associated with cell growth. This allows us to interrogate (and potentially manipulate) the chemical microenvironment while the biosynthetic machinery is operating, potentially at a fine time resolution. Cell-free protein synthesis (CFPS) systems are arguably the most prominent examples of cell-free systems used today [2]. However, CFPS in crude *E. coli* extracts has been used previously to explore fundamental biological questions. For example, Matthaei and Nirenberg used *E. coli* cell-free extracts to decipher the genetic code [3,4]. Later, Spirin and coworkers continuously exchanged reactants and products in a CFPS reaction, which improved protein production. However, while these extracts could run for up to tens of hours, they could only synthesize a single product and were likely energy limited [5]. More recently, energy and cofactor regeneration in CFPS has been significantly improved; for example, ATP can be regenerated using substrate level phosphorylation [6] or even oxidative phosphorylation [2]. Today, cell-free systems are used in a variety of applications ranging from therapeutic protein production [7] to synthetic biology [1,8]. There are also several CFPS technology platforms, such as the PANOx-SP (PEP, Amino Acids, NAD, Oxalic Acid, Spermidine, and Putrescine) and Cytomin platforms developed by Swartz and coworkers [2,9], and the transcription and translation (TX-TL) platform of Noireaux [10]. Taken together, CFPS is a promising technology for protein production. However, if CFPS is to become a mainstream technology for applications such as point of care biomanufacturing, we must first understand the performance limits of these systems, and eventually optimize their yield and productivity. Toward this unmet need, we have developed a dynamic constraint-based modeling that can be used to interrogate cell-free systems.

Genome scale stoichiometric reconstructions of microbial metabolism popularized by static, constraint-based modeling techniques such as flux balance analysis (FBA) have become standard tools [11]. Since the first genome scale stoichiometric model of *E. coli*, developed by Edwards and Palsson [12], well over 100 organisms, including industrially important prokaryotes such as *E. coli* [13] or *B. subtilis* [14], are now available [15]. Stoichiometric models rely on a pseudo-steady-state assumption to reduce unidentifiable genome-scale kinetic models to an underdetermined linear algebraic system, which can be solved efficiently even for large systems using linear programming. Traditionally, stoichiometric models have also neglected explicit descriptions of metabolic regulation and control mechanisms, instead opting to describe the choice of pathways by prescribing an objective function on metabolism. Interestingly, similar to early cybernetic models, the most common metabolic objective function has been the optimization of biomass formation [16], although other metabolic objectives have also been estimated [17]. Recent advances in constraint-based modeling have overcome the early shortcomings of the platform, including describing metabolic regulation and control [18] and incorporating genome sequence into the model [19,20]. Dynamic constraint-based methods have also been developed in which the metabolic flux is computed over short-time intervals subject to time-varying constraints [21]. These methods are common, have been used in varied applications, [22–25], and there are open source packages to support this class of calculation [26–28]. Thus, constraint-based approaches, and their dynamic extensions, have proven useful in the discovery of metabolic engineering strategies and represent the state of the art in metabolic modeling [29,30]. However, while constraint-based tools have been used extensively to analyze whole cells systems, they have not yet been widely applied to study cell-free reactions.

In this study, we constructed a dynamic constraint-based model of cell-free protein expression. This approach avoids the pseudo-steady-state assumption found in traditional constraint-based approaches, which allowed for the direct integration of dynamic metabolite measurements into the flux estimation problem, along with the accumulation or depletion of network metabolites. We adapted the sequence specific constraint-based model of Vilkhovoy and coworkers [31] into a dynamic constraint-based model of cell-free *E. coli* metabolism and protein production, and leveraged the kinetic model of Horvath and coworkers [32] to provide synthetic data to inform metabolite constraints. CFPS synthesis is often (but not always) conducted in small scale batch reactors. Thus, the concentration of the components of the reaction mixture, and the associated rates of the metabolic processes in the reaction are not always constant. The Vilkhovoy et al. study considered only the first hour of the CFPS reaction producing the model protein, chloramphenicol acetyltransferase (CAT). During this initial phase, the metabolic rates were approximately constant and a classical sequence specific flux balance analysis approach was sufficient to describe protein synthesis [31]. However, after this initial phase, there was a significant shift in productivity following the exhaustion of glucose (which occurred at approximately 1.5 h). Horvath and coworkers developed a fully kinetic model that described the complete three hour reaction time course, including the shift in productivity following glucose exhaustion [32]. While this model described the CFPS dynamics, and the decrease in productivity following the exhaustion of glucose, the identification of the model from 37 measured metabolite trajectories was difficult. Thus, there was an unmet need for a tool that could describe the dynamics of a CFPS reaction, without the burden of identifying a full kinetic model. Toward this need, we developed a dynamic constraint-based modeling approach for CFPS reactions which directly incorporated metabolite measurements (as constraints) into the flux calculation. The dynamic constraint-based model satisfied time-dependent metabolite constraints, while predicting the concentration of the CAT protein and unconstrained metabolite concentrations. Model interrogation suggested the most important metabolite constraint was glucose, as excluding glucose yielded the greatest metabolite prediction error, and the greatest uncertainty in the estimated metabolic flux. Furthermore, we evaluated metabolite constraint sets with one more and one fewer metabolites than the base case (the 37 measured metabolites) to explore the impact of measurement selection on model performance. The single addition of metabolites yielded no significant improvement in the predictive power, while the single exclusion suggested glucose to be the most important measured metabolite in the base case. Next, we selected measurement species based on the results of singular value decomposition on the stoichiometric matrix. The top 36 species from the singular value decomposition (SVD) analysis with the addition of glucose improved the predictive power and reduced flux uncertainty compared with the base case. Finally, we developed a heuristic optimization approach to estimate the optimal list of metabolite measurements. This approach significantly improved metabolite prediction compared to the base case. However, both the base measurement set and the heuristically optimized experimental design poorly characterized flux uncertainty. Taken together, this suggested that dynamic constraint-based modeling can aid in experimental design and measurement selection for metabolite prediction, but the flux estimation problem remains challenging. Furthermore, while we modeled the cell-free production of only a single protein in this study, the sequence specific dynamic constraint-based modeling approach presented here could be extended to multi-protein synthetic circuits, RNA circuits [33] or even small molecule production.

2. Results

2.1. Cell-Free E. coli Metabolic Network

We constructed the cell-free stoichiometric network by removing growth associated reactions from the *i*AF1260 reconstruction of K-12 MG1655 *E. coli* [13], and removing reactions not present in the cell-free system (see Materials and Methods). We then added the transcription and translation template reactions of Allen and Palsson for the CAT protein [19]. Thus, our stoichiometric network described the material and energetic demands for transcription and translation at sequence specific level. The metabolic network consisted of 264 reactions and 146 species; a schematic of the central carbon metabolism is shown in Figure 1. The network described the major carbon and energy pathways and amino acid biosynthesis and degradation pathways. Lastly, we removed genes from the network that were knocked out in the *E. coli* host strain used to make the cell-free extract (A19 ΔtonA ΔtnaA ΔspeA ΔendA ΔsdaA ΔsdaB ΔgshA); see Jewett et al. for further details of the host strain and the cell-free extract preparation [34]. Using this network, we simulated time-dependent cell-free production of the model protein CAT. We used dynamic modified flux balance analysis, a stoichiometric modeling technique that does not make the pseudo steady state assumption and allows the accumulation and depletion of metabolite species. Horvath and coworkers predicted time-dependent cell-free production of CAT using a fully kinetic model trained against an experimental dataset of 37 metabolites, including the substrate glucose, the protein product CAT, organic acids, amino acids, and energy species [32]. This model was used to generate the metabolite constraints used in this study. Transcription and translation rates were subject to resource constraints encoded by the metabolic network, and transcription and translation model parameters were largely derived from literature (Table 1). In this study, we did not explicitly consider protein folding. However, the addition of chaperone or other protein maturation steps could easily be accommodated within the approach by updating the template reactions (see Palsson and coworkers [20]). The cell-free metabolic network and all model code and parameters can be downloaded under an MIT software license from the Varnerlab website [35].

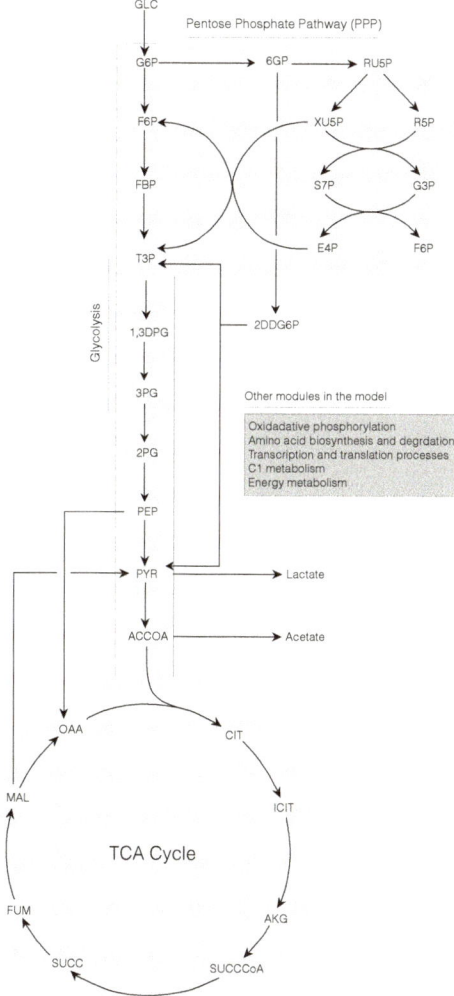

Figure 1. Schematic of the core portion of the cell-free *E. coli* metabolic network. The network consisted of 264 reactions and 146 metabolites. Metabolites of glycolysis, pentose phosphate pathway, Entner-Doudoroff pathway, and the tricarboxylic acid (TCA) cycle are shown. Metabolites of oxidative phosphorylation, amino acid biosynthesis and degradation, transcription/translation, chorismate metabolism, and energy metabolism are not shown.

Table 1. Reference values for transcription, translation, and mRNA degradation from literature. Transcription rate calculated from elongation rate, mRNA length, and promoter activity level. Translation rate calculated from elongation rate, protein length, and polysome amplification constant. The mRNA degradation rate calculated from a characteristic mRNA half-life. CAT: chloramphenicol acetyltransferase.

Description	Parameter	Value	Units	Reference
T7 RNA polymerase concentration	R_T	1.0	μM	
Ribosome concentration	R_X	2	μM	[10]
CAT mRNA length	l_G	660	nt	[36]
CAT protein length	l_P	219	aa	[36]
Transcription saturation coefficient	K_T	100	nM	estimated
Transcription elongation rate	$\dot{v}_T$	25	nt/s	[10]
Translation saturation coefficient	K_X	45	μM	estimated
Translation elongation rate	$\dot{v}_X$	1.5	aa/s	[10]
T7 Promoter activity level	u	0.9		estimated
Transcription rate	$k_{cat}^T = \left(\dfrac{\dot{v}_T}{l_G}\right) u$	123	h^{-1}	calculated
Polysome amplification constant	K_P	10		estimated
Translation rate	$k_{cat}^X = \left(\dfrac{\dot{v}_X}{l_P}\right) K_P$	247	h^{-1}	calculated
mRNA degradation time	$t_{1/2}$	8	min	BNID 106253
mRNA degradation rate	$k_{deg} = \dfrac{\ln(2)}{t_{1/2}}$	5.2	h^{-1}	calculated

2.2. Dynamic Constrained Simulation of cell-free Protein Synthesis

Cell-free synthesis of the CAT protein showed two production phases, an initial fast production phase before glucose exhaustion (at approximately 1.5 h) and a slow production phase following glucose exhaustion. The metabolite profile varied significantly between these phases; for example, pyruvate and lactate were produced during the first phase but consumed during the second. Thus, a static pseudo steady state flux balance approach was not possible for this system. However, a central advantage of cell-free systems is direct access to metabolite measurements, and the biosynthetic machinery during production. If we could directly integrate dynamic metabolite and protein concentration measurements into the flux estimation problem, we could potentially get a better estimate of the flux distribution. Toward this question, we developed a dynamic modeling approach in which metabolic fluxes were estimated so that all metabolites were non-negative and the simulated metabolites were constrained to lie within a bounded range of the measured value. Using this technique, we simulated the cell-free production of CAT subject to dynamic metabolite measurements.

We explored the influence of uncertainty in the transcription (TX) and translation parameters (TL) by sampling different values for the abundance and elongation rates of RNA polymerases and ribosomes, the polysome amplification constant, the mRNA degradation rate and other kinetic parameters appearing in the transcription and translation bounds. The base values for the TX/TL parameters are given in Table 1, and the uniform sampling procedure is described in the Materials and Methods. Central carbon metabolites (Figure 2), amino acids (Figure 3), and energy species (Figure 4) in the synthetic measurement set were captured, within experimental error, by an ensemble of dynamic constraint-based simulations. The synthetic metabolite constraints (blue regions) shown in each of the simulation figures was derived from the kinetic model of Horvath et al. [32], which was trained on experimental measurement of the 37 metabolites shown in Figures 2–4. Thus, the metabolite constraints used in this study were calculated based upon the kinetic model, which shows high fidelity with the experimental measurements. The flux estimation problem converged in greater than 99%

of the simulation time intervals, given these metabolite constraints. This suggested there were not gross measurement errors in the measurement constraints, as the stoichiometric constraints were satisfied. Moreover, it suggested the error introduced by the time discretization scheme did not lead to inconsistent metabolite estimates. The ensemble of models captured the time evolution of protein biosynthesis, and the consumption and production of organic acid, amino acid and energy species. Arginine and glutamate were excluded from the constraint set, but were still largely captured by the ensemble of dynamic constraint-based models, although with wide variance than the synthetic measurement set. During the first hour, glucose was consumed as the primary carbon source for ATP, amino acids, and protein synthesis. After glucose was depleted, lactate and pyruvate were consumed as alternate substrates for energy production and CAT synthesis. Taken together, we captured the 37 metabolite measurements in the base synthetic data set, and captured the biphasic behavior of CAT production, although we significantly over-predicted the translation rate for some elements of the ensemble. This suggested that there was excess capacity in the metabolic network, which could be used to enhance protein production.

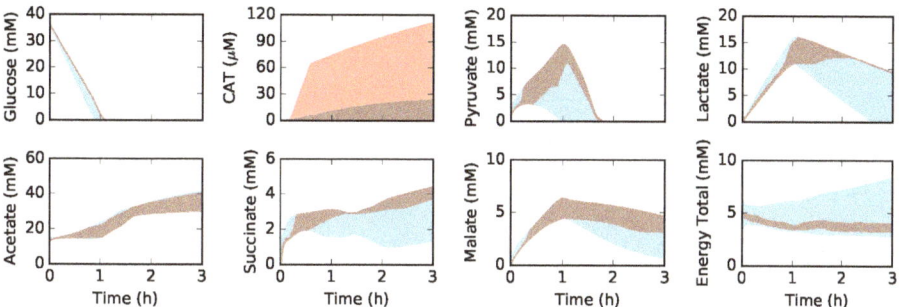

Figure 2. Simulated metabolite concentration versus synthetic data as a function of time. Central carbon metabolism, including glucose (substrate), chloramphenicol acetyltransferase (CAT) (product), and intermediates, as well as total concentration of energy species (energy total). The energy total denotes the summation of all energy species in the model (all bases and all phosphate states). The 95% confidence interval for the simulation conducted over the ensemble of transcription/translation parameter sets is shown in the orange shaded region, while the 95% confidence interval for the synthetic constraint data is shown in the blue shaded region. The synthetic data constraints were generated from the kinetic model of Horvath et al., which was trained using experimental measurements of the system simulated in this study [32].

We quantified the uncertainty in the estimated metabolic flux distribution, given constrained CAT production using flux variability analysis (FVA) for the base synthetic data across the three hours of measurement (Tables 2 and 3). The analysis was divided into two phases: phase 1 where glucose was consumed as the carbon source, and phase 2 when glucose was depleted and lactate and pyruvate were utilized. The reactions associated with protein synthesis (translation initiation, translation, tRNA charging, mRNA degradation) were unsurprisingly the most constrained, as CAT production was forced to remain the same. Transcription was not varied in this analysis. On the other hand, glycolytic, pentose phosphate, and Entner-Doudoroff reactions were not highly constrained, indicating the robustness of substrate utilization. However, one exception to this was the net reaction through *zwf* reaction, which was tightly constrained, suggesting that glycolysis alone cannot support protein production. Interestingly, although the two phases consumed different carbon sources, the flux variability remained similar. Taken together, these results suggested that there was significant flexibility in the ability of the metabolic network to meet the carbon and energy demands of protein synthesis. Next, we explored alternative measurement sets to constrain the simulation of cell-free protein synthesis.

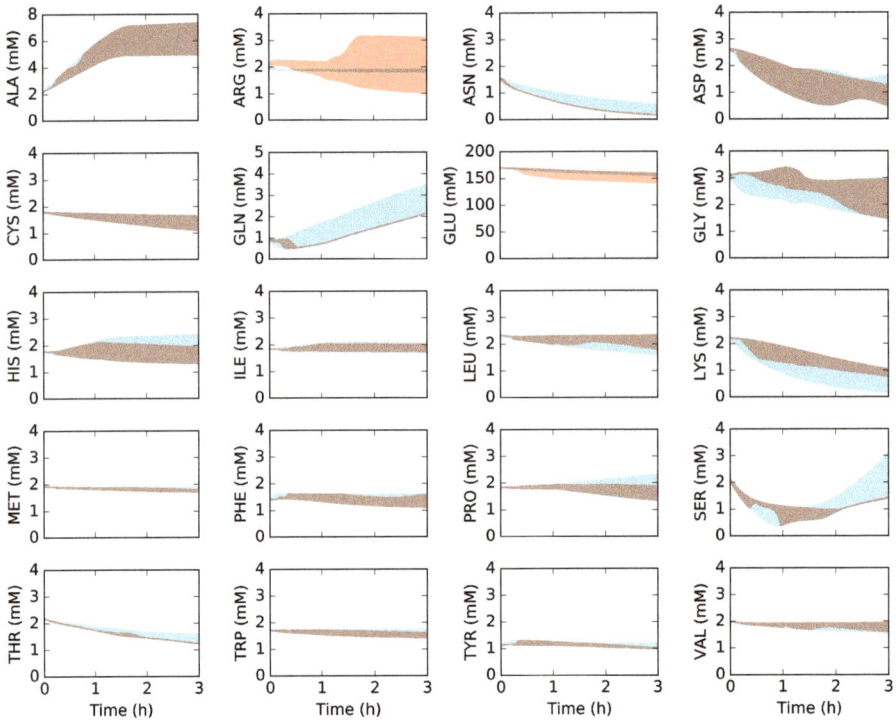

Figure 3. Simulation of amino acid concentration versus synthetic data as a function of time. The 95% confidence interval for the simulation conducted over the ensemble of transcription/translation parameter sets is shown in the orange shaded region, while the synthetic constraint data is shown in the blue shaded region. Arginine and glutamate were excluded from the constraint set. The synthetic data constraints were generated from the kinetic model of Horvath et al., which was trained using experimental measurements of the system simulated in this study [32].

Table 2. Flux uncertainty calculated using flux variability analysis for the base synthetic dataset during the first production phase (0 h to 1.5 h), normalized to the glucose consumption rate.

Enzyme/Pathway	Reaction	Uncertainty
RNA polymerase	Translation	<0.01
RNA polymerase	Translation initiation	<0.01
tRNA charging of alanine	tRNA charging (ALA)	<0.01
tRNA charging of cysteine	tRNA charging (CYS)	<0.01
tRNA charging of aspartate	tRNA charging (ASP)	<0.01
tRNA charging of histidine	tRNA charging (HIS)	<0.01
tRNA charging of serine	tRNA charging (SER)	<0.01
tRNA charging of tyrosine	tRNA charging (TYR)	<0.01
tRNA charging of phenylalanine	tRNA charging (PHE)	<0.01
tRNA charging of arginine	tRNA charging (ARG)	<0.01
tRNA charging of glutamate	tRNA charging (GLU)	<0.01

Table 2. *Cont.*

Enzyme/Pathway	Reaction	Uncertainty
mRNA degradation	mRNA degradation	<0.01
tRNA charging of tryptophan	tRNA charging (TRP)	<0.01
tRNA charging of proline	tRNA charging (PRO)	<0.01
tRNA charging of asparagine	tRNA charging (ASN)	<0.01
tRNA charging of isoleucine	tRNA charging (ILE)	<0.01
tRNA charging of glycine	tRNA charging (GLY)	<0.01
tRNA charging of glutamine	tRNA charging (GLN)	<0.01
tRNA charging of lysine	tRNA charging (LYS)	<0.01
tRNA charging of threonine	tRNA charging (THR)	<0.01
tRNA charging of valine	tRNA charging (VAL)	<0.01
tRNA charging of methionine	tRNA charging (MET)	<0.01
tRNA charging of leucine	tRNA charging (LEU)	<0.01
Step 6 of AMP synthesis	R_A_syn_6	<0.01
Orotate synthase 1	R_or_syn_1	<0.01
Metionine biosynthesis	R_met	0.01
Valine biosynthesis	R_val	0.01
Leucine biosynthesis	R_leu	0.01
Aldhyde-alcohol dehydrogenase	R_adhE_net	0.01
Malate dehydrogenase	R_mdh_net	0.01
Glycine biosynthesis	R_gly_deg	0.02
Threonine degradation 2	R_thr_deg2	0.02
Acetate kinase	R_ackA_net	0.02
Alanine biosynthesis	R_alaAC_net	0.02
Isoleucine biosynthesis	R_ile	0.03
Tyrosine biosynthesis	R_tyr	0.03
Histidine biosynthesis	R_his	0.03
Methylglyoxal degradation	R_mglx_deg	0.03
Transaldolase	R_talAB_net	0.04
Glycine cleavage system	R_gly_fol_net	0.04
Ribulose-phosphate 3-epimerase	R_rpe_net	0.04
Phosphate acetyltransferase	R_pta_net	0.04
Phosphoglycerate kinase	R_pgk_net	0.05
Glyceraldehyde-3-phosphate dehydrogenase	R_gapA_net	0.05
Fructose 1,6-bisphosphate aldolase	R_fbaA_net	0.05
Enolase	R_eno_net	0.05
Phenylalanine biosynthesis	R_phe	0.05
Transketolase 2	R_tkt2_net	0.06
Fumarate hydratase	R_fum_net	0.06
Transketolase 1	R_tkt1_net	0.07
Orotate synthase 2	R_or_syn_2	0.07
Phosphoglycerate mutase	R_gpm_net	0.08
Ribose-5-phosphate isomerase	R_rpi_net	0.09
CTP synthetase 1	R_ctp_1	0.09
CTP synthetase 2	R_ctp_2	0.09
Triosephosphate isomerase	R_tpiA_net	0.1
Step 7 of AMP synthesis	R_A_syn_7	0.15
Step 12 of AMP synthesis	R_A_syn_12	0.17
Lactate dehydrogenase	R_ldh_net	0.17
Step 5 of AMP synthesis	R_A_syn_5	0.2
Methylenetetrahydrofolate reductase	R_mthfr2a	0.2
Glucose-6-phosphate dehydrogenase	R_zwf_net	0.21
Methylenetetrahydrofolate dehydrogenase	R_mthfd_net	0.21
UMP synthesis	R_ump_syn	0.22
OMP synthesis	R_omp_syn	0.22
Lysine degradation	R_lys_deg	0.23
Lysine biosynthesis	R_lys	0.23
Isocitrate dehydrogenase	R_icd_net	0.23

Table 2. *Cont.*

Enzyme/Pathway	Reaction	Uncertainty
Threonine degradation 3	R_thr_deg3	0.24
Step 8 of AMP synthesis	R_A_syn_8	0.26
Tryptophan degradation	R_trp_deg	0.27
Methylenetetrahydrofolate dehydrogenase	R_mthfc_net	0.28
Tryptophan biosynthesis	R_trp	0.28
Aconitase	R_acn_net	0.33
Phosphoglucose isomerase	R_pgi_net	0.33
Step e of folate synthesis	R_fol_e	0.34
Step 4 of AMP synthesis	R_A_syn_4	0.4
GMP synthetase	R_gmp_syn	0.44
Step 9 of AMP synthesis	R_A_syn_9	0.48
XMP synthase	R_xmp_syn	0.53
Step 3 of AMP synthesis	R_A_syn_3	0.6
Step 10 of AMP synthesis	R_A_syn_10	0.63
Step 2b of folate synthesis	R_fol_2b	0.63
Glutamate dehydrogenase	R_gdhA_net	0.71
Step 3 of folate synthesis	R_fol_3	0.74
Step 4 of folate synthesis	R_fol_4	0.79
Pyruvate formate lyase	R_pflAB	0.8
Step 2 of AMP synthesis	R_A_syn_2	0.81
Step 2a of folate synthesis	R_fol_2a	0.81
Step 1 of folate synthesis	R_fol_1	0.99
Glucokinase	R_glk_atp	1
Step 1 of AMP synthesis	R_A_syn_1	1
Arginine degradation	R_arg_deg	1.23
Glycine biosynthesis	R_glyA	1.33
Phosphoribosylpyrophosphate synthase	R_prpp_syn	1.34
Chorismate synthesis	R_chor	1.35
Succinate thiokinase	R_sucCD	1.55
2-Ketoglutarate dehydrogenase	R_sucAB	1.55
GABA degradation 1	R_gaba_deg1	1.56
GABA degradation 2	R_gaba_deg2	1.56
Glutamate degradation	R_glu_deg	1.56
Arginine biosynthesis	R_arg	1.68
Pyruvate dehydrogenase	R_pdh	2.06
Malate synthase	R_aceB	2.3
Threonine degradation 1	R_thr_deg1	2.32
Isocitrate lyase	R_aceA	2.36
Threonine biosynthesis	R_thr	2.48
Citrate synthase	R_gltA	2.62
6-Phosphogluconate dehydrogenase	R_gnd	2.62
Cysteine biosynthesis	R_cysEMK	4.59
Cysteine degradation	R_cys_deg	4.6
Proline biosynthesis	R_pro	5.45
Proline degradation	R_pro_deg	5.47
6-Phosphogluconate dehydrase	R_edd	5.96
2-Keto-3-deoxy-6-phospho-gluconate aldolase	R_eda	5.96
Serine degradation	R_ser_deg	6.43
Nucleotide diphosphatase (ATP)	R_atp_amp	6.53
Nucleotide diphosphatase (UTP)	R_utp_ump	6.53
Nucleotide diphosphatase (GTP)	R_gtp_gmp	6.53
Nucleotide diphosphatase (CTP)	R_ctp_cmp	6.53
Cytidylate kinase	R_atp_cmp	6.56
Guanylate kinase	R_atp_gmp	6.6
UMP kinase	R_atp_ump	6.62
6-Phosphogluconolactonase	R_pgl	6.67
Serine biosynthesis	R_serABC	6.72

Table 2. *Cont.*

Enzyme/Pathway	Reaction	Uncertainty
NADH:ubiquinone oxidoreductase	R_nuo	7.39
NADH dehydrogenase 1	R_ndh1	7.39
NADH dehydrogenase 2	R_ndh2	7.39
Fumurate reductase	R_frd	7.44
Succinate dehydrogenase	R_sdh	7.93
Malic enzyme A	R_maeA	7.99
Malic enzyme B	R_maeB	8.01
Cytochrome oxidase bo	R_cyo	8.03
Cytochrome oxidase bd	R_cyd	8.03
ATP synthase	R_atp	11.71
PEP synthase	R_pps	12.98
Fructose-1,6-bisphosphate aldolase	R_fdp	12.98
Adenosinetriphosphatase	R_atp_adp	12.98
PEP carboxykinase	R_pck	12.98
Asparagine biosynthesis	R_asnB	13
Glutamate biosynthesis	R_gltBD	13
Glutamine degradation	R_gln_deg	13
Glutamine biosynthesis	R_glnA	13.05
Acetyl-CoA synthetase	R_acs	13.06
Inorganic pyrophosphatase	R_ppa	13.08
Adenylate kinase	R_adk_atp	13.36
PEP carboxylase	R_ppc	14.22
Phosphofructokinase	R_pfk	14.9
Pyruvate kinase	R_pyk	15.89
Transhydrogenase	R_pnt2	22.45
Transhydrogenase	R_pnt1	25.5
Aspartate degradation	R_asp_deg	25.5
Aspartate biosynthesis	R_aspC	25.83
Asparagine biosynthesis	R_asnA	247.53
Asparagine degradation	R_asn_deg	247.53

Table 3. Flux uncertainty calculated using flux variability analysis for the base synthetic dataset during the second production phase (1 h to 3 h), normalized to the glucose consumption rate.

Enzyme	Reaction	Uncertainty
Step 6 of AMP synthesis	R_A_syn_6	<0.01
Orotate synthase 1	R_or_syn_1	<0.01
Orotate synthase 2	R_or_syn_2	<0.01
Aldhyde-alcohol dehydrogenase	R_adhE_net	<0.01
RNA polymerase	Translation	<0.01
tRNA charging of phenylalanine	tRNA charging (PHE)	<0.01
tRNA charging of alanine	tRNA charging (ALA)	<0.01
tRNA charging of glutamine	tRNA charging (GLN)	<0.01
tRNA charging of threonine	tRNA charging (THR)	<0.01
tRNA charging of aspartate	tRNA charging (ASP)	<0.01
tRNA charging of glutamate	tRNA charging (GLU)	<0.01
tRNA charging of histidine	tRNA charging (HIS)	<0.01
tRNA charging of lysine	tRNA charging (LYS)	<0.01
tRNA charging of tyrosine	tRNA charging (TYR)	<0.01
tRNA charging of asparagine	tRNA charging (ASN)	<0.01
tRNA charging of serine	tRNA charging (SER)	<0.01
tRNA charging of methionine	tRNA charging (MET)	<0.01
tRNA charging of isoleucine	tRNA charging (ILE)	<0.01
tRNA charging of valine	tRNA charging (VAL)	<0.01
tRNA charging of proline	tRNA charging (PRO)	<0.01

Table 3. *Cont.*

Enzyme	Reaction	Uncertainty
tRNA charging of leucine	tRNA charging (LEU)	<0.01
tRNA charging of arginine	tRNA charging (ARG)	<0.01
tRNA charging of tryptophan	tRNA charging (TRP)	<0.01
tRNA charging of cysteine	tRNA charging (CYS)	<0.01
tRNA charging of glycine	tRNA charging (GLY)	<0.01
RNA polymerase	Translation initiation	<0.01
mRNA degradation	mRNA degradation	<0.01
Glycine cleavage system	R_gly_fol_net	0.01
Transaldolase	R_talAB_net	0.02
Transketolase 1	R_tkt1_net	0.02
Ribulose-phosphate 3-epimerase	R_rpe_net	0.02
Ribose-5-phosphate isomerase	R_rpi_net	0.03
Transketolase 2	R_tkt2_net	0.04
Valine biosynthesis	R_val	0.05
Leucine biosynthesis	R_leu	0.05
Malate dehydrogenase	R_mdh_net	0.06
Triosephosphate isomerase	R_tpiA_net	0.07
Fructose 1,6-bisphosphate aldolase	R_fbaA_net	0.07
Glycine biosynthesis	R_gly_deg	0.09
Threonine degradation 2	R_thr_deg2	0.09
Phosphoglucose isomerase	R_pgi_net	0.09
Methylglyoxal degradation	R_mglx_deg	0.09
Methylenetetrahydrofolate dehydrogenase	R_mthfd_net	0.13
Glucose-6-phosphate dehydrogenase	R_zwf_net	0.18
Tyrosine biosynthesis	R_tyr	0.2
Enolase	R_eno_net	0.2
Phosphoglycerate mutase	R_gpm_net	0.2
Phosphoglycerate kinase	R_pgk_net	0.21
Glyceraldehyde-3-phosphate dehydrogenase	R_gapA_net	0.21
Methylenetetrahydrofolate dehydrogenase	R_mthfc_net	0.28
Metionine biosynthesis	R_met	0.34
Phosphate acetyltransferase	R_pta_net	0.4
Acetate kinase	R_ackA_net	0.41
Fumarate hydratase	R_fum_net	0.43
Phenylalanine biosynthesis	R_phe	0.53
Glucokinase	R_glk_atp	1
Step 5 of AMP synthesis	R_A_syn_5	1
Step 7 of AMP synthesis	R_A_syn_7	1
OMP synthesis	R_omp_syn	1.09
Isoleucine biosynthesis	R_ile	1.13
Alanine biosynthesis	R_alaAC_net	1.49
Step 8 of AMP synthesis	R_A_syn_8	1.76
Methylenetetrahydrofolate reductase	R_mthfr2a	1.85
Isocitrate dehydrogenase	R_icd_net	1.87
Histidine biosynthesis	R_his	1.95
Step 4 of AMP synthesis	R_A_syn_4	2.02
Threonine degradation 3	R_thr_deg3	2.08
CTP synthetase 1	R_ctp_1	2.09
CTP synthetase 2	R_ctp_2	2.09
Lysine biosynthesis	R_lys	2.13
Lysine degradation	R_lys_deg	2.13
Lactate dehydrogenase	R_ldh_net	2.27
Tryptophan degradation	R_trp_deg	2.55
Tryptophan biosynthesis	R_trp	2.7
Aconitase	R_acn_net	2.82
UMP synthesis	R_ump_syn	2.85
Step 3 of AMP synthesis	R_A_syn_3	3.07

Table 3. *Cont.*

Enzyme	Reaction	Uncertainty
XMP synthase	R_xmp_syn	4.04
GMP synthetase	R_gmp_syn	4.04
Step 12 of AMP synthesis	R_A_syn_12	4.13
Step e of folate synthesis	R_fol_e	4.37
Step 2 of AMP synthesis	R_A_syn_2	4.74
Step 9 of AMP synthesis	R_A_syn_9	4.83
Step 10 of AMP synthesis	R_A_syn_10	4.83
Step 1 of AMP synthesis	R_A_syn_1	5.02
Glutamate dehydrogenase	R_gdhA_net	5.14
Phosphoribosylpyrophosphate synthase	R_prpp_syn	6.11
Step 4 of folate synthesis	R_fol_4	6.24
Step 3 of folate synthesis	R_fol_3	6.24
Step 2b of folate synthesis	R_fol_2b	7.03
Step 2a of folate synthesis	R_fol_2a	8.04
Step 1 of folate synthesis	R_fol_1	9.04
Glycine biosynthesis	R_glyA	9.58
Chorismate synthesis	R_chor	13.98
Arginine degradation	R_arg_deg	14.97
Succinate thiokinase	R_sucCD	17.94
GABA degradation 1	R_gaba_deg1	17.95
GABA degradation 2	R_gaba_deg2	17.95
Glutamate degradation	R_glu_deg	17.95
2-Ketoglutarate dehydrogenase	R_sucAB	18.06
Arginine biosynthesis	R_arg	18.14
Threonine degradation 1	R_thr_deg1	19.06
Pyruvate formate lyase	R_pflAB	19.97
Threonine biosynthesis	R_thr	22.23
Malate synthase	R_aceB	28.81
Isocitrate lyase	R_aceA	28.81
Pyruvate dehydrogenase	R_pdh	29.86
6-Phosphogluconate dehydrogenase	R_gnd	31.9
Citrate synthase	R_gltA	32.2
Cysteine biosynthesis	R_cysEMK	48.93
Cysteine degradation	R_cys_deg	49.31
Proline biosynthesis	R_pro	61.2
2-Keto-3-deoxy-6-phospho-gluconate aldolase	R_eda	61.86
6-Phosphogluconate dehydrase	R_edd	61.86
Proline degradation	R_pro_deg	62.62
Serine degradation	R_ser_deg	68.5
6-Phosphogluconolactonase	R_pgl	70.12
Serine biosynthesis	R_serABC	72.05
Nucleotide diphosphatase (ATP)	R_atp_amp	74.22
Nucleotide diphosphatase (UTP)	R_utp_ump	74.22
Nucleotide diphosphatase (GTP)	R_gtp_gmp	74.22
Nucleotide diphosphatase (CTP)	R_ctp_cmp	74.22
Cytidylate kinase	R_atp_cmp	74.74
Guanylate kinase	R_atp_gmp	76.11
UMP kinase	R_atp_ump	76.78
NADH dehydrogenase 1	R_ndh1	86.63
NADH:ubiquinone oxidoreductase	R_nuo	86.63
Malic enzyme A	R_maeA	86.77
NADH dehydrogenase 2	R_ndh2	87.01
Fumurate reductase	R_frd	87.01
Malic enzyme B	R_maeB	88.17
Succinate dehydrogenase	R_sdh	90.66
Cytochrome oxidase bo	R_cyo	92.08
Cytochrome oxidase bd	R_cyd	92.08

Table 3. *Cont.*

Enzyme	Reaction	Uncertainty
ATP synthase	R_atp	134.61
PEP synthase	R_pps	148.45
Asparagine biosynthesis	R_asnB	148.45
Glutamine degradation	R_gln_deg	148.45
Glutamate biosynthesis	R_gltBD	148.45
Acetyl-CoA synthetase	R_acs	148.45
Adenosinetriphosphatase	R_atp_adp	148.45
Fructose-1,6-bisphosphate aldolase	R_fdp	148.45
PEP carboxykinase	R_pck	148.45
Glutamine biosynthesis	R_glnA	149.67
Inorganic pyrophosphatase	R_ppa	150.79
Adenylate kinase	R_adk_atp	152.88
PEP carboxylase	R_ppc	158.68
Phosphofructokinase	R_pfk	161.64
Pyruvate kinase	R_pyk	170.95
Transhydrogenase	R_pnt2	258.03
Aspartate degradation	R_asp_deg	286.46
Transhydrogenase	R_pnt1	286.46
Aspartate biosynthesis	R_aspC	290.86
Asparagine biosynthesis	R_asnA	4421.48
Asparagine degradation	R_asn_deg	4421.48

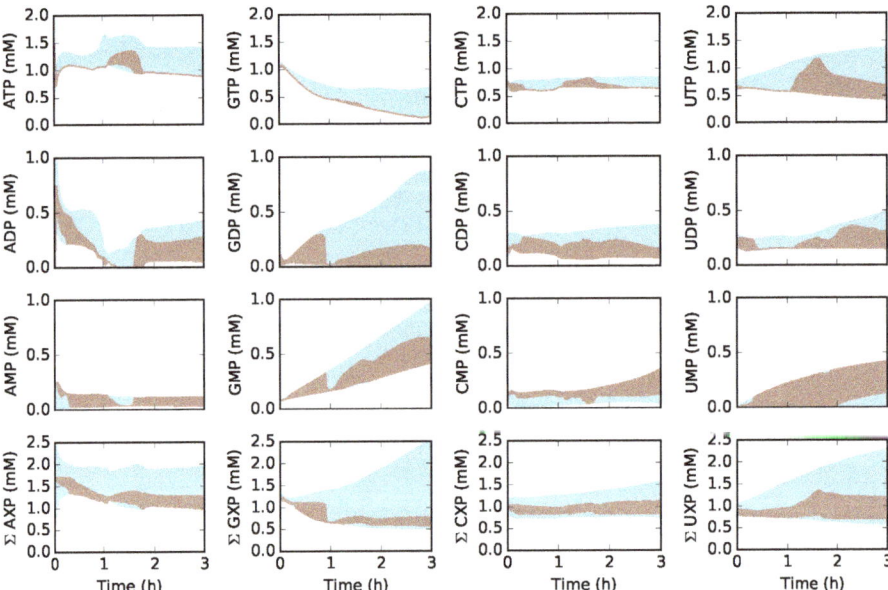

Figure 4. Simulation of energy species and energy totals by base versus synthetic data as a function of time. The 95% confidence interval for the simulation conducted over the ensemble of transcription/translation parameter sets is shown in the orange shaded region, while the synthetic constraint data is shown in the blue shaded region. The synthetic data constraints were generated from the kinetic model of Horvath et al., which was trained using experimental measurements of the system simulated in this study [32].

2.3. Alternative Measurement Sets

The base synthetic data set, consisting of 36 metabolite time series and the protein product CAT, was the measurement set used to train the kinetic model of Horvath et al. [32]. Thus, the confidence intervals on the synthetic data used in this study as constraints on the flux estimation problem were informed by experimental measurements of glucose, organic and amino acids, energy species and the protein product CAT. However, we have no a priori reason to suppose that this experimental design was optimal. Toward this question, we performed simulations and flux variability analysis (FVA) for alternative synthetic data sets to understand the importance of measurement selection when characterizing CFPS (Figures 5 and 6). In all cases, we assumed the same sampling frequency as the base synthetic dataset, but we varied which species were measured. First, we removed each of the 37 metabolites from the base set, one at a time, to create 37 measurement exclusion sets, consisting of 36 metabolites each (Figure 5, light gray dots). For each set, the state the dynamic model was used to calculate a value of error against the synthetic data, and FVA was used to calculate a value of flux uncertainty. Most of the exclusion sets clustered around the base case, with error values between 75% and 110%, and flux uncertainties between 93% and 103%, of the base case. The exception to this was the glucose exclusion set, which showed 89% higher error and 7% greater flux uncertainty. Within the primary cluster, a slight pattern emerged: the sets in which an organic acid were removed tended to result in increased error, while the removal of an amino acid tended to reduce error; however, this was not true across all metabolites. We also performed the analysis on several inclusion sets to determine which additional metabolites could improve predictive power (Figure 5, black dots). In particular, we added unmeasured central carbon metabolites to the base case, which resulted in 23 inclusion sets, consisting of 38 metabolites each. As with the exclusion sets, most of the inclusion sets clustered around the base case, with error values between 72% and 103%, and flux uncertainties between 94% and 102%, of the base case. Considering all exclusion and inclusion sets, there was generally no correlation between the metabolite prediction error and flux uncertainty. Taken together, these suggested central carbon metabolites, especially glucose, were important to characterize the network, but performing single additional measurements was not enough to significantly increase predictive power. Next, we explored whether measurement selection could be based upon the structural features of a metabolic network. Toward this question, we used singular value decomposition (SVD) of the stoichiometric matrix to suggest which metabolites should be measured.

Singular value decomposition (SVD) measurement selection outperformed the base case, with a prediction error improvement of 11% and similar flux variability (Figure 5, open square). SVD was used to decompose the stoichiometric matrix into 105 modes. The top 36 metabolites that had the greatest weighted sum across the modes that accounted for 95% of the network were estimated. Since our exclusion analysis identified glucose as the single most important metabolite, we added it to the top metabolites as determined by SVD to obtain a 37-metabolite constraint set, consisting of: GTP, GDP, GMP, ATP, ADP, AMP, UTP, UMP, CTP, CMP, GLN, GLU, ASP, LYS, LEU, HIS, THR, PHE, ALA, VAL, TYR, GLY, SER, H, ASN, ILE, MET, AKG, PYR, ARG, CYS, NH3, FUM, SUCC, TRP, ACCOA, and glucose. The SVD measurement set suggested that energy, and amino acid species carried the most information compared to central carbon species, which made up a relatively smaller fraction of the list. Surprisingly, the measurement set selected by SVD was approximately 80% similar to the original synthetic data generated by hand. However, the 20% difference was enough to improve the prediction error by approximately 11%. Taken together, measurements selected by SVD decomposition of the stoichiometric matrix improved the prediction of metabolite abundance, but SVD-based measurement selection did not improve flux variability.

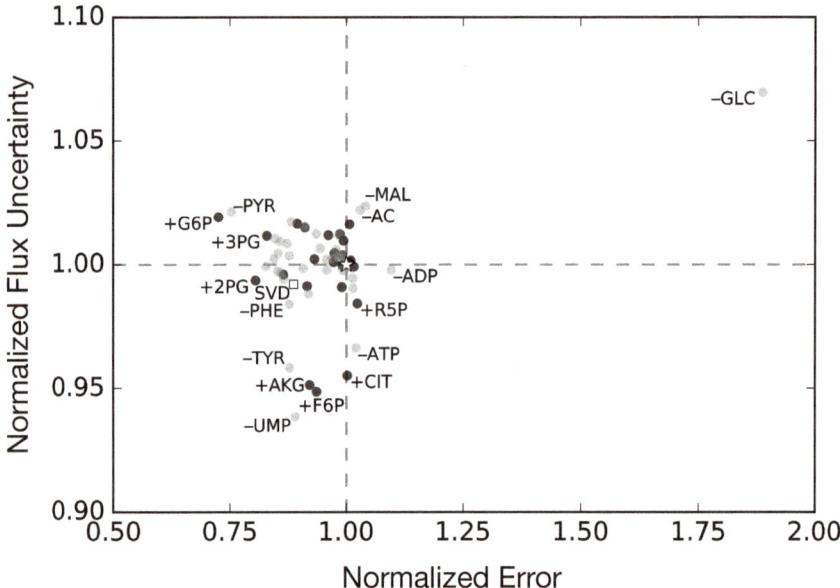

Figure 5. Flux uncertainty versus metabolite prediction error against synthetic data, normalized to the base case (white star), for exclusion (gray) and inclusion (black) metabolite constraint sets. The performance of the singular value decomposition (SVD)-determined metabolite constraint set is shown by the white square.

Next, we used heuristic optimization to systematically investigate the effect of changing the dimension and identify the measurement constraints (Figure 6). In particular, we minimized the error and flux variability of model predictions by varying the metabolites that appeared in the synthetic constraint set. We used a binary simulated annealing algorithm to switch metabolite membership in the constraint set on or off, and thus generated an ensemble of >200 measurement constraint sets (Figure 6A). While there was no strict error threshold, the simulated annealing algorithm was less likely to accept high-error sets into the ensemble; thus, the error of most sets in the ensemble was less than that of the base case. Specifically, the error varied from just over double to less than one ten-thousandth of the base case. Flux uncertainty was also a component of the objective function, but was only improved by 7%, suggesting this performance metric was tightly constrained; network flux values were not well characterized, even with comprehensive training datasets. As expected, there was an inverse relationship between the number of metabolite constraints and the prediction error (Figure 6B). However, the slope of that trend was striking; error was improved by three to four orders of magnitude, simply by increasing the number of constraints by 11 or fewer. Furthermore, the base synthetic measurement set was outperformed by the majority of the ensemble; often, the simulated annealing approach achieved the same error with fewer constraints, or much lower error with the same, or even fewer, constraints. This suggests that, while comprehensive, the original synthetic dataset was not optimal in terms of predictive power per measurement. However, the base case was one of the best in terms of reducing flux uncertainty.

Lastly, we investigated which metabolites were most effective at improving predictive power by considering how often it appeared in the ensemble (Table 4). Glucose unsurprisingly appeared most often (tied with G6P), but interestingly was not in every constraint set. Those that did not contain glucose had some of the highest errors, but also some of the smallest constraint set sizes (Figure 6B, black dots). The most frequent metabolites from the heuristic method were largely from glycolysis, pentose phosphate, and the TCA cycle (compared to the SVD analysis which gave greater consideration to energetic and amino acid species). To further understand the species selection, we calculated the frequency of appearance in the 57 best sets, those with error of least three orders of magnitude lower than the base case (Table 5). Nineteen metabolites appeared in all of these sets, and all but Alanine were central carbon metabolites (defined here as glycolysis, pentose phosphate, TCA). Taken together, measurement selection made a significant difference in capturing dynamic metabolite abundance in cell-free protein synthesis. Although the error decreased with increasing measurement number overall, the specific combination of metabolites was arguably even more important. Metabolic fluxes, however, remained unknown despite the large number of measurements taken.

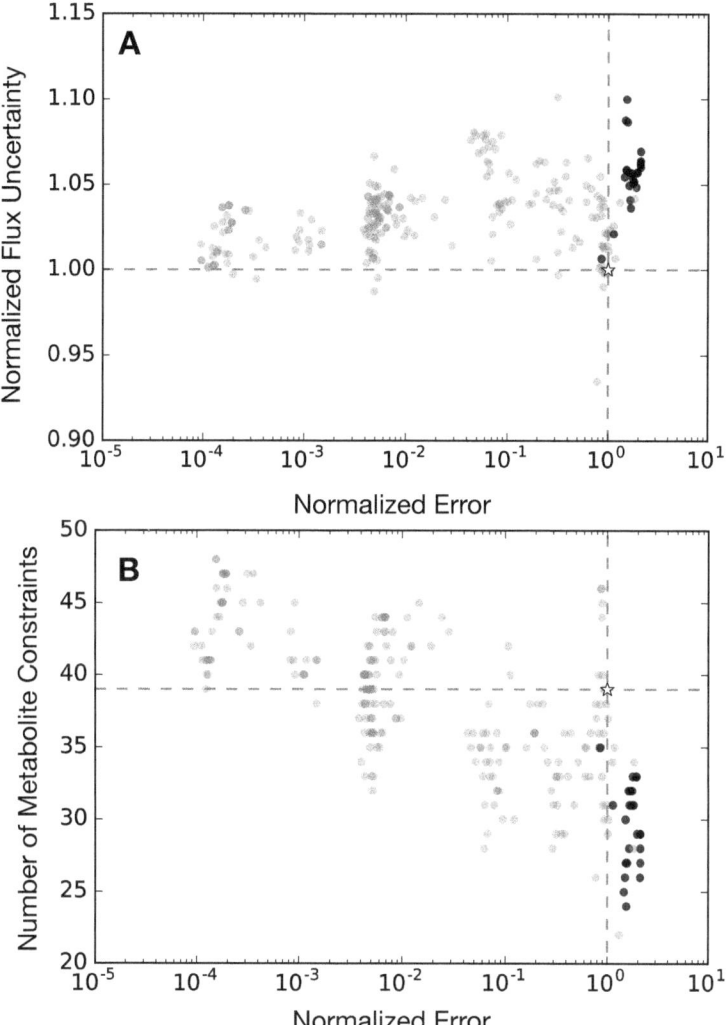

Figure 6. Flux uncertainty and metabolite prediction error for the simulated annealing experimental design approach. (**A**) normalized flux uncertainty versus normalized metabolite prediction error; (**B**) number of metabolite constraints versus normalized metabolite prediction error. Error was computed for the synthetic experimental designs normalized to the base synthetic dataset (white star). Sets that include glucose are show as gray circles, while those that do not are represented with black circles.

Table 4. Metabolites by frequency of appearance in the simulated annealing constraint sets.

Metabolite	Symbol	Frequency
alpha-D-Glucose	GLC	89.9%
Glucose 6-phosphate	G6P	89.9%
Citrate	CIT	85.7%
Isocitrate	ICIT	84.9%
Fumarate	FUM	84.5%
Fructose 6-phosphate	F6P	83.6%
6-Phospho-D-glucono-1,5-lactone	6PGL	81.5%
sedo-Heptulose 7-phosphate	S7P	79.4%
Alanine	ALA	77.7%
Guanosine triphosphate	GTP	77.3%
Malate	MAL	75.6%
D-Ribulose 5-phosphate	RU5P	74.8%
Erythrose 4-phosphate	E4P	73.1%
Adenosine diphosphate	ADP	72.3%
alpha-Ketoglutarate	AKG	71.8%
Uridine diphosphate	UDP	70.2%
Succinate	SUCC	69.7%
Cytidine monophosphate	CMP	69.3%
Guanosine diphosphate	GDP	67.2%
6-Phospho-D-gluconate	6PGC	67.2%
Arginine	ARG	66.8%
Ribose 5-phosphate	R5P	66.4%
Methionine	MET	65.5%
Glyoxylate	GLX	65.5%
Glutamine	GLN	63.9%
Phenylalanine	PHE	63.4%
Valine	VAL	62.6%
Glyceraldehyde 3-phosphate	G3P	62.2%
Adenosine monophosphate	AMP	62.2%
Proline	PRO	60.1%
Fructose 1,6-diphosphate	FDP	60.1%
Dihydroxyacetone phosphate	DHAP	60.1%
Histidine	HIS	59.7%
Glycine	GLY	59.7%
Oxaloacetate	OAA	58.4%
2-Dehydro-3-deoxy-D-gluconate 6-phosphate	2DDG6P	58.4%
Chloramphenicol acetyltransferase	CAT	56.3%
Cysteine	CYS	55.5%
Acetate	AC	54.2%
Succinyl coenzyme A	SUCCOA	53.8%
Uridine monophosphate	UMP	52.9%
Tryptophan	TRP	52.5%
Lactate	LAC	52.5%
Uridine triphosphate	UTP	52.1%
Aspartate	ASP	51.7%
Guanosine monophosphate	GMP	50.8%
Asparagine	ASN	50.4%
Cytidine diphosphate	CDP	50.0%
Phosphoenolpyruvate	PEP	48.3%
3-Phosphoglycerate	3PG	47.9%
2-Phosphoglycerate	2PG	47.1%
Lysine	LYS	43.3%
Threonine	THR	42.0%
Glutamate	GLU	38.7%
Tyrosine	TYR	37.0%
Adenosine triphosphate	ATP	37.0%
D-Xylulose 5-phosphate	XU5P	35.7%
Acetyl coenzyme A	ACCOA	34.9%
Cytidine triphosphate	CTP	33.2%
Serine	SER	31.9%
Isoleucine	ILE	29.8%
Leucine	LEU	26.5%
Pyruvate	PYR	21.8%

Table 5. Metabolites by frequency of appearance in the 57 best simulated annealing constraint sets, those with error at least three orders of magnitude lower than the base synthetic dataset.

Metabolite	Symbol	Frequency
D-Xylulose 5-phosphate	XU5P	100%
sedo-Heptulose 7-phosphate	S7P	100%
D-Ribulose 5-phosphate	RU5P	100%
Ribose 5-phosphate	R5P	100%
Oxaloacetate	OAA	100%
Isocitrate	ICIT	100%
alpha-D-Glucose	GLC	100%
Glucose 6-phosphate	G6P	100%
Glyceraldehyde 3-phosphate	G3P	100%
Fumarate	FUM	100%
Fructose 1,6-diphosphate	FDP	100%
Fructose 6-phosphate	F6P	100%
Erythrose 4-phosphate	E4P	100%
Dihydroxyacetone phosphate	DHAP	100%
Citrate	CIT	100%
Alanine	ALA	100%
6-Phospho-D-glucono-1,5-lactone	6PGL	100%
6-Phospho-D-gluconate	6PGC	100%
2-Dehydro-3-deoxy-D-gluconate 6-phosphate	2DDG6P	100%
Uridine triphosphate	UTP	98.2%
alpha-Ketoglutarate	AKG	98.2%
Succinate	SUCC	94.7%
Arginine	ARG	94.7%
3-Phosphoglycerate	3PG	94.7%
Guanosine triphosphate	GTP	91.2%
Uridine monophosphate	UMP	89.5%
Glutamine	GLN	87.7%
Cytidine monophosphate	CMP	87.7%
Tyrosine	TYR	82.5%
Threonine	THR	82.5%
Aspartate	ASP	80.7%
Cytidine diphosphate	CDP	75.4%
Uridine diphosphate	UDP	71.9%
Valine	VAL	70.2%
Methionine	MET	68.4%
Guanosine monophosphate	GMP	68.4%
Glycine	GLY	68.4%
Adenosine diphosphate	ADP	68.4%
Tryptophan	TRP	61.4%
Phenylalanine	PHE	59.6%
Acetate	AC	57.9%
Malate	MAL	52.6%
Phosphoenolpyruvate	PEP	50.9%
Leucine	LEU	49.1%
Proline	PRO	47.4%
Isoleucine	ILE	47.4%
Histidine	HIS	47.4%
Adenosine monophosphate	AMP	45.6%
Glyoxylate	GLX	43.9%
Lactate	LAC	40.4%
Chloramphenicol acetyltransferase	CAT	38.6%
Cysteine	CYS	36.8%
Asparagine	ASN	35.1%
Pyruvate	PYR	33.3%
Glutamate	GLU	31.6%
Succinyl coenzyme A	SUCCOA	22.8%
Guanosine diphosphate	GDP	22.8%
2-Phosphoglycerate	2PG	22.8%
Cytidine triphosphate	CTP	14%
Acetyl coenzyme A	ACCOA	14%
Lysine	LYS	7%
Adenosine triphosphate	ATP	5.3%
Serine	SER	0%

3. Discussion

In this study, we presented a dynamic constraint-based model of cell-free protein expression. This approach avoids the pseudo-steady-state assumption found in traditional constraint-based approaches, which allowed for the direct integration of metabolite measurements into the flux estimation problem, and the the accumulation or depletion of network metabolites. The approach used the *E. coli* cell-free protein synthesis metabolic network from Vilkhovoy and coworkers [31], and the simulated metabolite trajectories from the kinetic model of Horvath et al. [32] as constraints on the CFPS flux calculation. The dynamic constraint-based model satisfied time-dependent metabolite measurement constraints, predicted unconstrained metabolite concentrations as well as the concentration of a model protein, chloramphenicol acetyltransferase (CAT). Model interrogation suggested the most important metabolite measurement within the dataset to be glucose, as excluding the glucose yielded the greatest metabolite prediction error, and the greatest uncertainty in the estimated metabolic flux. Furthermore, we evaluated metabolite constraint sets with one more and one fewer metabolites than the base case (37 metabolites) to explore the impact of measurement selection on model performance. The single addition of metabolites yielded no significant improvement in the predictive power, while the single exclusion suggested glucose to be the most important measured metabolite in the base case. Next, we selected measurement species based on the results of singular value decomposition on the stoichiometric matrix. The top 36 species from the SVD analysis with the addition of glucose improved the predictive power and reduced flux uncertainty compared with the base case. Finally, we described a heuristic optimization approach to estimate the optimal list of metabolite measurements. Measurement sets determined by heuristic optimization vastly outperformed the accuracy of the base synthetic dataset; model precision, meanwhile, was virtually unchanged despite comprehensive measurement sets. Taken together, model interrogation showed that, even with a comprehensive dataset, there still exists a great amount of uncertainty associated with metabolic fluxes. This highlights the need for fluxomic data to fully understand biological networks.

Despite synthetic datasets consisting of greater than 30 metabolite time series, estimates of metabolic flux were largely uncertain. Flux variability analysis suggested that the metabolite constraints could be met with a wide range of different flux distributions. For instance, an open question in cell-free systems is the balance between glycolytic versus pentose phosphate pathway flux. In previous studies of *E. coli* cell-free protein synthesis, the kinetic model of Horvath and coworkers suggested that glucose was consumed primarily by glycolytic reactions, with minimal flux into the pentose phosphate pathway. However, Vilkhovoy et al. estimated, using sequence specific flux balance analysis with the same experimental dataset, that the CAT production was unaffected by the choice of pentose phosphate pathway versus glycolysis; deletion of either pathway did not change protein productivity. To answer this discrepancy, model analysis showed, during the first phase when glucose was being consumed, glycolytic and pentose phosphate fluxes (pgi and zwf, respectively) exhibited large uncertainty, as either could be utilized to satisfy CAT production. The measurement selection analysis was conducted by excluding or including a metabolite from the constraint set. The exclusion sets were dominated by the removal of glucose, and to a lesser extent the organic acids, suggesting that measurements of central carbon metabolism intermediates were more important than energetic and amino acid measurements. However, the inclusion sets showed no significant effect on error and flux uncertainty. There was generally no correlation between the error and flux uncertainty of a model constrained to a particular metabolite set, except with respect to the outlier glucose. Model calculations showed that, even with a comprehensive data set of 37 metabolite measurements, there was significant flux uncertainty. This suggested that there were many flux combinations that could give rise to the same set of time course measurements. This phenomenon was further supported by analyzing the ensemble of constraint sets determined by heuristic optimization. Although the optimization algorithm reduced the objective function by four degrees of magnitude, the flux variability remained stagnant in comparison. An ensemble of measurement sets ranging from 22 to 48 metabolite constraints was only able to reduce flux uncertainty by 7% from the base synthetic data set. The dynamic constraint-based model showed high flux variability in important branch points, including the glucose-6-phosphate split between glycolysis and pentose phosphate, the 6PGC split into pentose phosphate and Entner–Doudoroff, and the pyruvate split into TCA cycle versus lactate production. This may be why the high overall flux variability was robust to the varying of metabolite constraints. Using three different sampling approaches (single additions/exclusions, singular value decomposition, simulated annealing) coupled with the dynamic constraint-based model, we estimated key metabolites that could be prioritized in measurement selection, such as glucose. Although measuring central carbon metabolites and amino acids is the intuition of most researchers, model interrogation was able to provide the importance of certain species over others; for instance, measuring G6P, G3P, and F6P would be more fruitful than measuring PEP. Interestingly, many of the most valuable measurements were involved in upper glycolysis and pentose phosphate, such as glucose, G6P, and 6PGL. This may be because upstream metabolites have an effect on more of the network; any error or uncertainty in these metabolites will cascade down the rest of the network and magnify throughout. Taken together, the dynamic constraint-based model quantitatively affirmed the robustness of metabolism, and illustrated the complexity of inferring flux information from metabolite concentrations. Ultimately, to determine the metabolic flux distribution occurring in a cell-free system, we need to add additional constraints to the flux estimation calculation. This study suggested that metabolite measurements alone were not sufficient. However, these are not the only experimentally realizable types of constraints. For example, thermodynamic feasibility constraints may result in a better depiction of the flux distribution [37,38], and ^{13}C labeling constraints in cell-free systems could provide significant insight. However, while ^{13}C labeling techniques are well established for in vivo processes [39], application of these techniques to cell-free systems remains an active area of research.

The use of cell-free systems as a personalized point of care biomanufacturing tools, as platforms for vaccine development, or as the basis for portable pathogen detection are promising research directions [1,40–43]. cell-free systems have significant advantages in these application areas compared to in vivo systems. For example, as there is no longer a cell wall, we can experimentally observe the system and intervene if need be. Moreover, cell-free synthetic circuitry is highly portable. For example, it can be dried onto paper, easily transported, and potentially stored indefinitely [44]. Thus, after development and testing of the circuitry, and manufacturing of the devices, there is no need for large, bulky and expensive equipment usually associated with in vivo bioprocesses. However, to move beyond proof of concept and into industrial or medical practice will require extensive optimization. Mathematical modeling is an important tool in this regard. In this study, we explored the relationship between measurement selection and the ability to estimate metabolic flux in a cell-free system. One of the central advantages of cell-free systems is the ability to measure the concentration of metabolic intermediates. However, ultimately, we need to transform these measurements into testable hypothesis about the performance of a cell-free system. The connection between flux estimation and optimal measurement selection has been well studied for in vivo systems; for example, see the classic work of Savinell and Palsson [45,46]. Moreover, the robust quantification of metabolic flux in in vivo systems is also well developed, with both mature experimental and computational tools available (see [39,47,48]). However, quantification of metabolic flux in cell-free systems from metabolite measurements remains an open area of research. Certainly, techniques developed for the identification and quantification of in vivo systems could be applied to cell-free. For example, Lucks and coworkers used the D-optimal experimental design approach of Doyle and coworkers [49] to parameterize a kinetic model of RNA-based cell-free circuits [33]. However, Lucks and coworkers did not consider the resources required for the RNA circuits to function. Quantification of metabolic flux, and the associated resource production and consumption in cell-free applications, is not common. Instead, the synthetic biology community has focused on designing circuits and circuit components with specific behaviors e.g., [50–52], assuming the resources required to express these components will be available. At proof of concept scales, this is a reasonable assumption. However, as we move toward industrial practice, careful attention must be paid to resource generation and consumption—for example, optimizing the expressed proteome for cell-free extract production, or balancing cofactor utilization during the cell-free reaction [53,54]. Resource management has a direct impact on the performance and industrial viability of cell-free applications—for example, potentially limiting the rate of production and yield of circuit components, the size and complexity of possible synthetic circuits, the operational lifetime of synthetic devices, or the titer and rate of production of protein and small molecule products. Thus, as we move beyond technology development to realistic industrial or medical applications, the performance of synthetic devices will become increasingly important. Toward this need, we expect mathematical modeling will play an important role.

4. Materials and Methods

4.1. Formulation and Solution of the Model Equations

We modeled the time evolution of the ith metabolite concentration (x_i), the scaled activity of network enzymes (ϵ_i), transcription processes generating the mRNA m and translation processes generating the protein $\mathcal{P}$ in an *E. coli* cell-free metabolic network as a system of ordinary differential equations:

$$\dot{x}_i = \sum_{j=1}^{\mathcal{R}} \sigma_{ij} r_j\left(\mathbf{x}, \epsilon, \mathbf{k}\right) \qquad i = 1, 2, \ldots, \mathcal{M}, \tag{1}$$

$$\dot{\epsilon}_i = -\lambda_i \epsilon_i \qquad i = 1, 2, \ldots, \mathcal{N}, \tag{2}$$

$$\dot{m} = r_T - \lambda m, \tag{3}$$

$$\dot{\mathcal{P}} = r_X. \tag{4}$$

The quantity $\mathcal{R}$ denotes the number of metabolic reactions, $\mathcal{M}$ denotes the number of metabolites and $\mathcal{N}$ denotes the number of metabolic enzymes in the model. The quantity $r_j\left(\mathbf{x}, \epsilon, \mathbf{k}\right)$ denotes the rate of reaction j. Typically, reaction j is a nonlinear function of metabolite and enzyme abundance, as well as unknown kinetic parameters $\mathbf{k}$ ($\mathcal{K} \times 1$). The quantity σ_{ij} denotes the stoichiometric coefficient for species i in reaction j. If $\sigma_{ij} > 0$, metabolite i is produced by reaction j. Conversely, if $\sigma_{ij} < 0$, metabolite i is consumed by reaction j, while $\sigma_{ij} = 0$ indicates metabolite i is not connected with reaction j. Lastly, λ_i denotes the scaled enzyme activity decay constant. The system material balances were subject to the initial conditions $\mathbf{x}\left(t_o\right) = \mathbf{x}_o$ and $\epsilon\left(t_o\right) = \mathbf{1}$ (initially, we have 100% cell-free enzyme activity).

The cell-free model equations were solved using a dynamic constraint-based approach in which the rates of the metabolic fluxes, transcription and translation processes were estimated by solving an optimization subproblem from t to $t + \Delta t$. In particular, the biochemical fluxes $r_1, r_2, \ldots, r_{\mathcal{R}}$ which appear in the balance equations were calculated from t to $t + \Delta t$ by solving a constrained optimization subproblem with (potentially nonlinear) objective $\mathcal{O}\left(x_1, x_2, \ldots, x_{\mathcal{M}}\right)$:

$$\max_{r_1, r_2, \ldots, r_{\mathcal{R}}} \mathcal{O}\left(x_1, x_2, \ldots, x_{\mathcal{M}}\right) \tag{5}$$

subject to species constraints and flux bounds:

$$\left(\sum_{j=1}^{\mathcal{R}} \sigma_{ij} r_j - \dot{x}_i\right) \geq 0 \qquad i = 1, 2, \ldots, \mathcal{M}, \tag{6}$$

$$0 \leq r_j \leq \mathcal{U}_j\left(x_1, x_2, \ldots, x_{\mathcal{M}}, \kappa\right) \qquad j = 1, 2, \ldots, \mathcal{R}. \tag{7}$$

In this study, we maximized the rate of translation r_X unless otherwise specified. We discretized the derivative term for each species using a constant width h forward different approximation (however, this was done for convenience and more sophisticated techniques could have been used). The reaction bounds $\mathcal{U}_j\left(x_1, x_2, \ldots, x_{\mathcal{M}}, \kappa\right)$ are *potentially* non-linear functions of the system state, and can be updated during each time step. Here, we modeled the upper bound for flux j as $\hat{V}_{max} \epsilon_j(k)$, where $\hat{V}_{max}$ denotes a characteristic maximum reaction velocity, and $\epsilon_j(k)$ denotes the scaled enzyme activity catalyzing reaction j at time step k. The characteristic maximum reaction velocity was set to 600 mM/h (which corresponds to an average $k_{cat} \simeq 1000$ s^{-1} and and enzyme concentration of approximately 0.2 µM) unless otherwise specified. Additional species constraints can be added to directly incorporate metabolomic, proteomic or transcriptomic measurements into the flux calculation. In this study, we incorporated metabolite measurement constraints of the form:

$$\chi_{m,k+1}^{L} \leq x_{m,k+1} \leq \chi_{m,k+1}^{U} \qquad m = 1, 2, \ldots, \Xi, \tag{8}$$

where $\chi_{m,k+1}^{L}$ and $\chi_{m,k+1}^{U}$ denote the lower and upper measurement bound for metabolite m at time step $k+1$, where Ξ metabolites were measured over the time course of the cell-free reaction. Lastly, we imposed a user-configurable bound $\mathcal{B}_i$ on the maximum rate of change for metabolite i:

$$|\dot{x}_i| \leq \mathcal{B}_i \qquad i = 1, 2, \ldots, \mathcal{M} \tag{9}$$

and non-negativity constraints $x_i \geq 0$ for all metabolites and all time steps.

The bounds on the transcription rate ($\mathcal{L}_T = r_T = \mathcal{U}_T$) were modeled as:

$$r_T = V_T^{max} \left(\frac{G_\mathcal{P}}{K_T + G_\mathcal{P}} \right), \tag{10}$$

where $G_\mathcal{P}$ denotes the concentration of the gene encoding the protein of interest, and K_T denotes a transcription saturation coefficient. The maximum transcription rate V_T^{max} was formulated as:

$$V_T^{max} \equiv \left[R_T \left(\frac{\dot{v}_T}{l_G} \right) u\left(\kappa \right) \right], \tag{11}$$

where R_T denotes the RNA polymerase concentration (nM), $\dot{v}_T$ denotes the RNA polymerase elongation rate (nt/h), l_G denotes the gene length (nt). The term $u\left(\kappa \right)$ (dimensionless, $0 \leq u\left(\kappa \right) \leq 1$) is an effective model of promoter activity, where κ denotes promoter specific parameters. The general form for the promoter models was taken from Moon et al. [55], which was based on earlier studies from Bintu and coworkers [56], and similar to the genetically structured modeling approach of Lee and Bailey [57]. In this study, we considered only the T7 promoter model:

$$u_{T7} = \frac{K_{T7}}{1 + K_{T7}}, \tag{12}$$

where K_{T7} denotes a T7 RNA polymerase binding constant. The values for all promoter parameters are given in Table 1.

The translation rate (r_X) was bounded by:

$$0 \leq r_X \leq V_X^{max} \left(\frac{m}{K_X + m} \right), \tag{13}$$

where m denotes the mRNA abundance and K_X denotes a translation saturation constant.

The maximum translation rate V_X^{max} was formulated as:

$$V_X^{max} \equiv \left[K_P R_X \left(\frac{\dot{v}_X}{l_P} \right) \right]. \tag{14}$$

The term K_P denotes the polysome amplification constant, $\dot{v}_X$ denotes the ribosome elongation rate (amino acids per hour), and l_P denotes the number of amino acids in the protein of interest. The mRNA abundance m was estimated as:

$$m_{k+1} = m_k + (r_T - m_k\lambda)h, \tag{15}$$

where λ denotes the mRNA degradation rate constant (h^{-1}). All translation parameters are given in Table 1.

Metabolic fluxes were estimated at each time step using the GNU Linear Programming Kit (GLPK) v4.55 [58]. The objective of the optimization subproblem was to maximize the translation rate, subject to the stoichiometric and metabolite constraints. The model code, parameters and initial conditions used in this study are available under an MIT software license at the Varnerlab website [35]. The model code is written in the Julia programming language [59]. Default transcription and translation parameters are stored in `TXTLDictionary.jl`, while specific simulations are described in the `Solve_*.jl` files. Lastly, the figures for this study were produced using the `Plot_*.jl` scripts.

4.2. Sampling of Transcription and Translation Parameters

The influence of the uncertainty in the transcription (TX) and translation (TL) parameters was estimated by sampling the expected physiological ranges for these parameters as determined from literature. We generated uniform random samples between an upper (u) and lower (l) parameter bound of the form:

$$p^* = l + (u - l) \times \mathcal{U}(0, 1), \tag{16}$$

where $\mathcal{U}(0, 1)$ denotes a uniform random number between 0 and 1. The T7 RNA polymerase concentration was sampled between 800 and 1200 nM, ribosome levels between 1.5 and 3.0 μM, polysome amplification between 5 and 15, the RNA polymerase elongation rate between 20 and 30 nt/s, and the ribosome elongation rate between 1.0 and 3.0 aa/s [10,60]; see `TXTL_ensemble.jl` for a complete list of parameter ranges.

4.3. Generation and Evaluation of Alternative Measurement Sets

The measurement sets consisted of the base (one set of 37 metabolites), inclusion sets (23 sets of 38 metabolites each), exclusion sets (37 sets of 36 metabolites each), SVD-guided (one set of 37 metabolites), and simulated annealing samples (238 sets of varying length). In all cases, we assumed the same sampling frequency as the base synthetic dataset, but we varied which species were measured. The exclusion or inclusion measurement sets were constructed by removing or adding a metabolite to the base set, while the SVD-guided measurement set was constructed from high importance metabolites; the top 36 metabolites (plus glucose) that had the greatest singular value weighted sum across the SVD-modes, accounting for 95% of the network structure, were designated the SVD measurement set. Lastly, we used simulated annealing to generate potentially optimal measurements sets, where the objective was to minimize the product of the prediction error, and flux uncertainty. The prediction error, $\mathcal{E}$, was computed by comparing the simulated versus the measured value of a metabolite, for a $\mathcal{M}_{\text{core}}$ set of metabolites. On the other hand, the flux variability was computed using flux variability analysis (FVA) [61], subject to constraints on the CAT production rate, and the selected metabolite trajectories. In particular, the metabolite prediction error was calculated from the time-dependent state array:

$$\mathcal{E} = \sum_{i=1}^{\mathcal{M}_{\text{core}}} \sum_{t=t_i}^{\mathcal{T}} \left(\max\left(x_i(t) - y_i^U(t), 0\right) + \max\left(y_i^L(t) - x_i(t), 0\right) \right),$$

where $x_i(t)$ denotes the simulated value of metabolite i at time t, $y_i^U(t)$ denotes the upper bound of the 95% confidence interval on the synthetic data for metabolite i at time t, $y_i^L(t)$ denotes the lower bound of the 95% confidence interval on the synthetic data for metabolite i at time t, and $\mathcal{M}_{core}$ denotes the subset of metabolites in the core metabolism. For this calculation, the entire time course was considered ($t_i = 0$ h, $\mathcal{T} = 3$ h). The flux uncertainty was calculated from the maximal and minimal flux arrays:

$$\sigma_{overall} = \sum_{r_j \in \mathcal{R}_{core}} \sum_{t=t_i}^{\mathcal{T}} \left(r_j^{max}(t) - r_j^{min}(t) \right)^2,$$

where $r_j^{max}(t)$ denotes the maximum value of flux j, while $r_j^{min}(t)$ denotes the value of flux j at time t, calculated using flux variability analysis. The quantity $\mathcal{R}_{core}$ denotes the subset of reactions that constitute the core metabolism. For the flux uncertainty calculations, either the entire reaction time course was considered ($t_i = 0$ h, $\mathcal{T} = 3$ h), or the uncertainty was calculated separately for each phase (phase 1: $t_i = 0$ h, $\mathcal{T} = 1$ h; phase 2: $t_i = 1$ h, $\mathcal{T} = 3$ h).

The simulated annealing algorithm began by evaluating the error and flux uncertainty of the base case and multiplying these to obtain a cost function:

$$\text{cost} = \mathcal{E} \cdot \sigma_{overall}. \tag{17}$$

Then, each metabolite that was considered measurable was added to or removed from the constraint set with a certain probability p_{switch}:

$$\theta_i^{new} = \begin{cases} 1 - \theta_i, & \mathcal{U}(0,1) < p_{switch}, \\ \theta_i, & \mathcal{U}(0,1) > p_{switch}, \end{cases} \qquad i = 1, 2, \ldots, \mathcal{M}_{measurable}, \tag{18}$$

where $\theta_i \in \{0,1\}$ denotes a binary parameter encoding whether or not metabolite i is in the constraint set, $\mathcal{U}(0,1)$ denotes a uniform random number taken from a distribution between 0 and 1, and $\mathcal{M}_{measurable}$ denotes the set of metabolites deemed to be measurable. For each newly generated constraint set, we re-solved the dFBA and FVA problems, and re-calculated the cost function. All sets with a lower cost were accepted into the ensemble. Sets with a higher cost were also accepted into the ensemble, if they satisfied the acceptance constraint:

$$\mathcal{R}_{0,1}^{uniform} < exp\left(-\alpha \cdot \frac{\text{cost}_{new} - \text{cost}}{\text{cost}} \right), \tag{19}$$

where $\mathcal{R}_{0,1}^{uniform}$ denotes a random number taken from a uniform distribution between 0 and 1, cost denotes the cost of the current parameter set, cost_{new} denotes the cost of the new parameter set, and α denotes an adjustable parameter to control the tolerance to high-error sets. A total of 238 samples were accepted into the ensemble, of which there were 219 unique sets. Both $\mathcal{M}_{core}$ and $\mathcal{R}_{core}$ and user-configurable are defined in the model code repository available from the Varnerlab website [35].

5. Conclusions

In summary, we used a dynamic constraint-based modeling approach to simulate cell-free metabolism, and to study how measurement selection impacts model performance. We extended sequence specific flux balance analysis, by removing the pseudo steady state assumption, and adding synthetic metabolite measurement constraints to the flux calculation. Using this method, we simulated the cell-free synthesis of a model protein, chloramphenicol acetyltransferase, we identified the most important measured species in the cell-free system, and additional species that yielded the lowest metabolite prediction error and flux uncertainty. Only synthetic metabolite measurements were used in this study; however, this work built a foundation to rationally design experimental measurement protocols that could be implemented with a variety of analytical techniques. Taken together, these findings represent a novel tool for dynamic cell-free simulations, measurement selection and pathway analysis, not only for *E. coli*, but potentially for align variety of metabolic networks, whether in vivo or cell-free. However, while this first study was promising, there were several issues to consider in future work. First, while we described transcription and translation at a sequence specific level, we have not considered the complexities of protein folding, or post-translational modifications such as protein glycosylation. A more detailed description of transcription and translation reactions, including the role of chaperones in protein folding, has been used in in-vivo genome scale ME models; e.g., see O'Brien et al. [20]. These template reactions could easily be adapted to a cell-free system, thereby providing a potentially higher fidelity description protein synthesis and folding. Next, the inclusion of post-translational modifications such as protein glycosylation in the next generation of models will be important to describe the cell-free synthesis of therapeutic proteins. DeLisa and coworkers recently showed that glycoproteins can be synthesized in a cell-free system, using extract generated from modified *E. coli* cells capable of asparagine-linked protein glycosylation [62]. Simulation of the generation and attachment of glycans to protein targets could be an important step to optimizing cell-free glycoprotein production. Lastly, while we modeled the cell-free production of a only single protein in this study, sequence specific dynamic constraint models could be developed for multi-protein synthetic circuits, RNA circuits or even small molecule production. Thus, this approach offers a unique tool to model and potentially optimize a wide variety of application areas in synthetic biology.

Author Contributions: J.V. directed the modeling study. D.D., N.H. and J.V. developed the cell-free protein synthesis mathematical model, and parameter ensemble. Simulations were conducted by D.D. and N.H. The manuscript was prepared and edited for publication by D.D., N.H., and J.V.

Funding: This study was supported by a National Science Foundation Graduate Research Fellowship (DGE-1333468) to N.H. Research reported in this publication was also supported by the Systems Biology Coagulopathy of Trauma Program with support from the US Army Medical Research and Materiel Command under award number W911NF-10-1-0376. Lastly, this work was also supported by the Center on the Physics of Cancer Metabolism through Award Number 1U54CA210184-01 from the National Cancer Institute. The content is solely the responsibility of the authors and does not necessarily represent the official views of the National Cancer Institute or the National Institutes of Health.

Conflicts of Interest: The authors declare no conflict of interest. The funding sponsors had no role in the design of the study; in the collection, analyses, or interpretation of data; in the writing of the manuscript, and in the decision to publish the results.

References

1. Pardee, K.; Slomovic, S.; Nguyen, P.Q.; Lee, J.W.; Donghia, N.; Burrill, D.; Ferrante, T.; McSorley, F.R.; Furuta, Y.; Vernet, A.; et al. Portable, On-Demand Biomolecular Manufacturing. *Cell* **2016**, *167*, 248–259. [CrossRef] [PubMed]
2. Jewett, M.C.; Calhoun, K.A.; Voloshin, A.; Wuu, J.J.; Swartz, J.R. An integrated cell free metabolic platform for protein production and synthetic biology. *Mol. Syst. Biol.* **2008**, *4*, 220. [CrossRef] [PubMed]
3. Matthaei, J.H.; Nirenberg, M.W. Characteristics and stabilization of DNAase-sensitive protein synthesis in *E. coli* extracts. *Proc. Natl. Acad. Sci. USA* **1961**, *47*, 1580–1588. [CrossRef] [PubMed]
4. Nirenberg, M.W.; Matthaei, J.H. The dependence of cell-free protein synthesis in *E. coli* upon naturally occurring or synthetic polyribonucleotides. *Proc. Natl. Acad. Sci. USA* **1961**, *47*, 1588–1602. [CrossRef] [PubMed]
5. Spirin, A.; Baranov, V.; Ryabova, L.; Ovodov, S.; Alakhov, Y. A continuous cell-free translation system capable of producing polypeptides in high yield. *Science* **1988**, *242*, 1162–1164. [CrossRef] [PubMed]
6. Kim, D.M.; Swartz, J.R. Regeneration of adenosine triphosphate from glycolytic intermediates for cell-free protein synthesis. *Biotechnol. Bioeng.* **2001**, *74*, 309–316. [CrossRef] [PubMed]
7. Lu, Y.; Welsh, J.P.; Swartz, J.R. Production and stabilization of the trimeric influenza hemagglutinin stem domain for potentially broadly protective influenza vaccines. *Proc. Natl. Acad. Sci. USA* **2014**, *111*, 125–130. [CrossRef] [PubMed]

8. Hodgman, C.E.; Jewett, M.C. Cell-free synthetic biology: thinking outside the cell. *Metab. Eng.* **2012**, *14*, 261–219. [CrossRef] [PubMed]

9. Jewett, M.C.; Swartz, J.R. Mimicking the *Escherichia coli* cytoplasmic environment activates long-lived and efficient cell-free protein synthesis. *Biotechnol. Bioeng.* **2004**, *86*, 19–26. [CrossRef] [PubMed]

10. Garamella, J.; Marshall, R.; Rustad, M.; Noireaux, V. The All *E. coli* TX-TL Toolbox 2.0: A Platform for Cell-Free Synthetic Biology. *ACS Synth. Biol.* **2016**, *5*, 344–355. [CrossRef] [PubMed]

11. Lewis, N.E.; Nagarajan, H.; Palsson, B.Ø. Constraining the metabolic genotype-phenotype relationship using a phylogeny of in silico methods. *Nat. Rev. Microbiol.* **2012**, *10*, 291–305. [CrossRef] [PubMed]

12. Edwards, J.S.; Palsson, B.Ø. The *Escherichia coli* MG1655 in silico metabolic genotype: Its definition, characteristics, and capabilities. *Proc. Natl. Acad. Sci. USA* **2000**, *97*, 5528–5533. [CrossRef] [PubMed]

13. Feist, A.M.; Henry, C.S.; Reed, J.L.; Krummenacker, M.; Joyce, A.R.; Karp, P.D.; Broadbelt, L.J.; Hatzimanikatis, V.; Palsson, B.Ø. A genome-scale metabolic reconstruction for *Escherichia coli* K-12 MG1655 that accounts for 1260 ORFs and thermodynamic information. *Mol. Syst. Biol.* **2007**, *3*, 121. [CrossRef] [PubMed]

14. Oh, Y.K.; Palsson, B.Ø.; Park, S.M.; Schilling, C.H.; Mahadevan, R. Genome-scale reconstruction of metabolic network in Bacillus subtilis based on high-throughput phenotyping and gene essentiality data. *J. Biol. Chem.* **2007**, *282*, 28791–28799. [CrossRef] [PubMed]

15. Feist, A.M.; Herrgård, M.J.; Thiele, I.; Reed, J.L.; Palsson, B.Ø. Reconstruction of biochemical networks in microorganisms. *Nat. Rev. Microbiol.* **2009**, *7*, 129–143. [CrossRef] [PubMed]

16. Ibarra, R.U.; Edwards, J.S.; Palsson, B.Ø. *Escherichia coli* K-12 undergoes adaptive evolution to achieve in silico predicted optimal growth. *Nature* **2002**, *420*, 186–189. [CrossRef] [PubMed]

17. Schuetz, R.; Kuepfer, L.; Sauer, U. Systematic evaluation of objective functions for predicting intracellular fluxes in *Escherichia coli*. *Mol. Syst. Biol.* **2007**, *3*, 119. [CrossRef] [PubMed]

18. Hyduke, D.R.; Lewis, N.E.; Palsson, B.Ø. Analysis of omics data with genome-scale models of metabolism. *Mol. Biosyst.* **2013**, *9*, 167–174. [CrossRef] [PubMed]

19. Allen, T.E.; Palsson, B.Ø. Sequence-based analysis of metabolic demands for protein synthesis in prokaryotes. *J. Theor. Biol.* **2003**, *220*, 1–18. [CrossRef] [PubMed]

20. O'Brien, E.J.; Lerman, J.A.; Chang, R.L.; Hyduke, D.R.; Palsson, B.Ø. Genome-scale models of metabolism and gene expression extend and refine growth phenotype prediction. *Mol. Sys. Biol.* **2013**, *9*, 693. [CrossRef]

21. Mahadevan, R.; Edwards, J.S.; Doyle, F.J., 3rd. Dynamic flux balance analysis of diauxic growth in *Escherichia coli*. *Biophys. J.* **2002**, *83*, 1331–1340. [CrossRef]

22. Hjersted, J.L.; Henson, M.A.; Mahadevan, R. Genome-scale analysis of Saccharomyces cerevisiae metabolism and ethanol production in fed-batch culture. *Biotechnol. Bioeng.* **2007**, *97*, 1190–1204. [CrossRef] [PubMed]

23. Hjersted, J.L.; Henson, M.A. Steady-state and dynamic flux balance analysis of ethanol production by Saccharomyces cerevisiae. *IET Syst. Biol.* **2009**, *3*, 167–179. [CrossRef] [PubMed]

24. Hanly, T.J.; Henson, M.A. Dynamic flux balance modeling of microbial co-cultures for efficient batch fermentation of glucose and xylose mixtures. *Biotechnol. Bioeng.* **2011**, *108*, 376–385. [CrossRef] [PubMed]

25. Henson, M.A.; Hanly, T.J. Dynamic flux balance analysis for synthetic microbial communities. *IET Syst. Biol.* **2014**, *8*, 214–229. [CrossRef] [PubMed]

26. Höffner, K.; Harwood, S.M.; Barton, P.I. A reliable simulator for dynamic flux balance analysis. *Biotechnol. Bioeng.* **2013**, *110*, 792–802. [CrossRef] [PubMed]

27. Gomez, J.A.; Höffner, K.; Barton, P.I. DFBAlab: A fast and reliable MATLAB code for dynamic flux balance analysis. *BMC Bioinform.* **2014**, *15*, 409. [CrossRef] [PubMed]

28. Gomez, J.A.; Barton, P.I. Dynamic Flux Balance Analysis Using DFBAlab. *Methods Mol. Biol.* **2018**, *1716*, 353–370. [CrossRef] [PubMed]

29. McCloskey, D.; Palsson, B.Ø.; Feist, A.M. Basic and applied uses of genome-scale metabolic network reconstructions of *Escherichia coli*. *Mol. Syst. Biol.* **2013**, *9*, 661. [CrossRef] [PubMed]

30. Zomorrodi, A.R.; Suthers, P.F.; Ranganathan, S.; Maranas, C.D. Mathematical optimization applications in metabolic networks. *Metab. Eng.* **2012**, *14*, 672–686. [CrossRef] [PubMed]

31. Vilkhovoy, M.; Horvath, N.; Shih, C.H.; Wayman, J.; Calhoun, K.; Swartz, J.; Varner, J. Sequence Specific Modeling of *E. coli* Cell-Free Protein Synthesis. *bioRxiv* **2017**. [CrossRef]

32. Horvath, N.; Vilkhovoy, M.; Wayman, J.A.; Calhoun, K.; Swartz, J.; Varner, J. Toward a Genome Scale Sequence Specific Dynamic Model of Cell-Free Protein Synthesis in *Escherichia coli*. *bioRxiv* **2017**. [CrossRef]

33. Hu, C.Y.; Varner, J.D.; Lucks, J.B. Generating Effective Models and Parameters for RNA Genetic Circuits. *ACS Synth. Biol.* **2015**, *4*, 914–926. [CrossRef] [PubMed]

34. Jewett, M.; Voloshin, A.; Swartz, J., Prokaryotic systems for in vitro expression. In *Gene Cloning and Expression Technologies*; Weiner, M., Lu, Q., Eds.; Eaton Publishing: Westborough, MA, USA, 2002; pp. 391–411.

35. Varnerlab. Available online: http://www.varnerlab.org/downloads/ (accessed on 16 August 2018).

36. Kigawa, T.; Muto, Y.; Yokoyama, S. Cell-free synthesis and amino acid-selective stable isotope labeling of proteins for NMR analysis. *J. Biomol. NMR* **1995**, *6*, 129–134. [CrossRef] [PubMed]

37. Henry, C.S.; Broadbelt, L.J.; Hatzimanikatis, V. Thermodynamics-Based Metabolic Flux Analysis. *Biophys. J.* **2006**, *92*, 1792–1805. [CrossRef] [PubMed]

38. Hamilton, J.J.; Dwivedi, V.; Reed, J.L. Quantitative Assessment of Thermodynamic Constraints on the Solution Space of Genome-Scale Metabolic Models. *Biophys. J.* **2013**, *105*, 512–522. [CrossRef] [PubMed]

39. Zamboni, N.; Fendt, S.M.; Sauer, U. ^{13}C-based metabolic flux analysis. *Nat. Protoc.* **2009**, *4*, 878–892. [CrossRef] [PubMed]

40. Slomovic, S.; Pardee, K.; Collins, J.J. Synthetic biology devices for in vitro and in vivo diagnostics. *Proc. Natl. Acad. Sci. USA* **2015**, *112*, 14429–14435. [CrossRef] [PubMed]

41. Andries, O.; Kitada, T.; Bodner, K.; Sanders, N.N.; Weiss, R. Synthetic biology devices and circuits for RNA-based 'smart vaccines': A propositional review. *Expert Rev. Vaccines* **2015**, *14*, 313–331. [CrossRef] [PubMed]

42. Rustad, M.; Eastlund, A.; Marshall, R.; Jardine, P.; Noireaux, V. Synthesis of Infectious Bacteriophages in an *E. coli*-based Cell-free Expression System. *J. Vis. Exp.* **2017**. [CrossRef] [PubMed]

43. Moore, S.J.; MacDonald, J.T.; Freemont, P.S. Cell-free synthetic biology for in vitro prototype engineering. *Biochem. Soc. Trans.* **2017**, *45*, 785–791. [CrossRef] [PubMed]

44. Pardee, K.; Green, A.A.; Ferrante, T.; Cameron, D.E.; DaleyKeyser, A.; Yin, P.; Collins, J.J. Paper-based synthetic gene networks. *Cell* **2014**, *159*, 940–954. [CrossRef] [PubMed]

45. Savinell, J.M.; Palsson, B.O. Optimal selection of metabolic fluxes for in vivo measurement. I. Development of mathematical methods. *J. Theor. Biol.* **1992**, *155*, 201–214. [CrossRef]

46. Savinell, J.M.; Palsson, B.O. Optimal selection of metabolic fluxes for in vivo measurement. II. Application to *Escherichia coli* and hybridoma cell metabolism. *J. Theor. Biol.* **1992**, *155*, 215–242. [CrossRef]

47. Link, H.; Kochanowski, K.; Sauer, U. Systematic identification of allosteric protein-metabolite interactions that control enzyme activity in vivo. *Nat. Biotechnol.* **2013**, *31*, 357–361. [CrossRef] [PubMed]

48. Buescher, J.M.; Antoniewicz, M.R.; Boros, L.G.; Burgess, S.C.; Brunengraber, H.; Clish, C.B.; DeBerardinis, R.J.; Feron, O.; Frezza, C.; Ghesquiere, B.; et al. A roadmap for interpreting (13)C metabolite labeling patterns from cells. *Curr. Opin. Biotechnol.* **2015**, *34*, 189–201. [CrossRef] [PubMed]

49. Gadkar, K.G.; Varner, J.; Doyle, F.J., III. Model identification of signal transduction networks from data using a state regulator problem. *Syst. Biol.* **2005**, *2*, 17–30. [CrossRef]

50. Mutalik, V.K.; Qi, L.; Guimaraes, J.C.; Lucks, J.B.; Arkin, A.P. Rationally designed families of orthogonal RNA regulators of translation. *Nat. Chem. Biol.* **2012**, *8*, 447–454. [CrossRef] [PubMed]

51. Westbrook, A.M.; Lucks, J.B. Achieving large dynamic range control of gene expression with a compact RNA transcription-translation regulator. *Nucleic Acids Res.* **2017**, *45*, 5614–5624. [CrossRef] [PubMed]

52. Hu, C.Y.; Takahashi, M.K.; Zhang, Y.; Lucks, J.B. Engineering a Functional Small RNA Negative Autoregulation Network with Model-Guided Design. *ACS Synth. Biol.* **2018**, *7*, 1507–1518. [CrossRef] [PubMed]

53. Chen, X.; Li, S.; Liu, L. Engineering redox balance through cofactor systems. *Trends Biotechnol.* **2014**, *32*, 337–343. [CrossRef] [PubMed]

54. Yang, L.; Yurkovich, J.T.; Lloyd, C.J.; Ebrahim, A.; Saunders, M.A.; Palsson, B.O. Principles of proteome allocation are revealed using proteomic data and genome-scale models. *Sci. Rep.* **2016**, *6*, 36734. [CrossRef] [PubMed]

55. Moon, T.S.; Lou, C.; Tamsir, A.; Stanton, B.C.; Voigt, C.A. Genetic programs constructed from layered logic gates in single cells. *Nature* **2012**, *491*, 249–253. [CrossRef] [PubMed]

56. Bintu, L.; Buchler, N.E.; Garcia, H.G.; Gerland, U.; Hwa, T.; Kondev, J.; Phillips, R. Transcriptional regulation by the numbers: Models. *Curr. Opin. Genet. Dev.* **2005**, *15*, 116–124. [CrossRef] [PubMed]

57. Lee, S.B.; Bailey, J.E. Genetically structured models forlac promoter–operator function in the *Escherichia coli* chromosome and in multicopy plasmids: Lac operator function. *Biotechnol. Bioeng.* **1984**, *26*, 1372–1382. [CrossRef] [PubMed]

58. *GNU Linear Programming Kit (GLPK)*, Version 4.52. GNU Project, 2016; Available online: https://www.gnu. org/software/glpk/ (accessed on 16 August 2018).

59. Bezanson, J.; Edelman, A.; Karpinski, S.; Shah, V.B. Julia: A Fresh Approach to Numerical Computing. *SIAM Rev.* **2017**, *59*, 65–98. [CrossRef]

60. Underwood, K.A.; Swartz, J.R.; Puglisi, J.D. Quantitative polysome analysis identifies limitations in bacterial cell-free protein synthesis. *Biotechnol. Bioeng.* **2005**, *91*, 425–435. [CrossRef] [PubMed]

61. Müller, A.C.; Bockmayr, A. Fast thermodynamically constrained flux variability analysis. *Bioinformatics* **2013**, *29*, 903–909. [CrossRef] [PubMed]

62. Jaroentomeechai, T.; Stark, J.C.; Natarajan, A.; Glasscock, C.J.; Yates, L.E.; Hsu, K.J.; Mrksich, M.; Jewett, M.C.; DeLisa, M.P. Single-pot glycoprotein biosynthesis using a cell-free transcription-translation system enriched with glycosylation machinery. *Nat. Commun.* **2018**, *9*, 2686. [CrossRef] [PubMed]

Article

Effect of the Length-to-Width Aspect Ratio of a Cuboid Packed-Bed Device on Efficiency of Chromatographic Separation

Guoqiang Chen and Raja Ghosh *

Department of Chemical Engineering, McMaster University, 1280 Main Street West,
Hamilton, ON L8S 4L7, Canada; cheng38@mcmaster.ca
* Correspondence: rghosh@mcmaster.ca; Tel.: +1-905-525-9140 (ext. 27415)

Received: 22 August 2018; Accepted: 4 September 2018; Published: 6 September 2018

Abstract: In recent papers we have discussed the use of cuboid packed-bed devices as alternative to columns for chromatographic separations. These devices address some of the major flow distribution challenges faced by preparative columns used for process-scale purification of biologicals. Our previous studies showed that significant improvements in separation metrics such as the number of theoretical plates, peak shape, and peak resolution in multi-protein separation could be achieved. However, the length-to-width aspect ratio of a cuboid packed-bed device could potentially affect its performance. A systematic comparison of six cuboid packed-bed devices having different length-to-width aspect ratios showed that it had a significant effect on separation performance. The number of theoretical plates per meter in the best-performing cuboid packed-bed device was about 4.5 times higher than that in its equivalent commercial column. On the other hand, the corresponding number in the worst-performing cuboid-packed bed was lower than that in the column. A head-to-head comparison of the best-performing cuboid packed bed and its equivalent column was carried out. Performance metrics compared included the widths and dispersion indices of flow-through and eluted protein peaks. The optimized cuboid packed-bed device significantly outperformed its equivalent column with regards to all these attributes.

Keywords: chromatography; chromatography box; cuboid packed-bed; bioseparation; protein; separation efficiency

1. Introduction

Column chromatography is widely used for both analytical and preparative purification of biopharmaceuticals [1–4]. Analytical columns, which have very large bed-height to diameter ratios, and are packed with fine chromatography media (in the 3–10 micron range) give excellent resolution in multi-protein separation [3,5]. On the other hand, preparative columns, especially those used for large-scale bioseparation (which have small bed-height to diameter ratios) and are packed with larger resin particles (>30 microns) do not [3,6]. While these columns can be loaded with large volumes of feed material, their separation efficiencies are drastically lower than analytical columns.

In an analytical column, flow distribution in the radial direction is relatively insignificant compared to that in the axial direction. Also, the volume of sample injected in an analytical column is very small, and therefore solutes move through the column in a "plug-like" manner. When liquid flows into a preparative column, it is distributed in radial and axial directions as shown in Figure 1 (on the left). In large-scale preparative columns, the diameter is frequently comparable, or even larger than the bed height. For example, in the 20,000 L scale purification process for monoclonal antibodies, the typical column diameter is 100 cm and the typical bed height is 20 cm [7]. Therefore, efficient radial flow distribution in the column headers is critically important. Column flow maldistribution and consequent poor separation efficiency has been investigated by many researchers [8–13]. Poor separation has been mainly attributed to dispersion in peripherals, non-uniform packing and maldistribution in headers [8,9]. In addition, radial temperature gradients in the large-diameter columns could affect local superficial velocity and thereby retention parameters [14]. Non-uniformity of flow within preparative columns results in peak broadening and poor resolution, which decrease recovery, increase operating time and buffer usage, and necessitate subsequent concentration steps [8–14].

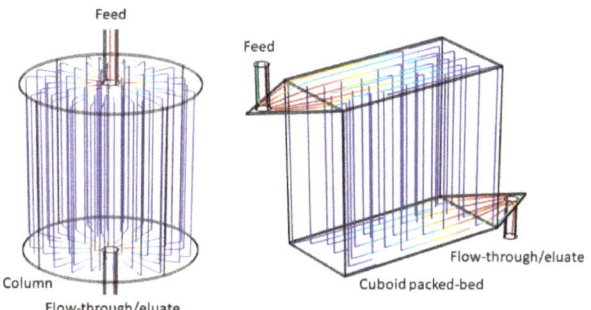

Figure 1. Idealized flow paths within a column and a cuboid packed-bed device.

Researchers have tried to address the problem of column flow maldistribution primarily by modifying and improving the headers [15–23]. This includes the use of ribbed, hyperbolic- and parabolic-shaped headers, the use of collimators and manifolds, the use of bifurcating, radially interconnecting and fractal flow distributors, and the use of frits with appropriate porosity. Other approaches include control of column packing [24–26], adjustment of the column inlet temperature [27], and the use of parallel flow [28], curtain flow [29], and radial flow [30] columns.

In recent papers, we have proposed a radically different approach i.e., the use of box-shaped or cuboid packed-bed devices for reducing flow maldistribution [31,32]. The design of the cuboid packed bed is inspired by that of lateral-fed membrane chromatography (LFMC) devices [33–38]. The cuboid packed-bed device (see Figure 1, on the right) consists of a box-shaped resin-bed into which liquid is distributed using an upper lateral channel, and from which liquid is collected using a lower lateral channel [31,32]. This flow arrangement makes the flow path lengths within the packed bed uniform and narrows down the solute residence time distribution (RTD). Using cuboid packed-bed devices containing different types of ion-exchange media, we were able to demonstrate their superior separation performances relative to their equivalent columns (i.e., containing the same media and having the same bed-height and bed-volume) [31,32]. The cuboid packed-bed devices consistently gave sharper and more symmetrical flow-through and eluted peaks, and higher resolution in multi-protein separations. The length-to-width aspect ratios of a cuboid packed-bed device of a given bed height and bed volume could be adjusted in a flexible manner. That brought us to a pertinent question, i.e., was this ratio important? Within a longer device, the velocity gradient in the lateral channels could be expected to be greater. In our earlier study [32], we hypothesized that the greater the velocity gradient, the poorer the separation. On the other hand, increasing the width of the device and thereby the width of the lateral channels could potentially introduce some non-uniformity of flow along the channel width and thereby affect separation efficiency [32].

In this paper, we have attempted to optimize the performance of a 5 mL cuboid packed-bed device by systematically examining 6 different cuboid devices having identical bed height and bed volume, but different length-to-width ratios. The performance of these devices was compared based on the number of theoretical plates per meter bed height. A head-to-head comparison of the best-performing cuboid packed-bed device and its equivalent column was also carried out. Performance metrics such as the widths, and dispersion indices of flow-through and eluted protein peaks were compared.

2. Materials and Methods

Trizma base (T1503), trizma hydrochloride (T3253), sodium hydroxide (795429), hydrochloric acid (258148), bovine serum albumin (BSA, isoelectric point 4.8, A7906), lysozyme (isoelectric point 11, L6876) were purchased from Sigma-Aldrich (St. Louis, MO, USA). Sodium chloride (SOD002.205) was purchased from Bioshop (Burlington, ON, Canada). Strong anion exchange Capto Q medium (17-5316-03) and HiTrap Capto Q column (5 mL bed volume, 16 mm diameter, 25 mm bed height, 11-0013-03) were purchased from GE Healthcare Biosciences, QC, Canada. All the buffers and the solutions were prepared using water obtained from a SIMPLICITY 185 water purification unit Millipore (Molsheim, France). Prior to use, buffers and solutions were filtered through a 0.1 μm VVLP membrane (Millipore, MA, USA) and degassed.

The different 5 mL cuboid packed-bed devices (see Figure 2), which had the same bed height and cross-sectional area were fabricated in-house (material: acrylic for the transparent top and bottom plates, polyvinyl chloride for the central frame). The dimensions were as follows: 14.14 mm (length) × 14.14 mm (width) × 25 mm (height), 20 mm × 10 mm × 25 mm, 25 mm × 8 mm × 25 mm, 33.3 mm × 6 mm × 25 mm, 50 mm × 4 mm × 25 mm, and 80 mm × 2.5 mm × 25 mm. The basic design of a 5 mL cuboid packed-bed device has been described in detail in our previous paper [32]. Briefly, it consists of 3 parts: an upper and a lower plate with engraved lateral channel for flow distribution/collection, and a central frame with a rectangular slot for housing the cuboid packed bed. A photograph of the assembled 20 mm × 10 mm × 25 mm device is also shown in Figure 2 (inset). Figure 2 also shows the bed dimensions of the equivalent 5 mL HiTrap column (material: polypropylene).

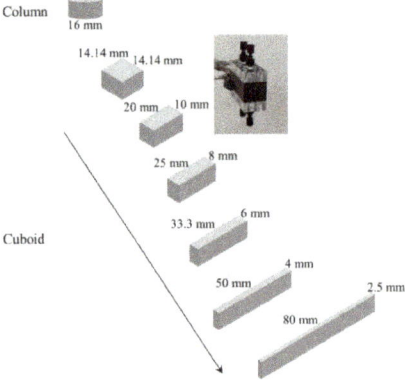

Figure 2. Dimensions of the different cuboid packed-bed devices and their equivalent column. Inset photograph: assembled 20 mm × 10 mm × 25 mm device.

The cuboid packed-bed devices were packed with the Capto Q media using the protocol described in our previous paper [32]. The chromatography experiments were carried out using an AKTA prime system (GE Healthcare Biosciences, Baie d'Urfe, QC, Canada). The number of theoretical plates in the cuboid packed-bed devices and the column was determined at different flow rates by injecting 100 μL 0.8 M NaCl solution as tracer (2% of the bed volume), using 0.4 M NaCl solution as the mobile phase. The best performing cuboid packed-bed device was selected based on the number of theoretical plates. This device was then compared head-to-head with the equivalent commercial column in terms of widths, asymmetry and tailing factors, and dispersion indices of flow through and eluted protein peaks obtained at different flow rates, using different sample injection loops. In these experiments 50 mM Tris-HCl buffer, pH 8 was used as the binding buffer and 50 mM Tris-HCl buffer + 1 M NaCl, pH 8 was used as the eluting buffer. Flow-through peaks were obtained by injecting different volumes of 1 mg/mL lysozyme solution, while eluted peaks were obtained using 2 mg/mL BSA solution.

All experiments with the cuboid devices and their equivalent column were run in duplicate and average values are reported.

3. Results and Discussion

The data for the number of theoretical plates (N), tailing factor (TF) and asymmetry factor (AF) obtained with the six cuboid packed-bed devices and their equivalent column are summarized in Table 1. N was calculated as follows [39]:

$$N = 5.545 \left(\frac{V_R}{w_{0.5}} \right)^2 \tag{1}$$

where V_R is the residence volume, and $w_{0.5}$ is the peak width at half height.

TF was calculated as follows [39]:

$$TF = \left(\frac{w_{0.05}^L + w_{0.05}^R}{2 w_{0.05}^L} \right) \tag{2}$$

where $w_{0.05}^L$ is the front peak width at 5% height, $w_{0.05}^R$ is the rear peak width, also at 5% height.

AF was calculated as follows [39]:

$$AF = \left(\frac{w_{0.1}^R}{w_{0.1}^L} \right) \tag{3}$$

where $w_{0.1}^L$ is the front peak width at 10% height, while $w_{0.1}^R$ is the rear peak width at 10% height.

Table 1. The number of theoretical plates and attributes of salt peaks obtained with the different cuboid packed-bed devices and their equivalent column at different flow rates (media: Capto Q anion exchange, bed volume: 5 mL, mobile phase: 0.4 M NaCl, tracer: 0.8 M NaCl, loop: 0.1 mL, v_S = 15, 60, 150 and 300 cm/h at flow rates of 0.5, 2, 5 and 10 mL/min respectively).

Efficiency Metrics	Flow Rate (mL/min)	Column	Cuboid—Length × Width (mm × mm)					
			14.14 × 14.14	20 × 10	25 × 8	33.3 × 6	50 × 4	80 × 2.5
Aspect ratio			1:1	2:1	25:8	33.3:6	25:2	32:1
Number of theoretical plates per meter, $\bar{N}_S$ (/m)	0.5	2296	8923	10,327	7381	6465	6886	710
	2	2251	7106	7690	6347	5663	5575	633
	5	1968	4403	4620	3867	3684	3402	612
	10	1781	2915	3000	2705	2450	2237	602
Tailing factor	0.5	1.61	1.04	0.92	0.96	0.92	0.95	1.03
	2	1.54	1.05	0.95	0.96	0.93	0.99	0.99
	5	1.29	1.07	0.99	1.00	0.97	0.99	1.00
	10	1.18	1.07	1.03	1.03	0.98	1.01	1.03
Asymmetry factor	0.5	2.15	1.02	0.84	0.87	0.89	0.89	1.02
	2	2.07	1.04	0.93	0.88	0.90	0.94	0.96
	5	1.57	1.09	0.99	0.98	0.95	0.97	0.97
	10	1.35	1.08	1.06	1.00	1.00	1.03	1.03

Five out of the six cuboid packed-bed devices out-performed the column in terms of the number of theoretical plates and other peak attributes. Only the performance of the 80 mm × 2.5 mm cuboid packed-bed device was worse than that of the column. With all the chromatography devices examined, i.e., the column and the six cuboid packed beds, the number of theoretical plates per meter ($\bar{N}$) decreased when the flow rate was increased above the optimum flow rate, as expected based on the van Deemter equation. With the column, the number of theoretical plates increased slightly from 1781/m (at 10 mL/min, v_S = 300 cm/h) to 2296/m (at 0.5 mL/min, v_S = 15 cm/h). With the best performing cuboid packed-bed device (i.e., 20 mm × 10 mm × 25 mm) the number increased very significantly from 3000/m at a flow rate of 10 mL/min (v_S = 300 cm/h) to 10327/m at a flow rate of 0.5 mL/min (v_S = 15 cm/h). However, the number of plates did not increase further when the flow rate was decreased below 0.5 mL/min. For instance, the value was 10,207/m at 0.4 mL/min (v_S = 12 cm/h) and 8165/m at 0.2 mL/min (v_S = 6 cm/h) (data not shown in Table 1), indicating that 0.5 mL/min was the optimal flow rate for the best-performing cuboid packed-bed device. The tailing and asymmetry factor of the salt peaks obtained with the column increased with decrease in flow rate. However, the corresponding values for the cuboid packed-bed device remained consistently close to 1 at all the flow rates examined due to consistently uniform flow distribution.

Figures 3 and 4 show salt peaks obtained with the six cuboid packed-bed devices and the column during the theoretical plate measurement experiments at the flow rates of 0.5 and 10 mL/min respectively. In these experiments, 0.4 M NaCl solution was used as the running buffer while 0.8 M NaCl solution was used as tracer. The volume of tracer injected was 100 μL (i.e., 2% of the bed volume).

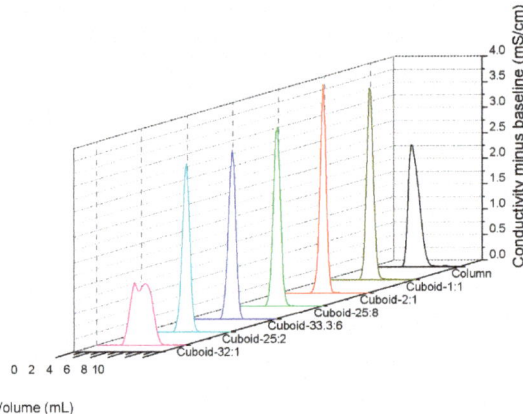

Figure 3. Salt peaks obtained with the six cuboid packed-bed devices and their equivalent column at low flow rate (media: Capto Q anion exchange, bed volume: 5 mL, mobile phase: 0.4 M NaCl, tracer: 0.8 M NaCl, loop: 0.1 mL, conductivity baseline was normalized for comparison, flow rate: 0.5 mL/min, v_S = 15 cm/h).

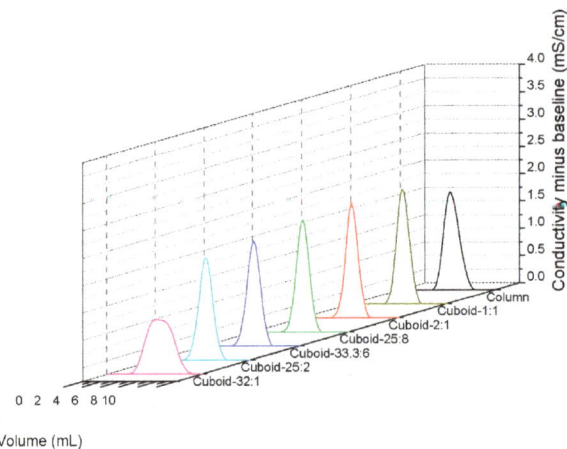

Figure 4. Salt peaks obtained with the six cuboid packed-bed devices and their equivalent column at high flow rate (media: Capto Q anion exchange, bed volume: 5 mL, mobile phase: 0.4 M NaCl, tracer: 0.8 M NaCl, loop: 0.1 mL, conductivity was normalized for comparison, flow rate: 10 mL/min, v_S = 300 cm/h).

In the chromatograms shown in the figure, the conductivity baseline was normalized for ease of comparison by deduction of the conductivity corresponding to 0.4 M NaCl. Consistent with the numerical results summarized in Table 1, the peaks obtained with the best cuboid packed-bed device (dimension: 20 mm × 10 mm × 25 mm) were sharper than those obtained with the other cuboid packed-bed devices and the column, especially at the lower flow rate i.e., 0.5 mL/min.

The difference between the maximum and minimum residence time within a cuboid packed-bed device would depend primarily on the magnitude of the channel velocity gradient [32]:

$$|\tau_{max} - \tau_{min}| = \left[\frac{2l}{2v_0 - l\left|\frac{dv_U}{dz}\right|}\right] - \left[\frac{2l}{2v_0 - \frac{l}{2}\left|\frac{dv_U}{dz}\right|}\right] \tag{4}$$

where τ_{max} is the maximum residence time, τ_{min} is the minimum residence time, l is the length of the cuboid packed-bed device, v_0 is the channel inlet velocity, and v_U is the liquid velocity in the upper channel at location z.

v_0 could be obtained by:

$$v_0 = \left(\frac{V}{hw}\right) \tag{5}$$

where V is the volumetric flow rate, h is the channel height, and w is the width of the channel (and therefore the width of the cuboid packed-bed device).

In an earlier paper it has been shown that [31]:

$$\left|\frac{dv_U}{dz}\right| = \frac{v_S}{h} \tag{6}$$

where v_S is the superficial velocity which could be obtained by:

$$v_S = \left(\frac{V}{lw}\right) \tag{7}$$

From Equations (4)–(7):

$$|\tau_{max} - \tau_{min}| = \left(\frac{2hlw}{3V}\right) \tag{8}$$

Therefore, the difference between the maximum and the minimum residence time within a cuboid packed-bed device would depend on the volumetric flow rate, the channel height, and the area of cross-section of the cuboid packed-bed device. For the six cuboid packed-bed devices studied, the channel height and area of cross-section were maintained the same. Therefore, based on Equation (8), the difference between the maximum and the minimum residence time within a cuboid packed-bed device could be expected to be the same, i.e., they would have the same efficiency. However, Table 1 shows that this was clearly not the case, i.e., the efficiency depended on the aspect ratio, with the best performance being observed with the device having 2:1 aspect ratio. The efficiency decreased with an increase in this ratio, particularly severely when the ratio was 32:1. This discrepancy with the expectations based on Equation (8) was probably due to the effect of aspect ratio on the pressure drop within the channel. One of the design criteria for obtaining high efficiency in separation with a cuboid packed-bed device is that the channel pressure drop should be significantly lower than the pressure drop across the packed bed [31,32]. An equation based on the resistance model used in membrane process [34] could be used to estimate the pressure drop in the packed bed (ΔP_v):

$$\Delta P_v = v_S R_v \tag{9}$$

where R_v is the hydraulic resistance offered by the packed bed which would depend on the property of the packed bed and on the flow rate but would be independent of the channel aspect ratio.

In a similar way, the pressure drop in the lateral channel (ΔP_l) could be estimated as follows:

$$\Delta P_l = \overline{v_U} R_l \tag{10}$$

where $\overline{v_U}$ is the average velocity in the upper channel, R_l is the lateral resistance in the channel. $\overline{v_U}$ is given by [32]:

$$\overline{v_U} = \left(2v_0 - z \left| \frac{dv_U}{dz} \right| \right) \Big/ 2 \tag{11}$$

From Equations (5)–(7) and (11):

$$\overline{v_U} = v_0/2 \tag{12}$$

Thus:

$$\Delta P_l = v_0 R_l/2 \tag{13}$$

The magnitude of both v_0 and R_l would increase with aspect ratio. v_0 would increase due to a decrease in the area of cross-section of the channel while R_l would increase due to an increase in channel length. Therefore, the greater the aspect ratio, the greater would be the net pressure differential between the channel extremities. When the channel is long and narrow (i.e., when the aspect ratio is very high), ΔP_l might become comparable in magnitude to ΔP_v and thereby the basic design criteria for cuboid packed beds, i.e., the channel pressure drop should be significantly lower than the pressure drop across the packed bed, would not hold good any more. Under such conditions, the higher channel pressure closer to the entrance of a narrow channel would result in more liquid going through the regions of the packed bed closer to the inlet than through the regions of the packed bed further down, i.e., v_S would no longer be constant. This would also result in a non-linear channel velocity gradient, making Equation (8) invalid for such a situation. Interestingly, the cuboid packed-bed device with an aspect ratio of 1:1 performed marginally poorer than the one with 2:1 ratio. A lower aspect ratio implies a wider channel. The channels in each of the cuboid packed-bed devices were fed from a narrow inlet and drained through a narrow outlet using a tapered distributor. Figure 5 shows the diagrams for the tapered distributors corresponding to the 1:1 and 2:1 aspect ratio cuboid packed-bed devices. While such tapered flow distributors worked well at the scale examined in this paper, i.e., 5 mL, they are likely to be less effective with larger devices with wider channels. For such larger devices, a better flow distributor could potentially be required.

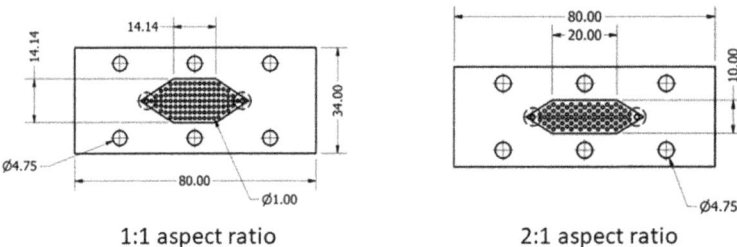

Figure 5. Tapered channel distributors used in cuboid packed-bed devices.

The above discussion highlights an efficiency trade-off between the length and the width of a cuboid packed-bed device. Increasing the length would increase the pressure in the channel which could decrease the efficiency of the device. On the other hand, increasing the width of the device could result in flow non-uniformity along the width of the channel which in turn could reduce the efficiency. In this study, the 2:1 aspect ratio cuboid packed-bed device represents the efficiency "sweet spot" for the 5 mL cuboid packed-bed devices examined. However, it should be noted that the difference in efficiency between the 2:1 and 1:1 aspect ratio devices was not very big, particularly at the higher flow rates. For larger devices, non-uniformity along the width of the channel could potentially become a significant factor and advanced flow distribution features to reduce such variability would need to be incorporated.

The 20 mm × 10 mm × 25 mm cuboid packed-bed device which had approximately 4.5 times as many theoretical plates as the column was used in all subsequent head-to-head comparisons with the column. This cuboid packed-bed device is henceforth referred to as the best cuboid packed-bed device to avoid any confusion.

Figure 6 and Figure S1 (supplementary information) show the unbound protein (i.e., lysozyme) breakthrough peaks obtained with the column and the best cuboid packed-bed device at different flow rates (1 mL/min and 5 mL/min, i.e., at v_S = 30 cm/h and 150 cm/h respectively). In these experiments, 50 mM Tris-HCl, pH 8 was used as the running buffer and 5 mL of 1 mg/mL lysozyme solution prepared in the running buffer was injected after ultraviolet (UV) absorbance equilibration. The large lysozyme sample volume ensured that the UV absorbance plateaued for a significant duration before decaying back to the baseline. The unbound protein breakthrough curves obtained with the best cuboid packed-bed device were sharper at both flow rates, clearly indicating its superior flow distribution attribute.

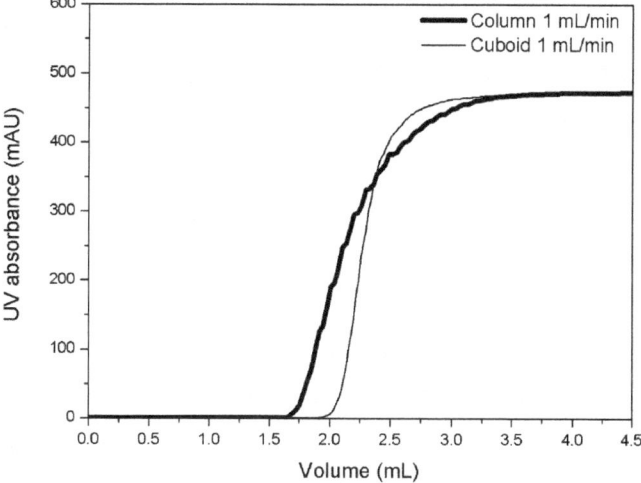

Figure 6. Unbound lysozyme breakthrough curves obtained with the best cuboid packed-bed device and its equivalent column (media: Capto Q anion exchange, bed volume: 5 mL, buffer: 50 mM Tris-HCl, pH 8.0, sample: 1 mg/mL lysozyme, flow rate: 1 mL/min, v_S = 30 cm/h).

Figure 7 and Figure S2 (supplementary information) show lysozyme flow through peaks obtained with the best cuboid packed-bed device and the column, at different flow rates (1 mL/min and 5 mL/min, i.e., at v_S = 30 cm/h and 150 cm/h respectively) by injecting 0.5 mL of 1 mg/mL lysozyme solution. Consistent with the salt peaks shown in Figures 3 and 4, the lysozyme flow-through peaks obtained with the best cuboid packed-bed device were significantly sharper and more symmetrical than those obtained with the column. In order to compare the above flow through peaks, a parameter termed dispersion index (*DI*) was used, and is defined below:

$$DI = \left(\frac{2w_{0.5}}{V_{inj}} \right) \tag{14}$$

where V_{inj} is the volume injected. The calculated values are summarized in Table 2.

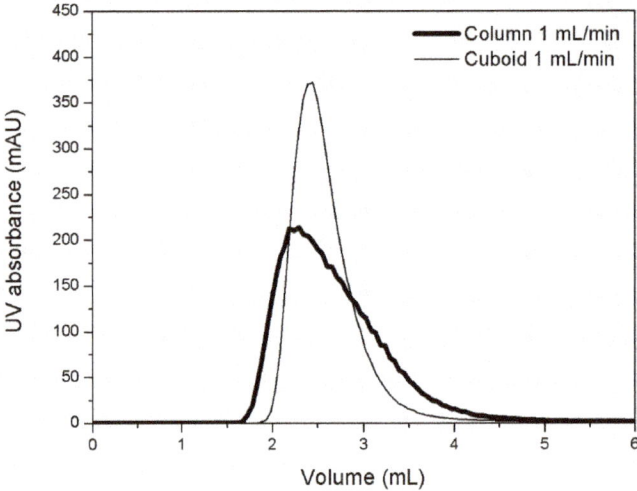

Figure 7. Lysozyme flow-through peaks obtained with the best cuboid packed-bed device and its equivalent column (media: Capto Q anion exchange, bed volume: 5 mL, buffer: 50 mM Tris-HCl, pH 8.0, sample: 1 mg/mL lysozyme, loop: 0.5 mL, flow rate: 1 mL/min, v_S = 30 cm/h).

Table 2. Comparison of the attributes of the lysozyme flow-through peaks obtained with the best cuboid packed-bed device and its equivalent column (media: Capto Q anion exchange, bed volume: 5 mL, buffer: 50 mM Tris-HCl, pH 8.0, sample: 1 mg/mL lysozyme, v_S = 30 and 150 cm/h for the flow rate of 1 and 5 mL/min respectively, injection volume: 0.5 mL)

Flow Rate (mL/min)	Column		Cuboid	
	Width at Half Height (mL)	Dispersion Index	Width at Half Height (mL)	Dispersion Index
5	0.98	3.94	0.62	2.50
1	1.1	4.42	0.6	2.38

The peak widths at half height and *DI* values of the flow through peaks obtained with the best cuboid packed-bed device by injecting 0.5 mL of lysozyme sample were significantly lower than those obtain with the column at the different flow rates. Moreover, the peaks were more symmetrical, the difference in symmetry being greater at the lower flow rate. Once again, these results demonstrate the role of flow distribution in the performance of chromatographic devices.

Figure 8 and Figure S3 (supplementary information) show the BSA elution peaks obtained with the best cuboid packed-bed device and the column. The binding buffer used in this experiment was 50 mM Tris-HCl buffer, pH 8, while 50 mM Tris-HCl buffer, 1M NaCl, pH 8 was used as the eluting buffer. These experiments were carried out at two different flow rates (1 mL/min and 5 mL/min, i.e., at v_S = 30 cm/h and 150 cm/h respectively) by injecting 5 mL of 2 mg/mL BSA solution. Figures S4 and S5 (supplementary information) show the results obtained from similar experiments carried out using 0.5 mL of 2 mg/mL BSA solution. In each case, a step change from the binding buffer to 100% eluting buffer was made immediately after sample injection. At all the different conditions examined (i.e., shown in Figure 8 and Figure S3–S5), the BSA elution peaks obtained with the best cuboid packed-bed device were sharper than those obtained with the column. Earlier in the paper, the parameter *DI* was used for quantifying the spread of flow through peaks. The corresponding metric for an eluted peak could be referred to as the apparent dispersion index (*ADI*), quantified in a similar manner, i.e., using Equation (14). The qualifier "apparent" is necessary for an eluted peak as its spread depends not only on column hydraulics but also on the gradient used for elution. Therefore, for comparison of two eluted peaks using *ADI*, the elution gradient has to be identical. The calculated

ADI data obtained from the peaks shown in Figure 8 and Figure S3–S5 (supplementary information) are summarized in Table 3 and Table S1 (supplementary information). At each experimental condition, the peak width at half height and corresponding *ADI* obtained with the best cuboid packed-bed device was significantly lower than that obtained with the column.

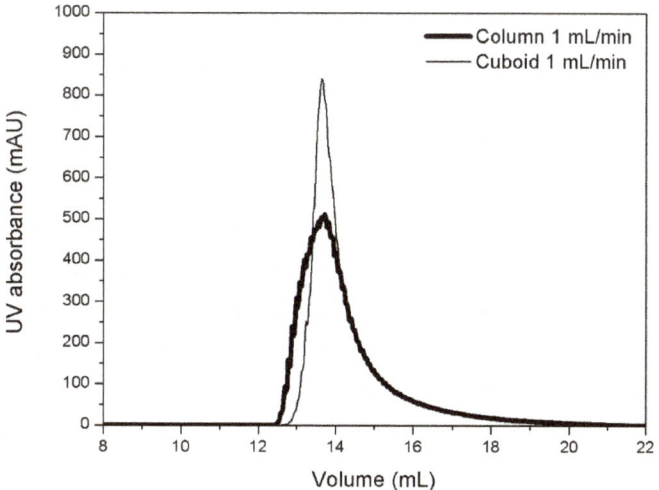

Figure 8. Eluted BSA peaks obtained with the best cuboid packed-bed device and its equivalent column (media: Capto Q anion exchange, bed volume: 5 mL, binding buffer: 50 mM Tris-HCl, pH 8.0, eluting buffer: 50 mM Tris-HCl, 1 M NaCl, pH 8.0, sample: 2 mg/mL BSA, loop: 5 mL, flow rate: 1 mL/min, v_S = 30 cm/h).

Table 3. Comparison of the attributes of the eluted BSA peaks obtained with the best cuboid packed-bed device and its equivalent column (media: Capto Q anion exchange, bed volume: 5 mL, binding buffer: 50 mM Tris-HCl, pH 8.0, eluting buffer: 50 mM Tris-HCl, 1 M NaCl, pH 8.0, sample: 2 mg/mL BSA, v_S = 30 and 150 cm/h for the flow rate of 1 and 5 mL/min respectively, injection volume: 0.5 mL)

Flow Rate (mL/min)	Column		Cuboid	
	Width at Half Height (mL)	Apparent Dispersion Index	Width at Half Height (mL)	Apparent Dispersion Index
5	1.82	7.28	0.9	3.6
1	1.51	6.04	0.6	2.4

Chromatography columns used for large-scale purification of biologicals such as monoclonal antibodies have low bed height to diameter ratios. For instance affinity chromatography, anion exchange chromatography, and cation exchange chromatography columns having ratios of 0.15, 0.18 and 0.08, respectively, have been used for 20,000 L scale production of monoclonal antibodies [7]. The bed height-to-diameter ratio of the cuboid packed-bed devices and colums used in this study is 1.57. Therefore, it may be anticipated that the cuboid packed-bed devices would show even better relative performances when the bed height-to-diameter ratio is lower. The amount of material used to fabricate the wall of a chromatography device is an important factor determining equipment cost. For a given wall thickness, a column would use the least amount of material as the circular cross-section ensures the lowest perimeter for a given cross-sectional area. For the cuboid packed-bed device, the amount of material used to fabricate the wall would depend on the length-to-width aspect ratio, as the perimeter would depend on it, the minimum being when the aspect ratio is one. In terms of wall material cost, the cuboid packed-bed device with an aspect ratio of one would be more expensive

Processes **2018**, *6*, 160

than the column by a factor of 1.128. This factor would increase with increase in the aspect ratio. In summary, a cuboid packed-bed device shows significant promise in applications such as large-scale, high-resolution purifications of biologicals.

4. Conclusions

Preparative columns used in large-scale biopharmaceutical purification processes typically have low bed height-to-diameter ratios. Ensuring adequate flow distribution in such columns is challenging. Cuboid packed-bed devices examined in this study were designed to provide superior flow distribution and thereby better separation than equivalent preparative columns. The length-to-width aspect ratio of a cuboid packed-bed device had a significant effect on performance. The cuboid packed-bed device with a 2:1 aspect ratio showed the best performance while the one with an aspect ratio of 32:1 performed the worst. The number of theoretical plates in the best-performing cuboid packed-bed device was approximately 4.5 times of that in the equivalent column. The performance of the worst-performing cuboid packed-bed device was poorer than that of the column. If the length-to-width ratio of the cuboid packed-bed device is too high, increase in channel back pressure could decrease the efficiency of the device. On the other hand, increasing the width of the device could potentially result in non-uniformity along the width of the channel which, in turn, could also somewhat reduce the efficiency of separation. The best-performing cuboid packed-bed device significantly outperformed the column in terms of all the protein separation metrics examined. The results discussed in this paper show that precice optimization of the design of a cuboid packed-bed device is essential to ensure that it performs efficiently.

5. Patents

CHROMATOGRAPHY DEVICE AND METHOD FOR FILTERING A SOLUTE FROM A FLUID R Ghosh—US Patent App. US20170349626A1

Supplementary Materials: The following are available online at www.mdpi.com/xxx/s1, Figure S1: Unbound lysozyme break through curves obtained with the best cuboid packed-bed device and its equivalent column (media: Capto Q anion exchange, bed volume: 5 mL, buffer: 50 mM Tris-HCl, pH 8.0, sample: 1 mg/mL lysozyme, flow rate: 5 mL/min, v_S = 150 cm/h), Figure S2: Lysozyme flow through peaks obtained with the best cuboid packed-bed device and its equivalent column (media: Capto Q anion exchange, bed volume: 5 mL, buffer: 50 mM Tris-HCl, pH 8.0, sample: 1 mg/mL lysozyme, loop: 0.5 mL, flow rate: 5 mL/min, v_S = 150 cm/h), Figure S3: Eluted BSA peaks obtained with the best cuboid packed-bed device and its equivalent column (media: Capto Q anion exchange, bed volume: 5 mL, binding buffer: 50 mM Tris-HCl, pH 8.0, eluting buffer: 50 mM Tris-HCl, 1 M NaCl, pH 8.0, sample: 2 mg/mL BSA, loop: 5 mL, flow rate: 5 mL/min, v_S = 150 cm/h), Figure S4: Eluted BSA peaks obtained with the best cuboid packed-bed device and its equivalent column (media: Capto Q anion exchange, bed volume: 5 mL, binding buffer: 50 mM Tris-HCl, pH 8.0, eluting buffer: 50 mM Tris-HCl, 1 M NaCl, pH 8.0, sample: 2 mg/mL BSA, loop: 0.5 mL, flow rate: 1 mL/min, v_S = 30 cm/h), Figure S5: Eluted BSA peaks obtained with the best cuboid packed-bed device and its equivalent column (media: Capto Q anion exchange, bed volume: 5 mL, binding buffer: 50 mM Tris-HCl, pH 8.0, eluting buffer: 50 mM Tris-HCl, 1 M NaCl, pH 8.0, sample: 2 mg/mL BSA, loop: 0.5 mL, flow rate: 5 mL/min, v_S = 150 cm/h), Table S1: Comparison of the attributes of the eluted BSA peaks obtained with the best cuboid packed-bed device and its equivalent column (media: Capto Q anion exchange, bed volume: 5 mL, binding buffer: 50 mM Tris-HCl, pH 8.0, eluting buffer: 50 mM Tris-HCl, 1 M NaCl, pH 8.0, sample: 2 mg/mL BSA, v_S = 30 and 150 cm/h for the flow rate of 1 and 5 mL/min respectively, injection volume: 5 mL).

Author Contributions: G.C. made the research plan, carried out the experiments, and wrote the manuscript. R.G. supervised the experiments and revised the manuscript.

Funding: This research was funded by the Natural Sciences and Engineering Research Council (NSERC) of Canada, and the Ontario Research Fund, Research Excellence Program.

Acknowledgments: We would like to thank the Natural Science and Engineering Research Council (NSERC) of Canada and the Ontario Research Fund, Research Excellence (ORF RE) program for funding this study. We thank Paul Gatt for fabricating the cuboid packed-bed devices used in this work based on the design provided by R.G. We thank Umatheny Umatheva for help with preparing Figure 1 of this paper.

Conflicts of Interest: The author declares no conflict of interest.

References

1. Marchand, D.H. Chromatography in analytical biotechnology. *Curr. Opin. Biotechnol.* **1994**, *5*, 72–76. [CrossRef]

2. Cramer, S.M.; Jayaraman, G. Preparative chromatography in biotechnology. *Curr. Opin. Biotechnol.* **1993**, *4*, 217–225. [CrossRef]

3. Freitag, R.; Horvath, C. Chromatography in the downstream processing of biotechnological products. *Adv. Biochem. Eng. Biotechnol.* **1996**, *53*, 17–59. [PubMed]

4. Hanke, A.T.; Ottens, M. Purifying biopharmaceuticals: Knowledge-based chromatographic process development. *Trends Biotechnol.* **2014**, *32*, 210–220. [CrossRef] [PubMed]

5. Chisti, Y.; Moo-Young, M. Large scale protein separations: Engineering aspects of chromatography. *Biotechnol. Adv.* **1990**, *8*, 699–708. [CrossRef]

6. Jungbauer, A. Chromatographic media for bioseparation. *J. Chromatogr. A* **2005**, *1065*, 3–12. [CrossRef] [PubMed]

7. Aldington, S.; Bonnerjea, J. Scale-up of monoclonal antibody purification processes. *J. Chromatogr. B* **2007**, *848*, 64–78. [CrossRef] [PubMed]

8. Lode, F.G.; Rosenfeld, A.; Yuan, Q.S.; Root, T.W.; Lightfoot, E.N. Refining the scale-up of chromatographic separations. *J. Chromatogr. A* **1998**, *796*, 3–14. [CrossRef]

9. Moscariello, J.; Purdom, G.; Coffman, J.; Root, T.W.; Lightfoot, E.N. Characterizing the performance of industrial-scale columns. *J. Chromatogr. A* **2001**, *908*, 131–141. [CrossRef]

10. Guiochon, G.; Farkas, T.; Guan-Sajonz, H.; Koh, J.-H.; Sarker, M.; Stanley, B.J.; Yun, T. Consolidation of particle beds and packing of chromatographic columns. *J. Chromatogr. A* **1997**, *762*, 83–88. [CrossRef]

11. Shalliker, R.A.; Broyles, B.S.; Guiochon, G. Physical evidence of two wall effects in liquid chromatography. *J. Chromatogr. A* **2000**, *888*, 1–12. [CrossRef]

12. Park, J.C.; Raghavan, K.; Gibbs, S.J. Axial development and radial non-uniformity of flow in packed columns. *J. Chromatogr. A* **2002**, *945*, 65–81. [CrossRef]

13. Farkas, T.; Guiochon, G. Contribution of the radial distribution of the flow velocity to band broadening in hplc columns. *Anal. Chem.* **1997**, *69*, 4592–4600. [CrossRef]

14. Dapremont, O.; Cox, G.B.; Martin, M.; Hilaireau, P.; Colin, H. Effect of radial gradient of temperature on the performance of large-diameter high-performance liquid chromatography columns. *J. Chromatogr. A* **1998**, *796*, 81–99. [CrossRef]

15. Johnson, C.; Natarajan, V.; Antoniou, C. Evaluating two process scale chromatography column header designs using cfd. *Biotechnol. Progr.* **2014**, *30*, 837–844. [CrossRef] [PubMed]

16. Wu, Y.X.; Ching, C.B. Theoretical study of the effect of frit quality on chromatography using computational fluid dynamics. *Chromatographia* **2003**, *57*, 329–337. [CrossRef]

17. Shalliker, R.A.; Broyles, B.S.; Guiochon, G. Visualization of sample introduction in liquid chromatographic columns: Contribution of a flow distributor on the sample band shape. *J. Chromatogr. A* **1999**, *865*, 83–95. [CrossRef]

18. Davydova, E.; Deridder, S.; Eeltink, S.; Desmet, G.; Schoenmakers, P.J. Optimization and evaluation of radially interconnected versus bifurcating flow distributors using computational fluid dynamics modelling. *J. Chromatogr. A* **2015**, *1380*, 88–95. [CrossRef] [PubMed]

19. Vangelooven, J.; De Malsche, W.; De Beeck, J.O.; Eghbali, H.; Gardeniers, H.; Desmet, G. Design and evaluation of flow distributors for microfabricated pillar array columns. *Lab Chip* **2010**, *10*, 349–356. [CrossRef] [PubMed]

20. Yuan, Q.S.; Rosenfeld, A.; Root, T.W.; Klingenberg, D.J.; Lightfoot, E.N. Flow distribution in chromatographic columns1. *J. Chromatogr. A* **1999**, *831*, 149–165. [CrossRef]

21. Fee, C.; Nawada, S.; Dimartino, S. 3d printed porous media columns with fine control of column packing morphology. *J. Chromatogr. A* **2014**, *1333*, 18–24. [CrossRef] [PubMed]

22. Kearney, M.M.; Petersen, K.R.; Vervloet, T.; Mumm, M.W. Fluid Transfer System with Uniform Fluid Distributor. U.S. Patent 5354460 A, 11 October 1994.

23. Kearney, M. Control of fluid dynamics with engineered fractals-adsorption process applications. *Chem. Eng. Commun.* **1999**, *173*, 43–52. [CrossRef]

24. Johnson, T.F.; Levison, P.R.; Shearing, P.R.; Bracewell, D.G. X-ray computed tomography of packed bed chromatography columns for three dimensional imaging and analysis. *J. Chromatogr. A* **2017**, *1487*, 108–115. [CrossRef] [PubMed]

25. Guiochon, G. Preparative liquid chromatography. *J. Chromatogr. A* **2002**, *965*, 129–161. [CrossRef]

26. Hofmann, M. Use of ultrasound to monitor the packing of large-scale columns, the monitoring of media compression and the passage of molecules, such as monoclonal antibodies, through the column bed during chromatography. *J. Chromatogr. A* **2003**, *989*, 79–94. [CrossRef]

27. Ching, C.B.; Wu, Y.X.; Lisso, M.; Wozny, G.; Laiblin, T.; Arlt, W. Study of feed temperature control of chromatography using computational fluid dynamics simulation. *J. Chromatogr. A* **2002**, *945*, 117–131. [CrossRef]

28. Camenzuli, M.; Ritchie, H.J.; Shalliker, R.A. Gradient elution chromatography with segmented parallel flow column technology: A study on 4.6 mm analytical scale columns. *J. Chromatogr. A* **2012**, *1270*, 204–211. [CrossRef] [PubMed]

29. Foley, D.; Pereira, L.; Camenzuli, M.; Edge, T.; Ritchie, H.; Shalliker, R.A. Curtain flow chromatography ('the infinite diameter column') with automated injection and high sample through-put: The results of an inter-laboratory study. *Microchem. J.* **2013**, *110*, 127–132. [CrossRef]

30. Besselink, T.; van der Padt, A.; Janssen, A.E.M.; Boom, R.M. Are axial and radial flow chromatography different? *J. Chromatogr. A* **2013**, *1271*, 105–114. [CrossRef] [PubMed]

31. Ghosh, R. Using a box instead of a column for process chromatography. *J. Chromatogr. A* **2016**, *1468*, 164–172. [CrossRef] [PubMed]

32. Ghosh, R.; Chen, G. Mathematical modelling and evaluation of performance of cuboid packed-bed devices for chromatographic separations. *J. Chromatogr. A* **2017**, *1515*, 138–145. [CrossRef] [PubMed]

33. Madadkar, P.; Wu, Q.; Ghosh, R. A laterally-fed membrane chromatography module. *J. Membr. Sci.* **2015**, *487*, 173–179. [CrossRef]

34. Ghosh, R.; Madadkar, P.; Wu, Q. On the workings of laterally-fed membrane chromatography. *J. Membr. Sci.* **2016**, *516*, 26–32. [CrossRef]

35. Madadkar, P.; Ghosh, R. High-resolution protein separation using a laterally-fed membrane chromatography device. *J. Membr. Sci.* **2016**, *499*, 126–133. [CrossRef]

36. Madadkar, P.; Nino, S.L.; Ghosh, R. High-resolution, preparative purification of pegylated protein using a laterally-fed membrane chromatography device. *J. Chromatogr. B* **2016**, *1035*, 1–7. [CrossRef] [PubMed]

37. Madadkar, P.; Sadavarte, R.; Butler, M.; Durocher, Y.; Ghosh, R. Preparative separation of monoclonal antibody aggregates by cation-exchange laterally-fed membrane chromatography. *J. Chromatogr. B* **2017**, *1055–1056*, 158–164. [CrossRef] [PubMed]

38. Madadkar, P.; Umatheva, U.; Hale, G.; Durocher, Y.; Ghosh, R. Ultrafast separation and analysis of monoclonal antibody aggregates using membrane chromatography. *Anal. Chem.* **2017**, *89*, 4716–4720. [CrossRef] [PubMed]

39. Moldoveanu, S.C.; David, V. *Essentials in Modern HPLC Separations*, 1st ed.; Elsevier: Amsterdam, The Netherlands, 2013.

Article

Glycosylation Flux Analysis of Immunoglobulin G in Chinese Hamster Ovary Perfusion Cell Culture

Sandro Hutter [1,2,†], **Moritz Wolf** [1,†], **Nan Papili Gao** [1,2], **Dario Lepori** [1], **Thea Schweigler** [1], **Massimo Morbidelli** [1] and **Rudiyanto Gunawan** [1,2,3,*]

1   Institute for Chemical and Bioengineering, Department of Chemistry and Applied Biosciences, ETH Zurich, 8093 Zurich, Switzerland; sandro.hutter@outlook.com (S.H.); moritz.wolf@chem.ethz.ch (M.W.); nan.papili-gao@chem.ethz.ch (N.P.G.); dario.lepori@bluewin.ch (D.L.); thea-schweigler@web.de (T.S.); massimo.morbidelli@chem.ethz.ch (M.M.)
2   Swiss Institute of Bioinformatics, 1015 Lausanne, Switzerland
3   Department of Chemical and Biological Engineering, University at Buffalo, The State University of New York, Buffalo, NY 14260, USA
*   Correspondence: rudiyant@buffalo.edu
†   These authors contributed equally to this work.

Received: 30 May 2018; Accepted: 18 September 2018; Published: 1 October 2018

Abstract: The terminal sugar molecules of the N-linked glycan attached to the fragment crystalizable (Fc) region is a critical quality attribute of therapeutic monoclonal antibodies (mAbs) such as immunoglobulin G (IgG). There exists naturally-occurring heterogeneity in the N-linked glycan structure of mAbs, and such heterogeneity has a significant influence on the clinical safety and efficacy of mAb drugs. We previously proposed a constraint-based modeling method called glycosylation flux analysis (GFA) to characterize the rates (fluxes) of intracellular glycosylation reactions. One contribution of this work is a significant improvement in the computational efficiency of the GFA, which is beneficial for analyzing large datasets. Another contribution of our study is the analysis of IgG glycosylation in continuous perfusion Chinese Hamster Ovary (CHO) cell cultures. The GFA of the perfusion cell culture data indicated that the dynamical changes of IgG glycan heterogeneity are mostly attributed to alterations in the galactosylation flux activity. By using a random forest regression analysis of the IgG galactosylation flux activity, we were further able to link the dynamics of galactosylation with two process parameters: cell-specific productivity of IgG and extracellular ammonia concentration. The characteristics of IgG galactosylation dynamics agree well with what we previously reported for fed-batch cultivations of the same CHO cell strain.

Keywords: N-linked glycosylation; perfusion cell culture; CHO cells; constraint-based modeling; monoclonal antibody

---

1. Introduction

Therapeutic recombinant monoclonal antibodies constitute the most important class of drugs in the biopharmaceutical industry, making up approximately half of the total revenue of biopharmaceutical products in 2013 [1]. The production of mAb drugs typically employs batch or fed-batch cultivations of mammalian cells. The state-of-the-art batch/fed-batch cell cultures are able to meet the large production volume requirement of mAbs, with reactor sizes of up to 25,000 L [2]. Nevertheless, the increasing number of mAb products entering different stages of clinical trials and the burgeoning market of biosimilars driven by impending patent expirations of blockbuster mAb drugs, give a strong motivation for the development of new production technology that is more robust and cost effective [3]. The US Food and Drug Administration (FDA) initiative on quality by design and process analytical technology put further emphasis on implementing quantitative approaches for process improvements in biopharmaceutical manufacturing [4,5].

Continuous manufacturing technology offers an effective and flexible way for the large scale and robust production of drug compounds [6]. In the biopharmaceutical industry, the application of continuous cell culture technology has thus far been limited to the production of unstable products that require constant recovery [7]. Nevertheless, continuous perfusion cell cultures have previously been demonstrated to be capable of producing antibodies at a volumetric rate that matches or exceeds that of fed-batch cultures [8]. In addition to the high productivity, the stable steady state operation mode in conjunction with the short residence time of perfusion cultures translates to a tight maintenance of product quality. Whether or not product qualities and their key controlling parameters can be directly translated from the traditional batch/fed-batch cell cultures to the perfusion cell cultivations is still unresolved. Past work on converting mAb production from batch/fed-batch to perfusion cell cultures gave conflicting reports on product qualities, where a few studies demonstrated consistent product qualities [9,10], and many others showed differences between the two production modes [11–15].

Among the key critical quality attributes (CQAs) of therapeutic mAbs is the glycan structures of the Fc domain [16]. The N-linked glycosylation is a common post-translational modification of proteins, a process that occurs in the endoplasmic reticulum (ER) and Golgi apparatuses. The Fc glycan structure has been shown to impact protein folding [17], clearance [18], bioactivity [19] and efficacy [20], as well as immunogenicity [21]. Moreover, there exists naturally-occurring heterogeneity in the glycosylation of mAbs. Thus, the FDA approval of mAb drugs is given for a particular composition of mAb glycoforms [5]. Because of the importance of the N-linked glycosylation, the majority of recombinant mAb drugs are produced using mammalian host cells in order to achieve human-like glycan structures [22]. In particular, Chinese Hamster Ovary (CHO) cells have become the major expression host for the biopharmaceutical production of therapeutic mAbs [23].

The N-linked glycosylation of mAbs has been shown to depend on the host genetic background [24], expression vector [25], media composition [16], media supplements [26], and bioprocess parameters [27]. However, the mechanism of the above dependence remains to be established [28]. Toward closing this gap, mathematical models of the glycosylation network have previously been developed [29–36]. Many of these models have a large number of unknown and system-specific kinetic parameters that need to be fitted to experimental data [29–33]. A number of parameter-free models have also been proposed to study the N-linked glycosylation process based on the constraint-based modeling approach (i.e., stoichiometric models) [34–36]. Recently, we proposed a flux analysis method, called glycosylation flux analysis, which provides predictions of the rate or flux of intracellular glycosylation reactions using the stoichiometric model of the glycosylation network and the cell secretion rates of mAb glycoforms [36]. Similar to the well-known metabolic flux analysis (MFA) [37], the GFA is based on the molar balance equations of glycoforms involved in the glycosylation reaction network under the pseudo steady state assumption. The GFA further makes use of the relatively small number of enzymes involved in the glycosylation process to reduce the degrees of freedom in the flux estimation. More specifically, we assumed that glycosylation fluxes vary with time according to a (global) cell-specific factor and a (local) enzyme-specific factor. When applied to study the IgG production of CHO cells in fed-batch cultivations, the GFA was able to give insights on how changes in the media affect the intracellular IgG glycosylation reactions [36].

The formulation of the flux estimation in the GFA involves a nonlinear least-square regression, which requires computationally intensive global optimizations. One contribution of this work is a reformulation of the flux estimation in the GFA into an iterative procedure involving two linear regression problems, which provides higher computational efficiency. In this study, we applied the improved GFA to analyze the N-linked glycosylation of IgG in perfusion CHO cell cultures undergoing setpoint changes in the viable cell densities and perfusion rates. In the post-analysis of the GFA results, we employed a random forest regression analysis to elucidate the key process parameters that affect intracellular IgG glycosylation fluxes. Finally, we compared the characteristics of IgG glycosylation between the traditional fed-batch and continuous perfusion cell cultures and identified similarities and differences.

2. Materials and Methods

Figure 1 illustrates the analysis of IgG glycosylation in our study. The perfusion cell culture experiments are summarized in Section 2.1. In the data preprocessing step, we computed the cell-specific secretion fluxes of IgG glycoforms, as well as other molecules such as glucose, lactate and ammonia, from cell culture measurements. The data preprocessing is described in Section 2.2. We applied the GFA to the secretion fluxes of IgG glycoforms calculated above using a constraint-based model of the glycosylation network. The GFA generates estimates of enzyme-specific glycosylation activities at different measurement time points. An improved iterative procedure for the GFA is detailed in Section 2.3. Finally, we performed a random forest regression analysis, as outlined in Section 2.4, to identify the process parameters that are able to explain the dynamic behavior of intracellular glycosylation activities.

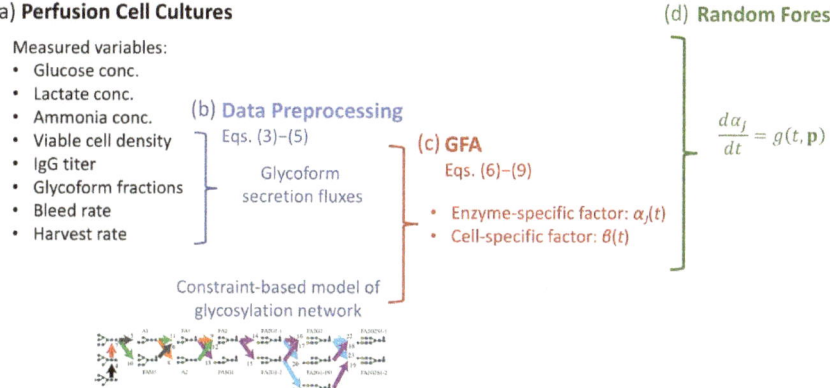

Figure 1. Analysis procedure of IgG glycosylation in CHO perfusion cell cultures. (**a**) Four perfusion cell cultures were performed (see Section 2.1). Measurements of glucose/lactate/ammonia concentrations, viable cell density, IgG titer, glycoform fractions, bleed rates and harvest rates were taken from each cell culture run. (**b**) In the data preprocessing step (see Section 2.2), the cell-specific secretion fluxes of IgG glycoforms, as well as the cell-specific secretion/uptake rates of IgG, glucose, lactate and ammonia, were computed. (**c**) The glycosylation flux analysis (GFA) was applied to the secretion fluxes of IgG and its glycoforms (see Section 2.3). The GFA is based on a constraint-based model of the glycosylation network where the glycosylation fluxes vary with time according to two multiplicative factors: the enzyme-specific factor $\alpha_J(t)$ and the cell-specific factor $\beta(t)$. The factor $\alpha_J(t)$ captures the dynamic changes in the enzyme-specific glycosylation activities, while the factor $\beta(t)$ describes the dynamic changes in the cell-specific IgG productivity. (**d**) A random forest regression analysis was carried out to rank the process parameters (**p**) based on their ability to predict the dynamic changes of $\alpha_J(t)$.

2.1. Continuous Perfusion Cell Cultures

The detailed procedure of the perfusion cell culture is available elsewhere [38]. Here, we provide a brief summary of the experiments. A proprietary CHO cell line expressing IgG1 was cultured using a previously developed 1.5 L perfusion cell culture setup [39], as depicted in Figure 2. In total, four perfusion cell culture experiments were performed with different viable cell density (VCD) and perfusion rate (PR) set-point profiles, as shown in Figure 3. Each of the experiments was inoculated following the same procedure. CHO cells were held back in the reactor by a cell retention device so that only cell free reaction mixture left the reactor through the harvest stream. The feed flowrate (F) of fresh media into the reactor and the bleed (B) and harvest (H) flowrates out of the reactor were balanced to keep the reactor volume constant. The perfusion rate (PR) is given by the flowrate through the reactor, as follows:

$$PR = H + B = F \tag{1}$$

The perfusion rate represents the rate of fresh media supplied to the cells. We define the cell-specific perfusion rate (CSPR) as the rate of fresh media per cell:

$$CSPR = \frac{PR}{VCD} \tag{2}$$

During the experiments, the following measurements were collected on a daily basis: cell counts and cell viability, concentrations of glucose, lactate and ammonia, IgG titer, and protein glycan distribution.

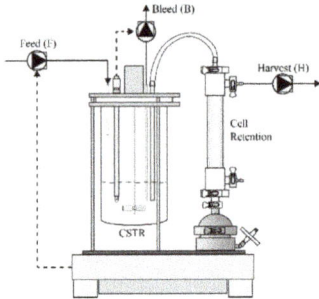

Figure 2. A schematic of perfusion cell culture reactor (Figure adapted from [39] with the permission from the authors). CHO cells were cultivated in suspension in a continuous stirred tank reactor with continuous feeding of fresh nutrients. Cell-free spent media was constantly collected in the harvest stream, while cells remained in the stirred tank reactor. A bleed stream removed a small fraction of the reactor mixture, including biomass, which was used to regulate viable cell density.

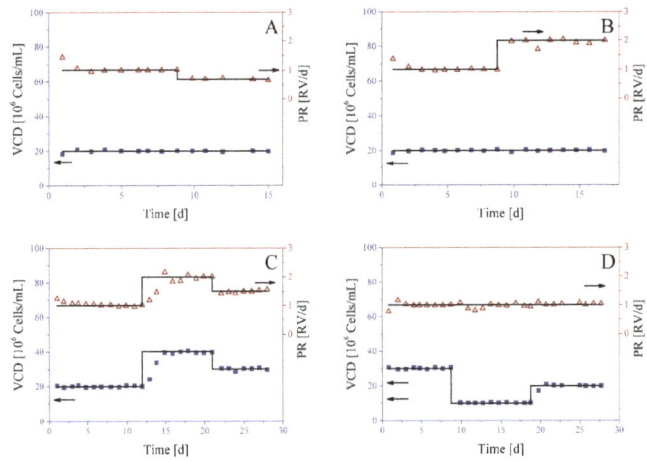

Figure 3. Perfusion cell culture experiments. Four perfusion cell culture experiments, labeled (**A**), (**B**), (**C**) and (**D**), were conducted with varying VCD and PR given set-points (solid lines). The experimental VCD and PR are shown as blue filled squares and red empty triangles, respectively.

2.2. Estimation of Secretion and Uptake Fluxes

As inputs to the GFA, the cell-specific secretion flux of each IgG glycoform ($v_{E,i}$) was determined based on the molar balance in the reactor, as follows

$$V\frac{dc_{E,i}(t)}{dt} = Vv_{E,i}(t)\text{VCD}(t) - (\text{H}(t) + \text{B}(t))c_{E,i}(t) \tag{3}$$

where $c_{E,i}$ denotes the concentration of the i-th IgG glycoform and V denotes the reactor volume. Since the reactor volume is kept (approximately) constant, Equation (3) can be rearranged to give:

$$v_{E,i}(t) = \frac{\left(\frac{dc_{E,i}(t)}{dt} + (\text{h}(t) + \text{b}(t))c_{E,i}(t)\right)}{\text{VCD}(t)} \tag{4}$$

The variables $\text{h}(t) = \frac{\text{H}(t)}{V}$ and $\text{b}(t) = \frac{\text{B}(t)}{V}$ represent the specific harvest and bleed rate, respectively. The concentration of the IgG glycoform was calculated as the product of the measured glycoform fraction (f_i) and the IgG titer (T) according to:

$$c_{E,i}(t_k) = f_i(t_k)\text{T}(t_k) \tag{5}$$

where t_k denotes the k-th measurement time point.

For the estimation of the secretion fluxes in Equation (4), the computed concentration of IgG glycoforms $c_{E,i}(t_k)$ was first smoothened (as a function of time) using spline fitting. The time derivative $\frac{dc_{E,i}(t)}{dt}$ was then evaluated by taking the first order derivative of the spline curve. Equation (4) can be used to compute the cell-specific uptake/secretion fluxes of IgG, glucose, lactate and ammonia by replacing $c_{E,i}(t)$ with the concentration measurements of IgG, glucose, lactate and ammonia, respectively.

2.3. Glycosylation Flux Analysis

The glycosylation flux analysis is based on a constraint-based model of the protein glycosylation reaction network. Like other flux analysis methods, the GFA is developed for evaluating the intracellular (glycosylation) fluxes using the cellular (mAb glycoforms) secretion rates. In the GFA, a pseudo steady state assumption is taken to derive the stoichiometric model of the glycosylation network, as follows:

$$\frac{d\mathbf{c}_I(t)}{dt} = \mathbf{S}\mathbf{v}_I(t) - \mathbf{v}_E(t) = 0 \tag{6}$$

where $\mathbf{c}_I$ denotes the vector of m intracellular IgG glycoform concentrations, $\mathbf{v}_I$ denotes the vector of n intracellular IgG glycosylation reaction fluxes (rates), $\mathbf{v}_E$ denotes the vector of secretion fluxes of IgG glycoforms estimated above, and $\mathbf{S}$ denotes the $m \times n$ stoichiometric matrix. The (i,j)-th element of $\mathbf{S}$ gives the number of the i-th glycoform molecule produced (if positive) or consumed (if negative) by the j-th glycosylation reaction. Since the number of reaction fluxes (i.e., the number of unknowns) typically exceeds that of glycoforms (i.e., the number of equations), the estimation of $\mathbf{v}_I$ from $\mathbf{v}_E$ in Equation (6) is underdetermined. In other words, there exist many $\mathbf{v}_I$ for the same experimentally determined $\mathbf{v}_E$. In the method Flux Balance Analysis, the most plausible $\mathbf{v}_I$ is set to the vector that maximizes a cellular objective, such as the biomass production [37]. However, the appropriate cellular objective to use for glycosylation networks is not immediately obvious.

Instead of assuming a cellular objective, the GFA enforces constraints on how groups of glycosylation fluxes vary together. More specifically, in the GFA each glycosylation flux $v_{I,j}(t)$ is computed as the product of a reference flux value $v_{I,j}{}^{\text{ref}}$, an enzyme-specific factor $\alpha_J(t)$ and a cell-specific factor $\beta(t)$, as follows:

$$v_{I,j}(t) = \alpha_J(t)\beta(t)v_{I,j}{}^{\text{ref}} \tag{7}$$

The variables $\alpha_J(t)$ and $\beta(t)$ represent the fold-change amplification or attenuation and can therefore be normalized to 1 at a chosen reference time point t_{ref}.

The cell-specific factor $\beta(t)$ captures the (global) influence of the cell metabolism on the glycosylation process, more specifically the total amount of mAb entering/leaving the glycosylation network. For this reason, $\beta(t)$ is represented by the ratio of the cell-specific productivity (q_{mAb}) between time t and the reference time t^{ref}, as follows:

$$\beta(t) = \frac{q_{mAb}(t)}{q_{mAb}(t^{ref})} \tag{8}$$

On the other hand, the factor $\alpha_J(t)$ describes the (local) influence of enzymatic processing capacity, which captures the dependence of glycosylation on factors such as enzyme expression and activity, as well as co-factor and nucleotide sugar availability. Note that the number of enzymes involved in the glycosylation network is typically much smaller than the number of reactions, as an enzyme catalyzes multiple reactions. Thus, the estimation of $\mathbf{v}_I(t)$ can be reformulated to fitting $\alpha_J(t)$ and $v_{I,j}^{ref}$ to the secretion fluxes, as follows:

$$\min_{\alpha_J(t), v_{I,j}^{ref}} \Phi = \|\mathbf{S}\mathbf{v}_I(t) - \mathbf{v}_E(t)\|_2^2 \tag{9}$$

For a more detailed derivation of the GFA, we refer interested readers to the original publication [36].

The formulation in Equation (9) is a nonlinear programming problem that requires a global optimization algorithm to solve. In the following, we describe an alternative and more computationally efficient procedure for solving the regression problem in the GFA. The procedure is based on decomposing the nonlinear least square regression above into two linear regression problems. First, for a given reference flux vector $\mathbf{v}_I^{ref}$, one can formulate the following linear regression problem to obtain the least square estimate of α_J:

$$\mathbf{v}_E(t_k) = \beta(t_k)\mathbf{S}\mathbf{\Psi}_I\alpha(t_k) = \beta(t_k)\mathbf{S}\begin{bmatrix} v_{I,1}^{ref} & & \\ \vdots & 0 & \\ v_{I,n_{J_1}}^{ref} & & \\ & v_{I,n_{J_1}+1}^{ref} & 0 \\ 0 & \vdots & \\ & v_{I,n_{J_1}+J_2}^{ref} & \\ & & \ddots \end{bmatrix}\begin{bmatrix} \alpha_{J_1}(t_k) \\ \alpha_{J_2}(t_k) \\ \vdots \end{bmatrix} \tag{10}$$

where $\mathbf{\Psi}_I$ is an $n \times e$ matrix with e being the number of enzymes involved in the glycosylation network, and n_{J_l} is the number of fluxes catalyzed by the enzyme J_l. $\mathbf{\Psi}_I$ is constructed by grouping the reference fluxes ($v_{I,j}^{ref}$) according the enzyme that catalyzes the reactions and stacking each group (block-)diagonally. In this manner, each $v_{I,j}^{ref}$ is multiplied by the corresponding enzyme-specific factor (α_J). In the second linear regression, given $\alpha_J(t_k)$, one obtains $v_{I,j}^{ref}$ by solving for the least square estimates of the following problem:

$$\mathbf{v}_E(t_k) = \beta(t_k)\mathbf{S}\mathbf{\Omega}(t_k)v_{I,j}^{ref}$$

$$= \beta(t_k)\mathbf{S}\begin{bmatrix} \alpha_{J_1}(t_k) & & & \\ & \ddots & & 0 \\ & & \alpha_{J_1}(t_k) & \\ & & & \alpha_{J_2}(t_k) \\ & & & & \ddots \\ 0 & & & & \alpha_{J_2}(t_k) \\ & & & & & \ddots \end{bmatrix}\begin{bmatrix} v_{I,1}^{ref} \\ \vdots \\ v_{I,n_{J_1}}^{ref} \\ v_{I,n_{J_1}+1}^{ref} \\ \vdots \\ v_{I,n_{J_1}+n_{J_2}}^{ref} \\ \vdots \end{bmatrix} \tag{11}$$

where $\mathbf{\Omega}$ is an $n \times n$ diagonal matrix with the enzyme-specific factor (α_J) as its diagonal elements.

Given the two linear regressions in Equations (10) and (11), we estimated α_J and $\mathbf{v}_I^{ref}$ following an iterative procedure as follows:

(1) generate a uniformly distributed random vector of $\mathbf{v_I}^{ref}$ within a biologically feasible range ($v_{I,j}^{ref} \in [0, 25]$; unit : pg/cell/day),

(2) given $\mathbf{v_I}^{ref}$ from step (1), solve for or update α_J using Equation (10),

(3) given α_J from step (2), solve for or update $\mathbf{v_I}^{ref}$ using Equation (11), and

(4) repeat steps (2) and (3) until the change of Φ as described in Equation (9) becomes smaller than a threshold (default 10^{-10}).

Equations (10) and (11) were solved following the iterative procedure above using the MATLAB subroutine lsqlin constraining α_J to values between zero and 20 and $\mathbf{v_I}^{ref}$ to be positive. In order to improve the chance of obtaining the global minimum of Φ, we adopted a multi-start strategy and ran the aforementioned iterative estimation for multiple random initial vectors of $\mathbf{v_I}^{ref}$. Among the results of the multi-start runs, we took the best α_J and $\mathbf{v_I}^{ref}$ values that correspond to the lowest Φ value. Note that the multi-start strategy is embarrassingly parallel and can be implemented on a multiprocessor computing platform.

2.4. Random Forest for Regression

For identifying explanatory variables of the dynamical changes in the galatosyltransferase activity $\alpha_{J,GalT}$, we formulated a regression problem using the change in the GalT enzyme-specific factors over time $d\alpha_{J,GalT}/dt$ as the response variable, and the experimental parameters of perfusion cell cultures as predictor variables. More specifically, we are interested the following regression problem:

$$\frac{d\alpha_{J,GalT}}{dt} = g\left(t, \alpha_{J,GalT}, \mathbf{p}\right) \tag{12}$$

which describes the dynamic change in $\alpha_{J,GalT}$ using the function $g\left(t, \alpha_{J,GalT}, \mathbf{p}\right)$ that depends on time, $\alpha_{J,GalT}$, and process parameters $\mathbf{p}$. Note that the function $g\left(t, \alpha_{J,GalT}, \mathbf{p}\right)$ is likely nonlinear in nature. Here, we employed Random Forest (RF) [40] to build the above regression model using data from all four perfusion cell culture experiments. RF regression involves building an ensemble of unpruned regression trees, in which each regression tree is created using a bootstrap sample of the original dataset. At each node of a tree, a subset of predictors is selected randomly to determine the best decision split of the samples. The final prediction of the regression trees in RF is obtained by averaging the predictions of the entire ensemble. Notably, RF regression is able to capture nonlinear dependencies of the response variable on the predictors.

In our work, we applied RF using normalized data of each variable, in which the data were centered and divided by the standard deviation. We created a RF regression model using an ensemble of 100 trees and employed one third of the total predictors for the decision split at each node. Predictor variables (features) were subsequently ranked based on their average impurity gain over all splits and all trees. Predictor variables with higher impurity gains contribute more to the variability in the prediction of the response variable, and hence are considered more important.

3. Results

3.1. Perfusion Cell Culture Experiments

We performed four perfusion cell cultures (Experiment A, B, C and D) using different VCD and PR set-point profiles, as illustrated in Figure 3. In Experiments A and B, we kept the VCD set-point constant for the entire duration of the cell cultures but shifted the PR set-point between day nine and ten (down in Experiment A and up in Experiment B). In Experiment C, we changed the VCD and PR set-points in a manner that maintained the CSPR at the same value. Finally, in Experiment D, we varied the VCD set-point while leaving the PR set-point constant. As shown in Figure 3, the VCD and PR followed the set-points very well with only minor deviations (also see Table S1). As the control of VCD was done only by bleeding (i.e., removal of cells), the VCD unsurprisingly tracked a decrease in the set-points better than an increase. Supplementary Figures S1–S4 give a summary of the cell culture parameters, including VCD, PR, CSPR, the concentrations of glucose (Glc), lactose (Lac), ammonia (Amm) and IgG, and the cell-specific growth rate (μ), for all experiments. As shown in Figure 4, the cell-specific productivities of IgG, i.e., the secretion rate of IgG divided by the VCD, in all experiments follows a decreasing trend over the course of the cell cultivation. Correspondingly, the experimental secretion rates of the IgG glycoforms computed in the data pre-processing decrease with time (see Figure 5 and Supplementary Figure S5).

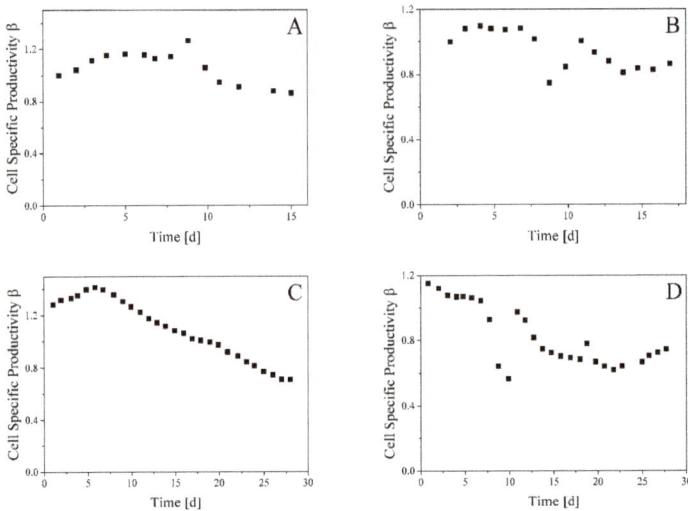

Figure 4. Cell-specific productivity. The cell-specific productivity generally decreases over the course of the four perfusion cell culture experiments (**A**), (**B**), (**C**) and (**D**).

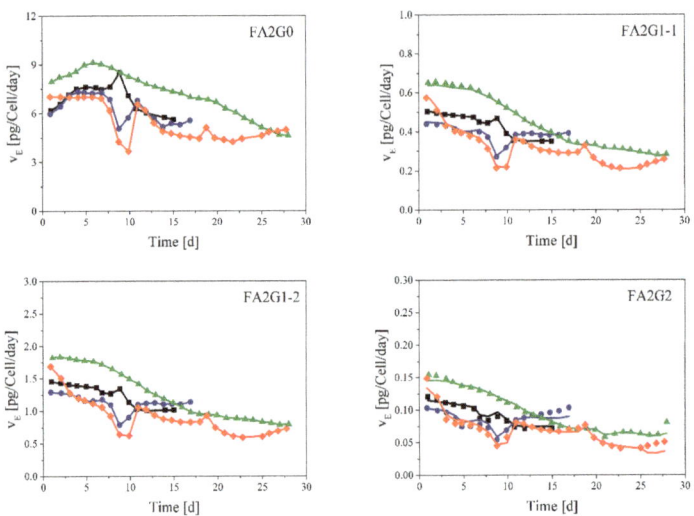

Figure 5. Secretion fluxes of the main IgG glycoforms. The solid symbols show the experimental secretion fluxes computed in the data preprocessing step, as outlined in Section 2.2 (Experiment A: black squares, Experiment B: blue circles, Experiment C: green triangle, Experiment D: red diamonds). The lines show the secretion fluxes from the fitting of α_j in the GFA, as outlined in Section 2.3.

3.2. Glycosylation Flux Analysis

For the GFA, we employed the IgG glycosylation reaction network depicted in Figure 6, which consists of 19 IgG glycoforms and 25 IgG glycosylation reactions. The glycosylation reaction network was based on a previously published network [41], where we omitted glycosylation reactions and molecules corresponding to IgG glycoforms that are not detected in our experiments and do not participate as intermediate species. Since each perfusion cell culture was started in the same manner, we used the same reference IgG glycosylation flux vector $\mathbf{v}_I^{ref}$ in the GFA of all experiments, more specifically using day one of experiment A as the reference time sample point (i.e., $\mathbf{v}_I^{ref}$ of day one in Experimental A is set to the vector of 1 s). Figure 5 shows the fitting of IgG glycoform secretion fluxes in the GFA for the major glycoforms (see Supplementary Figure S5 for the rest of IgG glycoforms).

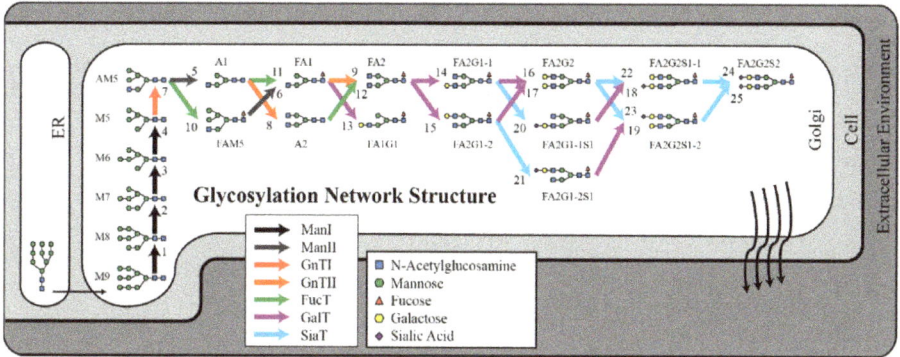

Figure 6. Glycosylation network for the GFA of immunoglobulin G in CHO-S. The enzyme names are abbreviated as follows: α-Mannosidase I and II (Man I/II), *N*-Acetylglucosaminyltransferase I and II (GnT I/II) and Fucosyltransferase (FucT), Galactosyltransferase (GalT) and Sialyltransferase (SiaT). The glycan labels are provided in Supplementary Table S2.

We compared the computational speed of the iterative GFA in analyzing the perfusion cell culture datasets with that of the original implementation. On a test computer (3.33 GHz Intel Xeon W3680, 18 GB RAM), the iterative GFA converged within 5 min using 250 random starting points. Of note, among the 250 random starts, 40% converged to the same minimum Φ value. Meanwhile, the original implementation of GFA using the global optimization toolbox MEIGO [42] required at least 42 min to converge. However, among ten repeated independent runs of MEIGO, only one converged to the same minimum Φ value as the iterative GFA, while the other runs converged to higher Φ values.

Figure 7 gives the time profiles of the enzyme-specific factors $\alpha_j(t)$ for each of the enzymes in the IgG glycosylation network. Most of the enzyme-specific factors, particularly those of α-Mannosidase I and II (Man I/II), *N*-Acetylglucosaminyltransferase I and II (GnT I/II) and Fucosyltransferase (FucT), maintain a constant activity level throughout the cell cultivation (i.e., $\alpha_j(t) \approx 1$). Meanwhile, the galactosyltransferase (GalT)-specific factor decreases during the beginning of the four experiments and varies with changes in the VCD and PR set-points. Since the fluxes catalyzed by sialyltransferase (SiaT) are close to zero, the estimate of the corresponding $\alpha_j(t)$ becomes unreliable due to high sensitivity to experimental noise, and thus is omitted from further analysis.

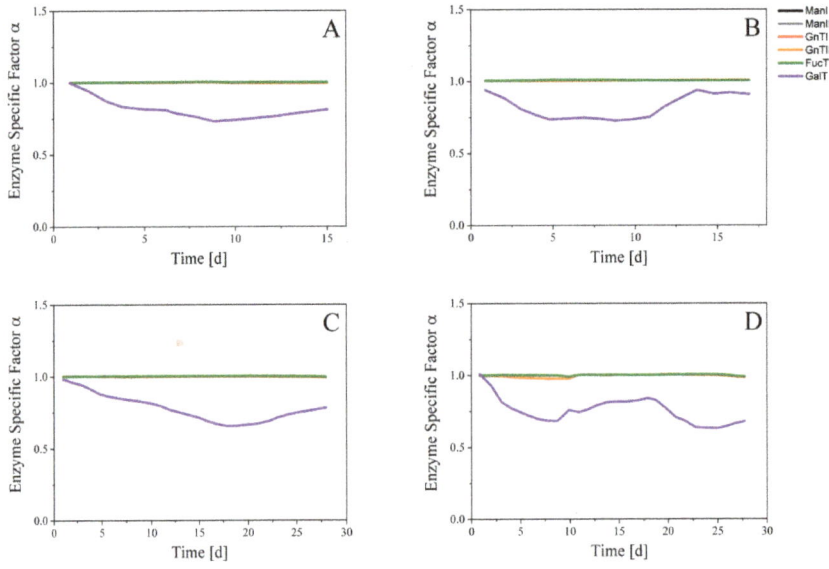

Figure 7. Predicted enzyme-specific factors. The activity of fluxes catalyzed by Man I (black), Man II (grey), GnT I (red), GnT II (orange) and FucT (green) remain relatively constant in all experiments. However, the fluxes catalyzed by GalT (purple) shows significant variation over the course of the four perfusion cell culture experiments (**A**), (**B**), (**C**) and (**D**).

3.3. Effects of Process Parameters on Glycosylation

As mentioned above, the enzyme-specific factor of IgG galactosylation fluxes displays the most dynamical change during the perfusion cell culture. However, the relationship between the changes in the enzyme-specific factor $\alpha_{GalT}(t)$ and the other process parameters is difficult to discern by a simple observation of the experimental data. For this reason, we employed a random forest regression analysis using the change of galactosyltransferase-specific factor over time $\frac{d\alpha_{GalT}(t)}{dt}$ as the response variable and using fourteen process parameters as the predictor variables (see Section 2.4). A RF regression model should be able to capture any nonlinear dependencies between the response and predictor variables. Here, we considered the following predictors: $\alpha_{GalT}(t)$, VCD, PR, B, CSPR, time and the concentrations of IgG, Glc, Lac and Amm, and the specific productivities of IgG, Glc, Lac and Amm (i.e., q_{IgG}, q_{Glc}, q_{Lac} and q_{Amm}, respectively). Furthermore, we excluded data from the startup period of the cell culture (i.e., days one to three of each experiment), as we were more interested in the regulation of IgG glycosylation during the steady state operations and setpoint changes of the perfusion cell culture. Finally, we ranked the predictor variables in decreasing magnitudes of the impurity gains. A higher impurity gain points to a predictor variable with higher importance in explaining the response variable.

Figure 8 gives the ranking of the predictor variables in decreasing impurity gains. The specific productivity of IgG (q_{IgG}) and the concentration of ammonia (Amm) are the two most important predictors of the dynamical changes in the galactosyltransferase specific activity. Indeed, when we repeated the RF regression using only q_{IgG} and Amm as the predictor variables, we observed a similar quality of data fitting to the response variables (see Supplementary Figure S6). Kolmogorov-Smirnov (KS) test and Wilcoxon rank sum test further confirmed that the residuals of the RF regression models using all 14 predictors and those using only q_{IgG} and Amm, are not statistically different (KS test *p*-value = 0.857; Wilcoxon rank sum test *p*-value = 0.824). Following q_{IgG} and Amm in the ranking are the glucose uptake rate (q_{GLC}) and bleed rate (B), indicating that the specific growth rate has a moderate contribution to the changes in the IgG galactosylation activity. The influence of growth rate on IgG galactosylation is somewhat expected, as the nucleotide sugar UDP-Galactose is used for different forms of cellular glycosylation, not solely for IgG [43].

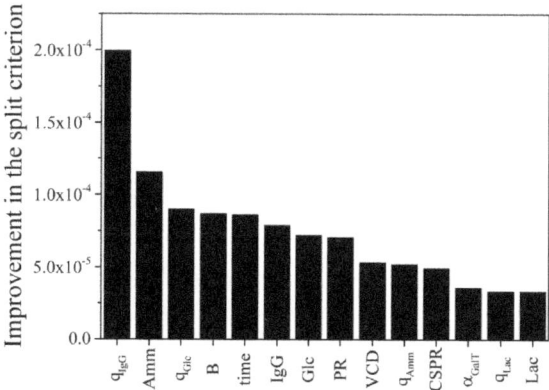

Figure 8. Ranking of predictors based on importance. The predictors are sorted in decreasing impurity gains as measured by the improvement in the split criterion.

4. Discussion

In this work, we improved the computational efficiency of the GFA and applied the GFA to analyze the IgG glycosylation of four perfusion CHO cell culture experiments. The GFA relies on a stoichiometric model of the IgG glycosylation network to estimate IgG intracellular glycosylation reaction fluxes from the secretion fluxes of IgG glycoforms [36]. The GFA assumes that the intracellular IgG glycosylation fluxes vary with time according to two multiplicative factors: α_J and β_J. The enzyme-specific factor α_J represents the relative change of the fraction of IgG that can processed by a specific enzyme, while the cell-specific factor β_J describes the relative change of the specific productivity of IgG. In particular, α_J captures dynamical alterations in the intracellular IgG glycosylation fluxes that cannot be explained by increases or decreases in the amount of IgG entering the glycosylation network.

We caution against making an inference on the intracellular IgG glycosylation fluxes directly from the cell culture data—i.e., without using the appropriate mass and flux balances. For example, a larger fraction of an IgG glycoform could result from a higher secretion rate of the specific IgG glycoform or from lower secretion rates of the other glycoforms. Furthermore, as noted in Section 2.3, the flux balance around the cells in Equation (6) is underdetermined, implying that multiple solutions there exist for the intracellular IgG glycosylation fluxes for the same observed secretion fluxes. The GFA gets around such an issue by assuming that reactions catalyzed by the same enzyme vary with time commensurately. As demonstrated in the study, by quantifying dynamic intracellular glycosylation activities in an enzyme-specific manner, the GFA enables the identification of key process parameters that are indicative of and potentially contribute to the changes in the intracellular IgG glycosylation process.

In contrast to other approaches that rely on kinetic modeling of the glycosylation network, the GFA does not require specifying a large number of kinetic parameters, whose values are uncertain and likely system specific. However, the GFA requires time-series data for computing the secretion fluxes of different glycoforms, which include viable cell density, IgG titer and glycoform fractions. Such data may not always be available during the production of IgG. The GFA is also unable to provide mechanistic insights for the dynamic changes in the enzyme-specific factors α_J, at least not directly. In this study, we performed a post-analysis using a random forest regression to identify the key process parameters that are able to explain the variability in $\alpha_J(t)$.

The four perfusion cell cultures in this study differed among each other in the VCD and PR set-point profiles, which were designed with the goal of understanding how IgG productivity and glycosylation vary with the set-points and process variables during steady-state operations. The cell culture conditions (i.e., media composition, seeding density, pH, temperature) of the four experiments were kept being as similar as possible, so that the effect of changes in operating set-points VCD and PR on the glycosylation process could be examined.

In our previous study using the same CHO cells, we reported that the cell-specific productivity of IgG in fed-batch cultivations increases with time, and that such an increase is associated with lower α_Js for the upstream enzymes ManI/II, GnTI/II and FucT [36]. We attributed the decreasing α_Js to the inability of these enzymes to process the increasing amount of IgG entering the intracellular glycosylation network, beyond the processing capacity of each enzyme. In contrast to fed-batch cultivations, the cell-specific productivity of IgG generally decreases over the course of the perfusion cell culture, as shown in Figure 4. The GFA shows that the enzyme-specific factors of the aforementioned upstream enzymes stay approximately constant at roughly 1 throughout the experiments. With α_J staying near 1, the IgG intracellular glycosylation fluxes associated with these enzymes thus vary proportionally with the cell-specific productivity $\beta(t)$. This trend is not surprising, considering that the amount of IgG that needs to be processed decreases over time.

The enzyme-specific factor with the most dynamical changes in the perfusion cell cultures corresponds to galactosyltransferase (GalT), which parallels that in the fed-batch cultivations. However, unlike the reduction in the GalT-specific factor $\alpha_{GalT}(t)$ over time during the fed-batch cultivations [36], $\alpha_{GalT}(t)$ dynamics in perfusion cell cultures showed both up and downward trends during the different steady-state operation periods, as shown in Figure 7. Because of the rich dynamics of $\alpha_{GalT}(t)$ in the perfusion cell culture, the relationship among the GalT-specific factor $\alpha_{GalT}(t)$ and process variables was not immediately obvious. By applying a random forest regression analysis, we identified the specific productivity of IgG and the concentration of ammonia as the two variables with the highest importance in explaining the dynamic changes of IgG galactosylation (see Figure 8). The accumulation of ammonia and its inhibitory activity on GalT have previously been reported as one of the main reasons for decreasing IgG galactosylation in fed-batch cultivations [36,44]. Unlike ammonia, the connection between the IgG specific productivity and galactosylation is not immediately clear. The partial correlation between $\frac{d\alpha_{GalT}(t)}{dt}$ and q_{IgG} is small and negative (partial correlation = -0.037), suggesting that the relationship between the two variables as revealed by RF analysis is likely to be nonlinear. The negative partial correlation further implies that keeping all other process parameters the same, a higher q_{IgG} is concomitant with decreasing α_{GalT} with time. Such a trend is in general agreement with how α_J of the upstream enzymes vary with the cell-specific productivity of IgG as explained earlier.

Taken together, the key process variables that are associated with IgG galactosylation in our perfusion cell cultures resemble those in the fed-batch cultivations of the same CHO cells. Note that the aforementioned similarity between fed-batch and perfusion cell cultures cannot be concluded by a direct observation of the process data and is only apparent through the combination of the GFA and random forest regression analysis. Such similarity suggests the possibility of using extensive data and knowledge from traditional batch/fed-batch production to inform or predict the IgG glycosylation in continuous perfusion cell cultures, which is a topic of interest in our research groups. In a separate publication, we noted comparable IgG glycoform patterns between our perfusion cell culture process at steady state and the fed-batch cultivation beyond the stationary growth phase [15]. In the literature, past studies produced contradictory conclusions on the IgG glycosylation when comparing the two cell culture modes, where some studies demonstrated similarities [9,10] and others showed discrepancies in the IgG glycan structures [13,14]. A recent report of a head-to-head comparison between perfusion and fed-batch cell cultures noted a strong similarity in the IgG glycosylation dynamics [45], an observation that aligns well with the result of our analysis.

5. Conclusions

The application of glycosylation flux analysis to IgG production data from four perfusion CHO cell culture experiments shed light on the possible controlling factors of IgG glycosylation. The activity of galactosyltransferase and thus intracellular IgG galactosylation fluxes displayed the most time-varying changes during the perfusion cell cultivation. The GFA, coupled with a random forest regression analysis, pointed to the cell-specific productivity of IgG and the concentration of ammonia in the media as the most important process parameters for explaining the dynamical alterations in GalT activity. These observations parallel our previous report on IgG glycosylation in fed-batch production. The similarity between fed-batch and perfusion cell culture, especially in how IgG galactosylation flux activity varies with the cell culture parameters, suggests the possibility of using extensive data and knowledge from batch/fed-batch production to predict the dynamics of IgG glycosylation in the continuous perfusion cell cultures.

Supplementary Materials: The following are available online at http://www.mdpi.com/2227-9717/6/10/176/s1, Table S1: Summary of the reactor set points and the corresponding measured values, Table S2: Glycan structures in glycosylation network, Figure S1: Process measurements of Experiment A, Figure S2: Process measurements of Experiment B, Figure S3: Process measurements of Experiment C, Figure S4: Process measurements of Experiment D, Figure S5: Secretion flux fitting of other IgG glycoforms, Figure S6: Residuals of random forest regression.

Author Contributions: S.H., M.W., M.M. and R.G. conceived and designed the study; S.H. performed the computational analysis; M.W. performed the wet lab experiments; N.P.G. performed the random forest regression analysis; S.H., D.L., T.S. and R.G. developed the improved GFA; M.M. supervised the wet lab experiments; S.H., M.W. and R.G. wrote the paper.

Funding: This research was funded by Swiss Economic Ministry KTI (CTI) program grant number 19190.2 PFIW—IW and by ETH Zurich.

Acknowledgments: The authors would like to thank Ernesto Scibona and Thomas Villiger for their help and explanations regarding the glycosylation process.

Conflicts of Interest: No conflicts of interest are declared.

References

1. Ecker, D.M.; Jones, S.D.; Levine, H.L. The therapeutic monoclonal antibody market. *MAbs* **2015**, *7*, 9–14. [CrossRef] [PubMed]
2. Kelley, B. Industrialization of MAb production technology. *MAbs* **2009**, *1*, 443–452. [CrossRef] [PubMed]
3. Li, F.; Vijayasankaran, N.; Shen, A.; Kiss, R.; Amanullah, A. Cell culture processes for monoclonal antibody production. *MAbs* **2010**, *2*, 466–479. [CrossRef] [PubMed]
4. Rathore, A.S. Roadmap for implementation of quality by design (QbD) for biotechnology products. *Trends Biotechnol.* **2009**, *27*, 546–553. [CrossRef] [PubMed]
5. Food and Drug Administration. *FDA Guidance for Industry. PAT—A Framework for Innovative Pharmaceutical Development, Manufacturing, and Quality Assurance*; Food and Drug Administration: Rockville, MD, USA, 2004.
6. Reay, D.; Ramshaw, C.; Harvey, A. *Process Intensification: Engineering for Efficiency, Sustainability and Flexibility*, 2nd ed.; Elsevier: New York, NY, USA, 2013.
7. Boedeker, B.G.D. Recombinant factor VIII (Kogenate®) for the treatment of Hemophilia A: The first and only world-wide licensed recombinant protein produced in high-throughput perfusion culture. In *Modern Biopharmaceuticals*; Wiley-VCH Verlag GmbH & Co. KGaA: Weinheim, Germany, 2013; pp. 429–443.
8. Clincke, M.F.; Molleryd, C.; Zhang, Y.; Lindskog, E.; Walsh, K.; Chotteau, V. Very high density of CHO cells in perfusion by ATF or TFF in WAVE bioreactorTM, Part I: Effect of the cell density on the process. *Biotechnol. Prog.* **2013**, *29*, 754–767. [CrossRef] [PubMed]
9. Meuwly, F.; Weber, U.; Ziegler, T.; Gervais, A.; Mastrangeli, R.; Crisci, C.; Rossi, M.; Bernard, A.; von Stockar, U.; Kadouri, A. Conversion of a CHO cell culture process from perfusion to fed-batch technology without altering product quality. *J. Biotechnol.* **2006**, *123*, 106–116. [CrossRef] [PubMed]

10. Lee, S.-Y.; Kwon, Y.-B.; Cho, J.-M.; Park, K.-H.; Chang, S.-J.; Kim, D.-I. Effect of process change from perfusion to fed-batch on product comparability for biosimilar monoclonal antibody. *Process Biochem.* **2012**, *47*, 1411–1418. [CrossRef]

11. Ryll, T.; Dutina, G.; Reyes, A.; Gunson, J.; Krummen, L.; Etcheverry, T. Performance of small-scale CHO perfusion cultures using an acoustic cell filtration device for cell retention: Characterization of separation efficiency and impact of perfusion on product quality. *Biotechnol. Bioeng.* **2000**, *69*, 440–449. [CrossRef]

12. Lüllau, E.; Kanttinen, A.; Hassel, J.; Berg, M.; Haag-Alvarsson, A.; Cederbrant, K.; Greenberg, B.; Fenge, C.; Schweikart, F. Comparison of Batch and Perfusion Culture in Combination with Pilot-Scale Expanded Bed Purification for the Production of Soluble Recombinant β-Secretase. *Biotechnol. Prog.* **2003**, *19*, 37–44. [CrossRef] [PubMed]

13. Lipscomb, M.L.; Palomares, L.A.; Hernández, V.; Ramírez, O.T.; Kompala, D.S. Effect of production method and gene amplification on the glycosylation pattern of a secreted reporter protein in CHO cells. *Biotechnol. Prog.* **2005**, *21*, 40–49. [CrossRef] [PubMed]

14. Zhuang, C.; Zheng, C.; Chen, Y.; Huang, Z.; Wang, Y.; Fu, Q.; Zeng, C.; Wu, T.; Yang, L.; Qi, N. Different fermentation processes produced variants of an anti-CD52 monoclonal antibody that have divergent in vitro and in vivo characteristics. *Appl. Microbiol. Biotechnol.* **2017**, *101*, 5997–6006. [CrossRef] [PubMed]

15. Karst, D.J.; Steinebach, F.; Morbidelli, M. Continuous integrated manufacturing of therapeutic proteins. *Curr. Opin. Biotechnol.* **2018**, *53*, 76–84. [CrossRef] [PubMed]

16. Berger, M.; Kaup, M.; Blanchard, V. Protein glycosylation and its impact on biotechnology. In *Genomics and Systems Biology of Mammalian Cell Culture*; Hu, W.S., Zeng, A.-P., Eds.; Springer: Berlin/Heidelberg, Germany, 2012; pp. 165–185.

17. Aebi, M. N-linked protein glycosylation in the ER. *Biochim. Biophys. Acta-Mol. Cell Res.* **2013**, *1833*, 2430–2437. [CrossRef] [PubMed]

18. Solá, R.J.; Griebenow, K. Glycosylation of therapeutic proteins: An effective strategy to optimize efficacy. *BioDrugs* **2010**, *24*, 9–21. [CrossRef] [PubMed]

19. Jefferis, R. Glycosylation as a strategy to improve antibody-based therapeutics. *Nat. Rev.* **2009**, *8*, 226–234. [CrossRef] [PubMed]

20. Goh, J.S.Y.; Liu, Y.; Liu, H.; Chan, K.F.; Wan, C.; Teo, G.; Zhou, X.; Xie, F.; Zhang, P.; Zhang, Y.; et al. Highly sialylated recombinant human erythropoietin production in large-scale perfusion bioreactor utilizing CHO-gmt4 (JW152) with restored GnT I function. *Biotechnol. J.* **2014**, *9*, 100–109. [CrossRef] [PubMed]

21. Harding, F.A.; Stickler, M.M.; Razo, J.; DuBridge, R.B. The immunogenicity of humanized and fully human antibodies: Residual immunogenicity resides in the CDR regions. *MAbs* **2010**, *2*, 256–265. [CrossRef] [PubMed]

22. Matasci, M.; Hacker, D.L.; Baldi, L.; Wurm, F.M. Protein therapeutics Recombinant therapeutic protein production in cultivated mammalian cells: Current status and future prospects. *Drug Discov. Today Technol.* **2008**, *5*, 37–42. [CrossRef] [PubMed]

23. Jayapal, K.P.; Wlaschin, K.F.; Hu, W.-S.; Yap, M.G.S. Recombinant protein therapeutics from CHO cells—20 years and counting. *Chem. Eng. Prog.* **2007**, *103*, 40–47.

24. Wright, A.; Morrison, S.L. Effect of glycosylation on antibody function: Implications for genetic engineering. *Trends Biotechnol.* **1997**, *15*, 26–32. [CrossRef]

25. Moremen, K.W.; Ramiah, A.; Stuart, M.; Steel, J.; Meng, L.; Forouhar, F.; Moniz, H.A.; Gahlay, G.; Gao, Z.; Chapla, D.; et al. Expression system for structural and functional studies of human glycosylation enzymes. *Nat. Chem. Biol.* **2017**, *14*, 156–162. [CrossRef] [PubMed]

26. Radhakrishnan, D.; Robinson, A.S.; Ogunnaike, B.A. Controlling the glycosylation profile in mAbs using time-dependent media supplementation. *Antibodies* **2018**, *7*, 1. [CrossRef]

27. Ivarsson, M.; Villiger, T.K.; Morbidelli, M.; Soos, M. Evaluating the impact of cell culture process parameters on monoclonal antibody N-glycosylation. *J. Biotechnol.* **2014**, *188*, 88–96. [CrossRef] [PubMed]

28. Varki, A. Biological roles of oligosaccharides: All of the theories are correct. *Glycobiology* **1993**, *3*, 97–130. [CrossRef] [PubMed]

29. Umaña, P.; Bailey, J.E. A Mathematical model of N-linked glycoform biosynthesis. *Biotechnol. Bioeng.* **1997**, *55*, 890–908. [CrossRef]

30. Krambeck, F.J.; Bennun, S.V.; Narang, S.; Choi, S.; Yarema, K.J.; Betenbaugh, M.J. A mathematical model to derive *N*-glycan structures and cellular enzyme activities from mass spectrometric data. *Glycobiology* **2009**, *19*, 1163–1175. [CrossRef] [PubMed]

31. Jimenez del Val, I.; Nagy, J.M.; Kontoravdi, C. A dynamic mathematical model for monoclonal antibody N-linked glycosylation and nucleotide sugar donor transport within a maturing Golgi apparatus. *Biotechnol. Prog.* **2011**, *27*, 1730–1743. [CrossRef] [PubMed]

32. Jiménez del Val, I.; Constantinou, A.; Dell, A.; Haslam, S.; Polizzi, K.M.; Kontoravdi, C. A quantitative and mechanistic model for monoclonal antibody glycosylation as a function of nutrient availability during cell culture. *BMC Proc.* **2013**, *7*, O10. [CrossRef]

33. Jedrzejewski, P.M.; Jiménez del Val, I.; Constantinou, A.; Dell, A.; Haslam, S.M.; Polizzi, K.M.; Kontoravdi, C. Towards controling the glycoform: A model framework linking extracellular metabolites to antibody glycosylation. *Int. J. Mol. Sci.* **2014**, *15*, 4492–4522. [CrossRef] [PubMed]

34. Spahn, P.N.; Hansen, A.H.; Henning, G.; Arnsdorf, J.; Kildegaard, H.F.; Lewis, N.E.; Arnsdorf, J.; Kildegaard, H.F.; Lewis, N.E.; Markov, A. A markov chain model for N-linked protein glycosylation—Towards a low-parameter tool for model-driven. *Metab. Eng.* **2015**, *33*, 52–66. [CrossRef] [PubMed]

35. Spahn, P.N.; Hansen, A.H.; Kol, S.; Voldborg, B.; Lewis, N.E. Predictive glycoengineering of biosimilars using a Markov chain glycosylation model. *Biotechnol. J.* **2017**, *12*, 1–8. [CrossRef] [PubMed]

36. Hutter, S.; Villiger, T.K.; Brühlmann, D.; Stettler, M.; Broly, H.; Soos, M.; Gunawan, R. Glycosylation flux analysis reveals dynamic changes of intracellular glycosylation flux distribution in Chinese hamster ovary fed-batch cultures. *Metab. Eng.* **2017**, *43*, 9–20. [CrossRef] [PubMed]

37. Antoniewicz, M.R. Methods and advances in metabolic flux analysis: A mini-review. *J. Ind. Microbiol. Biotechnol.* **2015**, *42*, 317–325. [CrossRef] [PubMed]

38. Wolf, M. Development and Optimization of Mammalian Cell Perfusion Cultures for Continuous Biomanufacturing. Ph.D. Thesis, ETH Zurich, Zurich, Switzerland, 2018.

39. Karst, D.J.; Serra, E.; Villiger, T.K.; Soos, M.; Morbidelli, M. Characterization and comparison of ATF and TFF in stirred bioreactors for continuous mammalian cell culture processes. *Biochem. Eng. J.* **2016**, *110*, 17–26. [CrossRef]

40. Breiman, L. Random forests. *Mach. Learn.* **2001**, *45*, 5–32. [CrossRef]

41. Villiger, T.K.; Scibona, E.; Stettler, M.; Broly, H.; Morbidelli, M.; Soos, M. Controlling the time evolution of mAb N-linkedglycosylation—Part II: Model-based predictions. *Biotechnol. Prog.* **2016**, *32*, 1135–1148. [CrossRef] [PubMed]

42. Egea, J.A.; Henriques, D.; Cokelaer, T.; Villaverde, A.F.; MacNamara, A.; Danciu, D.-P.; Banga, J.R.; Saez-Rodriguez, J. MEIGO: An open-source software suite based on metaheuristics for global optimization in systems biology and bioinformatics. *BMC Bioinform.* **2014**, *15*, 136. [CrossRef] [PubMed]

43. Del Val, I.J.; Polizzi, K.M.; Kontoravdi, C. A theoretical estimate for nucleotide sugar demand towards Chines Hamster Ovary cellular glycosylation. *Sci. Rep.* **2016**, *6*, 28547. [CrossRef] [PubMed]

44. Gawlitzek, M.; Ryll, T.; Lofgren, J.; Sliwkowski, M.B. Ammonium alters N-glycan structures of recombinant TNFR-IgG: Degradative versus biosynthetic mechanisms. *Biotechnol. Bioeng.* **2000**, *68*, 637–646. [CrossRef]

45. Walther, J.; Lu, J.; Hollenbach, M.; Yu, M.; Hwang, C.; McLarty, J.; Brower, K. Perfusion cell culture decreases process and product heterogeneity in a head-to-head comparison with fed-batch. *Biotechnol. J.* **2018**, e1700733. [CrossRef] [PubMed]

Review

Rotor-Stator Mixers: From Batch to Continuous Mode of Operation—A Review

Andreas Håkansson [1,2]

[1] Department of Food Technology, Engineering and Nutrition, Lund University, SE-221 00 Lund, Sweden; andreas.hakansson@food.lth.se; Tel.: +46-44-250-38-26

[2] Department of Food and Meal Science, Kristianstad University, SE-291 88 Kristianstad, Sweden

Received: 16 March 2018; Accepted: 30 March 2018; Published: 3 April 2018

Abstract: Although continuous production processes are often desired, many processing industries still work in batch mode due to technical limitations. Transitioning to continuous production requires an in-depth understanding of how each unit operation is affected by the shift. This contribution reviews the scientific understanding of similarities and differences between emulsification in turbulent rotor-stator mixers (also known as high-speed mixers) operated in batch and continuous mode. Rotor-stator mixers are found in many chemical processing industries, and are considered the standard tool for mixing and emulsification of high viscosity products. Since the same rotor-stator heads are often used in both modes of operation, it is sometimes assumed that transitioning from batch to continuous rotor-stator mixers is straight-forward. However, this is not always the case, as has been shown in comparative experimental studies. This review summarizes and critically compares the current understanding of differences between these two operating modes, focusing on shaft power draw, pumping power, efficiency in producing a narrow region of high intensity turbulence, and implications for product quality differences when transitioning from batch to continuous rotor-stator mixers.

Keywords: rotor-stator mixer; high shear mixer; inline; batch; continuous; emulsification; mixing

1. Introduction

A fully continuous mode of production is often desired in most types of industrial processing. Continuous production decreases the per unit cost of production and reduces the risk of quality differences between batches. However, batch processing is still commonly employed in many production lines, especially in the food, pharmaceutical, and cosmetics sectors.

Continuous production does introduce some additional difficulties. The first requirement is that each unit operation in the production process can be achieved in a continuous setup. Second, converting a given batch process into a continuous one requires an understanding of what (if any) differences there are between the batch and continuous versions of each unit operation. This second difficulty also applies to product development projects for continuous mode production, since laboratory testing is almost always performed in batch.

This review focus on liquid processing in rotor-stator mixers (RSMs), also known as high-shear mixers. RSMs, together with high-pressure homogenizers, are considered the standard tool for mixing and emulsification of liquid dispersions. High-pressure homogenizers are generally used for low to intermediate viscosity products and RSMs for products with higher viscosities [1]. Whereas high-pressure homogenization is an inherently continuous operation, RSMs can be operated in either batch or continuous mode. The same rotor-stator head is often used in both batch and continuous mode of operation, see illustrations in Figures 1 and 2. Continuous mode RSMs are sometimes referred to as inline (or in-line) RSMs.

A. **Continuous**
   mode of operation

B. **Batch**
   mode of operation

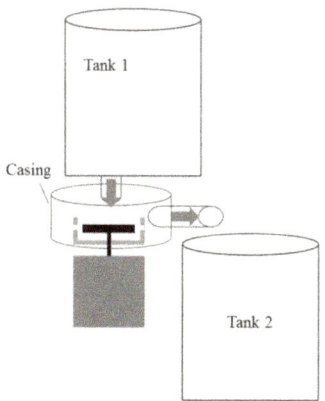
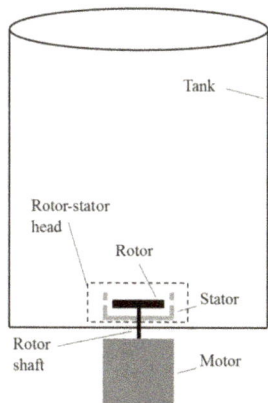

Figure 1. Schematic drawings of rotor-stator mixers (RSMs) operated in continuous (**A**) and batch (**B**) mode of operation.

A. **Continuous**
   mode of operation

B. **Batch**
   mode of operation

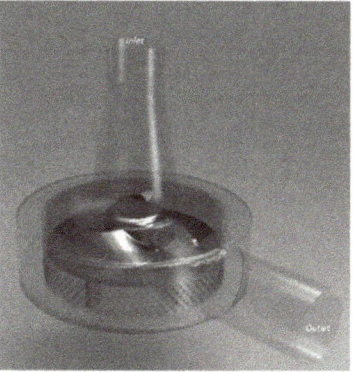
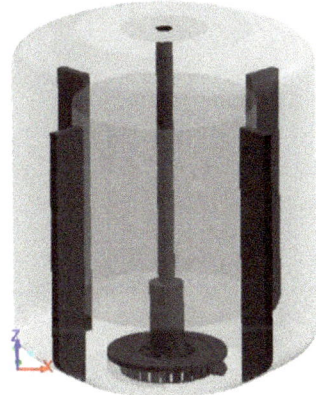

Figure 2. Schematics of RSMs for use in continuous (**A**) and batch (**B**) mode of operation.

From an industrial perspective, it is important to understand how the characteristics of a given batch mode RSM compare to those of a given continuous mode RSM. This understanding is crucial for both converting an existing batch production process to continuous production, and for generalizing results from laboratory (batch) experiments to pilot and production scale in a product development process. Since the rotor-stator heads are often very similar between batch and continuous modes of operation, it is tempting to assume that converting between the modes is straight-forward, both in terms of production economy (i.e., power draw of the rotor shaft) and in terms of obtained product quality (mixing or dispersing efficiency). However, as has become apparent in the scientific literature, this is not obviously the case [2,3]. Great care must therefore be taken when comparing RSMs run in batch to RSMs run in continuous mode of operation.

RSMs are used in many different applications, particularly in the food, pharmaceutical, and cosmetic processing industries. However, as pointed out in an editorial from 2001, despite their wide use, there has been a lack of fundamental understanding [4]. During the last 16 years there has been an increasing number of scientific research projects aimed at characterizing and understanding RSMs. Three major reviews have been published since then, summarizing many of these advances. In 2004, Atiemo-Obeng and Calabrese provided a comprehensive review of the mechanical designs of RSMs, focusing on power draw, flow profiles, and scale-up [5]. This review was also updated in 2016 [6]. Another fairly recent review has been provided by Zhang et al. [7], focusing on power draw and flow fields, but also providing an overview of the proposed emulsification scaling-laws and the mass and energy transfer correlations. However, none of these previous reviews provide a comprehensive discussion on the difference between batch and continuous mode of operation. Moreover, there has been a number of relevant studies on this in the last couple of years, after these reviews were written.

The objective of this contribution is to provide a more specific review on what is known about the similarities and differences between RSMs operated in batch and continuous mode, including the most recent advances. The intention is to provide an overview, both for engineering professionals struggling with the transition from batch to continuous rotor-stator mixing, and for the research community utilizing or studying RSMs. After a brief description of RSMs, this review will focus on four topics: The shaft power draw (power requirements) of batch and continuous mode RSMs (Section 3), the flowrate and pumping capacity of RSMs (Section 4), the pumping and turbulent dissipation efficiency (Section 5), and the implications of these differences on emulsion processing (Section 6). A summary of recommendations for further studies to resolve remaining issues is provided in Section 7, and the review is concluded in Section 8. Although RSMs can be operated under both laminar and turbulent conditions, this review will focus on turbulent RSMs, which are the most common RSMs employed in industrial applications.

2. The Rotor-Stator Mixer

The term RSM does not refer to a specific design but a range of mixer geometries [5]. RSMs are produced by several different manufacturers, and each have their own design, or more often several different designs for use with different applications, see References [5,7] for detailed overviews of different RSM geometries.

The common denominator of these RSMs is that they all consist of one or several high velocity rotors and one or several static stator screens separated by a short distance, the rotor-stator clearance, δ. The rotor accelerates the fluid tangentially and redirects it radially through the stator holes or slots. This gives rise to steep velocity gradients in the stator slot, or in the direct proximity of it, and creates a narrow region of high intensity hydrodynamics stresses [8–15], which give rise to high mixing and dispersing efficiency, characteristic of RSMs [14,15].

Although there are many different rotor-stator designs, they can broadly be classified into two groups based on the rotor: teeth-designs and blade-designs. A schematic view of the two design principles can be seen in Figure 3. As seen in the figure, the blade-design uses a rotor similar to that found in a centrifugal pump, either extending all the way from the shaft (as in the figure) or with shorter blades mounted on a plate attached to the rotor. The teeth-design uses a circular plate-mounted rotor, as seen in Figure 3. Both blade- and teeth-designs can use different stator screens (differing in the shape and size of the holes). Many rotor-stator heads also have multiple (i.e., 2–3) sets of concentric rotors and stators.

Figure 4 displays velocity profiles calculated with computational fluid dynamics (CFD) from two recently published investigations; one on blade-design [16] and one on teeth-design [11]. Looking at the general outline of the flow, there are many similarities. We can see how the fluid obtains a high tangential velocity in the rotor-stator clearance region, and how it is accelerated into a turbulent jet as it enters the (outer) stator slot. Note that the jet attaches to the leading edge of the stator and that the jet only fills a small portion of the slot [8,11,12]. This gives rise to a re-circulation region in the slots and, consequently, a "back-flow" of fluid that re-enters the slot from the bulk without passing the rotor [5,8,12,15,17,18].

The same rotor-stator heads are often used in both batch and continuous RSMs [3,19], the difference is primarily in how they are mounted, and how the product flow is subjected to the rotor-stator. In batch operation, the rotor-stator is mounted inside a mixing tank, either as an integrated part of the bottom of the tank as in Figure 1 (as is often the case in production-scale batch RSMs) or mounted on an impeller shaft lowered into the tank (more common for laboratory-scale batch RSMs). When used for continuous production, the rotor-stator head is mounted inside a narrow casing, similar to a centrifugal pump, with an inlet directing fluid towards the center of rotation and an outlet mounted at the periphery, see Figure 2.

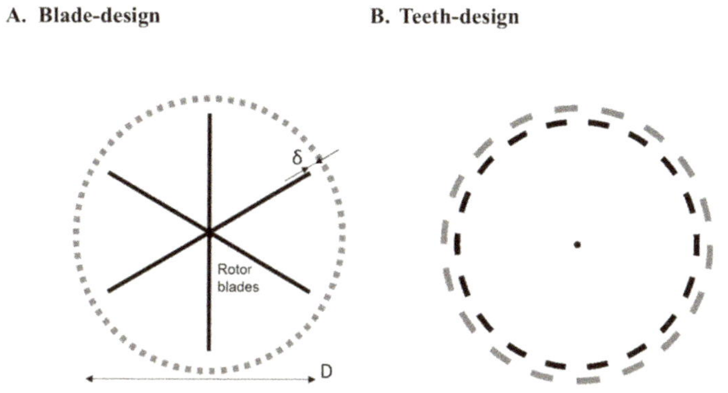

A. Blade-design   **B. Teeth-design**

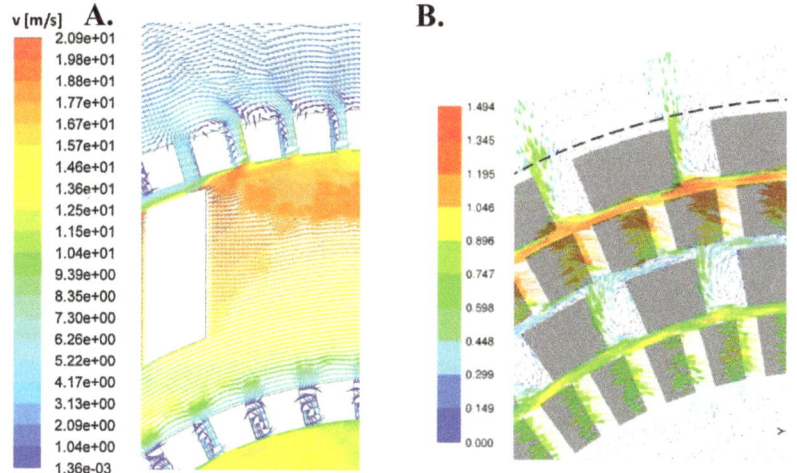

Figure 3. Schematic representation of a rotor-stator head with blade-design (**A**) and teeth-design (**B**). D denotes the rotor diameter, and δ denotes the rotor-stator clearance.

Figure 4. Velocity fields calculated using computational fluid dynamics (CFD) for two rotor-stator heads. (**A**) The flow field in a blade-design (Silverson RSM), reproduced with permission from [16], published by Elsevier, 2015; (**B**) The flow field in a teeth-design (Fluko RSM) from Xu et al. [11], reproduced with permission from [11]; published by Willey Online Library, 2013. Both show systems with two rows of rotor-stators.

3. Shaft Power Draw

The economic benefit of using an RSM depends on the power draw, P_{shaft}, required to operate the rotor shaft at a given speed. Much attention has been put into investigating and predicting the power draw for RSMs.

3.1. Batch Mode RSMs

A batch RSM is principally an impeller mixer, and the power draw of such systems have been under scientific investigation since the 1880s [20]. From dimensional analysis, it has been suggested that shaft power scales with rotor speed (N), rotor diameter (D), and fluid density (ρ) [21,22]:

$$N_P = \frac{P_{\text{shaft}}}{\rho N^3 D^5}. \tag{1}$$

Under turbulent conditions, the power number, N_P, is constant with respect to impeller speed and diameter, but depends on tank and impeller geometry [21–23]. Several studies have shown that Equation (1) is valid for batch RSMs [5,7,8,10,17,24,25]. Fully turbulent, Reynolds number independent, power numbers are obtained above a critical Reynolds number of approximately 10^4 [24–27], with Re defined based on the rotor tip-speed:

$$\text{Re} = \frac{\rho N D^2}{\mu} \tag{2}$$

where μ denotes fluid viscosity. The power number N_P is often found to be in the range 1–3, depending on the geometry of the rotor-stator head [5,7,10,17,24,27] and the tank design [25]. These values are often found to be in the same range or somewhat lower than the power numbers of Rushton type impeller mixers [22,28].

For impeller mixers, several experimental studies have also investigated and established mathematical relations for how the impeller and tank geometry influences N_P. Less information is available for batch RSMs. The tank geometry is expected to influence the power draw less than in tank agitators, since most of the pressure drop occurs in the stator. Regarding RSM geometry, several high-quality investigations have been published [17,24,25,29], but they often compare commercial designs and it has been difficult to establish which geometrical difference is responsible for the observed effects. However, there has been two recent exceptions. One study focused on the effects of stator hole width (keeping all other design parameters constant) and suggests that N_P decreases with increasing stator hole diameter for a single row Tetra Pak design [10]. Another study, conducted on a range of commercial available Silverson designs, suggests that the N_P is proportional to the square of the stator hole area [26]. The discrepancy between these two different systems is still not understood.

3.2. Continuous Mode RSMs

For continuous mode of operation, the power draw is more complex as the flow through the RSM is not set by the system, as in the batch case, but can be varied over a large range by the processing equipment of the continuous line. Experiments reveal that power draw depends on both rotor speed and flowrate. During the last decade several experimental investigations, using a number of different RSM designs, have found support for a three factor model that describes the power draw [7,27,30–34]:

$$P_{\text{shaft}} = \Pi_{\text{rot}} + \Pi_{\text{flow}} + \Pi_{\text{L}} = N_{P0} \rho N^3 D^5 + N_{P1} \rho Q N^2 D^2 + \Pi_{\text{L}}, \tag{3}$$

where Q is the flowrate through the RSM. Equation (3) can be understood by considering that the continuous RSM is something in between batch RSM and a centrifugal pump [5,34]. The first term, Π_{rot} is similar to that found for the batch design, and describes the effect of rotor speed. The second term, Π_{flow}, is similar to the power draw of a centrifugal pump and describes the effect of flowrate. The third term, Π_L is a loss-term, representing the energy lost due to vibrations and losses in bearings [33].

Equation (3) contains two constants, N_{P0} and N_{P1}. Just as N_P for the batch RSM, these are design dependent but do not change with flowrate or rotor speed. N_{P0} is often found to be on the order of 0.1 and N_{P1} on the order of 10. See Table 1 for specific values of these constants for a few different RSM designs. Although several studies have compared N_{P0} and N_{P1} values for different commercial designs [27,28,33,35], no systematic studies describing the effect of different design variables have yet been reported.

Although Equation (3) has received substantial experimental support, it should be remembered that RSMs can be operated under a wide set of conditions and such designs differ substantially. A correction including a fourth and a fifth term has been suggested by Jasinska et al. [16]. However, these extra terms only apply when operating the RSM at flowrates that are below what is typically used in commercial applications [16] (p. 47), and can therefore often be neglected.

Equation (3) has been used with success for describing power draw in continuous mode RSMs manufactured by Fluko [27], Silverson [28,33], Tetra Pak [35,36], and Ytron [15]. However, in one study [35], it was reported that for Conti TDS designs it is more appropriate to use a different correlation:

$$P = N_P^*(\text{Re}) \cdot \rho Q N^2 D^2 + \Pi_L. \tag{4}$$

Note that N^*_P in Equation (4) is Reynolds number dependent, in contrast to N_{P0} and N_{P1} in Equation (3) [27]. It is still not completely clear why this difference is observed. The Conti design has no obvious geometrical difference when compared to the Fluko, Silverson, Tetra Pak, and Ystral mixers—for which Equation (3) is valid. The reported Conti design is with a slotted stator and a rotor with both teeth and blades [37], whereas the designs supported by Equation (3) range from teeth-designs to blades, and with a variety of different stator designs [7,28,30–34].

Table 1. A comparison of power and pumping characteristics of some continuous mode rotor-stator mixers (RSMs).

RSM				Power Characteristics		Pumping Characteristics *		Ref.	Figure 5 **
Rotor	Stator	Manufacturer	D (m)	N_{P0}	N_{P1}	c_1	c_2		
Blade	Circular holes	Tetra Pak	0.20	0.11	9.2	0.46	−3.8	[36]	I
Blade	Square holes	Fluko	0.060	0.24	8.4	0.24	−1.1	[27]	II
Blade	Circular holes	Silverson	0.12	0.10	6.4	0.44	−3.2	[34]	III
Teeth (1 row)	Teeth	Ystral	0.12	0.13	9.7	0.23	−2.4	[34]	IV
Teeth (2 rows)	Teeth	Fluko	0.060	0.15	14.5	0.13	−4.5	[27]	V

* See Section 5.2 for definitions and a discussion on pumping characteristics. ** Relates the five mixers to the graphs in Figure 5.

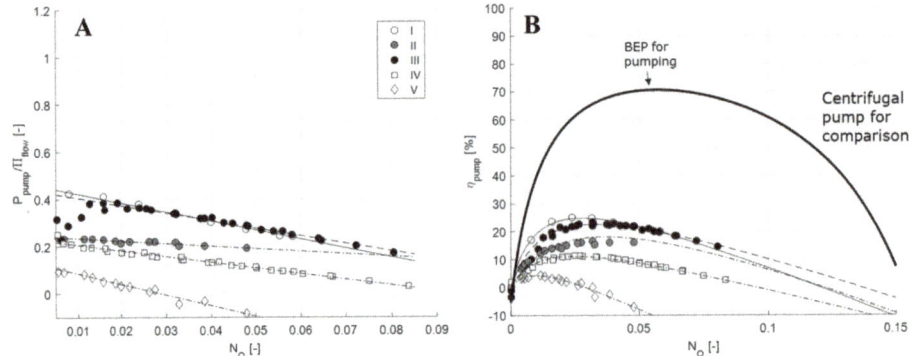

Figure 5. (**A**) Plot of Equation (11) to determine the pump constants c_1 and c_2 for the five mixers in Table 1. (P_{pump} denotes the pumping power of the mixer and Π_{flow} denotes the flow-term in the power-draw correlation, see Section 5.2.) (**B**) Pumping efficiency (η_{pump}) as a function of the flow number (N_Q) for the five mixers, compared to data for a typical centrifugal pump. Data from [27,34,36]. See Reference [35] for the methodology.

4. RSM Flowrate and Pumping

The net flowrate passing through the stator holes, Q, can be described in terms of a flow-number, N_Q, via:

$$N_Q = \frac{Q}{ND^3}. \tag{5}$$

This applies for both batch [8,9,11,12,38] and continuous modes of operation. However, it should be noted that the interpretation and controllability of this value differ substantially between the two modes. For a continuous RSM, flowrate is an externally set and easily measured parameter. Flowrate, and consequently N_Q, can be adjusted by varying what is referred to as the system curve in pump design; the total pressure loss of the system the mixer is connected to. In practice, N_Q can be decreased by using a valve downstream of the mixer or increased by adding a separate feed pump placed in series with the mixer.

When the mixer is operated in batch mode, however, N_Q is a constant that depends on the geometry of the mixing head [8,9,12,17,38], and to some degree on the tank geometry. The N_Q parameter is important for batch RSMs since it determines how fast the liquid is mixed. More specifically the expectation value for the time a fluid element spends in the tank between two passages of the rotor-stator head is [3,19,29]:

$$\tau = \frac{V_T}{Q} = \frac{V_T}{N_Q ND^3}, \tag{6}$$

where V_T is the tank liquid volume.

Another difference between the two modes of operation is that Q (and consequently N_Q) is difficult to measure for batch RSMs; it requires a non-intrusive experimental technique for measuring fluid velocities inside of and just outside of the stator slots, such as laser Doppler anemometry (LDA) [11,12,38] or particle image velocimetry (PIV) [17]. Alternatively, it can be determined by a CFD model that has been validated by one of the above-mentioned experimental techniques [17].

Table 2 compiles values of N_Q for a number of different batch RSMs and compares them to the N_Q span resulting from operating some different continuous mode RSMs under technically relevant flowrates [3,16,27,34,39]. As seen in Table 2, flow numbers are between 0.1 and 0.3 for batch RSMs. Systematic investigations are scarce, but based on a recent PIV investigation, it has been suggested that N_Q decreases with increasing stator slot width. This phenomenon has been linked to the increase in backflow obtained when increasing the slot width [9].

Table 2. Flow numbers (N_Q) for a number of batch and continuous mode RSMs.

RSM				Rotor Speed, U (m/s)	N_Q (-)	Method *	Ref.
Rotor	Stator	Manufacturer	D (m)				
Batch Mode of Operation							
Blade	Rectangular slots	Tetra Pak	0.20	3–14	0.11	PIV	[8]
Blade	Rectangular slots	Tetra Pak	0.20	3–14	0.11–0.15	PIV	[9]
Blade	Circular holes	Silverson	0.0028	3–5	0.22	LDA	[12,38]
Blade	Rectangular slots	Silverson	0.0028	6	0.18	CFD	[17]
Blade	Square holes	Silverson	0.0028	6	0.26	CFD	[17]
Inline Mode of Operation							
Blade	Circular holes	Tetra Pak	0.20	20	0.02–0.06	-	[3]
Blade	Circular holes	Silverson	0.040	6–22	0.0003–0.037	-	[39]
Blade	Circular holes	Silverson	0.022	5–12	0.0003–0.0095	-	[39]
Blade	Circular holes	Silverson	0.0038	6–10	0.002–0.04	-	[16]
Blade	Circular holes	Silverson	0.12	13–19	0.0005–0.08	-	[34]
Teeth	Teeth	Ystral	0.12	13–19	0.0005–0.08	-	[34]
Teeth	Teeth	Fluko	0.060	5–10	<0.05	-	[27]
Blade	Circular holes	Fluko	0.060	5–10	<0.05	-	[27]

* Method used for obtaining the flowrate through the stator slots in batch mode of operation: PIV (particle image velocimetry), LDA (laser Doppler anemometry) and CFD (computational fluid dynamics). (For continuous mode of operation, the flowrate is an externally measureable parameter).

For continuous RSMs, N_Q is substantially lower ($N_Q < 0.1$); it is not uncommon that continuous mode RSMs are operated at a flow number one or several decades below that of batch RSMs. This implies that the flowrates through the mixer are substantially lower for continuous mixers compared to those through batch mixers. Using the same rotor-stator head and operating it at the same rotor speed will therefore result in much lower radial velocities in the rotor-stator region. This difference can also be explained using a centrifugal pump analogy. The flowrate is determined by the properties of the pump (what in pump-theory is referred to as a pump curve) and the properties of the system (the system curve) [16]. Since tanks used with batch RSMs provide a much lower flow resistance than the pipes used for continuous RSMs, the flowrate becomes substantially higher.

5. Pumping and Turbulent Dissipation Efficiency

RSMs are designed to deliver a narrow region of high intensity shear and/or turbulence in order to achieve efficient mixing and emulsification. Under turbulent conditions, this corresponds to delivering a high local energy density or dissipation rate of turbulent kinetic energy (TKE) [13,14,16,40–42]. When comparing batch and continuous mode RSMs, one should therefore keep in mind that the proportion of the supplied energy which can be translated into intense turbulence differs between the two modes of operation.

5.1. Batch RSM

For a batch RSM run under turbulent conditions, all of the shaft energy (except the losses, Π_L) must ultimately be dissipated as heat. This implies that the loss-free shaft power (P'_{shaft}) equals the total dissipated power (P_{diss}):

$$P'_{shaft} = P_{shaft} - \Pi_L = P_{diss}. \tag{7}$$

Not all of this energy is available for mixing or emulsification since these phenomena occur in a narrow region in or around the rotor-stator head [13,40]. The energy that is dissipated outside of this high-intensity region will be an additional loss-term, seen from the point of view of the efficiency of converting energy into efficient mixing. Estimations of the local dissipation rates of TKE in a batch RSM have suggested that approximately 80% of the total dissipation occurs in the high intensity region (independent of rotor speed) for a one-row blade design [13]. This value is also close to the one reported from estimations from CFD simulations [12,17] for a similar geometry.

Assuming that the losses in bearings and those due to vibrations are negligible ($\Pi_L/P_{shaft} << 1$), the energy available for dispersion or mixing in a batch RSM, can be estimated directly by the power number:

$$P_{diss} = 0.8 \cdot N_P \rho N^3 D^5. \tag{8}$$

Since the pump number is relatively easy to determine [33] for a given batch RSM, it is also relatively easy to estimate how much energy is available for turbulent mixing and/or emulsification in a given batch design.

5.2. Continuous Mode RSMs

It is considerably more complicated to calculate how much energy is available for generating turbulence when running an RSM in a continuous mode of operation. In this case, energy provided by the shaft can take two routes: it can either be converted to turbulent fluctuations (and subsequently dissipated as heat), P_{diss}, or it can be used for pumping, P_{pump}. In the latter case, energy is transferred to increase the average velocity or the static pressure of the fluid, often referred to as the "head" across the RSM [35]. These effects are summarized via:

$$P\prime_{shaft} = P_{shaft} - \Pi_L = P_{diss} + P_{pump} <=> P_{diss} = P\prime_{shaft} - P_{pump}. \tag{9}$$

Reformulated using Equation (3):

$$P_{diss} = \Pi_{rot} + \Pi_{flow} - P_{pump}. \tag{10}$$

Note the difference between the Π-terms and the P-terms in Equations (9) and (10). Π_{flow} and Π_{rot} are the terms used in correlations to model power draw (Equation (3)), whereas P_{diss} and P_{pump} are the power associated with the two underlying mechanisms of turbulent dissipation and pumping. As seen below, experimental investigations reveal that the terms in the power draw correlations do not translate directly into terms in the energy balance (e.g., $\Pi_{flow} \neq P_{flow}$). Again, further insight on the continuous RSM can be gained by comparing it to a centrifugal pump, where the pumping power is proportional to the flow-term in Equation (3) (Π_{flow}) and a linear function of the flow number [35,43]:

$$P_{pump} = \Pi_{flow} \cdot (c_1 + c_2 N_Q), \tag{11}$$

where c_1 and c_2 are constants that depend on the design of the centrifugal pump. A similar relation has been shown to hold true for several mixer designs [35]. Figure 5A shows the linear fit of the right-hand side parenthesis in Equation (11), used to experimentally determined the pumping powers for five continuous RSMs using data from several different investigators, RSM manufacturers, and rotor-stator head designs [27,34,36]. See Table 1 for values and design specifications.

Combining Equations (10) and (11) allows one to determine how much of the energy fed into the system is used for pumping (η_{pump}) and how much is dissipated as turbulence (η_{turb}):

$$\eta_{pump} = \frac{P_{pump}}{P\prime_{shaft}} = \frac{(c_1 + c_2 N_Q) N_Q}{N_{P0}/N_{P1} + N_Q}, \tag{12a}$$

$$\eta_{turb} = \frac{P_{diss}}{P\prime_{shaft}} = \frac{N_{P0}/N_{P1} - (1 - c_1) N_Q - c_2 N_Q^2}{N_{P0}/N_{P1} + N_Q}. \tag{12b}$$

Note that the efficiency is given by four empirically determined constants: the power draw constants (N_{P0}, N_{P1}) and the pumping constants (c_1 and c_2). These four constants are relatively easy to determine experimentally by measuring the power draw and the increase in head across an RSM for a range of different rotor-speeds and flowrates (see Reference [33] for the power draw methodology and Reference [35] for the pump-constant methodology).

Figure 5B shows the pumping efficiencies for the five different continuous mode RSMs from Table 1. The pumping efficiency of a centrifugal pump (LKH50, Alfa Laval, Lund, Sweden) has also been inserted as a comparison. Just as for the centrifugal pump, the percentage of energy translated into pumping in a continuous mode RSM (η_{pump}) varies with flowrate (N_Q). For centrifugal pumps, the flowrate with the maximal pumping efficiency, often referred to as the best efficiency point (BEP) is always the desired operating point. However, for the continuous RSM, determining which flowrate is optimal will be considerably more complicated. The primary objective of the RSM is to transfer as much energy as possible into turbulence. This would suggest that a minimal η_{pump} (and hence a maximal η_{turb}) would be desired. However, continuous RSMs are often also designed to contribute to pumping the fluid; they are often used without external feed pumps. Hence, for most application a reasonable balance between pumping and turbulence efficiency is desired.

Note that the teeth-designs generally give rise to lower pumping efficiencies (see Table 1). One of the RSMs in Table 1 even gives a negative efficiency at high flowrates, implying that it is able to convert some of the power supplied from an external feed pump into turbulence. This suggest that teeth-designs are more desirable for applications where RSMs are not intended to contribute to pumping (i.e., when an external feed pump is used) and that blade-designs are more desirable for processes without an external pump.

6. Implications for Emulsification

From an industrial perspective, the most important question with regards transitioning from batch to continuous RSM is how to operate a continuous mode RSM in a way such that it results in the same product quality as batch RSM. In an emulsification context, this corresponds to the question of how to predict the resulting drop diameters in batch and continuous operation RSMs, and to the question of whether there are mechanistic differences between the two products.

Turbulent drop breakup is often explained in terms of Kolmogorov–Hinze theory [44–48], which suggests scaling relations between the largest drop diameter that can survive a given turbulent field and the dissipation rate of TKE of that field. Depending on the size of this limiting drop in relation to the size of the smallest turbulent structures (the Kolmogorov length-scale), different explicit scaling laws have been suggested, see References [7,41,47] for comprehensive reviews and some different explicit formulations.

However, as shown for impeller mixers [49] and high-pressure homogenizers [50], in order for this approach to be satisfactory, the dissipation rate of turbulent kinetic energy should be the local value in the most intense region (where breakup takes place). However, since the local dissipation rate of TKE is highly challenging to measure [49], practical application of Kolmogorov–Hinze theory to RSMs are often based on externally measurable quantities such as rotor speed [31,39,51], the total dissipation power (P_{diss}) [52,53], and the globally defined Reynolds and Weber numbers [41]. The global Weber number is defined as:

$$We = \frac{\rho N^2 D^3}{\sigma}, \tag{13}$$

where σ is the interfacial tension of the drop. The Kolmogorov–Hinze theory is very general, and not specific to design or mode of operation. However, there is some disagreement in the scientific literature when it comes to the question of if there are mechanistic differences between emulsification in the two modes of operation, and hence, if the same scaling law expressions can be used for both modes of operation.

Experimental studies have reported some systematic differences. Emulsions passed n times through a batch RSM do not always show the same drop size as an emulsion processed for a time $t = n\tau$, despite the fact that this would result in the same the number of passages though the RSM, at least in terms of expectation number [2,3]. Three different standpoints discussing this discrepancy and the emulsification implications can be found in the RSM literature.

6.1. Flowrate and Its Influence on Turbulence

As previously mentioned, there is a decisive difference in flowrate (and thus in N_Q) between RSMs operated in batch and continuous mode; batch mode RSMs show approximately ten times higher flowrates (Table 2). The difference between the two systems can be investigated by understanding the effect of flowrate on emulsification. Hall et al. [39] undertook a large systematic investigation of emulsification in continuous mode RSMs and suggested that the flowrate-based Reynold number:

$$\mathrm{Re}_Q = \frac{\rho Q d}{A_{\mathrm{tot}} \mu},$$ (14)

where d is the slot diameter and A_{tot} is the total flow-through area of the stator, has a small but significant effect on the resulting drop size (when tip-speed is kept constant). When keeping the geometry constant, it can be shown that Re_Q is proportional to the product between flow number and Reynolds number [54]:

$$\mathrm{Re}_Q \propto N_Q \mathrm{Re}.$$ (15)

This would suggest that $N_Q \mathrm{Re}$ would be an appropriate scaling law when comparing emulsification results from batch to inline mode of operation. However, it should be kept in mind that the variations in flowrate seen in this type of experiment are much smaller than that between continuous and batch RSMs.

6.2. Radial Flow and Dissiaption Profile Scaling

RSM mixing and emulsification ultimately depends on the hydrodynamic conditions created in the rotor-stator region. Thus, it is interesting to investigate if there are any differences to the flow fields between batch and continuous modes of operation for the same rotor speed and rotor-stator head geometry. Unfortunately, no such experimental studies have been reported. However, a recent CFD investigation might be used to shed some light on the situation [54]. The study [54] reports flow fields obtained with CFD for a continuous mode RSM run at different flowrates and rotor speeds. The lower flowrates correspond to those generally obtained for continuous mode of operation ($N_Q = 0.007$) and the higher to those obtained in batch mode of operation ($N_Q = 0.082$). The radial velocity profiles in the stator holes were compared and it was found that neither Re_Q nor $N_Q \mathrm{Re}_Q$ were appropriate scaling laws, in contrast to what was suggested in Section 6.1. Instead, it was found that both radial and tangential velocities (appropriately scaled with rotor speed) were determined by the flow number [54]. This has an important implication on the difference between the two modes of operation. Since transitioning from batch RSM to a continuous mode RSM decreases N_Q, the velocity profile in the rotor-stator head will undergo a substantial change. This change is not merely a scaling due to the reduction in flowrate but a shift into a fundamentally different turbulent flow [54]. Most notably the position of the highest local dissipation rate of TKE shifts from the turbulent jet formed downstream of the slot in the batch RSM flow number, to the rotor-stator clearance for the continuous RSM flow numbers [54]. This effect is illustrated in Figure 6.

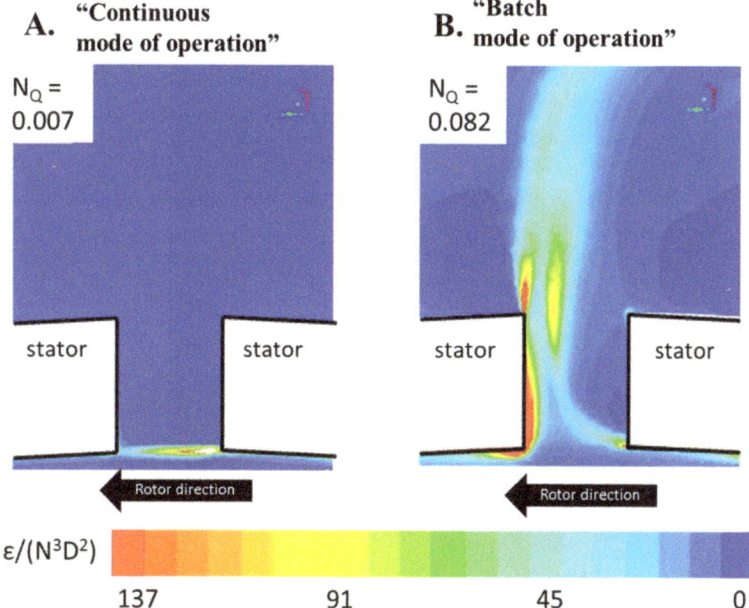

Figure 6. CFD-estimated dissipation rates of turbulent kinetic energy TKE in the stator slot region for the same rotor-stator head run at a flow number corresponding to an RSM run under continuous mode of operation (**A**) and run under batch mode of operation (**B**). Adapted from data obtained in Reference [54].

A shift in the position of highest local dissipation rate of TKE suggests a mechanistic difference between the two modes of operation. As seen in Figure 6, the dissipation volume is smaller for the low-N_Q (continuous mode) case than in the high-N_Q (batch mode) case. This also suggests that the average dissipation rate in the two regions will scale with different parameters: with the clearance length-scale for continuous RSMs and with the slot diameter for the batch RSM. However, this has not yet been experimentally verified.

Due to the lack of experimentally measured flow fields, this difference in flow pattern between modes of operation has not yet been experimentally verified, but a validation study has shown that the CFD model employed is able to capture the position of high intensity local dissipation at least for the batch RSM [55].

Further insight can be obtained by single drop breakup visualizations. However, only one such study on RSMs has yet been reported [42]. This study was conducted for a batch RSM ($N_Q = 0.11$) [8] and showed that drops are deformed and subsequently broken up just downstream of the stator hole [42], as suggested by the CFD simulations for the N_Q-values found in batch RSM (i.e., in Figure 6). However, no corresponding investigations on continuous RSMs (or low N_Q-systems) have yet been reported.

6.3. A Purely Stochastic Effect

An altogether different, but highly promising approach to describe the previously reported differences between batch and continuous modes of operation on emulsification results has recently been reported. Carrillo De Hert and Rodgers [19] suggest that the differences only apply if the wrong scaling is employed when comparing data from different modes of operation.

In the first step of their study, they conclude that the mode drop diameter, d_0, resulting from processing an emulsion n times through their continuous mode RSM at flowrate Q and rotor speed N is given by [19]:

$$d_0 = CQ^{-1/5}N^{-6/5}n^{-1/5}. \tag{16}$$

Note the $-6/5$ exponent, that corresponds to the scaling expected from Kolmogorov–Hinze breakup in the turbulent viscous regime [7,47].

The authors then continue by suggesting that the only difference between continuous and batch modes of operation is in the stochastic effect of the rotor-stator head passage [19]. After processing an emulsion for a time t in a batch system, the expectation number of the number of passages is:

$$n = \frac{t}{\tau}. \tag{17}$$

However, for each volume element of the emulsion, the actual number of passages is a stochastic property following a Poisson distribution. By assuming that Equation (16) (the model for continuous mode of operation) applies each time a volume element passes the rotor-stator head, they conclude that the corresponding model for a batch system after being processed for a time t would be [19]:

$$d_0 = CQ^{1/5}N^{-6/5} \cdot [\exp(t/\tau) - 1]^{-1} \cdot \sum_{n=1}^{\infty} \frac{1}{n^{1/5}n!} \left(\frac{t}{\tau}\right)^n. \tag{18}$$

Moreover, for $t/\tau > 2$, Equation (18) converges to [19]:

$$d_0 = CQ^{1/5}N^{-6/5} \left(\frac{t}{\tau}\right)^{-1/5}, \tag{19}$$

which was found to accurately describe their data for a wide range of properties [19].

These results suggest that there is no mechanistic difference between the modes of operation and that (at least when processing times are fairly large) emulsification results can be translated directly using Equation (17); one continuous mode RSM passage would then correspond directly to processing for t/τ in a batch RSM. However, this is not completely general. In a previous study on emulsification of mayonnaise, it was concluded that this scaling was inadequate to describe the experimental differences [3]. The reason behind this discrepancy is still not understood, but it is hypothesized that it is related to the higher volume fraction of oil in Reference [3] which increases the complexity of the process.

In summary, there is as of yet no consensus in literature on which (if any) of these theories (Sections 6.1–6.3) best describes the differences between emulsification efficiency in batch and continuous modes of operation. Some of the confusion can be explained by postulating that the underlying differences between batch and continuous modes—i.e., the higher flowrate in batch systems—are the different hydrodynamic effects of the rotor-stator head in RSMs with different designs. It might be that the Re_Q-scaling is appropriate for the design investigated by Hall et al. [39] (a pilot scale Silverson dual blade design), that the N_Q-scaling is appropriate for the design investigated by Håkansson et al. [54] (a production scale Tetra Pak blade design), and that there is no mechanistic difference for the design investigated by Carrillo De Hert and Rodgers [19] (a laboratory scale Silverson blade design). However, it is not clear why these different behaviors would occur, and how they could be linked to the design differences. Further investigations are needed in order to draw any definite conclusions on this matter.

7. Suggestion for Further Research

During the last couple of years, significant advances have been made into improving the fundamental understanding of RSMs in general and more specifically on the difference between batch and continuous modes of operation. However, there are still a number of issues that need further investigation, especially when it comes to the difference between continuous and batch modes of operation:

- Although general models for the scaling of power draw with operating parameters have now been obtained for both modes of operation (Equations (1) and (3)), there is still a lack of systematic investigation into how the model parameters (N_P, N_{P0}, N_{P1}) depend on the design parameters (rotor, stator and tank dimensions). A better understanding of this would be helpful in the mechanical design of both batch and continuous mode RSMs.

- As seen throughout this review, the continuous mode RSM could be seen as something between a batch RSM and a centrifugal pump. Further investigations on the relative pumping and turbulence producing properties of different mixer designs (i.e., determination of pumping constants, c_1 and c_2) would be helpful for choosing the right rotor-stator head for a given application.

- A large number of experimental studies correlating drop sizes to operating parameters have been published, but it has been difficult to use these studies to obtain a fundamental understanding of the breakup process or the underlying hydrodynamics of RSMs. The single drop breakup visualizations reported by Ashar et al. [42] shows a promising alternative approach where the breakup probabilities are measured directly and then linked to the local hydrodynamic conditions. Expanding these types of investigations into other RSM geometries and repeating it for a continuous mode RSM is needed to further our fundamental understanding.

- As seen in Section 6, there is some remaining uncertainty whether there exist mechanistic differences between breakup when using the same rotor-stator head in the batch or continuous mode of operation. Comparing the scaling suggested by Carrillo De Hert and Rodgers [19] to data from more RSM designs could be one way towards reaching a more definite conclusion; single drop breakup visualization in a continuous mode RSM at varying N_Q-values would be another interesting way forward.

8. Summary and Conclusions

The objective of this contribution was to review the current scientific based understanding of the differences between RSMs in batch or continuous mode of operation. Section 3 showed that correlations for shaft power draw are available for both modes of operation, allowing for accurate prediction of process economy in terms of energy expenditure. In Section 4, it was seen that the flow number (N_Q), and consequently the flow through the stator screen, Q, is considerably lower for continuous mode of operation (compared to batch mode) when using the same rotor-stator head and operating it at the same rotor speed. Section 5 showed that, in general, a much higher proportion of the energy fed to the shaft is converted into turbulence in the high-intensity region where mixing and emulsification takes place for a batch RSM than for an RSM operated in continuous mode. For a continuous mode RSM, more of the energy is used for pumping (i.e., increasing the head of the flow). Section 6 discussed what this implies when comparing emulsification efficiencies between the two modes of operation. Several different theories have been suggested, but there is of yet no clear consensus in the literature for how continuous mode RSMs should be operated in order to give the same emulsion as in a batch RSM.

Acknowledgments: This study received no external funding. Fredrik Innings at Lund University is acknowledged for valuable discussion and insightful comments on an earlier draft of the manuscript.

Conflicts of Interest: The author declares no conflict of interest.

References

1. Schultz, S.; Wagner, G.; Urban, K.; Ulrich, J. High-pressure homogenization as a process for emulsification. *Chem. Eng. Technol.* **2004**, *27*, 361–368. [CrossRef]
2. Bourne, J.R.; Studer, M. Fast reactions in rotor-stator mixers of different size. *Chem. Eng. Process.* **1992**, *31*, 285–296. [CrossRef]
3. Håkansson, A.; Chaudhry, Z.; Innings, F. Model emulsions to study the mechanism of industrial mayonnaise emulsification. *Food Bioprod. Process.* **2016**, *98*, 189–195. [CrossRef]
4. Calabrese, R. Research needs and opportunities in fluid mixing technology. *Chem. Eng. Res. Des.* **2001**, *79*, 111–112. [CrossRef]
5. Atiemo-Obeng, V.A.; Calabrese, R. Rotor-stator mixing devices. In *Handbook of Industrial Mixing*; Paul, E.L., Atiemo-Obeng, V.A., Kresta, S.M., Eds.; Wiley: Hoboken, NJ, USA, 2004; pp. 479–505.
6. Atiemo-Obeng, V.A.; Calabrese, R. Rotor-stator mixing devices. In *Advances in Industrial Mixing*; Kresta, S.M., Etchells, A.W., Dickey, D.S., Atiemo-Obeng, V.A., Eds.; Wiley: Hoboken, NJ, USA, 2016; pp. 255–258.
7. Zhang, J.; Xu, S.; Li, W. High shear mixers: A review of typical applications and studies on power draw, slot pattern, energy dissipation and transfer properties. *Chem. Eng. Process.* **2012**, *57–58*, 25–41. [CrossRef]
8. Mortensen, H.H.; Calabrese, R.V.; Innings, F.; Rosendahl, L. Characteristics of a batch rotor-stator mixer performance elucidated by shaft torque and angle resolved PIV measurements. *Can. J. Chem. Eng.* **2011**, *89*, 1076–1095. [CrossRef]
9. Mortensen, H.H.; Innings, F.; Håkansson, A. The effect of stator design on flowrate and velocity fields in a rotor-stator mixer—An experimental investigation. *Chem. Eng. Res. Des.* **2017**, *121*, 245–254. [CrossRef]
10. Mortensen, H.H.; Innings, F.; Håkansson, A. Local levels of dissipation rate of turbulent kinetic energy in a rotor-stator mixer with different stator slot widths—An experimental investigation. *Chem. Eng. Res. Des.* **2018**, *130*, 52–62. [CrossRef]
11. Xu, S.; Cheng, Q.; Li, W.; Zhang, J. LDA Measurements and CFD simulations of an in-line high shear mixer with ultrafine teeth. *AIChE J.* **2014**, *60*, 1143–1155. [CrossRef]
12. Utomo, A.; Baker, M.; Pacek, A.W. Flow pattern, periodicity and energy dissipation in a batch rotor-stator mixer. *Chem. Eng. Res. Des.* **2008**, *89*, 1397–1409. [CrossRef]
13. Håkansson, A.; Mortensen, H.H.; Andersson, R.; Innings, F. Experimental investigations of turbulent fragmenting stresses in a Rotor-Stator Mixer. Part 1. Estimation of turbulent stresses and comparison to breakup visualizations. *Chem. Eng. Sci.* **2017**, *171*, 625–637. [CrossRef]
14. Jasinska, M.; Baldyga, J.; Hall, S.; Pacek, A.W. Dispersion of oil droplets in rotor-stator mixers: Experimental investigations and modelling. *Chem. Eng. Process.* **2014**, *84*, 45–53. [CrossRef]
15. Özcan-Taskin, G.; Kubicki, D.; Padron, G. Power and flow characteristics of three rotor-stator heads. *Can. J. Chem. Eng.* **2011**, *89*, 1005–1017. [CrossRef]
16. Jasinska, M.; Baldyga, J.; Cooke, M.; Kowalski, A.J. Specific features of power characteristics of in-line rotor-stator mixers. *Chem. Eng. Process.* **2015**, *9*, 143–159. [CrossRef]
17. Utomo, A.; Baker, M.; Pacek, A.W. The effect of stator geometry on the flow pattern and energy dissipation rate in a rotor-stator mixer. *Chem. Eng. Res. Des.* **2009**, *87*, 533–542. [CrossRef]
18. Espinoza, C.J.U.; Simmons, M.J.H.; Albertini, F.; Mihailova, O.; Rothman, D.; Kowalski, A.J. Flow studies in an in-line Silverson 150/250 high shear mixer using PIV. *Chem. Eng. Res. Des.* **2018**, *132*, 989–1004. [CrossRef]
19. Carrillo De Hert, S.; Rodgers, T.L. Continuous, recycle and batch emulsification kinetics using a high-shear mixer. *Chem. Eng. Sci.* **2017**, *167*, 265–277. [CrossRef]
20. Unwin, W.C. On the friction of water against solid surfaces of different degrees of roughness. *Proc. R. Soc. Lond.* **1880**, *31*, 54–58. [CrossRef]
21. White, A.M.; Brenner, E. Studies in agitation vs. the correlation of power data. *Trans. Am. Inst. Chem. Eng.* **1934**, *30*, 555–597.
22. Rushton, J.H.; Costich, E.W.; Everett, H.J. Power characteristics of mixing impellers. Part 1. *Chem. Eng. Prog.* **1950**, *46*, 395–404.
23. Bates, R.L.; Fondy, P.L.; Corpstein, R.R. An examination of some geometric parameters of impeller power. *Ind. Eng. Chem. Process Des. Dev.* **1963**, *2*, 310–314. [CrossRef]
24. Padron, G.A. Measurement and Comparison of Power Draw in Batch Rotor-Stator Mixers. Master's Thesis, University of Maryland, College Park, MD, USA, 2001.

25. Myers, K.J.; Reeder, M.F.; Ryan, D. Power draw of a high-shear homogenizer. *Can. J. Chem. Eng.* **2001**, *79*, 94–99. [CrossRef]
26. James, J.; Cooke, M.; Trinh, L.; Hou, R.; Martin, P.; Kowalski, A.; Rodgers, T.L. Scale-up of batch rotor-stator mixers. Part 1–Power constants. *Chem. Eng. Res. Des.* **2017**, *124*, 313–320. [CrossRef]
27. Cheng, Q.; Xu, S.; Shi, J.; Li, W.; Zhang, J. Pump capacity and power consumption of two commercial in-line high shear mixers. *Ind. Eng. Chem. Res.* **2012**, *52*, 525–537. [CrossRef]
28. Cooke, M.; Rodgers, T.L.; Kowalski, A.J. Power consumption characteristics of an in-line Silverson high shear mixer. *AIChE J.* **2012**, *58*, 1683–1692. [CrossRef]
29. Nienow, A.W. On impeller circulation and mixing effectiveness in the turbulent flow regime. *Chem. Eng. Sci.* **1997**, *52*, 2257–2565. [CrossRef]
30. Baldyga, J.A.; Kowalski, A.J.; Cooke, M.; Jasinska, M. Investigations of micromixing in the rotor-stator mixer. *Chem. Process Eng.* **2007**, *28*, 867–877.
31. Hall, S.; Pacek, A.W.; Kowalski, A.J.; Cooke, M.; Rothman, D. The effect of scale and interfacial tension on liquid–liquid dispersion in-line Silverson rotor-stator mixers. *Chem. Eng. Res. Des.* **2013**, *91*, 2156–2168.
32. Kowalski, A.J. Power consumption of in-line rotor-stator devices. *Chem. Eng. Process.* **2009**, *48*, 581–585. [CrossRef]
33. Kowalski, A.J.; Cooke, M.; Hall, S. Expression for turbulent power draw of an inline Silverson high shear mixer. *Chem. Eng. Sci.* **2011**, *66*, 241–249. [CrossRef]
34. Sparks, T. Fluid Mixing in Rotor/Stator Mixers. Ph.D. Thesis, Cranfield University, Cranfield, UK, 1996.
35. Håkansson, A.; Innings, F. The dissipation rate of turbulent kinetic energy and its relation to pumping power in inline rotor-stator mixers. *Chem. Eng. Process.* **2017**, *115*, 46–55. [CrossRef]
36. Lindahl, A. Fluid Dynamics of Rotor Stator Mixers. Master's Thesis, Luleå University, Luleå, Sweden, 2013.
37. Schönstedt, B.; Jacob, H.-J.; Schilde, C.; Kwade, A. Scale-up of the power draw of inline-rotor-stator mixers with high throughput. *Chem. Eng. Res. Des.* **2015**, *93*, 12–20. [CrossRef]
38. Pacek, A.; Baker, M.; Utomo, A.T. Characterisation of flow pattern in a rotor stator high shear mixer. In Proceedings of the European Congress of Chemical Engineering, Copenhagen, Denmark, 16–20 September 2007.
39. Hall, S.; Cooke, M.; Pacek, A.W.; Kowalski, A.J.; Rothman, D. Scaling up of silverson rotor-stator mixers. *Can. J. Chem. Eng.* **2011**, *89*, 1040–1050. [CrossRef]
40. Jasinska, M.; Baldyga, J.; Cooke, M.; Kowalski, A. Application of test reactions to study micromixing in the rotor-stator mixer (test reactions for rotor-stator mixer). *Appl. Therm. Eng.* **2013**, *57*, 172–179. [CrossRef]
41. Rueger, P.E.; Calabrese, R.V. Dispersion of water into oil in a rotor-stator mixer. Part 1: Drop breakup in dilute systems. *Chem. Eng. Res. Des.* **2013**, *91*, 2122–2133. [CrossRef]
42. Ashar, M.; Arlov, D.; Carlsson, F.; Innings, F.; Andersson, R. Single droplet breakup in a rotor-stator mixer. *Chem. Eng. Sci.* **2018**, *181*, 186–198. [CrossRef]
43. White, F. *Fluid Mechanics*, 4th ed.; McGraw-Hill: Boston, MA, USA, 1998; ISBN 0070697167.
44. Kolmogorov, A.N. On the breakage of drops in a turbulent flow. *Dokl. Akad. Nauk. SSSR* **1949**, *66*, 825–828.
45. Hinze, J.O. Fundamentals of the Hydrodynamic Mechanism of Splitting in dispersion processing. *AIChE J.* **1955**, *1*, 289–295. [CrossRef]
46. Davies, J.T. Drop sizes of emulsions related to turbulent energy dissipation rates. *Chem. Eng. Sci.* **1985**, *40*, 839–842. [CrossRef]
47. Vankova, N.; Tcholakova, S.; Denkov, N.D.; Ivanov, I.B.; Vulchev, V.D.; Danner, T. Emulsification in turbulent flow 1. Mean and maximum drop diameters in inertial and viscous regimes. *J. Colloid Interface Sci.* **2007**, *312*, 363–380. [CrossRef] [PubMed]
48. Walstra, P. Emulsions. In *Fundamentals of Interface and Colloid Science*; Lyklema, J., Ed.; Elsevier: Amsterdam, The Netherland, 2005; Volume 5, pp. 1–94, ISBN 9780124605305.
49. Zhou, G.; Kresta, S.M. Correlation of mean drop size and minimum drop size with the turbulence energy dissipation and the flow in an agitated tank. *Chem. Eng. Sci.* **1998**, *53*, 2063–2079. [CrossRef]
50. Håkansson, A.; Fuchs, L.; Innings, F.; Revstedt, J.; Trägårdh, C.; Bergenståhl, B. High resolution experimental measurement of turbulent flow field in a high pressure homogenizer model and its implications on turbulent drop fragmentation. *Chem. Eng. Sci.* **2011**, *66*, 1790–1801. [CrossRef]
51. Hall, S.; Cooke, M.; El-Hamouz, A.; Kowalski, A.J. Droplet break-up by in-line Silverson rotor-stator mixer. *Chem. Eng. Sci.* **2011**, *66*, 2068–2079. [CrossRef]

52. Karbstein, H.; Schubert, H. Developments in the continuous mechanical production of oil-in-water macro-emulsions. *Chem. Eng. Process.* **1995**, *34*, 205–211. [CrossRef]

53. Rodgers, T.L.; Cooke, M. Rotor-stator devices: The role of shear and the stator. *Chem. Eng. Res. Des.* **2012**, *90*, 323–327. [CrossRef]

54. Håkansson, A.; Arlov, D.; Carlsson, F.; Innings, F. Hydrodynamic difference between inline and batch operation of a rotor-stator mixer head—A CFD approach. *Can. J. Chem. Eng.* **2017**, *95*, 806–816. [CrossRef]

55. Mortensen, H.H.; Arlov, D.; Innings, F.; Håkansson, A. A validation of commonly used CFD methods applied to Rotor Stator Mixers using PIV measurements of fluid velocity and turbulence. *Chem. Eng. Sci.* **2018**, *177*, 340–353. [CrossRef]

Article

A Reaction Database for Small Molecule Pharmaceutical Processes Integrated with Process Information

Emmanouil Papadakis [1], Amata Anantpinijwatna [2], John M. Woodley [1] and Rafiqul Gani [1,*]

[1]   Department of Chemical and Biochemical Engineering, Technical University of Denmark,
      DK-2800 Kgs. Lyngby, Denmark; empap@kt.dtu.dk (E.P.); jw@kt.dtu.dk (J.M.W.)
[2]   Faculty of Engineering, King Mongkut's Institute of Technology Ladkrabang, 10520 Bangkok, Thailand;
      amatana.dtu@gmail.com
*   Correspondence: rag@kt.dtu.dk; Tel.: +45-4525-2882

Received: 12 September 2017; Accepted: 2 October 2017; Published: 12 October 2017

Abstract: This article describes the development of a reaction database with the objective to collect data for multiphase reactions involved in small molecule pharmaceutical processes with a search engine to retrieve necessary data in investigations of reaction-separation schemes, such as the role of organic solvents in reaction performance improvement. The focus of this reaction database is to provide a data rich environment with process information available to assist during the early stage synthesis of pharmaceutical products. The database is structured in terms of reaction classification of reaction types; compounds participating in the reaction; use of organic solvents and their function; information for single step and multistep reactions; target products; reaction conditions and reaction data. Information for reactor scale-up together with information for the separation and other relevant information for each reaction and reference are also available in the database. Additionally, the retrieved information obtained from the database can be evaluated in terms of sustainability using well-known "green" metrics published in the scientific literature. The application of the database is illustrated through the synthesis of ibuprofen, for which data on different reaction pathways have been retrieved from the database and compared using "green" chemistry metrics.

Keywords: reaction database; pharmaceutical process engineering; organic solvents; "green" metrics analysis

---

1. Introduction

Organic chemistry has an important role to play in the development of synthetic routes for new drugs during early stage process development. To pursue synthesis at a high level, access to chemical information is needed, which can be provided by using knowledge databases, experience, literature review and/or computer-aided tools [1,2]. The retrieved data is used for similarity search, reaction data retrieval, synthesis route planning, drug discovery-development and prediction of physicochemical properties [3]. The development of methods, algorithms and tools to systematize data collection, retrieval of chemical information-data, and to assist the solution approach to many problems related to the synthesis of molecules in organic chemistry has been developed since the 1970s. The methods and tools for reaction synthesis are based on retrieving chemical information organized in chemical reaction databases where data for individual reactions and structural information for different components involved in the reaction are stored.

Computer-aided tools have been developed to solve problems related to "synthesis" and "retrosynthesis." The focus of these tools is to generate a number of possible chemical synthesis paths for possible precursors (synthesis tree) to achieve the synthesis of a given target compound.

In retrosynthesis, the process of generating the possible pathways starts from the given target compound and, by going backwards, the reactions necessary to synthesize the target compound are identified. In addition, the reactions to produce the reactants of identified reactions are generated. The process is repeated until commercially available reactants are identified. These approaches are based on heuristics and logical rules and all of them rely on knowledge databases [4–8]. Recently, computer-aided tools that are based on algorithmic approaches have been developed, such as The Route Designer [9], which automatically extracts rules that capture the essence of the reactions in the chemical reaction database [10]. The tool ICSYNTH utilizes a graph-based approach with available data from the literature to generate the reaction rules [9]. Many other computer-aided methods and tools for reaction synthesis have already been developed with different characteristics. For example, tools to perform combinatorial searches, to screen generated alternatives based on information retrieved from knowledge databases and to perform extensive reaction assessment calculations [11–15].

Searching for reactions and retrieving the relevant information is a complex problem because it involves searching for chemical structures (complete or partial), transformation information (reaction centers), description of the reactions (reaction type, general comments) and numerical data such as experimental reaction data (including conversion, yield, selectivity, reaction conditions etc.). Reaction databases that help to organize, store and retrieve data continue to be developed (Houben-Weyl [16] and Theillheimer [17]), but more recently, the field of reaction databases has evolved further and databases (see Table 1) such as CASREACT [18], ChemReact [17] and REAXYS (previously Beilstein plus Reactions) [19] have been established, while reaction databases such as ChemInform [20] have become well-known.

1.1. General Databases

In these types of database, the information included is focused on organic reactions and synthetic methods in general. The CASREACT reaction database [18] was started in 1840 and since then more than 74.9 million reactions have been added as it is updated daily. The information is related to organic synthesis including organometallics, total synthesis of natural products and biocatalytic (biotransformation) reactions. This database can be used to provide information on different ways to produce the same product (single step or multi-step reactions), used for applications of a particular catalyst and various ways to carry out specific functional group transformations. The REAXYS reaction database [19]—based on data from Elsevier's industry-leading chemistry databases (CrossFire Beilstein, CrossFire Gmelin and Patent Chemistry Database)—includes data for more than 40.7 million reactions, dating from 1771 to the present. It includes a large number of compounds (organic, inorganic and organometallic) and experimental reaction details (yield, solvents etc.). It is searchable for reactions, substances, formulas, and data such as physico-chemical properties data, spectra. Additionally, the REAXYS database can be used for synthesis route planning. The Current Chemical Reaction (CCR) database [21] includes over one million organic reactions together with reaction diagrams, critical conditions and bibliographic data. The Reference library of synthetic methodology (RefLib) covers reaction data from 1946 to 1992. The database contains information from different sources and the latest version has a comprehensive heterocyclic chemistry database [17].

The ChemReact reaction database [17] is a closed database that covers the period from 1974 to 1998 and includes over 3.5 million reactions. It is searchable by reaction type and provides information for the reaction transformation classified by type of reaction and relevant data (bibliographic, spectra and yield). Chemogenesis is a web-book [22], dealing with chemical reactions and chemical reactivity. It examines the rich science between the periodic table and the established disciplines of inorganic and organic chemistry. The Organic Synthesis database [23], includes more than 6000 organic reactions and is searchable by the reaction type or the structure of the compounds and it provides information for single and multi-step organic reaction together with reaction components, conditions and description. The reaction database-Chemical Synthesis [24] enables the user to find reactions related to reagents or target products and it also provides information with the necessary details of the reagents. The Synthetic Pages reaction database [25], covers 292 reactions and provides information for the optimized reaction procedure. It is searchable by reaction type and/or the structure of the reagent or the target product. The Chemical Thesaurus reaction database [26] contains 4000 reactions classified as organic, inorganic, organometallic, transition metal and biochemical.

The WebReaction reaction database [27] covers over 400,000 reactions; it can be searched by defining the structure of the reactant and the product and it performs search based on the reaction similarity with focus on reaction center. The Science of Synthesis database (previously Houben-Weyl) [16] covers information for organic and organometallic reactions with detailed experimental procedures, methodology evaluation and discussion of the field. Finally, the SPRESI reaction database [28] contains 4.6 million reactions and it enables searching of structures, references and reactions.

The Synthetic Reaction Updated (previously Methods in Organic Synthesis) lists many organic reactions (in graphical form) and is searchable by reaction type [29].

1.2. Specialized Databases

These databases are specialized in one class of reaction type. The ChemInform reaction database [20] includes more than 2 million reactions, including organic, enzymatic and microbial reactions. The available data can be used for the application of new reagents and also for catalysts as with the preparation of natural and pharmaceutical products. Other aspects that are covered by the ChemInform database include synthetic procedures, enantio-and diastereoselective syntheses and new protection/de-protection procedures. The Biotage Pathfinder reaction database [30] is specialized in the verified methods of microwave synthesis.

The e-EROS (Encyclopedia of Reagents for Organic Synthesis) [31] focuses on the reagents and catalysts used in organic chemistry for synthesis. The FlowReact Search [32] covers a range of over 2000 flow chemistry reactions adapted from publications on pharmaceutical, fine chemical and biotech companies. The Protecting Groups reaction database [33] provides information for protection, de-protection and trans-protection methods, stability, liability, and reaction conditions, and includes up-to-date information. Recently, a reaction library focused on generic reactions (88 reactions, ~20,000 reactants) with high reliability and reasonable yield has been developed by Masek et al. [34]. The objective of this library is to provide information on synthetically feasible design ideas for de novo drug design.

Representing chemical reactions in a structured way is a complex task. The reaction information contained in a database needs to fulfil several criteria and needs to be categorized with respect to their searchable reaction information. The criteria that a reaction database should fulfill are [17]:

(i)   **Each reaction is an individual record in the database (detailed and graphical).** The reaction must be able to be retrieved from the database as a detailed record (reagents, products, stoichiometry etc.). It can also be extracted as a graphical representation where the reaction scheme is shown. In many databases, the reaction is represented in a graphical form.

(ii)   **Structural information for target product as well as substrates.**

(iii) **Reaction centers.** The reaction center of a reaction is the collection of atoms and bonds that are changed during the reaction [3].

(iv)   **Reaction components must be searchable.** Information for the components involved in the reaction such as reagent, catalysts, solvents etc.

(v)   **Multistep reactions.** In the case of multistep reactions, all reactions (individual and whole pathway) must be searchable.

(vi)   **Reaction conditions.** Conditions such as pH, temperature, pressure etc. should be searchable by exact and a suitable range of values.

(vii) **Reaction classification.** The type of reaction (i.e., esterification) should be searchable.

(viii) **Post-processing of the database contents.** Export of the retrieved reaction data in other tools (i.e., MS Excel).

Many reaction databases have been developed over time—some of them have a large number of reactions available and others a smaller number, and some of the databases cover the whole range of the organic and/or inorganic reactions. There are also reaction databases that cover more specialized reactions such as solid reactions, flow reactions etc. It can also be seen that most of the databases cover the most important criteria as defined by Zass [17], such as the need for individual reaction records (criterion i, in Table 1). In Table 1, existing reaction databases are listed and have been classified based on the different presented criteria. The numbers of reactions, as well as online sources, have also been listed.

Table 1. Database review. All the databases have been summarized with respect to the number of reactions and the focus of the database.

Database	Number of Reaction	Criteria [17]	Reference
CASREACT	>74.9 million (1840–present)	i, iv, v, vi	[18]
REAXYS (previously CrossFire Beilstein)	40.7 million (1771–present)	i, ii, iv, vii	[19]
Theilheimer	>72200 (1946–1980)	i, v, vi, vii	[35]
ChemInform RX	>2 million (since 1990–present)	i, iv, vi	[20]
Current chemical reactions	1,083,758 (1840–present)	i, vi	[21]
Methods in organic synthesis	33,000 (1999–2014)	i, vii	[29]
Reference library of synthetic methodology	209.800 (1946–2001)	i	[17]
ChemReact	3.5 million reactions (1974–1998)	i, vii	[17]
Chemogenesis	-	ii, iii	[22]
Organic synthesis	>6000 (1921–present)	i, ii, v, vi, vii	[23]
Reaction Database-Chemical Synthesis	-	i, ii	[24]
Synthetic Pages	292	i, ii, vi, vii	[25]
The chemical thesaurus	4000	i, ii	[26]
WebReactions	>400,000	i, ii, iii	[27]
Biotage Pathfinder (reaction assisted with microwave technology)	>1000	i, vi, vii, viii	[30]
e-EROS Encyclopedia of Reagents for Organic Synthesis	>70,000 (4000 *)	i, ii	[31]
FlowReact Search	>2000	i (reaction in flow)	[32]
Protecting groups	-	i	[33]
Science of Synthesis (previously Houben-Weyl)	240,000 (early 1800s–present)	i, ii, iii	[16]
SPRESI	4.6 million	i, ii, iii	[28]

The main objective of this article is to assist pharmaceutical process development in the early stages of the synthesis route selection and development, by providing enhanced process understanding. To achieve this task, a data-rich environment where knowledge can be collected, stored and retrieved is a requirement. A database that covers reactions taking place in pharmaceutical processes covering information connected to the criteria listed by Zass [17] and additionally covering process information has been developed to create an environment where process knowledge is available. The connection of individual reactions to criteria like scalability, cost, expected yield, and reaction steps, ease of separation, safety and to parameters such as reaction conditions, experimental data and models, they can improve the process understanding and the decision making process during the synthesis route selection process. In addition to constraints of high product quality and process economics, a pharmaceutical process needs to fulfill the criteria for environmental issues. In particular, for pharmaceutical processes, the environmental sustainability evaluation must be performed during the early stage of process development [36] before the approval of the regulatory bodies as the re-approval of the process can be a very expensive process [37]. Constable et al. [38] has reviewed "green" metrics proposed in literature and these metrics are used to increase the awareness of generated waste sources from the reaction and to identify opportunities for further improvement. The reviewed "green" metrics are listed in Table 2, where for each metric an explanation and the equation to quantify the specific metric are given.

Table 2. List of metrics that have been proposed for "green" chemistry (reviewed by Constable et al. [38]).

Metric	Explanation	Equation
Effective Mass yield (EM)	The percentage of the mass of product over the overall mass of non-benign compounds used during the synthesis.	$EM(\%) = \dfrac{Mass\ of\ products\ (kg)}{Mass\ of\ non-benign\ reagents\ (kg)} \times 100\%$
E-factor	The mass of total waste produced for a given amount of produced product.	$E - factor = \dfrac{Total\ waste\ (kg)}{kg\ product}$
Atom Economy	How much of the reactants remain in the product.	$Atom\ Economy(\%) = \dfrac{MW\ P}{\sum (MW\ A,\ B,\ D,\ F,\ G,\ I)} \times 100$ Where A, B, D, F, G, I: reactants; P: product
Mass Intensity (MI)	Total mass used to produce the product.	$MI = \dfrac{Total\ mass\ used\ in\ a\ process\ or\ process\ step\ (kg)}{Mass\ of\ product\ (kg)}$
Carbon efficiency	Percentage of carbon of the reactants that remain in the final product.	$Carbon\ efficiency\ (\%) = \dfrac{amount\ of\ carbon\ in\ product}{Total\ carbon\ present\ in\ reactants} \times 100$
Reaction mass efficiency (RME)	Mass of reactants remaining in the product.	$RME\ (\%) = \dfrac{mass\ of\ product(kg)}{mass\ of\ reactants\ (kg)} \times 10$

This information, in combination with other knowledge databases and computer-aided synthesis design (CASD) tools developed earlier, provides an opportunity for an integrated approach to the solution of problems related to synthesis route selection and improvement, taking into account important process considerations such as the development time to establish the synthesis route, product quality, cost of manufacture that are often linked to "green" chemistry metrics and the final approval of regulatory agencies [1]. This process related information is not available in the reaction databases listed in Table 1, but is needed for plant-wide design, process-operation simulation and optimization in studies related to sustainability and the economics of processes producing active pharmaceutical ingredients [39–41].

In this article, the developed reaction database is presented with a specific focus on reactions (including multiple reactions) taking place in pharmaceutical processes within the pharmaceutical industry and connecting them with process information. The reactions in this database have been categorized according to the reaction type, the target product to be produced (when single-step or multistep reactions are considered), the reaction product and the effect of the solvent use on the reacting system. Reaction conditions (temperature, pressure etc.), reaction components (reagents, catalysts etc.), reaction data (conversion, selectivity, etc.), scaling information and finally batch or continuous processing is included in the developed database. For each reaction entry, a description of the process exists and the references are provided. A more detailed description of the database development and structure follows later in this article.

This reaction type database, more specifically, aims to:

1. Identify reactions that are used to produce different types of products (Active Pharmaceutical Ingredients (API), Intermediates).
2. Identify reactions to be utilized, for a given compound availability.
3. Investigate the function of different type of solvents in single/multiphase reactive systems.
4. Facilitate the choice of the reaction conditions.
5. Evaluate the reaction pathway in terms of yield, cost and sustainability metrics.
6. Facilitate the reactor design from available experimental data and kinetic models.

In addition, with the process information that is included in the database and has been mentioned in points 1–6 above, the database fulfills most of the criteria defined by Zass [17] (see Table 3). Table 3 provides a comparison of the available database with respect to the criteria given by Zass [17]. It can be noted that most of the available databases provide information for individual reactions (criterion i) and molecular structure information on reactants and products (criterion ii). However, the remaining criteria are covered only in some databases (see Table 3).

Table 3. List of available databases and the criteria [17] they fulfill. Criterion: (i) individual records of reactions, (ii) chemical structure information, (iii) reaction centers, (iv) searchable reaction components, (v) multistep reactions, (vi) searchable reaction conditions, (vii) reaction classification, (viii) post-processing information.

Criterion	CASREACT	REAXYS	Theilheimer	ChemInform RX	Current Chemical reactions	Synthetic Reaction Updates	Reference Library of Synthetic methodology	ChemReact	Chemogenesis	Organic Synthesis	Reaction Database-Chemical Synthesis	Synthetic Pages	The Chemical Thesaurus	Webreactions	Biotage Pathfinder	e-EROS Encyclopedia of Reagents for Organic Synthesis	FlowReact Search	Protecting Groups	Science of Synthesis	SPRESI	This Work
i.	✓	✓	✓	✓	✓	✓	✓	✓	✓	✓	✓	✓	✓	✓	✓	✓	✓	✓	✓	✓	✓
ii.		✓							✓	✓	✓	✓	✓	✓		✓			✓	✓	
iii.	✓													✓					✓	✓	
iv.	✓	✓	✓	✓		✓				✓					✓						✓
v.	✓									✓					✓						✓
vi.										✓		✓			✓						✓
vii.		✓	✓	✓	✓	✓		✓		✓		✓		✓							✓
viii.																					✓

2. Reaction Database

The data required to populate a reaction database to satisfy the abovementioned objectives has been acquired from numerous published articles and patents. The collected knowledge from these sources has been structured in the database according to a developed ontology (knowledge representation) and stored for easy data retrieval and re-use in different likely applications. The database consists of classes, sub-classes, instances and objects. A class is a representation for a conceptual grouping of similar terms. Classes are the focus of most ontology. A class describes concepts in the domain. A class can have subclasses that represent concepts that are more specific than a super class [42]. A simplified flow-diagram, which serves as a guide for the reaction database in terms of knowledge representation system, classes and instances of data and information on the available data, and where information can be found in the article, is shown in Figure 1.

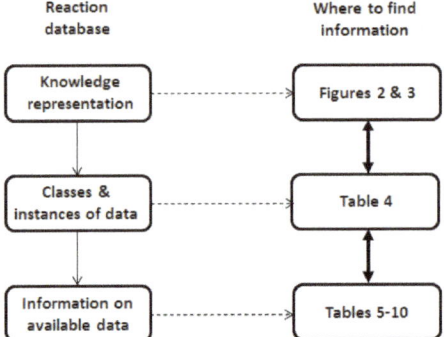

Figure 1. Simplified flow-diagram highlighting the contents of the reaction database. Figures 2 and 3 provide details of the knowledge representation system, Table 4 provides information on the classification of the data and Tables 5–10 provide information on the available data.

2.1. Knowledge Representation

For the development of the reaction database, classes have been used to represent the main knowledge categories such as the reaction type, the reaction, phases involved, how the phases are created, solvent use, solvent function, type of solvent, reaction conditions, available data and finally operation mode (listed in Table 4 and shown in Figure 2). The first knowledge class consists of different reaction types that are commonly found in pharmaceutical processes (i.e., hydrogenation). The set of these reaction types are called the instances of the class. The second class in the knowledge representation system (or data) is the reaction, which is divided in four sub-classes; the reactants, reaction products, and target product and reaction information (see Figure 3). The instances of the three first sub-classes of the second class are classified in terms of name of the compound, type of the compound and molecular structure while the fourth class summarizes information for the specific reaction. This type of information is important to identify the structural changes of the compounds during the reaction. The fourth class of data consists of instances describing the phases involved in the specific reaction. It is important to note that this class connects the reaction information with the reaction performance class, which will be described later, and it has an important role in the database since in this way, the advantages of using a multiphase or a single-phase system can be identified. The next two classes of the database consist of instances describing the solvent function, in case an organic solvent has been used in the reactive system, for example, the solvent function is *"creates a second phase and removes the reaction product,"* and the type and name of the used organic solvent. The last three classes of the data consist of instances describing the reaction performance under certain conditions. The reaction conditions class consists of instances, which have to do with the reaction variables such as reaction temperature, stoichiometric amount, catalyst (type and amount), pH, pressure and the need to use acid or base. The data class consists of four sub-classes, reaction data, dynamic data, kinetic model, and scale. The instances of the reaction data sub-classes are information related to reaction time (or residence time), conversion, selectivity, reaction yield and overall process yield (usually after isolation and purification). The instances of the dynamic data are sets of experimental data that can be used to fit or to develop a kinetic model. The next sub-class describes the availability of kinetic models that can be used either directly, or after fitting to the experimental data for reaction optimization studies. The last sub-class of the data is a super class that provides important information on the scale the reaction has been performed. Finally, the last class of the data is the operation mode, instances of this class can be different operational modes such as batch reaction or flow reaction.

2.2. Database Structure

Table 4 lists the classes of the data in the first column, the second column relates the classes to the instances that an individual class contains and in the third column, the instances are listed for different classes. The structure of the database is visually shown in Figure 2.

Table 4. Main classes of the reaction type database and the instances.

Main Classes	Relation with Instances	Instances
Reaction Type, T	$T = [T_1, T_2, ..., T_i, ..., T_n]$	T_i: reaction type in the knowledge base (i.e., acylation etc.)
Reaction, R	$R = [R_1, R_2, ..., R_i, ..., R_n]$	R_i: reaction of the ith reaction type; for each reaction information about the reactants and reaction products are provided as well as information for the target product and process (for example: 1st step for production of an API)
Phases involved, P	$P = [P_1, P_2, ..., P_i, ..., P_n]$	P_i: phase of the ith reaction (i.e., organic-aqueous, organic-gas etc.)
How phases are created, C	$C = [C_1, C_2, ..., C_i, ..., C_n]$	C_i: (i.e., solvent etc.)
Solvent function, F	$F = [F_1, F_2, ..., F_i, ..., F_n]$	F_i: (i.e., phase creation, carrier etc.)
Solvent type, ST	$ST = [ST_1, ST_2, ..., ST_i, ..., ST_n]$	ST_i: (i.e., ether, alcohol etc.)
Solvent, S	$S = [S_1, S_2, ..., S_i, ..., S_n]$	S_i: Solvents in ith reaction
Reaction condition, RC	$RC = [RC_1, RC_2, ..., RC_i, ..., RC_n]$	RC_i (i.e., Temperature, composition, cat, pH etc.)
Data, D	$D = [D_1, D_2, ..., D_i, ..., D_n]$	D_i (reaction data: conversion, selectivity, reaction time, and dynamic data: concentration vs. time, scale information and kinetic models etc.)
Operation Mode, OP	$OP = [OP_1, OP_2, ..., OP_i, ..., OP_n]$	OP_i: batch, continuous, fed batch

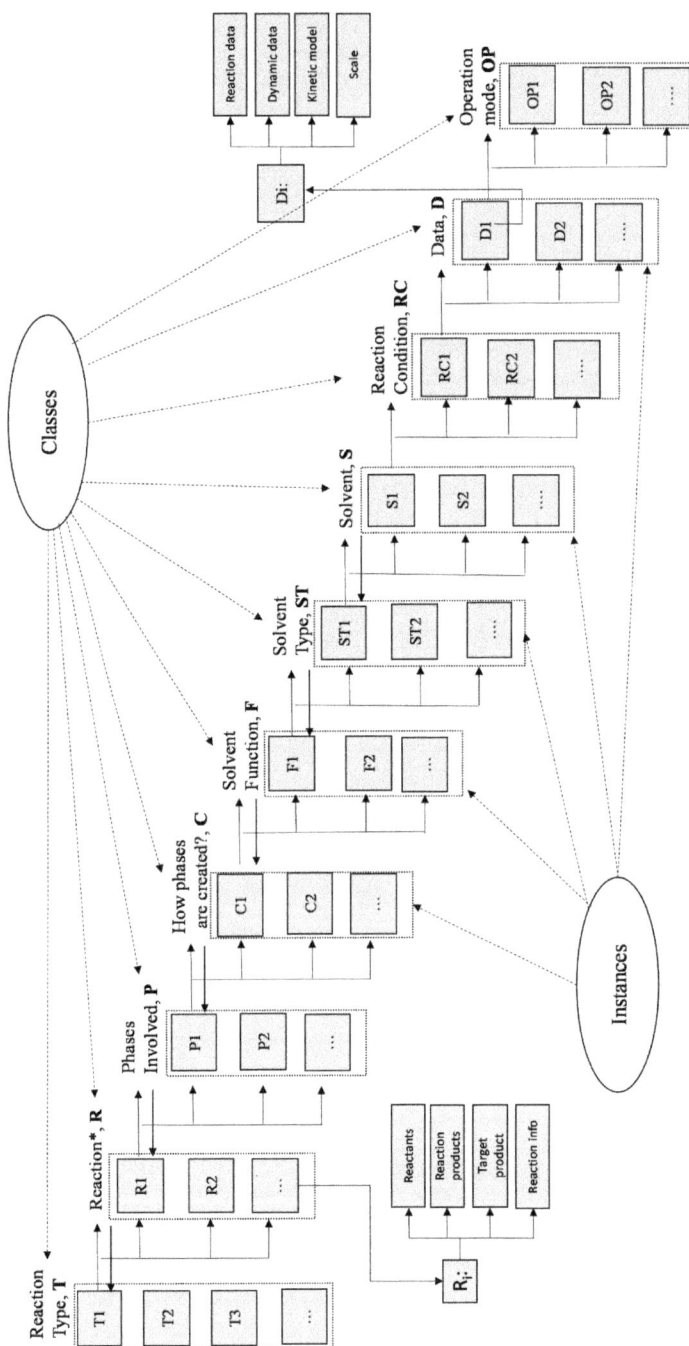

Figure 2. Knowledge representation system of the reaction database.

In Figure 3, the subclasses and the values of each instance in the "Reaction" class are illustrated. For example, each reaction has **reactants**—as well as **reaction products**—and can be used to eventually produce a **target product** (in case of multi-step reactions), each of the sub-classes take values such as the name of the compound (N), the type of the compound (T, for example, alcohol) and the molecular structure of the compound. The reaction info subclasses takes text values that can be used to give useful insights for the reaction.

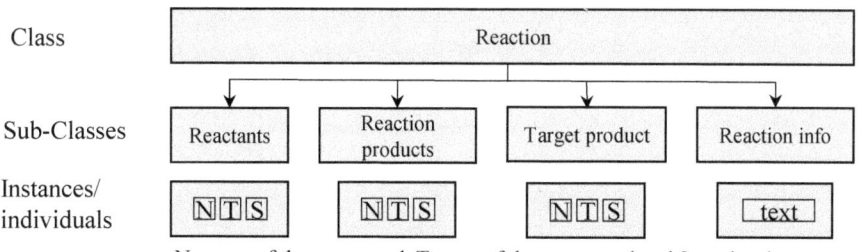

N: name of the compound, T: type of the compound and S: molecular structure

Figure 3. Sub-classes and instance/individuals for the reaction class of the database.

3. Statistics of the Reaction Database

To determine the range of applications and the capability of the reaction database, the statistics of the stored data within the database are needed. The statistics are given in terms of number of reactions, reaction types, list of APIs, reactions where the use of solvent improves the reaction performance and available kinetic models.

3.1. General Numbers

In this section, the general statistics of this database are given, for example, total number of reactions, total number of APIs, the number of the intermediates, reactions that require solvent, multiphase reactions, experimental data, type of reaction operation (batch or continuous, technology i.e., microwave technology). The general characteristics of the reaction type database are listed in Table 5.

Table 5. Summary of the information included in the database.

Category	Number
Total number of reactions	285
Types of reactions	44
Number of multiphase reactions	88
Number of reaction with solvents	226
Solvent	Dissolve, Phase creation, Substrate/catalyst carrier, compound extraction
Number of APIs (with total synthesis pathway)	21
Number of building blocks (type of compounds)	19
Number of experimental data	275 (conversion, selectivity, reaction yield, conditions), 32 (dynamic data), 11 (kinetic models)
Number of production mode data	96 (in flow), 203 (in batch)
Number of application examples	14 (chemicals), 16 (Fine chemicals), 251 (pharmaceuticals)

3.2. Reaction Types

The different reaction types included in the database are listed in Table 6, together with the number of the reactions, the catalyst need, the phases (usually) involved and the solvent function if it used.

Table 6. Reaction types included in database, phases involved and function of the used solvent.

Reaction Type	Catalyst	Phases	Solvent Function
1. Alkylation	Yes	Liquid (org.)	Dissolves reactants
		Liquid (org.)—Liquid (aq.)	Creates second phase
2. Hydrogenation	Yes	Liquid (org.)—Gas	Dissolves reactants
3. Epoxidation		Liquid (org.)—Liquid (aq.)	Creates second phase Reactant/catalyst carrier
4. Carbonylation	Yes	Liquid (org.)—Liquid (aq)—Gas	Creates phase Carrier for catalyst
	Yes	Liquid (org.)—Gas	Dissolves reactants
5. Hydroformulation	?		Creates second phase Catalyst carrier
6. Enzymatic reduction	Yes	Liquid (org.)—Liquid (aq.)	Reactant carrier Creates second phase
7. Arylation	Yes	Liquid (org.)—Liquid (aq.)	Creates second phase
	Yes	Liquid (org.)	Dissolves reactants
8. Oxidation	Yes	Liquid (org.)—Gas	Dissolves reactants
9. Transamination	yes	Liquid (org,)—Liquid (aq.)	Creates second phase Product removal
10. Saponification	No	Liquid (org.)	Dissolves reactants
11. Amidation	Yes/No	Liquid (org.)—liquid (aq.)	Creates second phase removes product
12. Amination	Yes	Liquid (org.)	Dissolves reactants
13. Esterification	Yes	Liquid (org.)	Solvent free
		Liquid (org.)	Dissolves reactant
14. Hydrolysis	Yes	Liquid (org.)—liquid (aq.)	Creates second phase removes product
			Dissolves reactants
15. Aminolysis	yes	Liquid (org.)	Dissolves reactants
16 .Condensation	No	Liquid (org.)	Dissolves reactant
17. Deprotection	No	Liquid (org.)	Dissolves reactant
18. Protection	Yes	Liquid (org.)	Dissolves reactant
19. Dehydration	Yes	Liquid (org.)—liquid (aq.)	Catalyst carrier Create second phase Product removal
20. Cyclization	No	Liquid (org.)—liquid (aq.)	Dissolves reactant Product separation
21. Lithiation	No	Liquid (org.)	Dissolves reactants

Note: aq.: aqueous and org.: organic.

3.3. Active Pharmaceutical Ingredients (APIs)

In Table 7, the list of the available APIs (or the final drug) in the database is given. The database includes at least one pathway for each API listed in Table 7. In some cases, more than two completely different published reaction pathways (for example, for Ibuprofen) exist, which are also listed in the database. Finally, in some cases efforts have been focused on improving a certain reaction within the reaction path that has also been included in the knowledge database.

Table 7. List of APIs and final drugs (*) in the database, of which complete reaction pathway and the reactions are provided in the database.

APIs	
1. 6-aminopenicillanic acid	12. Tramadol
2. Zuchopenthixol	13. Artemisinin
3. 6-Hydroxybuspirone	14. Saxagliptin
4. aliskiren hemifumarate	15. Atazanavir
5. Ibuprofen	16. PDE5 inhibitor *
6. Meclinertant *	17. Axitinib
7. Rufinamide	18. Olanzapine *
8. Ciprofloxacin	19. Amitriptyline
9. Naproxen	20. Tamoxifen
10. OZ439 * (antimalarial drug candidate)	21. Vildagliptin
11. Efavirenz *	

3.4. Reaction with Improved Reaction Performance When Solvent Is Used

Reaction improvements in terms of reaction time, reaction volume, yield, conversion and/or selectivity and post-processing improvement in the separation and purification steps related to solvent use are considered in database development. The functions of solvent and the possible process improvements are listed below and summarized in Table 8:

a. Reaction medium.

b. Separation of the main product in order to shift the equilibrium reaction towards the product side in order to increase the yield and/or reduce the separation steps required.

c. Separation of an inhibitory product to increase the productivity of the reaction.

d. Controlled released of substrate, it might improve the process safety in case of hazardous compounds or increase selectivity towards the desired product.

e. Reaction volume reduction.

f. Dissolves reactants to increase the reaction rate and/or to avoid process complications when the reaction involves compounds in solid phase at the reaction conditions.

Table 8. Solvent functions in reaction and their possible improvements.

		Possible Improvements			
		Productivity	Process safety	Separation steps	Waste reduction
Solvent Functions	Reaction medium	✔	✔	-	-
	Product removal (phase creation)	✔	-	✔	✔
	Substrate carrier (phase creation)	✔	✔	-	✔
	Catalyst carrier (Phase creation)	✔	✔	✔	✔

In Table 9 below, different reactive systems where solvent has been added in order to improve the reaction performance are listed. Table 9 has been classified based on the reaction type and the main product—it also gives the reaction phases, the solvent function and the reaction improvement.

Table 9. List of reactions where the use of the solvent has a specific function that leads in direct reaction performance improvement.

Reaction Type	Main Product	Phases	Solvent Function	Improvement
Amidation [1]	PDE5 inhibitor	Liquid(org.)—Solid	Product separation (Product not soluble in solvent)	Direct product separation
Enzymatic reduction [43]	Chiral alcohols	Liquid (aq.)—Ionic liquid	Substrate carrier	Increased productivity (82–92% yield)
		Liquid (aq.)	-	Productivity (42–46% yield)
		Liquid (aq.)—Organic solvent	Substrate carrier	Productivity (0% yield)
Alkylation [44]	Alyl azides	liquid (org. DMSO)	Dissolves reactants	High productivity (94% yield) but high waste generation
		Liquid (aq.)—liquid (org. DMSO)	Dissolves reactants	High in productivity (94% yield) and lower waste generation
		Liquid (aq.)—liquid (org. Isopropyl acetate)	Dissolves reactants	High productivity (96.5% yield) and lower waste generation
		Liquid (aq.)—liquid (org. Isooctane)	Dissolves reactants	High productivity (91.4% yield) and lower waste generation
Arylation [45]	3,3-disubstituted oxindoles	Liquid (aq.)—liquid (org, THF or Toluene)	Dissolve reactants	Increased reaction rate that leads to complete conversion and high yields compared to single phase systems
Arylation [46]	Arylation of Alkynes	Liquid (aq.)—liquid (org.)	Catalysts dissolved in aq. Phase	Catalyst recovery while maintaining high yields
Hydrolysis [47]	Naproxen	Liquid (aq.)—liquid (org.; Hexane or isooctane or toluene)	Product removal (in organic phase)	Increased yield, enzyme stability increases
Hydrolysis [48]	6-amino penicillanic acid	Liquid (aq.)—liquid (org.; butyl acetate)	Product removal in the organic phase	Productivity increases (product removal shifts the equilibrium reaction towards the product)
Transamination [49]	L-2 Aminobutyric acid	Liquid (aq.)—liquid (org.)	By-product inhibits the enzyme, removal in the organic phase	Increased conversion (96%)
		Liquid (aq.)	-	Conversion (~40%)
Transamination [50]	Chiral amines	Liquid (aq.)—Resin	Product removal	Equilibrium shifts towards product side
Enzymatic Reduction [51,52]	S-4-Chloro-3-hydroxybutyric acid ethyl ester	Liquid (aq.)—liquid (org.)	Substrate controlled release	Increased reaction productivity
Carbonylation [53,54]	Ibuprofen	Gas—Liquid (org.)—Liquid (aq.)	Dissolves catalyst (aq.)	Less waste generated, same productivity, slightly lower reaction rates, reduction in the separation steps
		Gas—Liquid (org.); MEK-Liquid (aq.)	Dissolves catalyst (aq.); Dissolves reactants (org.)	Increased reaction rates
Transamination [55,56]	Sitagliptin	Liquid (aq.); DMSO used as co-solvent	DMSO dissolves amine donor	Increased productivity; enantiomeric selectivity and less waste generated

Note: aq.: aqueous and org.: organic.

3.5. Kinetic Models Available

Table 10 lists the kinetic model availability (found through literature search) and their inclusion in the reaction database kinetic model library. Some of the available kinetic models in the literature have been analyzed, validated against experimental data and, if found acceptable, then been used for reaction optimization in order to establish the design space. In other cases a model has been used by taking it directly from the reported reference, for example, the model reported by Thakar et al. [57] for the second hydrogenation step of ibuprofen synthesis has been successfully used without any modification (of the kinetic parameters) to fit the dynamic experimental data published by Cho et al. [58].

Table 10. Kinetic models availability; * indicates those that are included in the kinetic model library.

Kinetic Models	Number	Reference
Dehydration	1	[59]
Enzymatic reduction	3	[51,60]
Esterification	1	[61]
Transamination	3*	[50,62,63]
Hydrolysis	2	[64,65]
Carbonylation	2*	[66–68]
Hydrogenation	2*	[57,69]

4. Reaction Database Application

The reaction database has multiple features that can assist in the creation of a data-rich environment in the early stage pharmaceutical process-product development. The knowledge stored in the database is searchable by forward or backward search options. As is illustrated in Figure 2, data can be retrieved for the specific search and the retrieved data is used for reaction improvement studies in subsequent calculation-analysis.

4.1. Reaction Data

Process improvements are usually related to resources such as development cost and time. The process of establishing the reactions, the experimental procedure, and the reaction conditions might require significant resources during the initial reaction screening that is required to identify the reaction pathway that leads to the production of the desired type of products (i.e., chiral alcohols). However, having an information-based system that can provide information for reaction identifications, reaction conditions and experimental procedures, can rapidly reduce the required time and cost of the initial screening process. The data-rich environment can also provide solution for reaction improvements related to the mass and heat transfer improvements by the use of new technologies such as flow reactions using for example new microwave technologies.

The use of experimental data (dynamic or end-points) can assist the improvement of the reaction system as the effect of reaction variable changes can be understood and quantified. Moreover, experimental data can be used to develop or to fit kinetic models that capture the behavior of the system under different conditions. These kinetic models can be used for validation studies, optimization studies to identify improved reaction conditions, evaluate different operation scenario and/or different reactor designs and networks.

4.2. Organic Solvents

Another class of process improvement is related to the solvent role during the synthesis step. There are cases where solvent use might enhance the reaction performance. Solvents might have different roles such as creating a second phase to remove an inhibitory product and shift the reaction equilibrium towards the product side, or simply it can create the second phase to remove the product in order to facilitate the following separation procedure. The solvent can also be used as a carrier for the controlled release of the substrate in the reaction mixture, which can minimize the amount of by-products produced when the concentration of substrate is high. The solvent can also have a role as the medium of the reaction and broaden the reaction conditions in order to improve reaction performance or satisfy other process concerns such as process safety. For example, if a reaction takes place at very low temperatures ($<-25\,°C$), the solvent should be liquid at this condition and have the ability to dissolve the reactants, products and catalyst [70].

4.3. Search Options

The search options of the database in terms of both the retrieved data and the use of that data for a defined process are given below.

1. Search for reaction types

    Different reaction types can be searched in the reaction database, the retrieved results provide information for the reaction (reactants, product and target product), the solvent role and how it improves the reaction, reaction conditions (i.e., temperature range, acid/base, different catalyst) and quantitative data (i.e., conversion, concentration vs. time), and finally applicability information such as scale or batch/continuous mode. The results can be used as similarity check, to identify reaction conditions, solvents and possibilities for improvement (i.e., equipment, production mode, technology) for quick reaction optimization.

2. Search for main products (such as APIs or intermediates or type of products like chiral alcohols)

    Searching for main products or type of products, reactions that are used to synthesize this type of compound can be retrieved. The results are used to identify different ways for synthesis and to evaluate them in terms of reaction performance, cost, scalability and sustainability.

3. Search for reactants

    The results obtained by searching reactants are used to identify ways for further utilizing them in case they have used or produced a product during a reaction.

4. Multiphase reactions

    Multiphase and single reactions where the solvent use has improved the reaction performance can be searched, the retrieved results are used to identify the role of the multiphase system, for example, solvent creates a second phase to remove inhibitory by-product and to quantify the improvement in reaction performance, for example, increased conversion.

    To summarize, the information retrieved from the reaction database can be used to:

a. Identify reaction pathways, reaction types, reactants, catalysts, solvents and base/acid.
b. Optimization reaction conditions.
c. Investigate the solvent role in process improvement.
d. Optimize the process development identified reactions in terms of cost, yield and time.
e. Improve the overall process performance in terms of separation process, overall yield, sustainability, safety, scalability, controllability and utilized mass.
f. Improve reactor design and evaluate different reactor designs.
g. Establish operation procedure for the reactors.
h. Assist in plant-wide design, simulation, and techno-economic optimization.
i. Enhance process understanding.

5. Application Example: Ibuprofen Synthesis and Evaluation

5.1. Problem Definition

To illustrate the applicability of the database, the synthesis of ibuprofen is selected as an example. The objectives of this example are:

1. To retrieve data relevant to the reaction pathway of Ibuprofen.
2. Collect data related to individual reactions.
3. Evaluate the alternatives based on green metrics.

Database Search: "Main Product = *'Ibuprofen'*".

5.1.1. Database Results

The main product sub-class is found in the "Reaction" class and from there information before (reactants, reaction types) and information forward (solvents, reaction conditions, data etc.) are retrieved. The database information as retrieved from the database is shown in Figure 4, which contains three screenshots for the purpose of illustration. Screenshot-1 connects the main product (ibuprofen) to the reaction type data; screenshot-2 connects the main product to the specific reaction information (temperature, pressure, solvent use, etc.); screenshot-3 connects the main product to modelling details (for example, kinetic model). The information is also given in the text as follows:

a. Summary of the findings (reaction pathways, reaction types, operation mode, available data and reference).
b. For each reaction pathway, each reaction is analyzed in terms of:

    i. Reactant, products, by-products, acids/base, solvents, catalysts.
    ii. Then the reaction conditions for each reaction is presented.
    iii. Finally, the reaction data is presented.

The retrieved information is used for the evaluation of different pathways to produce ibuprofen using the green chemistry metrics.

The database search gives three different reaction pathways. Pathway 1 consists of three reactive steps. It has been proposed by Elango et al. [71] and consists of three batch reaction steps—a Friedel craft acylation, a hydrogenation and finally, a carbonylation step. The first reactive step has been improved by Lindley et al. [72] using a continuous counter flow reaction-separation system which enables the recovery and recycle of the solvent and the unreacted reactants. The second reactive step is a hydrogenation step that takes place in a fed-batch reactor and the final step is a carbonylation step that also takes place in a fed-batch reactor. Pathway 2 consists of 3 reactive steps as well—a Friedel crafts acylation, an 1,2-aryl migration step and a saponification step—all the reactions are taking place in a continuous flow reactor and this reaction pathway that has been proposed by Snead et al. [73]. Finally, the third reaction pathway consists of the same three reactive steps, as the second pathway, although the intermediates and reactants are different Bogdan et al. [74]. Table 11 gives a summary of the reaction pathways retrieved from the database.

Table 11. Summary of the data retrieved from the database.

Pathway	Reaction Steps	Database Entries	Operation	Reference
1	1.1 Friedel Crafts acylation	67 and 74	Batch (67), continuous (74)	Elango et al. [71] Lindley et al. [72]
	1.2 Hydrogenation	68–71 and 92	Batch	Elango et al. [71]
	1.3 Carbonylation	9–13	Batch	Elango et al. [71]
2	3.1 Friedel Crafts	45	Continuous	Snead et al. [73]
	3.2 1–2 aryl migration	46	Continuous	Snead et al. [73]
	3.3 Saponification	44	Continuous	Snead et al. [73]
3	2.1 Friedel-Crafts	73	Continuous	Bogdan et al. [74]
	2.2 1–2,aryl migration	72	Continuous	Bogdan et al. [74]
	2.3 Saponification	44	Continuous	Bogdan et al. [74]

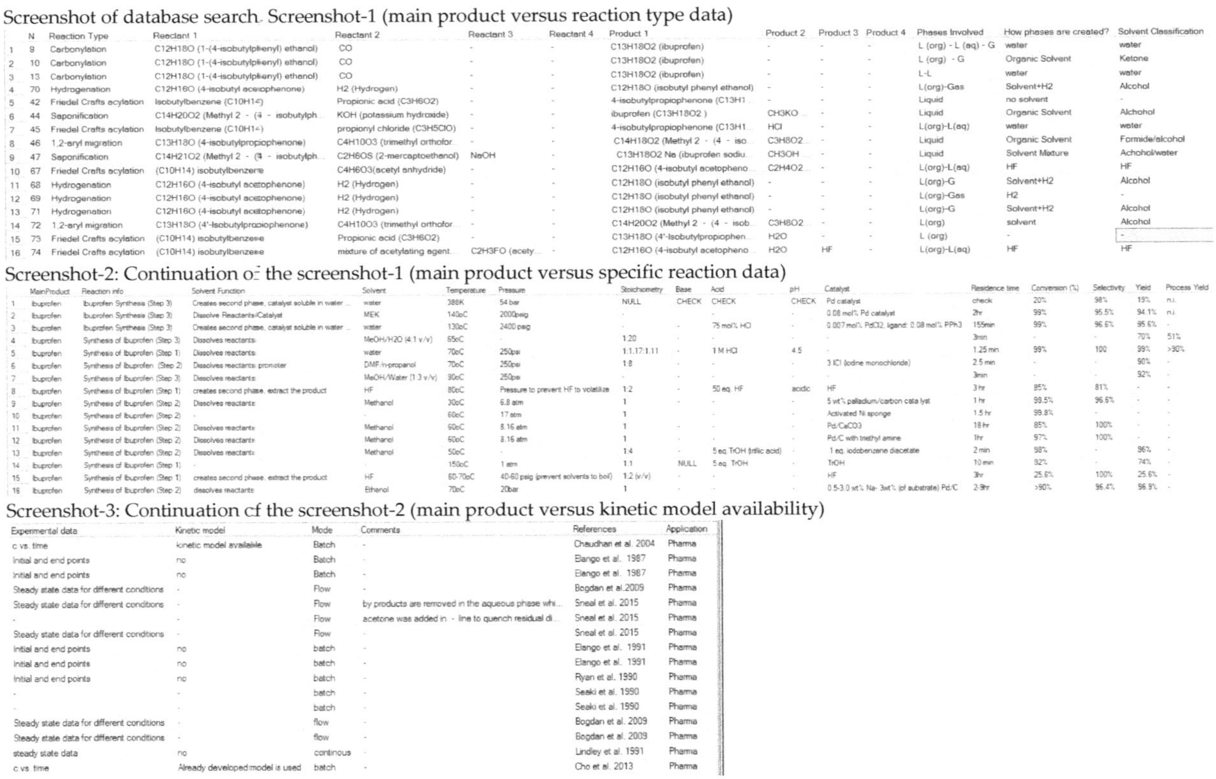

Screenshot of database search. Screenshot-1 (main product versus reaction type data)

	N	Reaction Type	Reactant 1	Reactant 2	Reactant 3	Reactant 4	Product 1	Product 2	Product 3	Product 4	Phases Involved	How phases are created?	Solvent Classification
1	9	Carbonylation	C12H18O (1-(4-isobutylphenyl) ethanol)	CO	-	-	C13H18O2 (ibuprofen)	-	-	-	L (org) - L (aq) - G	water	water
2	10	Carbonylation	C12H18O (1-(4-isobutylphenyl) ethanol)	CO	-	-	C13H18O2 (ibuprofen)	-	-	-	L (org) - G	Organic Solvent	Ketone
3	13	Carbonylation	C12H18O (1-(4-isobutylphenyl) ethanol)	CO	-	-	C13H18O2 (ibuprofen)	-	-	-	L-L	water	water
4	70	Hydrogenation	C12H16O (4-isobutyl acetophenone)	H2 (Hydrogen)	-	-	C12H18O (isobutyl phenyl ethanol)	-	-	-	L(org)-Gas	Solvent+H2	Alcohol
5	42	Friedel Crafts acylation	Isobutylbenzene (C10H14)	Propionic acid (C3H6O2)	-	-	4-isobutylpropiophenone (C13H1...	-	-	-	Liquid	no solvent	-
6	44	Saponification	C14H20O2 (Methyl 2 - (4 - isobutylph...	KOH (potassium hydroxide)	-	-	ibuprofen (C13H18O2)	CH3KO ...	-	-	Liquid	Organic Solvent	Alchohol
7	45	Friedel Crafts acylation	Isobutylbenzene (C10H14)	propionyl chloride (C3H5ClO)	-	-	4-isobutylpropiophenone (C13H1...	HCl	-	-	L(org)-L(aq)	water	water
8	46	1,2-aryl migration	C13H18O (4-isobutylpropiophenone)	C4H10O3 (trimethyl orthofor...	-	-	C14H18O2 (Methyl 2 - (4 - iso...	C3H8O2...	-	-	Liquid	Organic Solvent	Formide/alcohol
9	47	Saponification	C14H21O2 (Methyl 2 - (4 - isobutylph...	C2H6OS (2-mercaptoethanol)	NaOH	-	C13H18O2 Na (ibuprofen sodiu...	CH3OH ...	-	-	Liquid	Solvent Mixture	Achohol/water
10	67	Friedel Crafts acylation	(C10H14) isobutylbenzene	C4H6O3(acetyl anhydride)	-	-	C12H16O (4-isobutyl acetopheno...	C2H4O2...	-	-	L(org)-L(aq)	HF	HF
11	68	Hydrogenation	C12H16O (4-isobutyl acetophenone)	H2 (Hydrogen)	-	-	C12H18O (isobutyl phenyl ethanol)	-	-	-	L(org)-G	Solvent+H2	Alcohol
12	69	Hydrogenation	C12H16O (4-isobutyl acetophenone)	H2 (Hydrogen)	-	-	C12H18O (isobutyl phenyl ethanol)	-	-	-	L(org)-Gas	H2	-
13	71	Hydrogenation	C12H16O (4-isobutyl acetophenone)	H2 (Hydrogen)	-	-	C12H18O (isobutyl phenyl ethanol)	-	-	-	L(org)-G	Solvent+H2	Alcohol
14	72	1,2-aryl migration	C13H18O (4'-Isobutylpropiophenone)	C4H10O3 (trimethyl orthofor...	-	-	C14H20O2 (Methyl 2 - (4 - isob...	C3H8O2...	-	-	L(org)	solvent	Alcohol
15	73	Friedel Crafts acylation	(C10H14) isobutylbenzene	Propionic acid (C3H6O2)	-	-	C13H18O (4'-Isobutylpropiophen...	H2O	-	-	L (org)	-	-
16	74	Friedel Crafts acylation	(C10H14) isobutylbenzene	mixture of acetylating agent...	C2H3FO (acety...	-	C12H16O (4-isobutyl acetopheno...	H2O	HF	-	L(org)-L(aq)	HF	HF

Screenshot-2: Continuation of the screenshot-1 (main product versus specific reaction data)

	MainProduct	Reaction info	Solvent Function	Solvent	Temperature	Pressure	Stoichiometry	Base	Acid	pH	Catalyst	Residence time	Conversion (%)	Selectivity	Yield	Process Yield
1	Ibuprofen	Ibuprofen Synthesis (Step 3)	Creates second phase, catalyst soluble in water ...	water	388K	54 bar	NULL	CHECK	CHECK	CHECK	Pd catalyst	check	20%	98%	19%	n.i.
2	Ibuprofen	Ibuprofen Synthesis (Step 3)	Dissolve Reactants/Catalyst	MEK	140oC	2000psig	-		-	-	0.08 mol% Pd catalyst	2hr	99%	95.5%	94.1%	n.i.
3	Ibuprofen	Ibuprofen Synthesis (Step 3)	Creates second phase, catalyst soluble in water ...	water	130oC	2400 psig	-	-	75 mol% HCl	-	0.007 mol% PdCl2, ligand: 0.08 mol% PPh3	155min	99%	96.6%	95.6%	-
4	Ibuprofen	Synthesis of Ibuprofen (Step 3)	Dissolves reactants	MeOH/H2O (4:1 v/v)	65oC	-	1:20	-	-	-	-	3min	-	-	70%	51%
5	Ibuprofen	Synthesis of Ibuprofen (Step 1)	Dissolves reactants	water	70oC	250psi	1:1.17:1.11	-	1 M HCl	4.5	-	1.25 min	99%	100	99%	>90%
6	Ibuprofen	Synthesis of Ibuprofen (Step 2)	Dissolves reactants: promoter	DMF/n-propanol	70oC	250psi	1:8	-	-	-	3 ICl (iodine monochloride)	2.5 min	-	-	90%	-
7	Ibuprofen	Synthesis of Ibuprofen (Step 3)	Dissolves reactants	MeOH/Water (1:3 v/v)	90oC	250psi	-	-	-	-	-	3min	-	-	92%	-
8	Ibuprofen	Synthesis of Ibuprofen (Step 1)	creates second phase, extract the product	HF	80oC	Pressure to prevent HF to volatilize	1:2	-	50 eq. HF	acidic	HF	3 hr	85%	81%	-	-
9	Ibuprofen	Synthesis of Ibuprofen (Step 2)	Dissolves reactants	Methanol	30oC	6.8 atm	1	-	-	-	5 wt% palladium/carbon cata lyst	1 hr	99.5%	96.6%	-	-
10	Ibuprofen	Synthesis of Ibuprofen (Step 2)	-	-	60oC	17 atm	1		-	-	Activated Ni sponge	1.5 hr	99.8%	-	-	-
11	Ibuprofen	Synthesis of Ibuprofen (Step 2)	Dissolves reactants	Methanol	60oC	8.16 atm	1	-	-	-	Pd/CaCO3	18 hr	85%	100%	-	-
12	Ibuprofen	Synthesis of Ibuprofen (Step 2)	Dissolves reactants	Methanol	60oC	8.16 atm	1	-	-	-	Pd/C with triethyl amine	1hr	97%	100%	-	-
13	Ibuprofen	Synthesis of Ibuprofen (Step 2)	Dissolves reactants	Methanol	50oC	-	1:4	-	5 eq. TfOH (triflic acid)	-	1 eq. iodobenzene diacetate	2 min	98%	-	96%	-
14	Ibuprofen	Synthesis of Ibuprofen (Step 1)	-	-	150oC	1 atm	1:1	NULL	5 eq. TfOH	-	TfOH	10 min	92%	-	74%	-
15	Ibuprofen	Synthesis of Ibuprofen (Step 1)	creates second phase, extract the product	HF	60-70oC	40-60 psig (prevent solvents to boil)	1:2 (v/v)	-	-	-	HF	3hr	25.6%	100%	25.6%	-
16	Ibuprofen	Synthesis of Ibuprofen (Step 2)	dissolves reactants	Ethanol	70oC	20bar	1	-	-	-	0.5-3.0 wt% Na- 3wt% (of substrate) Pd/C	2-9hr	>90%	96.4%	96.9%	-

Screenshot-3: Continuation of the screenshot-2 (main product versus kinetic model availability)

Experimental data	Kinetic model	Mode	Comments	References	Application
c vs. time	kinetic model available	Batch	-	Chaudhari et al. 2004	Pharma
Initial and end points	no	Batch	-	Elango et al. 1987	Pharma
Initial and end points	no	Batch	-	Elango et al. 1987	Pharma
Steady state data for different conditions	-	Flow	-	Bogdan et al. 2009	Pharma
Steady state data for different conditions	-	Flow	by products are removed in the aqueous phase whi...	Sneal et al. 2015	Pharma
-	-	Flow	acetone was added in - line to quench residual di...	Sneal et al. 2015	Pharma
Steady state data for different conditions	-	Flow	-	Sneal et al. 2015	Pharma
Initial and end points	no	batch	-	Elango et al. 1991	Pharma
Initial and end points	no	batch	-	Elango et al. 1991	Pharma
Initial and end points	no	batch	-	Ryan et al. 1990	Pharma
-	-	batch	-	Seaki et al. 1990	Pharma
-	-	batch	-	Seaki et al. 1990	Pharma
Steady state data for different conditions	-	flow	-	Bogdan et al. 2009	Pharma
Steady state data for different conditions	-	flow	-	Bogdan et al. 2009	Pharma
steady state data	no	continous	-	Lindley et al. 1991	Pharma
c vs. time	Already developed model is used	batch	-	Cho et al. 2013	Pharma

Figure 4. Reaction database search results for "Main Product = Ibuprofen".

The details for reaction pathways 1 and 3 are given in the supplementary material (see Sections A.1 and A.2 for pathways 1 and 3 respectively) while the retrieved data for reaction pathway 2 are given and analyzed in the text below.

5.1.2. Pathway 2: Ibuprofen Synthesis

The individual reaction details for the reaction pathway proposed by Snead et al. [73] are presented in Table 12, where the reaction is given in terms of reactants and reaction product for each step and the overall reaction pathway is illustrated in Figure 5. The stoichiometric amounts of the reactants, the solvents, the catalyst, acid/base and by-products are also given in Table 12 for the three reaction steps involved in this pathway.

Table 12. Retrieved reaction information from the database.

Reaction Information	Reaction Step 1	Reaction Step 2	Reaction Step 3
	Friedel Crafts (Flow)	Aryl Migration (Flow)	Saponification (Flow)
Reaction	Isobutylbenzene + propionyl chloride → 4-isobutylpropiophenone + HCl	$C_{13}H_{18}O$ (4-isobutylpropiophenone) + $C_4H_{10}O_3$ (trimethyl orthoformate) → $C_{14}H_{18}O_2$ (Methyl 2-(4-isobutylphenyl) propanoate) + C3H8O2 (Dimethoxymethane)	$C_{14}H_{21}O_2$ (Methyl 2-(4-isobutylphenyl) propanoate) + C2H6OS (2-mercaptoethanol) + NaOH→ $C_{13}H_{18}O_2$ Na (ibuprofen sodium salt) + CH_3OH (MeOH)
Composition (Reactant A: Reactant B, in moles eq.)	1:1.17	1:8	1:8
Solvent	Water	DMF/1-propanol	MeOH/H_2O
Catalyst	$AlCl_3$	ICI	-
Acid/Base	HCl	-	-
By Products	-	-	-

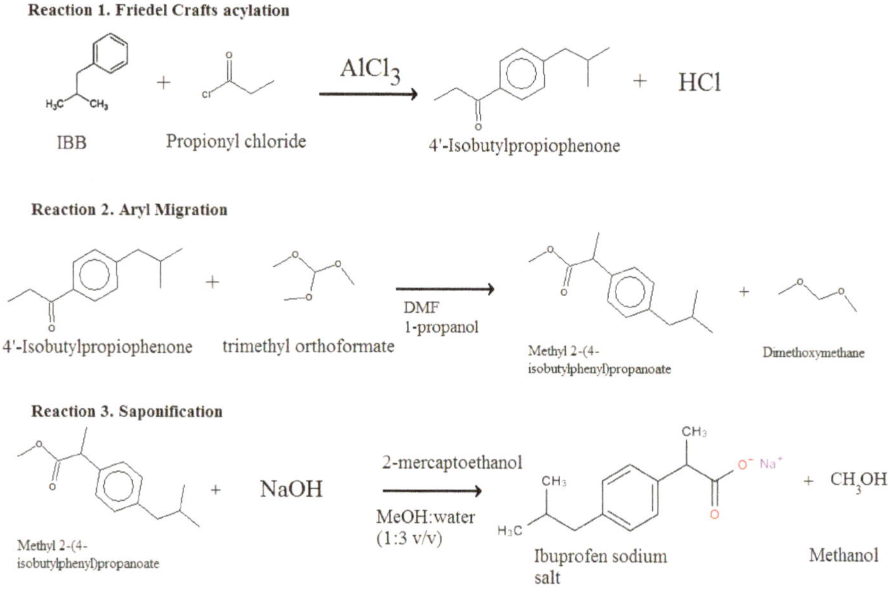

Reaction 1. Friedel Crafts acylation

IBB   Propionyl chloride   4'-Isobutylpropiophenone

Reaction 2. Aryl Migration

4'-Isobutylpropiophenone   trimethyl orthoformate   Methyl 2-(4-isobutylphenyl)propanoate   Dimethoxymethane

Reaction 3. Saponification

Methyl 2-(4-isobutylphenyl)propanoate   2-mercaptoethanol   MeOH:water (1:3 v/v)   Ibuprofen sodium salt   Methanol

Figure 5. Reaction pathway proposed by Snead et al. for the continuous flow synthesis of ibuprofen.

The reaction conditions in terms of temperature, pressure, residence time, catalyst amount and solvent amount are listed in Table 13 for all the reaction steps.

Table 13. Reaction Conditions for the three reactive steps.

Reaction Conditions	Reaction Step 1	Reaction Step 2	Reaction Step 3
	Friedel Crafts (Flow)	Aryl Migration (Flow)	Saponification (Flow)
Temperature	87 °C	90 °C	90 °C
Pressure	17 atm	14 atm	14 atm
Residence time	1.25 min	1min	1 min
Catalyst amount	1.11 eq. $AlCl_3$	3 eq. ICl	-
Solvent amount	-	0.25 eq. DMF/0.71 eq. n-propanol	MeOH/H2O (1:3 *v/v*)

The retrieved experimental data are given in Table 14 in terms of conversion, selectivity, overall reaction yield, experimental data and model availability.

Table 14. Available experimental data as retrieved from the database.

Type of Data	Reaction Step 1	Reaction Step 2	Reaction Step 3
	Friedel Crafts (Flow)	Aryl Migration (Flow)	Saponification (Flow)
Conversion	99%	90%	89%*(*yield)
Selectivity (main product; by-product)	-	-	-
Reaction Yield	-	-	-
Experimental	Steady state data for different residence times	Steady state data for different residence times	Steady state data for different residence times
Model	No	No	No

5.1.3. Reaction Pathways Evaluation through the "Green" Metrics

A simple evaluation based on green metrics [38] has been performed and the results are illustrated in Figure 6. For this analysis, pathway 1 (BHC pathway) with and without recycling of HF and IBB, pathway 3 proposed by Bogdan et al. [74], and pathway 2 proposed by Snead et al. [73] have been considered. The effective mass yield, which is a ratio of the produced product (in mass, kg) over the total amount of non-benign reactant, has been evaluated first. As shown in Figure 6, step 1 of the BHC synthesis requires larger amounts of non-benign reactants compared to pathways 2 and 3, whereas reaction steps 2 and 3 require much less non-benign reactants. Another metric that has been evaluated is the mass intensity (MI), which shows the total required mass for the reaction per kg of product. In Figure 6b, it can be seen that the first reaction steps of pathways 2 and 3 require fewer reactants than the amount required for the BHC pathway without considering the recycling. However, when recycle is considered, the MI metric has lower values for BHC pathway than the other two pathways where recycle is not possible. In addition, pathway 2 proposes much fewer reactants than are required by pathway 3.

The E-factor metric, which shows the generated waste per kg of product, has been evaluated for all the four cases (shown in (Figure 6c). The first step of the BHC pathway has been found to be the main contributor in the E-factor metric—even if step 1 produces a small amount of waste during the reaction, the large value of E-factor is caused by the large stoichiometric amounts of needed solvent and reactant. When the solvent and the reactant are recycled back into the reactor, the E-factor reduces dramatically and the small value of the E-factor is now caused by the small amount of waste and non-recovered solvent and reactant (~1%) [72]. The other two pathways (2 and 3) have relatively high E-factor values, which means that larger amounts of waste are generated through the synthesis steps.

The generated waste for pathway 2 has been found to be slightly lower compared to the reaction in pathway 3. Finally, the atom efficiency has been evaluated for the all pathways and is illustrated in Figure 6d. It can be seen that the atom efficiency for the BHC pathway is very high and therefore, most of the reactant atoms remain in the final product whereas the atom efficiencies are much lower for the two new pathways which means that pathways 2 and 3 might generate more waste than the batch process. Note that the interpretation and the analysis of each "green" metric should be performed individually for each reaction pathway as they represent different aspects of the process (for example, waste generation and total mass used per kg of product). Therefore, an overall conclusion about the "green extent" of the reaction pathways using weighted individual metrics cannot easily be made.

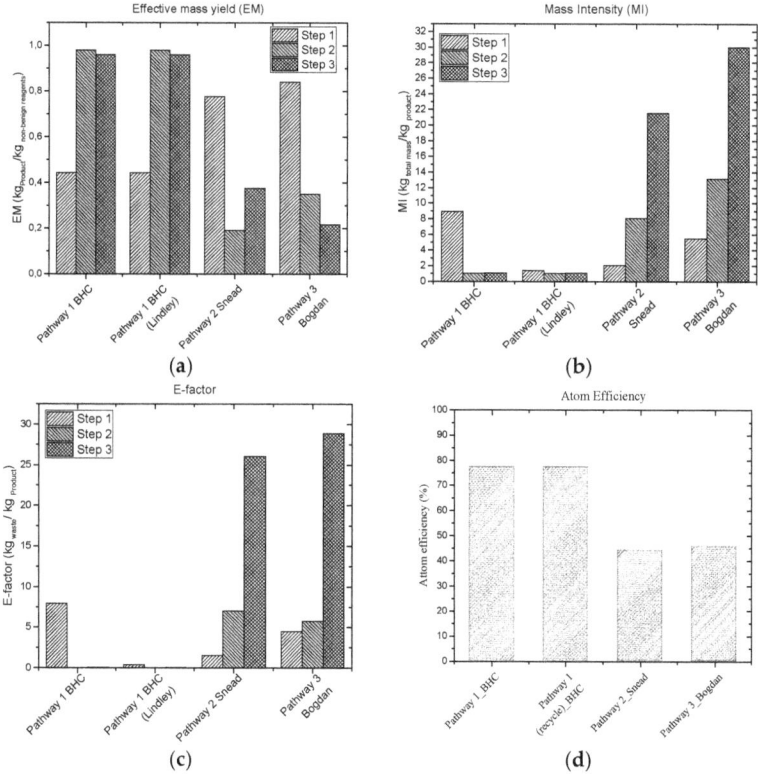

Figure 6. "Green" metrics evaluation for the reaction pathways found in the reaction database: (a) effective mass yield (EM) metric, (b) mass intensity (MI) metric, (c) E-Factor metric and (d) atom efficiency metric.

6. Conclusions

In this article, a reaction database has been developed to assist pharmaceutical process development during the early stages of the synthesis route selection and process-product development by providing enhanced process understanding. A data-rich environment is proposed for this task, where knowledge can be collected, stored and retrieved. The focus of this database is on the pharmaceutical processes and multiphase reactions taking place within them. The reactions in this database have been represented in terms of reaction type, target product to be produced (when single-step or multistep reactions are considered), reaction product and the effect of the solvent use in the reacting system. Information that is contained in the database includes: reaction conditions (temperature, pressure etc.), reaction components (reagents, catalysts etc.), reaction data (conversion, selectivity, dynamic data set, and kinetic models), scaling information and finally batch or continuous processing. For each reaction entry, a description of the process together with literature references are provided.

Reaction data collection is a crucial and very challenging task together with the development of an appropriate knowledge representation system. Also, verification of the consistency of the data is necessary but tests for consistency of data are not yet available, except for some phase equilibrium data.

The application of the database has been highlighted by retrieving data for the synthesis of ibuprofen and using the retrieved data to evaluate the identified reaction pathways using "green" metrics. This reaction database can be used to provide important information during the development of pharmaceutical processes at the early stages of process design. The reaction database covers chemical and biochemical reactions and the future aim is to extend it in terms of reactions and pathways to cover a wider range of reaction systems-products. Many multiphase reactions or single-phase reactions have been improved through the use of solvents available in the database. The solvents are either organic solvents or ionic solvents and in some cases, the extra phase is created by resin, especially for biochemical processes.

Supplementary Materials: Reaction data and reaction pathways for ibuprofen synthesis. The following are available online at www.mdpi.com/2227-9717/5/4/58/s1.

Author Contributions: This research has been carried out in collaboration with all authors. Papadakis and Anantpinijwatna collected reaction data. Papadakis designed the database, performed the analysis and drafted the manuscript, which is based on his PhD-thesis. John M. Woodley and Rafiqul Gani supervised the research work and revised the manuscript.

Conflicts of Interest: The authors declare no conflict of interest.

References

1. Caron, S.; Thomson, N.M. Pharmaceutical process chemistry: Evolution of a contemporary data-rich laboratory environment. *J. Org. Chem.* **2015**, *80*, 2943–2958. [CrossRef] [PubMed]

2. Gasteiger, J. Chemoinformatics: Achievements and challenges, a personal view. *Molecules* **2016**, *21*, 151. [CrossRef]

3. Warr, W.A. A short review of chemical reaction database systems, computer-aided synthesis design, reaction prediction and synthetic feasibility. *Mol. Inform.* **2014**, *33*, 469–476. [CrossRef]

4. Corey, E.J. General methods of synthetic analysis. Strategic bond disconnections for bridged polycyclic structures. *J. Am. Chem. Soc.* **1975**, *97*, 6116–6124. [CrossRef]

5. Bersohn, M.; Esack, A. Computers and organic synthesis. *Chem. Rev.* **1976**, *76*, 269–282. [CrossRef]

6. Corey, E.J.; Jorgensen, E.J. Computer-Assisted Synthetic Analysis. Synthetic Strategies Based on Appendages and the Use of Reconnective Transforms. *J. Am. Chem. Soc.* **1976**, *98*, 189–203. [CrossRef]

7. Agarwal, K.K.; Larsen, T.D.L.; Gelernter, H.L. Application of chemical transforms in synchem2, a computer program for organic synthesis route discovery. *Comput. Chem.* **1978**, *2*, 75–84. [CrossRef]

8. Wipke, T.W.; Ouchi, G.I.; Krishnan, S. Simulation and Evaluation of Chemical Synthesis—SECS: An Application of Artificial Intelligence Techniques. *Artif. Intell.* **1978**, *11*, 173–193. [CrossRef]

9. Ravitz, O. Data-driven computer aided synthesis design. *Drug Discov. Today Technol.* **2013**, *10*, 443–449. [CrossRef] [PubMed]

10. Bogevig, A.; Federsel, H.J.; Huerta, F.; Hutchings, M.G.; Kraut, H.; Langer, T.; Low, P.; Oppawsky, C.; Rein, T.; Saller, H. Route design in the 21st century: The IC SYNTH software tool as an idea generator for synthesis prediction. *Org. Process Res. Dev.* **2015**, *19*, 357–368. [CrossRef]

11. Salatin, T.D.; Jorgensen, W.L. Computer-assisted mechanistic evaluation of organic reactions. 1. Overview. *J. Org. Chem.* **1980**, *45*, 2043–2051. [CrossRef]

12. Chen, J.H.; Baldi, P. No electron left behind: A rule-based expert system to predict chemical reactions and reaction mechanisms. *J. Chem. Inf. Model.* **2009**, *49*, 2034–2043. [CrossRef] [PubMed]

13. Kayala, M.A.; Baldi, P. A Machine Learning Approach to Predict Chemical Reactions. *Adv. Neural Inf. Process. Syst.* **2011**, 747–755.

14. Kayala, M.A.; Azencott, C.A.; Chen, J.H.; Baldi, P. Learning to predict chemical reactions. *J. Chem. Inf. Model.* **2011**, *51*, 2209–2222. [CrossRef] [PubMed]

15. Gothard, C.M.; Soh, S.; Gothard, N.A.; Kowalczyk, B.; Wei, Y.; Baytekin, B.; Grzybowski, B.A. Rewiring chemistry: Algorithmic discovery and experimental validation of one-pot reactions in the network of organic chemistry. *Angew. Chem. Int. Ed.* **2012**, *51*, 7922–7927. [CrossRef] [PubMed]

16. Science of Synthesis. Thieme Chemistry. Available online: https://www.thieme.de/en/thieme-chemistry/science-of-synthesis-54780.html (accessed on 7 July 2016).

17. Zass, E. Databases of Chemical Reactions. In *Handbook of Chemoinformatics: From Data to Knowledge in 4 Volumes*; Gasteiger, J., Ed.; Wiley-VCH Verlag GmbH: Weinheim, Germany, 2003; pp. 667–699.

18. Reactions-CASREACT. Available online: http://www.cas.org/content/reactions (accessed on 1 July 2016).

19. Reaxys. Available online: https://www.elsevier.com/solutions/ (accessed on 7 July 2016).

20. ChemInform. Available online: http://www.cheminform.com/ (accessed on 7 July 2016).

21. Current Chemical Reaction. Available online: http://www.cheminform.com/ (accessed on 7 July 2016).

22. The Chemogenesis web book. Available online: http://www.meta-synthesis.com/webbook.html (accessed on 7 July 2016).

23. Organic Synthesis. Available online: http://www.orgsyn.org/ (accessed on 7 July 2016).

24. Reaction Database. Available online: http://www.sigmaaldrich.com/help-welcome/product-search/reaction-database.html (accessed on 7 July 2016).

25. Synthetic Pages. Available online: http://cssp.chemspider.com/ (accessed on 7 July 2016).

26. The Chemical Thesaurus. Available online: http://www.chemthes.com/ (accessed on 7 July 2016).

27. Webreactions. Available online: http://webreactions.net/index.html (accessed on 7 July 2016).

28. SPRECI InfoChem. Available online: http://www.infochem.de/products/databases/spresiweb.shtml (accessed on 7 July 2016).

29. Synthetic Reaction. Available online: http://pubs.rsc.org/lus/synthetic-reaction-updates (accessed on 7 July 2016).

30. Biotage PathFinder. Available online: www.biotagepathfinder.com/index.jsp (accessed on 7 July 2016).

31. e-EROS. Available online: http://onlinelibrary.wiley.com/book/10.1002/047084289X (accessed on 7 July 2016).

32. FlowReact. Available online: www.flowreact.com/index.php/text/about (accessed on 7 July 2016).

33. Protecting Groups. Available online: http://accelrys.com/products/datasheets/protecting-groups.pdf (accessed on 7 July 2016).

34. Masek, B.B.; Baker, D.S.; Dorfman, R.J.; DuBrucq, K.; Francis, V.C.; Nagy, S.; Richey, B.L.; Soltanshahi, F. Multistep Reaction Based De Novo Drug Design: Generating synthetically feasible design ideas. *J. Chem. Inf. Model.* **2016**, *56*, 605–620. [CrossRef] [PubMed]

35. Databases M. Theilheimer. Available online: http://akosgmbh.de/pdf/overview.pdf (accessed on 7 July 2016).

36. Butters, M.; Catterick, D.; Craig, A.; Curzons, A.; Dale, D.; Gillmore, D.; Green, S.P.; Marziano, I.; Sherlock, J.-P.; White, W. Critical assessment of pharmaceutical processes - A rationale for changing the synthetic route. *Chem. Rev.* **2006**, *106*, 3002–3027. [CrossRef] [PubMed]

37. Cervera-Padrell, A.E.; Skovby, T.; Kiil, S.; Gani, R.; Gernaey, K.V. Active pharmaceutical ingredient (API) production involving continuous processes–a process system engineering (PSE)-assisted design framework. *Eur. J. Pharm. Biopharm.* **2012**, *82*, 437–456. [CrossRef] [PubMed]

38. Constable, D.J.C.; Curzons, A.D.; Cunningham, V.L. Metrics to "green" chemistry-which are the best? *Green Chem.* **2002**, *4*, 521–527. [CrossRef]

39. Jolliffe, H.G.; Gerogiorgis, D.I. Plantwide design and economic evaluation of two continuous pharmaceutical manufacturing (CPM) cases: Ibuprofen and artemisinin. *Comput. Chem. Eng.* **2016**, *91*, 269–288. [CrossRef]

40. Schaber, S.D.; Gerogiorgis, D.I.; Ramachandran, R.; Evans, J.M.B.; Barton, P.I.; Trout, B.L. Economic Analysis of Integrated Continuous and Batch Pharmaceutical Manufacturing: A Case Study. *Ind. Eng. Chem. Res.* **2011**, *50*, 10083–10092. [CrossRef]

41. Jolliffe, H.G.; Gerogiorgis, D.I. Technoeconomic optimisation and comparative environmental impact evaluation of continuous crystallisation and antisolvent selection for artemisinin recovery. *Comput. Chem. Eng.* **2016**, *91*, 269–288. [CrossRef]

42. Singh, R.; Gernaey, K.V.; Gani, R. An ontological knowledge-based system for the selection of process monitoring and analysis tools. *Comput. Chem. Eng.* **2010**, *34*, 1137–1154. [CrossRef]

43. Pfruender, H.; Amidjojo, M.; Kragl, U.; Weuster-Botz, D. Efficient whole-cell biotransformation in a biphasic ionic liquid/water system. *Angew. Chem. Int. Ed.* **2004**, *43*, 4529–4531. [CrossRef] [PubMed]

44. Kopach, M.E.; Murray, M.M.; Braden, T.M.; Kobierski, M.E.; Williams, O.L. Improved Synthesis of 1-(Azidomethyl)-3,5-bis-(trifluoromethyl)benzene: Development of Batch and Microflow Azide Processes. *Org. Process Res. Dev.* **2009**, *13*, 152–160. [CrossRef]

45. Li, P.; Buchwald, S.L. Continuous-flow synthesis of 3,3-disubstituted oxindoles by a palladium-catalyzed α-arylation/alkylation sequence. *Angew. Chem. Int. Ed. Engl.* **2011**, *50*, 6396–6400. [CrossRef] [PubMed]

46. Domier, R.C.; Moore, J.N.; Shaughnessy, K.H.; Hartman, R.L. Kinetic Analysis of Aqueous-Phase Pd-Catalyzed, Cu-Free Direct Arylation of Terminal Alkynes Using a Hydrophilic Ligand. *Org. Process. Res. Dev.* **2013**, *17*, 1262–1271. [CrossRef]

47. Xin, J.Y.; Li, S.B.; Xu, Y.; Wang, L.L. Enzymatic resolution of (S)-(+)-naproxen in a trapped aqueous-organic solvent biphase continuous reactor. *Biotechnol. Bioeng.* **2000**, *68*, 78–83. [CrossRef]

48. Wang, Z.; Wang, L.; Xu, J.H.; Bao, D.; Qi, H. Enzymatic hydrolysis of penicillin G to 6-aminopenicillanic acid in cloud point system with discrete countercurrent experiment. *Enzyme Microb. Technol.* **2007**, *41*, 121–126. [CrossRef]

49. Shin, J.S.; Kim, B.G. Transaminase-catalyzed asymmetric synthesis of L-2-aminobutyric acid from achiral reactants. *Biotechnol. Lett.* **2009**, *31*, 1595–1599. [CrossRef] [PubMed]

50. Tufvesson, P.; Lima-Ramos, J.; Jensen, J.S.; Al-Haque, N.; Neto, W.; Woodley, J.M. Process considerations for the asymmetric synthesis of chiral amines using transaminases. *Biotechnol. Bioeng.* **2011**, *108*, 1479–1493. [CrossRef] [PubMed]

51. Houng, J.Y.; Tseng, J.C.; Hsu, H.F.; Wu, J.Y. Kinetic investigation on asymmetric bioreduction of ethyl 4-chloro acetoacetate catalyzed by baker's yeast in an organic solvent-water biphasic system. *Korean J. Chem. Eng.* **2008**, *25*, 1427–1433. [CrossRef]

52. Houng, J.Y.; Liau, J.S. Applying slow-release biocatalysis to the asymmetric reduction of ethyl 4-chloroacetoacetate. *Biotechnol. Lett.* **2003**, *25*, 17–21. [CrossRef] [PubMed]

53. Papadogianakis, G.; Maat, L.; Sheldon, R.A. Catalytic Conversions in Water. Part 5: Carbonylation of 1-(4-Isobutylphenyl) ethanol to Ibuprofen Catalysed by Water-Soluble Palladium-Phosphine Complexes in a Two-Phase System. *J. Chem. Technol. Biotechnol.* **1997**, *70*, 83–91. [CrossRef]

54. Chaudhari, R.V.; Mills, P.L. Multiphase catalysis and reaction engineering for emerging pharmaceutical processes. *Chem. Eng. Sci.* **2004**, *59*, 5337–5344. [CrossRef]

55. Savile, C.K.; Janey, J.M.; Mundorff, E.C.; Moore, J.C.; Tam, S.; Jarvis, W.R.; Colbeck, J.C.; Krebber, A.; Fleitz, F.J.; Brands, J.; et al. Biocatalytic asymmetric synthesis of chiral amines from ketones applied to sitagliptin manufacture. *Science* **2010**, *329*, 305–309. [CrossRef] [PubMed]

56. Dunn, P.J. The importance of green chemistry in process research and development. *Chem. Soc. Rev.* **2012**, *41*, 1452–1461. [CrossRef] [PubMed]

57. Thakar, N.; Berger, R.J.; Kapteijn, F.; Moulijn, J.A. Modelling kinetics and deactivation for the selective hydrogenation of an aromatic ketone over Pd/SiO$_2$. *Chem. Eng. Sci.* **2007**, *62*, 5322–5329. [CrossRef]

58. Cho, H.-B.; Lee, B.U.; Ryu, C.-H.; Nakayama, T.; Park, Y.-H. Selective hydrogenation of 4-isobutylacetophenone over a sodium-promoted Pd/C catalyst. *Korean J. Chem. Eng.* **2013**, *30*, 306–313. [CrossRef]

59. Cervera-Padrell, A.E. Moving from Batch towards Continuous Organic-Chemical Pharmaceutical Production. Ph.D. Thesis, Technical University of Denmark, Kgs. Lyngby, Denmark, 2011.

60. Bhatia, S.; Long, W.S.; Kamaruddin, A.H. Enzymatic membrane reactor for the kinetic resolution of racemic ibuprofen ester: Modeling and experimental studies. *Chem. Eng. Sci.* **2004**, *59*, 5061–5068. [CrossRef]

61. Shankar, S.; Agarwal, M.; Chaurasia, S.P. Study of reaction parameters and kinetics of esterification of lauric acid with butanol by immobilized Candida antarctica lipase. *Indian J. Biochem. Biophys.* **2013**, *50*, 570–576. [PubMed]

62. Shin, J.S.; Kim, B.G. Substrate inhibition mode of ω-transaminase from Vibrio fluvialis JS17 is dependent on the chirality of substrate. *Biotechnol. Bioeng.* **2002**, *77*, 832–837. [CrossRef] [PubMed]

63. Al-Haque, N.; Santacoloma, P.A.; Neto, W.; Tufvesson, P.; Gani, R.; Woodley, J.M. A robust methodology for kinetic model parameter estimation for biocatalytic reactions. *Biotechnol. Prog.* **2012**, *28*, 1186–1196. [CrossRef] [PubMed]

64. Diender, M.B.; Straathof, A.J.J.; van der Does, T.; Ras, C.; Heijnen, J.J. Equilibrium Modeling of Extractive Enzymatic Hydrolysis of Penicillin G with Concomitant 6-Aminopenicillanic Acid Crystallization. *Biotechnol. Bioeng.* **2002**, *78*, 395–402. [CrossRef] [PubMed]

65. Den Hollander, J.L.; Zomerdijk, M.; Straathof, A.J.J.; Van Der Wielen, L.A.M. Continuous enzymatic penicillin G hydrolysis in countercurrent water-butyl acetate biphasic systems. *Chem. Eng. Sci.* **2002**, *57*, 1591–1598. [CrossRef]

66. Jayasree, S.; Seayad, A.; Chaudhari, R.V. Novel palladium(II) complex containing a chelating anionic N-O ligand: efficient carbonylation catalyst. *Org. Lett.* **2000**, *2*, 203–206. [CrossRef] [PubMed]

67. Seayad, A.; Seayad, J.; Mills, P.L.; Chaudhari, R.V. Kinetic Modeling of Carbonylation of 1-(4-Isobutylphenyl)ethanol Using a Homogeneous PdCl2(PPh3)2/TsOH/LiCl Catalyst System. *Ind. Eng. Chem. Res.* **2003**, *42*, 2496–2506. [CrossRef]

68. Seayad, A.; Kelkar, A.A.; Chaudhari, R.V. Carbonylation of p-isobutyl phenylethanol to ibuprofen using palladium catalyst: activity and selectivity studies. *Stud. Surf. Sci. Catal.* **1998**, *113*, 883–889.

69. Rajashekharam, M.V.; Chaudhari, R.V. Kinetics of Hydrogenation of p-isobutyl acetophenone using a supported Ni catalyst in a slurry reactor. *Chem. Eng. Sci.* **1996**, *51*, 1663–1672. [CrossRef]

70. Watson, D.; Dowdy, E.D.; Depue, J.S.; Kotnis, A.S.; Leung, S.; Reilly, B.C.O. Development of a Safe and Scalable Oxidation Process for the Preparation of 6-Hydroxybuspirone: Application of In-Line Monitoring for Process Ruggedness and Product Quality. *Org. Process Res. Dev.* **2004**, *8*, 616–623. [CrossRef]

71. Elango, V.; Murphy, M.; Smith, B.L.; Davenport, K.G.; Mott, G.N.; Zey, E.G.; Moss, G.L. Method for Producing Ibuprofen. U.S. Patent 4,981,995 A, 1 January 1991.

72. Lindley, D.D.; Curtis, T.A.; Ryan, T.R.; de la Garza, E.M.; Hilton, C.B.; Kenesson, T.M. Process for the Production of 4-Isobutylacetophenone. U.S. Patent 5,068,448 A, 26 November 1991.

73. Snead, D.R.; Jamison, T.F. A three-minute synthesis and purification of ibuprofen: Pushing the limits of continuous-flow processing. *Angew. Chem. Int. Ed.* **2015**, *54*, 983–987. [CrossRef] [PubMed]

74. Bogdan, A.R.; Poe, S.L.; Kubis, D.C.; Broadwater, S.J.; McQuade, D.T. The continuous-flow synthesis of Ibuprofen. *Angew. Chem. Int. Ed.* **2009**, 8547–8550. [CrossRef] [PubMed]

Article

A Systematic Framework for Data Management and Integration in a Continuous Pharmaceutical Manufacturing Processing Line

Huiyi Cao [1], Srinivas Mushnoori [2], Barry Higgins [3], Chandrasekhar Kollipara [4], Adam Fermier [4], Douglas Hausner [1], Shantenu Jha [2], Ravendra Singh [1], Marianthi Ierapetritou [1] and Rohit Ramachandran [1,*]

[1] Department of Chemical and Biochemical Engineering, Engineering Research Center for Structured Organic Particulate Systems (C-SOPS), Rutgers, The State University of New Jersey, Piscataway, NJ 08854, USA; hc623@scarletmail.rutgers.edu (H.C.); douglas.hausner@rutgers.edu (D.H.); ravendra.singh@rutgers.edu (R.S.); marianth@soe.rutgers.edu (M.I.)
[2] Department of Electrical & Computer Engineering, Rutgers, The State University of New Jersey, Piscataway, NJ 08854, USA; scm177@scarletmail.rutgers.edu (S.M.); shantenu.jha@rutgers.edu (S.J.)
[3] Johnson & Johnson, Co., Cork, Ireland; bhiggin2@its.jnj.com
[4] Janssen Research & Development, Spring House, PA 19002, USA; ckollip8@its.jnj.com (C.K.); afermier@its.jnj.com (A.F.)
*   Correspondence: rohit.r@rutgers.edu; Tel.: +1-848-445-6278

Received: 28 March 2018; Accepted: 4 May 2018; Published: 10 May 2018

Abstract: As the pharmaceutical industry seeks more efficient methods for the production of higher value therapeutics, the associated data analysis, data visualization, and predictive modeling require dependable data origination, management, transfer, and integration. As a result, the management and integration of data in a consistent, organized, and reliable manner is a big challenge for the pharmaceutical industry. In this work, an ontological information infrastructure is developed to integrate data within manufacturing plants and analytical laboratories. The ANSI/ISA-88.01 batch control standard has been adapted in this study to deliver a well-defined data structure that will improve the data communication inside the system architecture for continuous processing. All the detailed information of the lab-based experiment and process manufacturing, including equipment, samples and parameters, are documented in the recipe. This recipe model is implemented into a process control system (PCS), data historian, as well as Electronic Laboratory Notebook (ELN) system. Data existing in the recipe can be eventually exported from this system to cloud storage, which could provide a reliable and consistent data source for data visualization, data analysis, or process modeling.

Keywords: data management; continuous pharmaceutical manufacturing; ISA-88; recipe; OSI Process Information (PI)

1. Introduction

For decades, the pharmaceutical industry has been dominated by a batch-based manufacturing process. This traditional method can lead to increased inefficiency and delay in time-to-market of product, as well as the possibility of errors and defects. Continuous manufacturing in contrast, is a newer technology in pharmaceutical manufacturing that can enable faster, cleaner and economical production. The US Food and Drug Association (FDA) has recognized the advancement of this manufacturing mode and has been encouraging its development as part of the FDA's QbD paradigm [1]. The application of process analytical technology (PAT) and control systems is a very useful effort to gain improved science-based process understanding [2]. One of the advantages of continuous pharmaceutical manufacturing process is that it provides the ability to monitor and rectify data/product in real time. Therefore, it has been considered a data rich manufacturing process. However, in the face of the enormous amount of data generated from a continuous process, a sophisticated data management system is required for the integration of analytical tools to the control systems, as well as the off-line measurement systems.

In order to represent, manage and analyze a large amount of complex information, an ontological informatics infrastructure will be necessary for the process and product development in the pharmaceutical industry [3]. The ISA-88 Batch Control Standard [4–6] is an international standard addressing batch process control, which has already been implemented in other industries for years. Therefore, adapting this industrial standard into pharmaceutical manufacturing could provide a design philosophy for describing equipment, material, personnel, as well as reference models [7,8]. This recipe-based execution could work as a hierarchical data structure for the assembly of data from the control system, process analytical technology PAT tools, and off-line measurement devices. The combination of the ISA-88 recipe model and the data warehouse informatics strategy [9] leads to the "recipe data warehouse" strategy [10]. This strategy could provide the possibility of the data management across multiple execution systems, as well as the ability for data analysis and visualization.

Applying the "recipe data warehouse" strategy to continuous pharmaceutical manufacturing provides a possible approach to handle the data produced via analytical experimentation and a process recipe execution. Not only the data itself but the context of data can also be well-captured and saved for documentation and reporting. However, unlike batch operations, continuous manufacturing is a complicated process containing series of interconnected unit operations with multiple execution layers. Therefore, it is quite challenging to integrate data across the whole system while maintaining an accurate representation of the complex manufacturing processes.

In addition to the data collection and the integration of the continuous manufacturing plant, the highly variable and unpredictable properties of raw materials are necessary to be captured and stored in a database because they could have an impact on the quality of the product [11]. These properties of relevance to continuous manufacturing are measured via many different analytical methods, including FT4 powder characterization [12], particle size analysis [13], Washburn technique [14], etc. The establishment of a raw material property database could be achieved by the "recipe data warehouse" strategy. Nevertheless, compared to the computer aided manufacturing used in production process, the degree of automation would vary a lot in different analytical platforms. Therefore, an easily accessible recipe management system will be highly desirable in the characterization laboratories.

Moreover, a cloud computing technology for data management and storage is adapted to deal with massive amount of data generated from the continuous manufacturing process. While traditional computational infrastructures involve huge investments on dedicated equipment, cloud computing offers a virtual environment for users to store or share infinite packets of information by renting hardware and storage systems for a defined time period [15]. Because of its many advantages including flexibility, security, and efficiency, cloud computing is suitable for data management in the pharmaceutical industry [16].

Objectives

In this work, a big data strategy following ISA-88 batch control standard is applied for data management in continuous pharmaceutical manufacturing. The ontology of recipe modeling and the design philosophy of a recipe is elaborated. A data management strategy is proposed for the data integration in both continuous manufacturing processes and the analytical platforms used for raw material characterization. This strategy has been applied to a pilot direct compaction continuous tablet manufacturing process to build up data flow from equipment to recipe database on the cloud. In the characterization of raw material and intermediate blends properties, experiment data is also captured and transferred via a web-based recipe management tool.

2. Materials and Methods

2.1. Ontology

An ontology is an explicit specification of a conceptualization where conceptualization refers to an abstract, simplified view of the world that we wish to represent for some purpose [17]. A more detailed definition of an ontology has been made where the ontology is described as a hierarchically structured set of terms for describing a domain that can be used as a skeletal foundation for a knowledge base [18]. Ontologies are created to support the sharing and reuse of formally represented knowledge among different computing systems, as well as human beings. They provide the shared and common domain structure for semantic integration of information sources. In this work, a conceptualization through the ISA-88 shows the advantage of building up a general conceptualization in pharmaceutical manufacturing and development domain.

Information technology has already developed the capabilities to support the implementation of such proposed informatics infrastructure. The language used in an ontology should be expressive, portable, and semantically defined, which is important for the future implementation and sharing of the ontology. The World Wide Web Consortium (W3C) has proposed several markup languages intended for web environment usage, and Extensive Markup Language (XML) is one of them [19]. XML is a metalanguage that defines a set of rules for encoding documents in a format which is both human readable and machine readable. The design of XML focuses on what data is and how to describe data. XML data is known as self-defining and self-describing, which means that the structure of the data is embedded with the data. There is no need to build the structure before storing the data when it arrives. The most basic building block of an XML document is an element, which has a beginning and ending tag. Since nested elements are supported in XML, it has the capability to embed hierarchical structures. Element names describe the content of the element while the structure describes the relationship between the elements.

As an XML Schema defines the structure of XML documents, an XML document can be validated according to the corresponding XML Schema. XML Schema language is also referred as XML Schema definition (XSD) which defines the constraints on the content and structure of documents of that type [20].

2.2. Recipe Model

In ISA-88, a recipe is defined as the minimum set of information that uniquely identifies the production requirements for a particular product [4]. However, there will still be a significant amount and different types of necessary information, which is required to describe products and to make products. Holding all the information in one recipe will be complicated and cumbersome for human beings. As a result, four types of the recipe are defined in ISA-88 to focus on different levels and accurate information.

2.2.1. Recipe Types

Table 1 shows the four recipe types defined in ISA-88 and their relationship. Each of the four recipe types is described in Table 1 according to the ISA-88. The different type of the recipe together with an example from pharmaceutical industry is shown in Figure 1.

Table 1. ISA-88 recipe types.

Type	Definition
General recipe	A type of recipe that defines raw materials and site independent processing requirements.
Site recipe	A type of recipe that usually derived from the general recipe to meet specific conditions or constraints of the site manufacturing the product.
Master recipe	A type of recipe that contains the information of an individual product and depends on equipment situation.
Control recipe	A type of recipe that defines the manufacturing process a single batch of product.

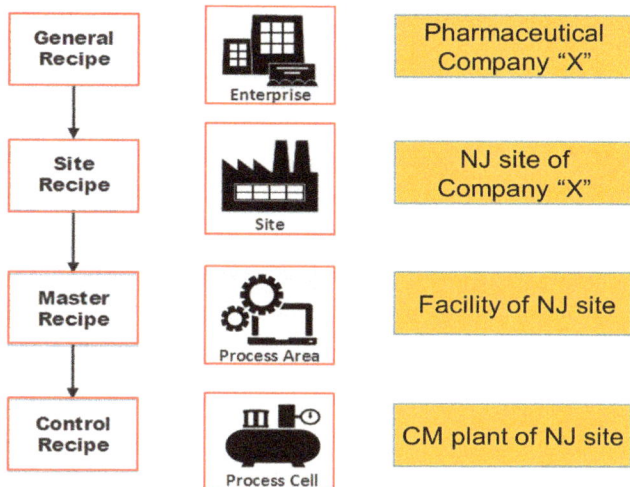

Figure 1. ISA-88 recipe model.

2.2.2. Process Model

The ISA-88 standard defines a process model that has four levels, including process, process stages, process operations, and process actions. The structure of this process model has been displayed in Figure 2. As shown in the figure, the higher level consists an ordered set of the lower levels. The information of equipment, parameter, and material can be included in the attributes of the process actions.

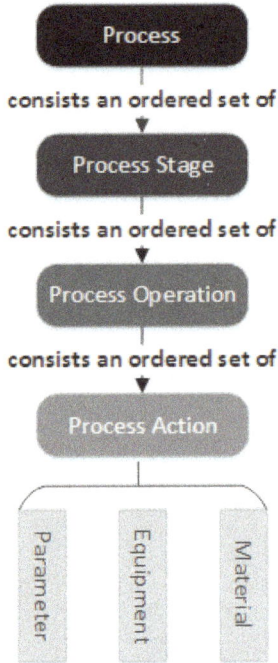

Figure 2. ISA-88 process model.

It is important to note that the reference model and guideline recommended in the ISA-88 standard are not to be strictly normative. This feature provides the sufficient level of flexibility to represent the current manufacturing process according to the ISA-88 standard, as well as the opportunity to expand the reference model to suit unusual manufacturing [21]. In this case, the ISA-88 recipe model is extended to continuous manufacturing, instead of the batch manufacturing process, to provide a hierarchical structure for data management and integration.

2.2.3. Recipe Model Implementation

The purpose of this part is to illustrate the proposed methodological method to assess them implementation of ISA-88 recipe model into continuous pharmaceutical manufacturing. Proven evidence has shown that developing a new product manufacturing process might be a difficult task. Although ISA-88 standard provides a practical reference model for sure, adapting a batch model to the continuous process will still be challenging. The implementation of the recipe model is divided into several steps to perform the activities in a structured way.

- Step 1. ISA-88 applicable area identification

The initial scope and project boundaries will be defined in this step-in order to discuss the main objectives. As mentioned above, the ISA-88 standard is not strictly defined and can be expanded to continuous manufacturing in contrast. Moreover, since analytical experiments are performed by each sample, one sample can be treated as a "batch" in the analytical process. Therefore, it is also reasonable to consider the offline characterization experiment of material to be applicable to the ISA-88 standard. In this situation, a technique-specific recipe containing detailed process steps could provide instructions for specific experiment type, as well as the designed structure for result documentation. As shown in Figure 3, after analytical tests are performed on samples according to the procedures in the recipes, experiment results will be recorded in the recipe as well.

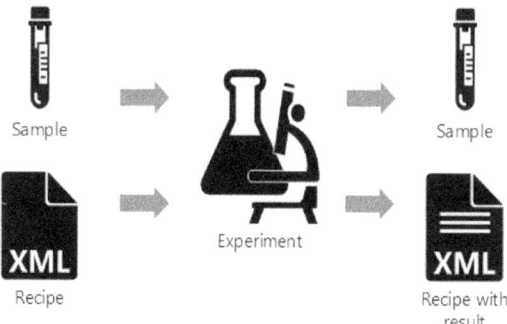

Figure 3. Material and data flow in an experiment process.

Preliminary data gathering is performed to provide a clear understanding of the stream process and the requirement of each particular case. The component may be modified in ISA-88 recipe model could be recipe type, recipe content structure, recipe representation, etc. The in-depth knowledge of process architecture, manufacturing execution systems, product and material information, and analytical platform will be necessary to the accomplishment of this step.

- Step 2. Recipe structure definition

An XML Schema is created for the implementation of an ISA-88 standard to provide the recipe model a structured, system-independent XML markup. This step needs to be completed based on the data collected from the previous step. An XML Schema document works as a master recipe standard. The goals of this document include setting restrictions to the structure and content of the recipe, maximizing the recipe model's expressive capability and realizing machine's ability of recipe validation.

- Step 3. Process analysis

Due to the master recipe's equipment dependent property, it will be complicated and inefficient to develop a single master recipe to describe the whole continuous pharmaceutical process. The process itself consists an ordered set of multiple unit operations, which may change from time to time, depending on the equipment used. This does happen a lot in the product process development stage, especially in the early time period. Another barrier is the transformation from research and development (R&D) to production. Usually, R&D performs the product development process on small-scale pilot plants, which could be quite different from the production plant in technology aspect. Moreover, the differences between technologies used in every plant are crucial to product manufacturing after the transformation. The entire problem mentioned above is critical to the implementation of ISA-88 to continuous pharmaceutical manufacturing.

A suitable approach to implementation of ISA-88 into continuous manufacturing might be developing a robust modular structure that could support a variety of equipment types and classes. In this case, a continuous process is divided into several unit operations, such as mixing, and blending. Each unit operation could be performed by different equipment types that correspond to a different recipe module. In other words, each equipment type might take place in a continuous process and will be mapped to a recipe module individually. The ordered combination of these recipe modules will form the master recipe of continuous manufacturing.

PAT tools used in continuous pharmaceutical manufacturing, Near-infrared NIR and Raman for example, are implemented in a continuous process to provide real-time monitoring of the material properties and are essential to manufacturing decision making. The PAT tools will also be treated as a unit operation that has a corresponding recipe module, in order to provide flexibility and reliability for the management of PAT.

- Step 4. Process mapping

XML-based master recipes will be developed in accordance with an XML Schema to represent the manufacturing process. Firstly, the equipment information, operation procedure, as well as process parameters of unit operations will be mapped carefully to the ISA-88 recipe model to generate respect recipe modules. These recipe modules could be transformed into the master recipe and used for the individual study and data capturing of unit operations.

2.3. Systematic Framework of Integration

The fast development of information and communication technology, such as big data and cloud computing, provides the capability to increase productivity, quality and flexibility across industries. The process industry is also seeking a possible approach to bringing together all data from different process levels and distributed manufacturing plants in a continuous and holistic way to generate meaningful information [22]. For continuous pharmaceutical manufacturing, a data flow across the whole process is proposed in Figure 1 following the traditional design pattern, automation pyramid [23], to create an information and communication infrastructure.

Enterprise resource planning (ERP) systems are placed at the top of structure shown in Figure 4. It provide the integrated view of the core business process, based on the data collected from various business activities [24]. Long-term resource planning—primarily human and material resources—will be its primary objective. Historian, on Level 4, is intended for collecting and organizing data to provide an information infrastructure. Meaningful information in different representations, ISA-88 recipe document, could be generated from the organizational assets, as well.

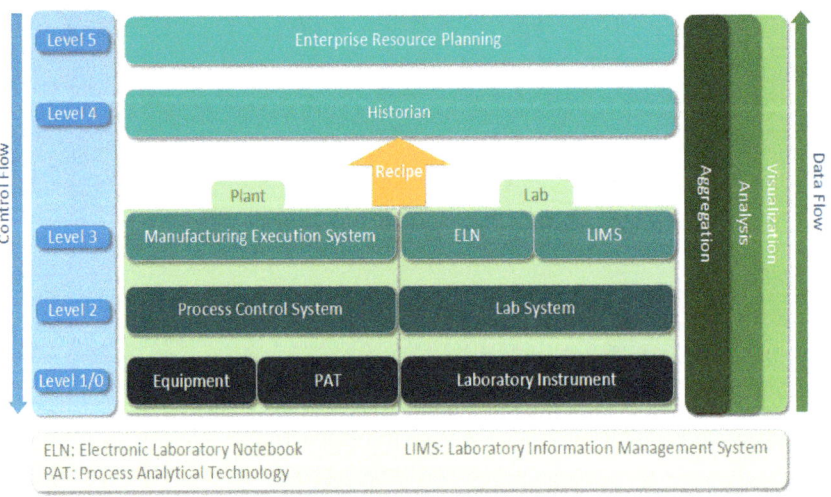

Figure 4. Architecture of data integration.

In terms of the levels below, they are separated into two parts, continuous manufacturing process, and analytical laboratory platform. Because of the distinct data acquisition method used these two areas, they will be elaborated separately in the following section.

2.3.1. Continuous Manufacturing

As shown in Figure 4, there are three more levels in the continuous manufacturing process. Manufacturing Execution System (MES) performs the real-time managing and monitoring work-in-process on the plant floor. During the manufacturing process, the process control system plays a significant role in controlling the system states and conditions to prevent severe problem during operations. Process equipment located in Level 1/0 contains process equipment and PAT tools. Process equipment is always consisted of the field device and embedded programmable logic controllers (PLC). PLC is an industrial computer control system which monitors input devices, such as machinery or sensors and makes the decision to control output devices according to its program [25]. The most powerful advantage of PLC is its capability to change the process or operation while collecting information. In terms of PAT tools, it has corresponding PAT data management tools that could communicate with the control platform. Above are the different levels of control systems across the manufacturing process.

Distinct from the control flows organized top-down, whereas the data capturing and information flow are bottom up. After generated from process equipment and PAT tools in the continuous process, various field data is fed into PCS to form a recipe structure which is accordance with ISA-88 recipe model. In other words, manufacturing data is collected, organized, interpreted and transformed into meaningful information starting from PCS level. This recipe structure will be mapped identically into historian for storage. The low-level network in this information architecture is mainly based on the communication over bus systems, such as field device to PLC. However, most of high-level systems use connections based on ethernet technology named object linking and embedding (OLE) for process control, which is also known as OPC. OPC is a software interface standard that enables the communication between Windows programs and industrial hardware devices to provide reliable and performable data transformation.

2.3.2. Laboratory Experiment

Unlike control systems used in continuous manufacturing process plant, the laboratory platform consists of light-weighted, but various types of instruments are generating analysis data in different formats. Therefore, on the third level located the electronic laboratory notebook (ELN) system or laboratory information management system (LIMS) that has the ability for support all kinds of analytical instruments. While material properties are measured by instruments and collected by laboratory systems, these data sets will be transformed into ELN or LIMS systems for further organization and interpretation. Meaningful information is generated in recipe form according to the ISA-88 standard in Level 3.

The basic workflow of using ELN to support experiment documentation is illustrated in Figure 5. The first step is to select the master recipe developed for this specific test and create a control recipe. If such master recipe doesn't exist, this issue needs to be reported to laboratory manager because only lab administrators have permission to create and import recipes. However, users are able to change the control recipe, adding/deleting steps or parameters, according to their particular experiment plan. After the execution of the control recipe, user will perform experiments according the steps in the recipe and recording test parameters at the same time. Eventually, all the data related to this single test—including operator, sample, instrument, and result—will be documented in the recipe and be transferred to cloud storage.

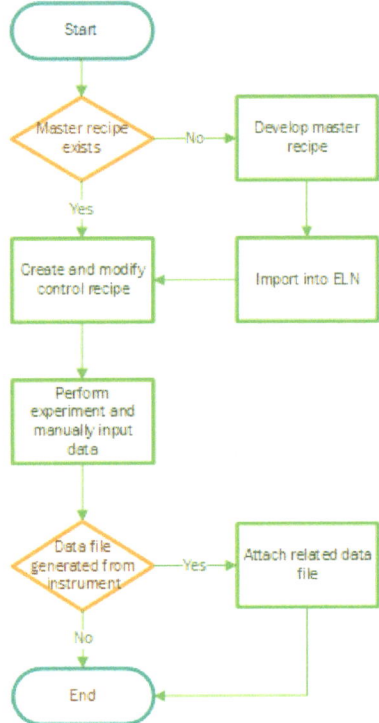

Figure 5. Workflow of using electronic lab notebook (ELN) to perform an experiment.

2.4. Transferring Data from Lab Computer to Cloud System

Transfer of data from a computing cloud to a storage cloud can be non-trivial, since cross platform support is not guaranteed to be available on the cloud service of choice. There is the option of using the application program interface (API) provided by the cloud service but the heterogeneity in the API's provided by different services makes this approach difficult. This issue is highlighted in our current case: the cloud platform of choice here is Box.com. The computing cloud, i.e., the Amazon Elastic Compute Cloud (EC2) runs a Linux distribution (Amazon Linux AMI) based on RHEL (Red Hat Enterprise Linux) and CentOS. Box.com, however, neither has support for, nor has any plans to include support for this platform. This puts the user in a situation where there is no direct method to easily and trivially push data to the cloud using a tool such as an app. Additionally, there is the difficulty in transitioning to a different platform should that need ever arise.

In order to address these issues, we have put in place a data transfer pipeline between the compute and storage clouds using davfs2 [26], a specific implementation of WebDAV (web-based Distributed Authoring and Versioning) [27], an extension of the hypertext transfer protocol (HTTP) that allows a user (client) to remotely create/edit web content. This method makes it possible to set up a directory on the Linux EC2 compute cloud that serves as a window into the cloud storage location. Now a rsync [28] may be set up to sync to this directory, and by extension, to the cloud.

3. Case Studies

Two different case studies have been performed to verify the proposed systematic framework in both pilot-plant and analytical laboratory.

3.1. Direct Compaction Continuous Tablet Manufacturing Process

A direct compaction continuous pharmaceutical tablet manufacturing process has been developed at the Engineering Research Center for Structured Organic Particulate Systems (C-SOPS), Rutgers University [29]. This schematic of the process is illustrated in Figure 6. Three gravimetric feeders are used to provide raw materials, including active pharmaceutical ingredient (API), lubricant and excipient. The flow rate of powder is controlled via manipulating the rotational screw speed of the feeder. If necessary, a co-mill can be used to delump the API and excipient while lubricant will be directly added into the blender to avoid over lubrication. After the continuous blender, the homogeneous powder mixture is sent to tablet compaction. Some of the tablets produced will be sent for dissolution testing. In order to use the gravity, the pilot plant is built in three levels high for better material flow. The feeders (K-Tron) are located on the top level. The second level is used for delumping and blending and the tablet compactor (FETTE) is located on the bottom floor.

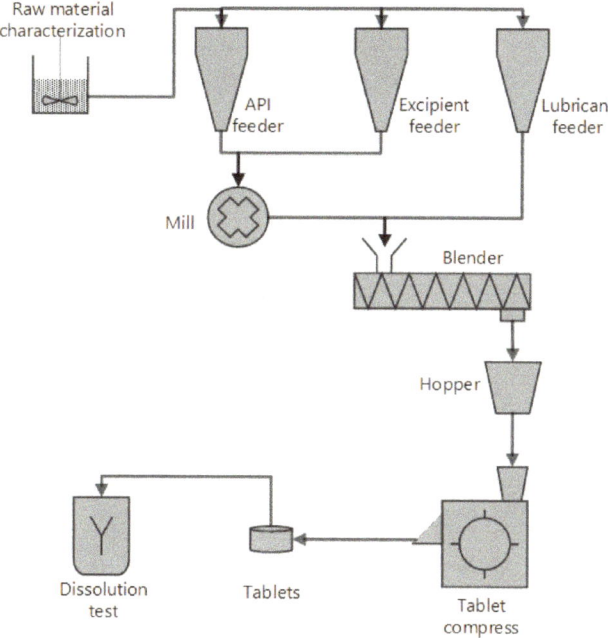

Figure 6. Continuous tablet compression process.

3.2. FT4 Powder Characterization Test

The input materials of continuous direct compaction tablet manufacturing process are powders whose physical and mechanical properties vary. Since this variation may have an impact on the performance of the final dosage, it is important to develop effective measurement methods on the critical properties. The FT4 Powder Rheometer of Freeman Technology developed over the last two decades is an ideal solution.

The FT4 Powder Rheometer (Figure 7) is designed to characterize powders under different conditions in ways that resemble large-scale production environments. It provides a comprehensive series of methods that allow powder behavior to be characterized across a whole range of process conditions. The methods include rheological, torsional shear, compressibility and permeability tests which can be performed using small bulk samples, such as 1, 10 or 25 mL.

Figure 7. FT4 powder rheometer.

4. Results and Discussions

4.1. Data Integration in Continuous Tablet Manufacturing

Figure 8 presents the data flow within the continuous direct compaction tablet manufacturing process. Data generated from both process equipment and PAT tools will be sent into DeltaV system and organized according to the ISA-88 recipe model. PI system, playing the role of historian, is able to receive data from PCS and build up recipe hierarchical structure using PI Event Frame. In the final step, cloud storage is chosen for permanent data storage and the portal of ERP.

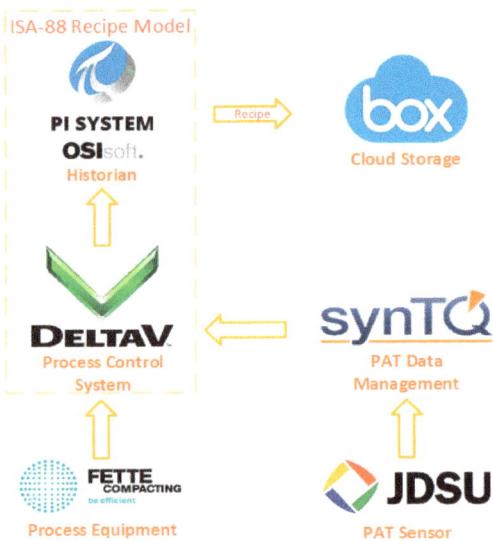

Figure 8. Data flow in continuous tablet manufacturing.

4.1.1. DeltaV Recipe Model

ISA-88 batch manufacturing standard is supported by DeltaV Distributed Control System. Recipes has been created and maintained by DeltaV's Recipe Studio application. Recipe Studio supports two types of recipes: master recipes and control recipes. A master recipe created and modified by process engineers, while control recipe can be modified and downloaded by operators. All recipes have four main parts: a header, a procedure (sequence or actions), the parameters, and the equipment. They are further elaborated in Table 2.

Table 2. Recipe parts.

Name	Function
Header	Contains information about the exact version and the author of the recipe
Procedure	Contains the procedural function chart (PFC) that defines the steps needed for the batch to run.
Parameters	Allow you to set values for different formulations of the product using the same recipe.
Equipment	Defined in DeltaV Explorer and associated with the recipe. Each operation in Recipe Studio is associated with a particular equipment unit.

Instead of the ISA-88 process model, the procedural control model can be implemented in DeltaV Recipe Studio. This model, shown in Figure 9, focuses on describing the process as it relates to physical equipment. Procedures, unit procedures, and operations of the continuous direct compaction process are constructed graphically using IEC 61131-3 compliant sequential function charts (SFCs) [30].

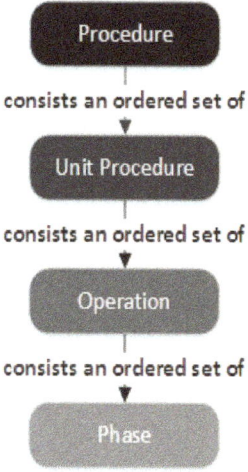

Figure 9. ISA-88 procedure model.

While the data generated from field devices is collected by DeltaV, PAT's spectrum data is captured and organized by the PAT data management platform, synTQ. synTQ has been designed to provide harmonization and integration of all plant-wide PAT data, including spectral data, configuration data, models, raw and metadata, and orchestrations (PAT methods). It is worth noting that synTQ is fully compatible with the Emerson process management system. This feature means data collected and managed in synTQ can be transferred into the DeltaV system for process control and data integration.

4.1.2. OSI PI Recipe Structure

PI system is an enterprise infrastructure for management of real-time data and events provided by OSIsoft. It is a suite of software products that are used for data collection, historicizing, finding, analyzing, delivering, etc. Figure 10 illustrates the structure and data flow of the PI system, in which PI data archive and PI asset framework (AF) are the keys parts.

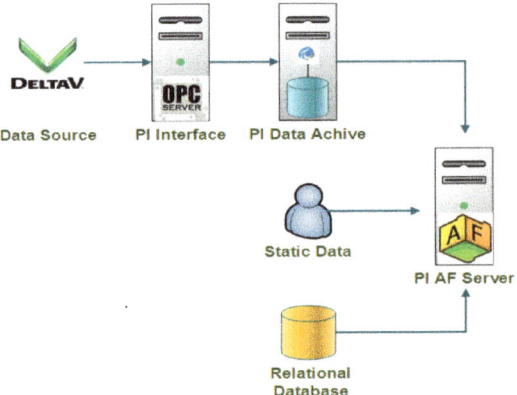

Figure 10. Structure and data flow of Process Information (PI) system.

After the PI interface receives data from a data source (DeltaV system in this case) via the OPC server, the PI Data Archive gets the data and routes it throughout the PI system, providing a common set of real-time data. PI AF is a single repository for objects, equipment, hierarchies, and models. It is designed to integrate, contextualize, and reference data from multiple sources including the PI data archive and non-PI sources such as human input and external relational databases. Together, these metadata and time series data provide a detailed description of equipment or assets. Figure 11 shows how equipment from the continuous tablet direct compaction process is mapped in PI System. The attributes of element representing equipment performance parameters are configured to get value from the corresponding PI data archive points.

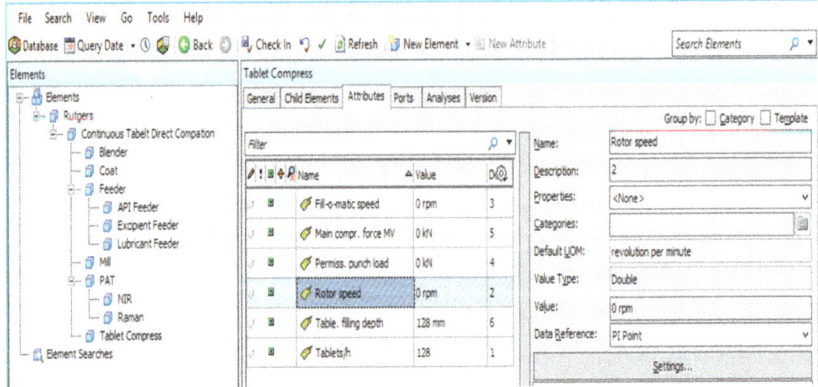

Figure 11. Include equipment into PI asset framework (AF) elements.

Moreover, PI AF supports the most significant function of implementing ISA-88 Batch Control Standard, PI Event Frame. Events are critical process time periods that represents something happening and impacting the manufacturing process or operations. The PI Event Frame is intended to capture critical event contexts which could be names, start/end time, and related information that are useful for analysis. Complex hierarchical events, shown in Figure 12, are designed in order to map the ISA-88 process model of continuous tablet press process.

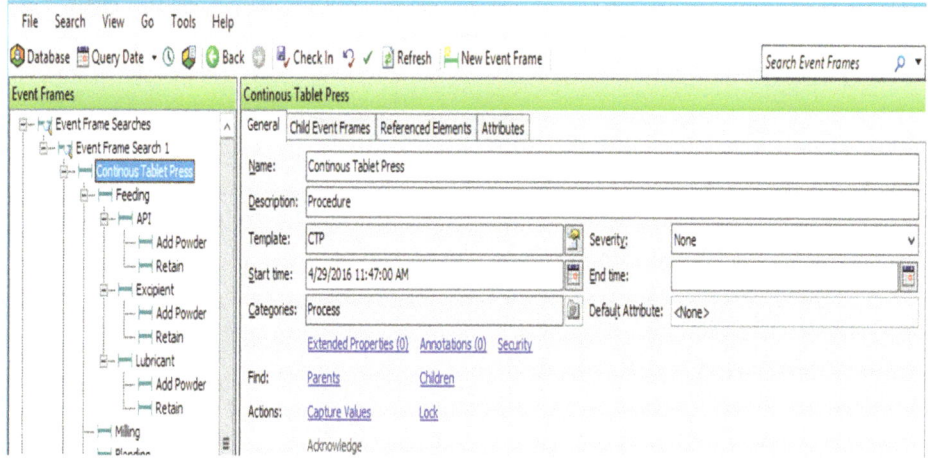

Figure 12. Using PI Event Frame to map ISA-88 recipe model.

4.2. Recipe-Based ELN System

A novel ELN system is developed and implemented for data management in raw material characterization laboratories to replace paper laboratory notebooks (PLN). Scientists could create, import, modify recipes within ELN following the ISA-88 standards, as well as input data and upload related data files. This system has sufficient capability of documenting experiment process and gathering data from various analytical platforms. It could also provide material property information that complies with recipe model.

4.2.1. Information Management

This recipe-based ELN system is developed as a custom module for Drupal, an open-source content management system (CMS) written in Hypertext Preprocessor (PHP). The stand release of Drupal, also referred as Drupal core, contains the essential features of CMS, including taxonomy, user account registration, menu management, and system administration. Beside this, Drupal provides many other features, such as high scalability, flexible content architecture, and supporting mobile devices. One widespread distribution of Drupal, Open Atrium is selected as the platform for laboratory data handling because of its advanced knowledge management and security features.

In this case, Drupal is installed and maintained on a cloud computing service, Amazon Web Service (AWS). AWS is an on-demand computing platform, instead of setting up actual workstations in computer rooms. AWS provides large computing capacity cheaper and quicker than actual physical servers. Drupal runs in a Linux operating system on one of AWS' service, Amazon Elastic Compute Cloud (EC2). The MySQL database which Drupal is connected to is supported by Amazon Relational Database Service (RDS). The adoption of cloud computing technology into laboratory data management benefits a lot, including economic computing resource expense, extensible infrastructure capacity, etc.

After installation and configuration, this Drupal system can be accessed in the form of the website via an Internet browser. Within Open Atrium, spaces can be created as the highest level of content structure. While the public space is open to all system users, the content of private space can only be accessed by website members that have been added to such space. After the ELN module is installed in Drupal, a notebook can be created as a section housed in every space. Via the section visibility widget, specific permissions for certain member group or people could be set for ELN section. Although the CMS is accessible across the world through Internet, the confidentiality of content in ELN is well protected.

The web interface of ELN system (Figure 13) contains two main parts: master recipe list and control recipe list. As mention in previous sections, master recipes are the templates for recipes used to perform experiments on individual samples, and they are analytical process dependent. Using ELN tool, master recipes can be created within the system or imported from external recipe document in XML format, as is shown in "Recipe Import" area. After clicking the green "Create Control" button, a control recipe will be generated as one copy of the master recipe with username and time in the name. It will work as the experiment note to document all the experiment process and data. Figure 14 is an example of control recipe for FT4 compressibility test.

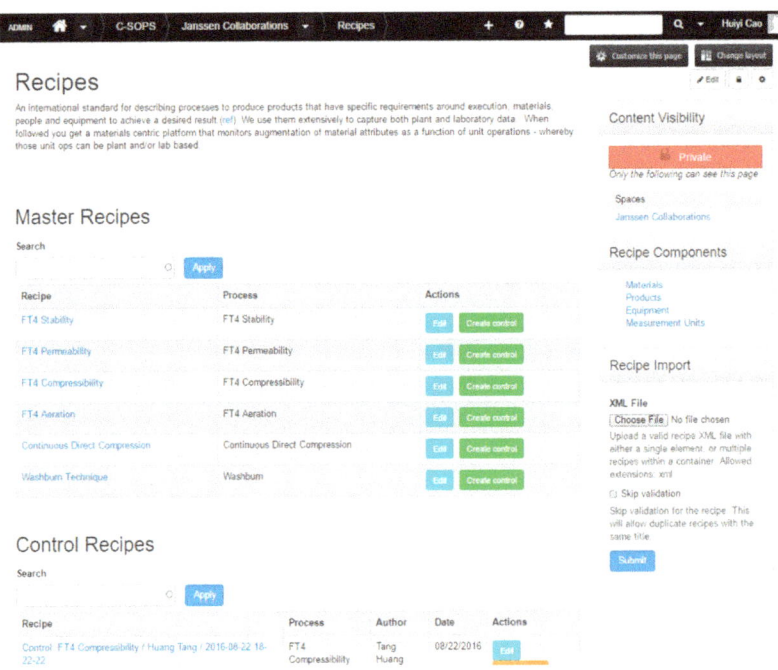

Figure 13. The web interface of ELN system.

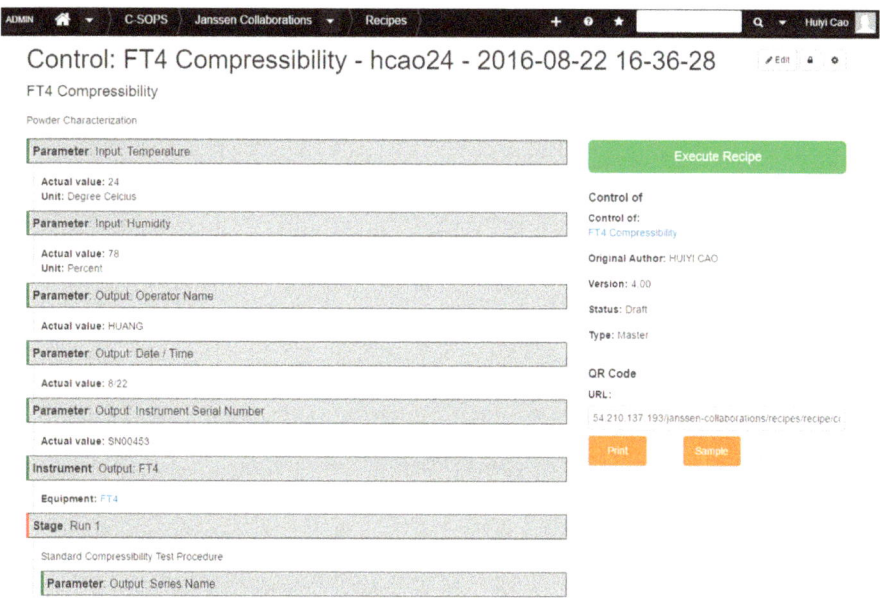

Figure 14. Example of control recipe.

The "Recipe Components" placed on the right sidebar of ELN front page contains four relevant information may be included in the recipe, and they are further introduced in Table 3. The list of each recipe component includes all the items, as well as their related information and attributes. Each item is linked to a page listing the recipes that use the particular piece of the component. Figures 15 and 16 show the equipment list and the information page of FT4 Powder Rheometer, for example. The QR code displayed on the top right of the page is another feature of ELN. Each equipment or material existing in ELN system will have a unique QR code, which has the corresponding Uniform Resource Locator (URL) address, encoded. By scanning the QR code attached to the material or equipment, its information, and related recipes could be accessed on the website.

Table 3. Recipe components.

Name	Definition	Attributes
Materials	Pharmaceutical ingredient used in manufacturing process	Material Name Role Aliases & Chemical Info
Products	Things produced from manufacturing process	Product Name Catalog
Equipment	The analytical instrument used in material characterization.	Equipment Name Equipment ID Equipment Description
Measurement Units	A definite magnitude of a quantity	Unit Name Unit Symbol

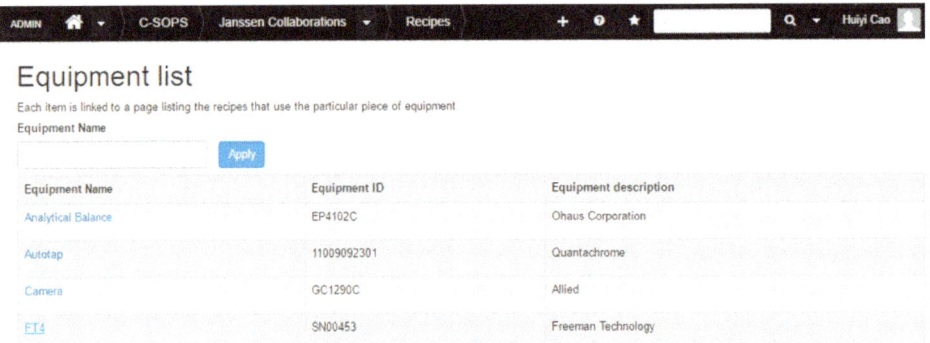

Figure 15. Equipment list in ELN system.

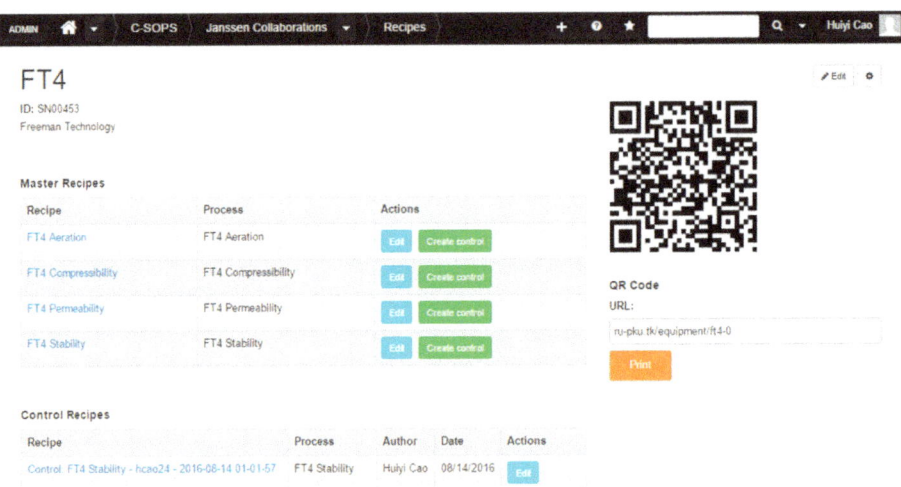

Figure 16. Equipment page of FT4 power rheometer.

As shown in Figure 17, the left side of this execution dialog display the recipe steps from process to process action. Part of the recipe steps is marked with a green pencil icon which means there are step parameters, step ingredients, or step equipment linked to such steps. Related step information will be presented on the right side of the prompt window. In terms of the step parameters, there are two types of them regarding of the actions need to be taken, "input" and "output". "Input" means this step parameter will be manually typed into ELN by a scientist. "Output" indicates that such parameter is recorded in the output data file from instrument system. Once the step parameters are saved, the green pencil icon will turn into a check mark. The last action in the workflow is to upload the data file generated by analytical device system if there is any.

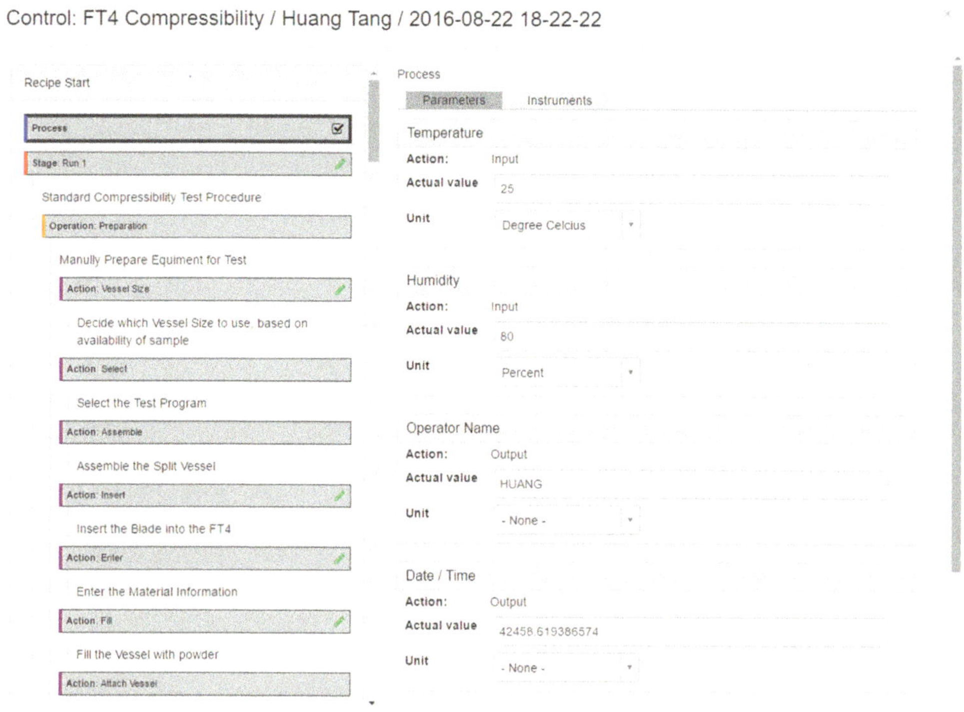

Control: FT4 Compressibility / Huang Tang / 2016-08-22 18-22-22

Figure 17. Execution dialog of control recipe.

4.2.2. Other Features

- QR code printing

It is worth mentioning that the ELN system has the capability of directly printing QR codes for recipe components via a barcode printer. Within this system, there always exists a unique QR code corresponded to the process material and equipment. A QR code sidebar can certainly be found on the material and equipment information web pages. After simply clicking the "Print" button, the QR code will be printed on a sticker from the barcode printer connected to the computer. Such convenient feature is enabled by the custom Drupal module "zprint". It is designed to send the command of printing current web URL to the printer via the Google Cloud Print service.

- Mobile device supporting

By reason of ELN system's web interface, this recipe portal could be accessed via all kinds of mobile devices not only desktops or laptops. Scientists can open up recipes through the internet browsing apps on their smartphones or tablets, as well as executing recipe. Moreover, mobile devices are more convenient for scanning QR codes to acquire information of material, equipment, and samples.

4.2.3. Data Flow

There are two primary data sources for the ELN system: human input and data file outputted by the instrument. Both will be loaded into the control recipe of the experiment as the actual value of step parameters. As mentioned before, manually inputting data can be done via the execution dialog shown in Figure 18. However, the extraction of data from instrument-generated data file will be performed by ELN system's Excel Parsing function. Considering the fact that most of the analytical device systems have the ability to export data into Excel supported documents (xlsx and csv for example), ELN's Excel Parsing capability is developed based on the contributed Drupal module "PHPExcel". After the data file is attached to the recipe, PHPExcel module will convert the data into an array. Once there is a match within the step parameters, the value of an array item will be updated to that step parameter. Therefore, ELN supports various of analytical instruments, as well as manually data input.

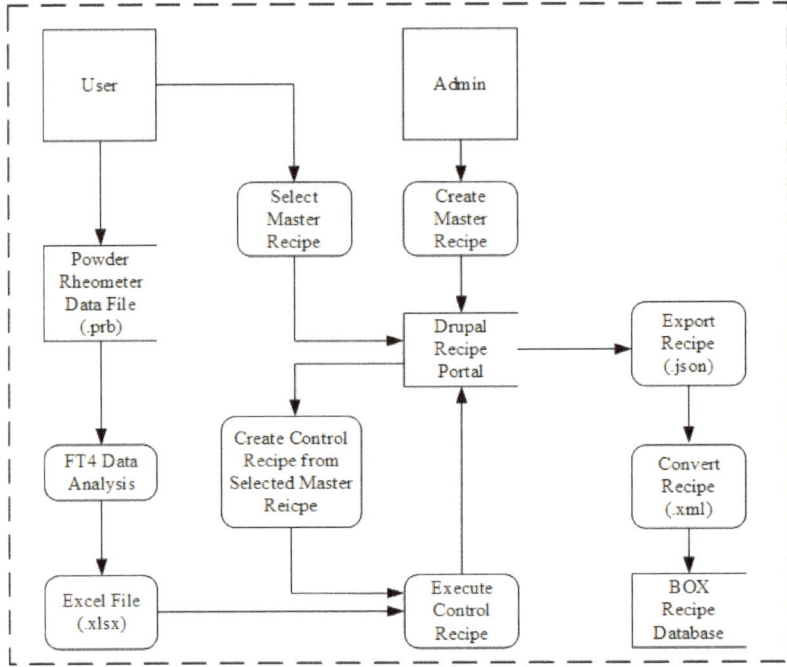

Figure 18. Data flow in performing FT4 experiment using the ELN system.

Containing the information of the analytical process, test results and instrument information, each control recipe is an integral documentation of experiment. The ELN system can export control recipe into XML document for archiving or data transformation. Drupal has the built-in functionality of dumping content into JavaScript Object Notation (JSON) format. JSON is an alternative to XML for storing and exchanging data. Thus, the recipe is transformed into JSON document at first. To complete a fast and strict conversion between JOSN and XML format, a small application is developed in Haskell language. After these two steps, an XML file of control recipe is generated from ELN system, which is suitable for archiving and sharing information.

Figure 8 shows data flow within ELN system using FT4 Powder Rheometer Characterization for example. In this case, the final step is uploading recipe document into BOX database, which is intended for data warehousing. BOX is a cloud computing business that provides content management and file sharing service. Keeping data in cloud storage is a secure and economical method of data archiving. This is also the cornerstone of the data analysis and visualization that also happens on cloud computing.

5. Conclusions and Future Directions

A systematic framework for data management for continuous pharmaceutical manufacturing has been developed. The ISA-88 Batch Control Standard is adapted to continuous manufacturing in order to provide a design philosophy, as well as reference models. The implementation of such a recipe model into the continuous process is well-summarized and assessed. The recipe data warehouse strategy is used for the purpose of data integration in both manufacturing plant and material characterization laboratory. The proper communication among PLC, PAT, PCS, and historian enable the data collection and transformation across the continuous manufacturing plant. A recipe-based ELN system is in charge of capturing data from various analytical platforms. Therefore, data from different process levels and distributed locations can be integrated and contextualized with meaningful information. Such knowledge plays an important role in supporting process control and decision making [31]. Future work includes validating the possibility of using this data management strategy to support the design of experiment (DOE). In terms of the laboratory platform, instrument configuration and calibration information will be included into experiment recipes, which are helpful for error analysis. The importance of this developed data management and integration is important since it will enable industrial practitioners to better monitor and control the process, identify risk, and mitigate process failures. This will enable the product to be produced with reduced time-to-market and increased quality, leading to potentially better and cheaper medicines.

Author Contributions: A.F., S.J., M.I., D.H and R.R. contributed theory and methods. B.H., C.K. and R.S. provided equipment and materials. H.C. and S.M. designed and performed the case studies. H.C., S.M. and R.S. wrote the paper.

Acknowledgments: This work is supported by the National Science Foundation Engineering Research Center on Structured Organic Particulate Systems, through Grant NSF-ECC 0540855, and Johnson and Johnson Company.

Conflicts of Interest: The authors declare no conflicts of interest.

References

1. Lee, S.L.; O'Connor, T.F.; Yang, X.; Cruz, C.N.; Chatterjee, S.; Madurawe, R.D.; Moore, C.M.V.; Yu, L.X.; Woodcock, J. Modernizing pharmaceutical manufacturing: From batch to continuous production. *J. Pharm. Innov.* **2015**, *10*, 191–199. [CrossRef]

2. Food and Drug Administration (FDA). *Guidance for Industry Pat—A Framework for Innovative Pharmaceutical Development, Manufacturing, and Quality Assurance*; Food and Drug Administration: Rockville, MD, USA, 2004.

3. Venkatasubramanian, V.; Zhao, C.; Joglekar, G.; Jain, A.; Hailemariam, L.; Suresh, P.; Akkisetty, P.; Morris, K.; Reklaitis, G.V. Ontological informatics infrastructure for pharmaceutical product development and manufacturing. *Comput. Chem. Eng.* **2006**, *30*, 1482–1496. [CrossRef]

4. Instrument Society of America (ISA). *Batch Control—Part 1: Models and Terminology*; ISA: Research Triangle Park, NC, USA, 1995.

5. Instrument Society of America (ISA). *Batch Control—Part 2: Data Structures and Guidelines for Languages*; SA: Research Triangle Park, NC, USA, 2001.

6. Instrument Society of America (ISA). *Batch Control—Part 3: General and Site Recipe Models and Representation*; SA: Research Triangle Park, NC, USA, 2003.

7. Dorresteijn, R.C.; Wieten, G.; Santen, P.T.E.V.; Philippi, M.C.; Gooijer, C.D.d.; Tramper, J.; Beuvery, E.C. Current good manufacturing practice in plant automation of biological production processes. *Cytotechnology* **1997**, *23*, 19–28. [CrossRef] [PubMed]

8. Verwater-Lukszo, Z. A practical approach to recipe improvement and optimization in the batch processing industry. *Comput. Ind.* **1998**, *36*, 279–300. [CrossRef]

9. Kimball, R.; Ross, M. *The Data Warehouse Toolkit: The Definitive Guide to Dimensional Modeling*; Wiley Computer Publishing: Indianapolis, IN, USA, 2013.

10. Fermier, A.; McKenzie, P.; Murphy, T.; Poulsen, L.; Schaefer, G. Bringing new products to market faster. *Pharm. Eng.* **2012**, *46*, 1–8.

11. Ierapetritou, M.; Muzzio, F.; Reklaitis, G. Perspectives on the continuous manufacturing of powder-based pharmaceutical processes. *AIChE J.* **2016**, *62*, 1846–1862. [CrossRef]

12. Vasilenko, A.; Glasser, B.J.; Muzzio, F.J. Shear and flow behavior of pharmaceutical blends—Method comparison study. *Powder Technol.* **2011**, *208*, 628–636. [CrossRef]

13. Ramachandran, R.; Ansari, M.A.; Chaudhury, A.; Kapadia, A.; Prakash, A.V.; Stepanek, F. A quantitative assessment of the influence of primary particle size polydispersity on granule inhomogeneity. *Chem. Eng. Sci.* **2012**, *71*, 104–110. [CrossRef]

14. Llusa, M.; Levin, M.; Snee, R.D.; Muzzio, F.J. Measuring the hydrophobicity of lubricated blends of pharmaceutical excipients. *Powder Technol.* **2010**, *198*, 101–107. [CrossRef]

15. Armbrust, M.; Stoica, I.; Zaharia, M.; Fox, A.; Griffith, R.; Joseph, A.D.; Katz, R.; Konwinski, A.; Lee, G.; Patterson, D.; et al. A view of cloud computing. *Commun. ACM* **2010**, *53*, 50–58. [CrossRef]

16. Subramanian, B. The disruptive influence of cloud computing and its implications for adoption in the pharmaceutical and life sciences industry. *J. Med. Mark. Device Diagn. Pharm. Mark.* **2012**, *12*, 192–203. [CrossRef]

17. Gruber, T.R. A translation approach to portable ontology specifications. *Knowl. Acquis.* **1993**, *5*, 199–220. [CrossRef]

18. Swartout, W.R.; Neches, R.; Patil, R. Knowledge sharing—Prospects and challenges. In Proceedings of the International Conference on Building and Sharing of Very Large-Scale Knowledge Bases 93, Tokyo, Japan, 1–4 December 1993.

19. W3C. Extensible Markup Language (XML) 1.0 (Fifth Edition). Available online: https://www.w3.org/TR/2008/REC-xml-20081126/ (accessed on 1 March 2018).

20. W3C. W3c XML Schema Definition Language (XSD) 1.1 Part 1: Structures. Available online: https://www.w3.org/TR/xmlschema11-1/ (accessed on 1 March 2018).

21. De Minicis, M.; Giordano, F.; Poli, F.; Schiraldi, M.M. Recipe development process re-design with ansi/isa-88 batch control standard in the pharmaceutical industry. *Int. J. Eng. Bus. Manag.* **2014**, *6*. [CrossRef]

22. Muñoz, E.; Capón-García, E.; Espuña, A.; Puigjaner, L. Ontological framework for enterprise-wide integrated decision-making at operational level. *Comput. Chem. Eng.* **2012**, *42*, 217–234. [CrossRef]

23. International Electrotechnical Commission (IEC). IEC 62264-1:2013. In *Enterprise-Control System Integration—Part 1: Models and Terminology*; IEC: Geneva, Switzerland, 2013.

24. Hwang, Y.; Grant, D. An empirical study of enterprise resource planning integration: Global and local perspectives. *Inf. Dev.* **2014**, *32*, 260–270. [CrossRef]

25. Alphonsus, E.R.; Abdullah, M.O. A review on the applications of programmable logic controllers (PLCS). *Renew. Sustain. Energy Rev.* **2016**, *60*, 1185–1205. [CrossRef]

26. Baumann, W. Davfs2. Available online: http://savannah.nongnu.org/projects/davfs2 (accessed on 1 March 2018).

27. Whitehead, J. Webdav: Versatile collaboration multiprotocol. *IEEE Internet Comput.* **2005**, *9*, 66–74. [CrossRef]

28. Tridgell, A. Efficient Algorithms for Sorting and Synchronization. Ph.D. Thesis, Australian National University, Canberra, Australia, 1999.

29. Singh, R.; Ierapetritou, M.; Ramachandran, R. An engineering study on the enhanced control and operation of continuous manufacturing of pharmaceutical tablets via roller compaction. *Int. J. Pharm.* **2012**, *438*, 307–326. [CrossRef] [PubMed]

30. Godena, G.; Lukman, T.; Steiner, I.; Bergant, F.; Strmčnik, S. A new object model of batch equipment and procedural control for better recipe reuse. *Comput. Ind.* **2015**, *70*, 46–55. [CrossRef]

31. Meneghetti, N.; Facco, P.; Bezzo, F.; Himawan, C.; Zomer, S.; Barolo, M. Knowledge management in secondary pharmaceutical manufacturing by mining of data historians-a proof-of-concept study. *Int. J. Pharm.* **2016**, *505*, 394–408. [CrossRef] [PubMed]

 *processes*

Article

On the Thermal Self-Initiation Reaction of *n*-Butyl Acrylate in Free-Radical Polymerization

Hossein Riazi [1,†], Ahmad Arabi Shamsabadi [1,†], Patrick Corcoran [1], Michael C. Grady [2], Andrew M. Rappe [3] and Masoud Soroush [1,*]

1   Department of Chemical and Biological Engineering, Drexel University, Philadelphia, PA 19104, USA; hr339@drexel.edu (H.R.); arabishamsabadi@gmail.com (A.A.S.); patandchris1@verizon.net (P.C.)
2   Axalta Coating Systems, Wilmington, DE 19803, USA; mcg26@drexel.edu
3   Department of Chemistry, University of Pennsylvania, Philadelphia, PA 19104, USA; rappe@sas.upenn.edu
*   Correspondence: soroushm@drexel.edu; Tel.: +1-(215)-895-1710
†   These authors contributed equally to this work.

Received: 13 December 2017; Accepted: 31 December 2017; Published: 4 January 2018

Abstract: This experimental and theoretical study deals with the thermal spontaneous polymerization of *n*-butyl acrylate (*n*-BA). The polymerization was carried out in solution (*n*-heptane as the solvent) at 200 and 220 °C without adding any conventional initiators. It was studied with the five different *n*-BA/*n*-heptane volume ratios: 50/50, 70/30, 80/20, 90/10, and 100/0. Extensive experimental data presented here show significant monomer conversion at all temperatures and concentrations confirming the occurrence of the thermal self-initiation of the monomer. The order, frequency factor, and activation energy of the thermal self-initiation reaction of *n*-BA were estimated from *n*-BA conversion, using a macroscopic mechanistic model. The estimated reaction order agrees well with the order obtained via our quantum chemical calculations. Furthermore, the frequency factor and activation energy estimates agree well with the corresponding values that we already reported for bulk polymerization of *n*-BA.

Keywords: *n*-butyl acrylate; thermal self-initiation; free-radical polymerization; reaction order

1. Introduction

Free-radical polymerization has been initiated by a thermal initiator like ammonium persulfate [1] and azobisisobutyronitrile [2], or a method such as plasma [3], ultrasound [4], UV-irradiation [5], ionizing-irradiation [6], and redox initiation [7]. A monomer can also initiate free-radical polymerization by itself. Examples are vinyl monomers that undergo self-initiation in bulk [8,9], gas [10], and (mini)emulsion media [11,12]. However, the latter needs a large amount of surfactants to stabilize monomer droplets and polymer particles [12]. The monomer self-initiation reactions can also be exploited as a radical generation source in other radical polymerization mechanisms like reversible addition-fragmentation chain transfer (RAFT) polymerization where a RAFT agent exists. Among the vinyl monomers, styrene has shown high tendency for thermal self-initiation. The proposed mechanism is the formation of two monoradicals from two styrene monomers via self-Diels-Alder cycloaddition [13,14]. Monomer self-initiation strongly depends on temperature; styrene needs more than one year to self-polymerize to reach 50% conversion at 29 °C, however, the same conversion will be obtained after only 4 h at 127 °C. In addition to styrene, thermal self-initiation of acrylate and methacrylate monomers have been studied [15–17]. For methyl methacrylate, the rate of thermal self-initiation is two orders of magnitude lower than styrene [18].

Purity is an important advantage of thermally self-initiated polymers. Pure polymers have many applications in medicine and food industries [19–21], because initiator and catalyst leftovers in polymers reduce their applicability in such areas. For example, molecularly imprinted polymers

(MIPs) as antibodies are synthesized through thermal self-initiation free-radical polymerization [22]. The absence of any kind of functional groups in the microstructure of self-polymerized polymers prevents them from participating in many kinds of side reactions. Thus, the so-called dormant chains can be used as a characterization tool in some living polymerizations [23]. In general, in any kind of application where fully-pure polymer chains are required, polymerization through thermal monomer self-initiation is a good choice. Another advantage of thermally self-polymerized polymers is their low molecular weight [24] especially where they are used as paints and coatings [25].

Environmental regulations constantly have forced resin manufacturers to reduce the level of organic solvents in their products. However, solvents are needed in coatings and paints to endow them with good flowability and brushability. Low molecular weight acrylate resins synthesized through thermal monomer self-initiation polymerization can obviate the need for high amount of solvent. Thermally self-polymerized acrylate polymers usually have Mw < 10,000 g/mol and thereby do not need a high amount of solvent to gain flowability [26]. In the synthesis of hybrid organic/inorganic nanoparticles, thermal monomer self-initiation may also contribute. Hui et al. [27] synthesized polystyrene/silica hybrid brushes via ATRP technique, however, they reported that some pure untethered polystyrene homopolymers are formed due to the thermal self-initiation of styrene. Their results showed that nearly 23 weight % of the styrene undergoes self-initiation reactions and these untethered homopolymer chains, though are formed unintentionally and are considered as impurity, improved elastic modulus and toughness of the polystyrene/silica brushes. Another advantage of thermally monomer-self-initiated polymerization comes into play when conventional thermal initiators are not able to polymerize some monomers. For example, it has been reported that benzoyl peroxide is not able to polymerize dimethylaminoethyl methacrylate, while this monomer undergoes thermal self-initiation polymerization [28]. Moreover, thermally monomer-self-initiated polymerization is a good method for the synthesis of macromonomers as the occurrence of a β-Scission reaction, which is a prevalent reaction at high temperatures, leads to the formation of an unsaturated carbon double bond at the backbone of the produced shortened chain [29]. Such macromonomers have wide applications in the synthesis of brush polymers and graft copolymers [30]. Similar to controlled/living polymerization, thermally monomer-self-initiated polymerization is capable of producing macromonomers without a need for an initiator or a mediating agent [29,30]. Monomer-self-initiated polymerization has also been utilized to synthesize graphene/polystyrene nanocomposites. Graphene was functionalized to improve its lipophilicity, while neither external initiator nor immobilization of initiating groups on the surface of the graphene were required to initiate the polymerization [31].

To fully exploit potentials of the monomer self-initiation reaction, a detailed understanding of its kinetics is needed [32]. Monomer self-initiation reaction affects polymer average molecular weights, molecular weight distribution, and consequently many properties of the polymers [10,15]. At high temperatures, styrene undergoes self-initiation reaction through a third order reaction [33]. For acrylate and methacrylate monomers, a kinetic study of the polymerization at high temperatures is more complicated as many side reactions take place at high temperatures. These side reactions are transfer to monomer, transfer to polymer, midchain radical propagation, midchain radical termination, backbiting, and β-Scission. In general, for non-styrenic monomers, the monomer self-initiation mechanism occurs through Flory's diradical mechanism [16,34,35]. For alkyl acrylate monomers, computational quantum chemistry studies [35–37] showed that two monomer molecules (M) react and form a singlet diradical (*MM*$_S$) (Figure 1):

$$M + M \rightarrow {}^*MM^*_S \tag{1}$$

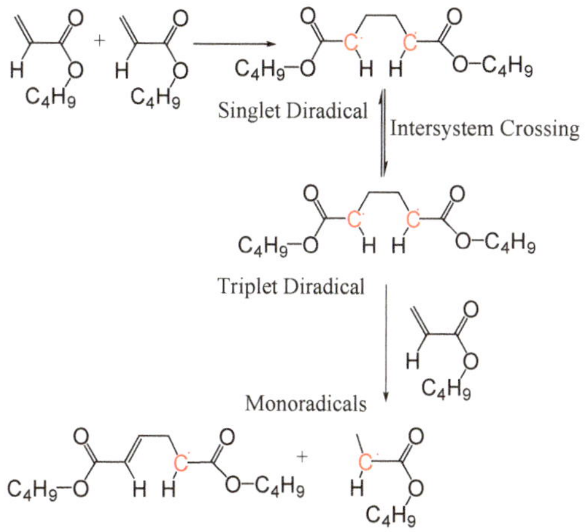

Figure 1. *n*-BA self-initiation mechanism [35].

The singlet diradical includes two monomer units. A singlet diradical has a pair of electrons, one spin-up and one spin-down (+1/2 and −1/2), in one orbital with the second, equal energy orbital, empty [38]. The singlet diradical then goes through a spin crossover reaction and changes into a triplet diradical (*MM*$_T$):

$$*MM*_S \rightarrow *MM*_T \tag{2}$$

The triplet diradical also includes two monomer units. A triplet diradical has two "spin-up" electrons in adjacent, degenerate (equal energy) orbitals (two electrons of the same spin, +1/2 and +1/2) [38]. Finally, a third monomer molecule abstracts a hydrogen atom from the triplet diradical and forms two monoradicals, one with two monomer units (MM*) and the other with one monomer unit (M*):

$$*MM*_T + M \rightarrow MM* + M* \tag{3}$$

Thus, the overall (apparent) monomer self-initiation reaction is:

$$3M \overset{k_{i,m}}{\rightarrow} MM* + M* \tag{4}$$

which according to quantum chemical calculations [35–37], is second order. In this paper, experimentally we investigate this theoretical finding (reaction order) via estimating the order of reaction from monomer conversion measurements made in solution and bulk polymerizations. In addition, we estimate the frequency factor and activation energy of the *n*-BA self-initiation reaction and compare them with those estimated in bulk polymerization of *n*-BA.

2. Materials and Methods

n-BA with purity of 98% containing 50 ppm of inhibitor (4-methoxyphenol) was purchased from Alfa-Aesar. Anhydrous *n*-heptane (99%) was supplied by Sigma-Aldrich, MO, USA. Stainless steel Swagelok tubes (Swagelok Inc., Huntingdon Valley, PA, USA) with length of 4-inch were used as batch reactors. Both ends of the reactors were capped with Swagelok stainless steel caps. These tubes provide 4.8 mL reaction volume and bear pressures as high as 3300 psig. Before polymerization, *n*-BA was passed through an inhibitor removal column to remove the manufacturer-added inhibitor from the monomer. The columns were DHR-4, supplied by Scientific Polymer Products of Ontario, New York, NY, USA. *n*-BA/*n*-heptane mixtures with different volume ratios of 50/50, 70/30, 80/20, 90/10, and 100/0 were then prepared by adding the inhibitor-free *n*-BA to *n*-heptane followed by magnet stirring. The mixtures were then transferred to a sealed round flask, where its inlet was blocked by a rubber cap, to be purged by ultra-high pure nitrogen. After 1 h purging, the nitrogen flow was stopped and the rubber cap was covered with an aluminum adhesive tape very quickly to make sure no oxygen diffused into the flask contents. Next, the flask, tubes and caps were transferred to a nitrogen-atmosphere glove box (LC Technology Solutions, Salisbury, MD, USA). Before transferring the items to the main chamber of the glove box, several vacuum-nitrogen purging cycles were repeated. When the concentration of oxygen in the glove box was lower than 0.1 ppm, the sealed flask was opened and around 3 mL of the flask contents was added to each tube. Afterwards, the caps were tightened and the tubes were removed from the glovebox and weighed, and their weights were recorded.

To carry out the polymerization, a fluidized sand bath was heated to 200 or 220 °C, and left in that temperature for a while to reach a steady state condition. Each polymerization reaction was started by fastening two tubes to a metal basket and putting it in the stabilized sand bath. The reaction times were 55, 110, 165, and 220 min. After staying in the sand bath for the desired time, the basket was removed from the sand bath and immersed in a cold-water bath to cool down the tubes and stop polymerization. The tubes were then detached from the basket, dried, and weighed again to compare their weights with the values recorded after removing them from the glove box. In case of weight difference, the tube was not considered for monomer conversion measurements. Finally, each tube's contents were discharged to an aluminum petri dish and left in a vacuum oven at 50 °C to allow the contents to dry for monomer conversion measurements. It is worth noting here that, our previous studies showed that nearly 1 min is required for the monomer inside the tubes to reach the sand bath temperature. Thus, the tubes were allowed to stay for one minute extra in the sand bath beyond each nominal reaction time. The monomer conversion was measured using the gravimetric method [39].

3. Results and Discussion

Figure 2 shows measured monomer conversion at 200 and 220 °C. These results agree with what one expects from free-radical polymerization; that is, the conversion increases with time, monomer concentration, and temperature. In addition, this figure shows that the *n*-BA self-initiation reaction in both solution and bulk media [39] is so strong that the monomer conversion can exceed 60% after only 220 min in the absence of any external initiators. Monomer self-initiation reactions of both acrylate and methacrylate families were investigated in our previous quantum chemical calculation studies [35–37]; it was found that the second reaction step, transition from a singlet diradical to a triplet one (Equation (2)), is the rate limiting step in the monomer self-initiation. We also found theoretically that the overall monomer self-initiation reaction is second order [35–37]. To validate the overall-reaction order obtained via the quantum chemical calculations, we estimate the overall-reaction order from monomer conversion measurements using a macroscopic mechanistic model.

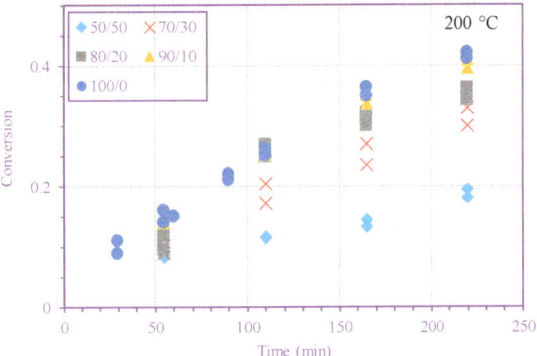

Figure 2. *Cont.*

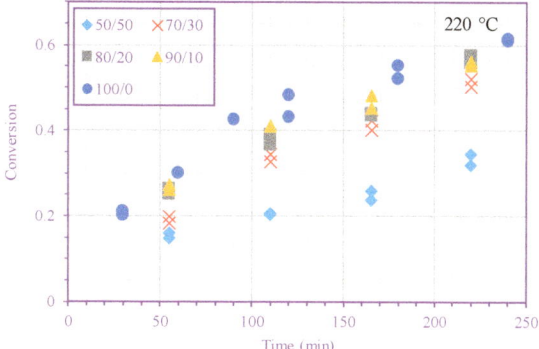

Figure 2. Measured *n*-BA conversion vs. time (different *n*-BA/*n*-heptane ratios) at 200 °C (**top**) and 220 °C (**bottom**).

In thermally monomer-self-initiated polymerization, various reactions happen simultaneously. A list of the most likely reactions is presented in Table A1 in the Appendix A; more details can be found in Ref. [39]. The reactions include, propagation by secondary and tertiary radicals, termination by secondary and tertiary radicals, monomer self-initiation, back-biting, β-Scission, inter-chain transfer to polymer, chain transfer to monomer, chain transfer to solvent, termination by combination, and termination by disproportion. Except for monomer self-initiation reaction, all other reactions are considered as elementary reactions and their reaction rate equations are derived by the method of moments. Writing component mass balances leads to 21 differential equations [39], describing monomer and solvent concentrations; concentrations of secondary radicals containing zero, one, two and three monomer unit(s); the zeroth, first and second moments of the secondary radical chain length distribution; the zeroth, first, second and third moments of the chain length distribution of tertiary radicals formed through intermolecular chain transfer to polymer reactions; the zeroth, first, second and third moments of the chain length distribution of tertiary radicals created by backbiting reactions; and the zeroth, first, second and third moments of the chain length distribution of live and dead polymer chains. Given the initial monomer concentration and the monomer concentration, $[M]_0$ and $[M]$, the monomer conversion, X, is calculated using:

$$X = \frac{[M]_0 - [M]}{[M]_0} \tag{5}$$

To simulate the model, we use the rate coefficients of all reactions except for the overall self-initiation reaction, reported in Ref. [39]. Although frequency factor and activation energy of *n*-BA self-initiation reaction in bulk had been estimated in our previous article [39], we re-estimate them here in solution to compare these results with our previous findings. In our previous bulk study [39], to simulate the macroscopic mechanistic model, the order of *n*-BA overall self-initiation reaction was set to 2, obtained from quantum chemical calculations [35–37]. Here, in addition to re-estimating the frequency factor and activation energy, we estimate the order of *n*-BA overall self-initiation reaction from the conversion measurements. We use ode15s command of MATLAB to integrate the differential equations of the macroscopic mechanistic model and use ga command of MATLAB to minimize the sum of squared residuals:

$$SSRs = \sum_{l=1}^{8} \sum_{j=1}^{5} \sum_{k=1}^{2} \left(X_e\left(t_l, C_{Sj}, T_k\right) - X_m\left(t_l, C_{Sj}, T_k, n, Z, E_a\right) \right)^2 \tag{6}$$

subject to:

$$5.145 \times 10^6 - 4.289 \times 10^6 \ \leq \ Z \leq 5.145 \times 10^6 + 4.289 \times 10^6 \ \left(\mathrm{M^{-1}s^{-1}} \right)$$

$$1.65 \times 10^5 - 0.05 \times 10^5 \leq E_a \leq 1.65 \times 10^5 + 0.05 \times 10^5 \ (\mathrm{J/mol})$$

$$1.5 \ \leq n \leq 2.5$$

where $5.145 \times 10^6 \pm 4.289 \times 10^6 \left(\mathrm{M^{-1}s^{-1}} \right)$ and $1.65 \times 10^5 \pm 0.05 \times 10^5$ (J/mol) are the frequency factor and activation energy values for n-BA bulk polymerization reported in [39], and 2 is the theoretical value of n reported in [35–37]. Here, X_e is the measured monomer conversion value at the reaction time t_l, the solvent concentration C_{Sj}, and the temperature T_k, while X_m is the model-predicted monomer conversion at the same time, solvent concentration and temperature. n, Z, and E_a are the order, frequency factor, and activation energy of the n-BA overall self-initiation reaction, respectively. The monomer conversion measurements are available at two different temperatures (200 and 220 °C), five different n-BA/n-heptane volume ratios (50/50, 70/30, 80/20, 90/10, and 100/0), and four different times (55, 110, 165, and 220 min) twice.

In high temperature free-radical polymerization, secondary reactions like transfer to solvent are possible. In this work, n-heptane was used as the solvent, because transfer reactions to this solvent are negligible. For other vinyl monomers such as styrene and methyl methacrylate, the transfer to n-heptane constant is also very low [40]. On the other hand, solvents like n-butanol, methyl ethyl ketone, and p-xylene participate in chain transfer reactions [41]. The macroscopic mechanistic model showed that the concentration of n-heptane does not change under the experimental conditions used in this study. For example, the highest transfer to n-heptane constant found in [40] is 90×10^{-4}, while values as low as 0.865×10^{-4} are also reported in the same reference. In the worst scenario when the transfer-to-solvent constant is 90×10^{-4}, n-BA/n-heptane ratio is 50/50 volume%, and the temperature is 220 °C, our model showed that the solvent concentration decreased by 0.39% from 3.4169 to 3.4033 (mol/L) after 250 min of the polymerization, indicating the insignificance of chain-transfer-to n-heptane reactions. Thus, it is logical to assume that transfer-to-n-heptane reactions are negligible.

The frequency factor, activation energy, and the order of n-BA overall self-initiation reaction estimated from the conversion measurements are: $1.069 \times 10^6 \pm 0.656 \times 10^6 \left(\mathrm{M^{-1}s^{-1}} \right)$, $1.61 \times 10^5 \pm 0.026 \times 10^5$ (J/mol), and 2.08 ± 0.28. A possible reason for this frequency factor value (estimated here for solution polymerization) being less than the value obtained for bulk polymerization [39] is the cage effect of the solvent. Figures 3 and 4 show the model prediction of n-BA conversion with these estimated values. They indicate that the model can predict the conversion well at all temperatures and solvent concentrations. The frequency factor and activation energy values reported here for the solution polymerization agree well with the values already reported for the bulk polymerization: $5.145 \times 10^6 \pm 4.289 \times 10^6 \left(\mathrm{M^{-1}s^{-1}} \right)$ and $1.65 \times 10^5 \pm 0.05 \times 10^5$ (J/mol) [39]. The order of the monomer self-initiation reaction found here also agrees well with the one obtained theoretically through quantum chemical calculations [35–37]. Figure 5 shows the sensitivity of our macroscopic mechanistic model to changes in the order, activation energy, and frequency factor. It indicates that the predicted conversion has the lowest sensitivity to the frequency factor (Z) and the highest sensitivity to the activation energy, E_a.

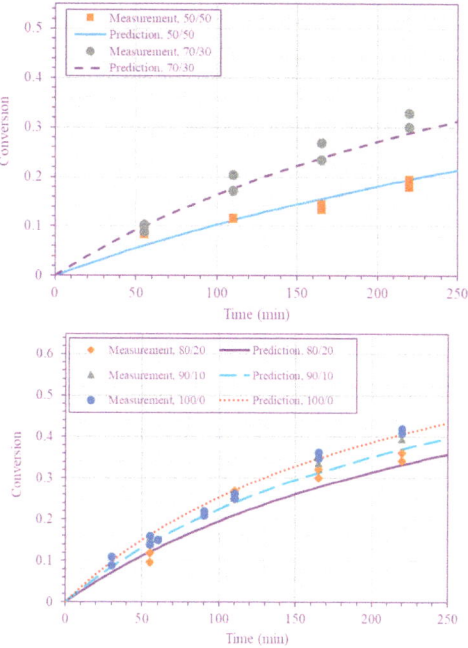

Figure 3. Measured and predicted *n*-BA conversion at 200 °C (*n*-BA/*n*-heptane volume ratios 50/50 and 70/30 (**top**), and 80/20, 90/10, and 100/0 (**bottom**)).

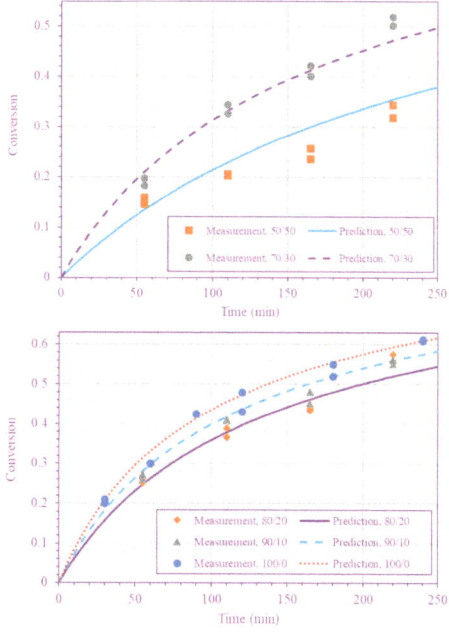

Figure 4. Measured and predicted *n*-BA conversion at 220 °C (*n*-BA/*n*-heptane volume ratios 50/50 and 70/30 (**top**), and 80/20, 90/10, and 100/0 (**bottom**)).

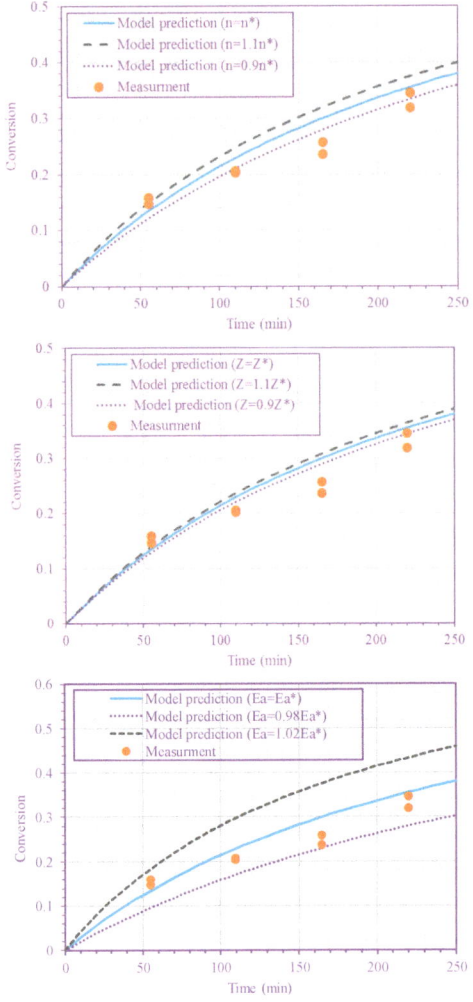

Figure 5. Sensitivity of the model predictions to changes in n (**top** figure), Z (**middle** figure), and E_a (**bottom** figure) (50/50 n-BA/n-heptane ratio; 220 °C). $Z^* = 1.069 \times 10^6 \left(M^{-1}s^{-1} \right)$, $E_a^* = 1.61 \times 10^5 (J/mol)$, and $n^* = 2.08$.

4. Conclusions

Experimental results showed that n-BA undergoes self-initiation reaction at high temperatures in the presence and absence of the solvent. Macroscopic mechanistic model predictions showed that n-heptane does not participate in the polymerization; that is, chain-transfer-to-n-heptane reactions are negligible under the experimental conditions. The monomer conversions predicted by the model agree well with the experimental data. In addition, the order of the n-BA overall self-initiation reaction estimated from conversion is 2.08 ± 0.28, which is in agreement with the theoretical value of 2 calculated using quantum chemical calculations [35–37]. Moreover, the frequency factor and activation energy of n-BA overall self-initiation reaction in solution polymerization are in agreement with those estimated for the same reaction in bulk. The sensitivity analysis showed that the macroscopic mechanistic model is adequately sensitive to the three estimated parameters.

Acknowledgments: The authors would like to thank Aaron Fafarman for his lab facility support.

Author Contributions: H.R. and A.A.S. jointly carried out the experiments and polymer analyses. H.R. prepared the initial draft of this paper. A.A.S., M.S., P.C., M.C.G., and A.M.R. reviewed and revised the manuscript. P.C., M.S., M.C.G. and A.M.R. conceived and designed the experiments. A.A.S. and M.S. analyzed the data.

Conflicts of Interest: The authors declare no conflicts of interest.

Appendix A

Table A1. *n*-BA polymerization reactions.

a.	Apparent monomer self-initiation reaction

$$3M \xrightarrow{k_i} R_1^* + R_2^*$$

b.	Propagation reactions

$$R_n^* + M \xrightarrow{k_p} R_{n+1}^*$$

$$R_n^{**} + M \xrightarrow{k_p^t} R_{n+1}^* \ (+\text{LCB})$$

$$R_n^{***} + M \xrightarrow{k_p^t} R_{n+1}^* \ (+\text{SCB})$$

$$R_n^* + U_m \xrightarrow{k_{mac}} R_{n+m}^{**}$$

c.	Backbiting reactions ($n > 2$)

$$R_n^* \xrightarrow{k_{bb}} R_n^{***}$$

d.	β-scission reactions ($n > 3$)

$$R_n^{***} \xrightarrow{k_\beta} U_3 + R_{n-3}^*$$

$$R_n^{***} \xrightarrow{k_\beta} R_{n-3}^* + U_3$$

$$R_n^{***} \xrightarrow{k_\beta} U_{n-2} + R_2^*$$

$$R_n^{***} \xrightarrow{k_\beta} R_2^* + U_{n-2}$$

$$R_n^{**} \xrightarrow{k_\beta} U_{n-m} + R_m^*$$

$$R_n^{**} \xrightarrow{k_\beta} R_m^* + U_{n-m}$$

e.	Intermolecular chain transfer to polymer reactions

$$R_n^* + D_m \xrightarrow{mk_{tr}^p} D_n + R_m^{**}$$

$$R_n^* + U_m \xrightarrow{mk_{tr}^p} D_n + R_m^{**}$$

f.	Chain transfer to monomer reactions

$$R_n^* + M \xrightarrow{k_{trM}} D_n + R_1^*$$

$$R_n^{**} + M \xrightarrow{k_{trM}^t} D_n + R_1^*$$

$$R_n^{***} + M \xrightarrow{k_{trM}^t} D_n + R_1^*$$

g.	Chain transfer to solvent reactions

$$R_n^* + S \xrightarrow{k_{trS}} D_n + R_0^*$$

$$R_n^{**} + S \xrightarrow{k_{trS}^t} D_n + R_0^*$$

$$R_n^{***} + S \xrightarrow{k_{trS}^t} D_n + R_0^*$$

h.	Termination by coupling reactions

$$R_n^* + R_m^* \xrightarrow{k_{tc}} D_{n+m}$$

$$R_n^* + R_m^{**} \xrightarrow{2k_{tc}^t} D_{n+m}$$

$$R_n^* + R_m^{***} \xrightarrow{2k_{tc}^t} D_{n+m}$$

$$R_n^{**} + R_m^{**} \xrightarrow{k_{tc}^{tt}} D_{n+m}$$

$$R_n^{**} + R_m^{***} \xrightarrow{2k_{tc}^{tt}} D_{n+m}$$

$$R_n^{***} + R_m^{***} \xrightarrow{k_{tc}^{tt}} D_{n+m}$$

Table A1. *Cont.*

i	Termination by disproportionation reactions
	$R_n^* + R_m^* \xrightarrow{k_{td}} D_n + U_m$
	$R_n^* + R_m^{**} \xrightarrow{k_{td}^t} D_n + U_m$
	$R_n^* + R_m^{**} \xrightarrow{k_{td}^t} D_m + U_n$
	$R_n^* + R_m^{***} \xrightarrow{k_{td}^t} D_n + U_m$
	$R_n^* + R_m^{***} \xrightarrow{k_{td}^t} D_m + U_n$
	$R_n^{**} + R_m^{**} \xrightarrow{k_{td}^{tt}} D_n + U_m$
	$R_n^{**} + R_m^{***} \xrightarrow{k_{td}^{tt}} D_n + U_m$
	$R_n^{**} + R_m^{***} \xrightarrow{k_{td}^{tt}} D_m + U_n$
	$R_n^{***} + R_m^{***} \xrightarrow{k_{td}^{tt}} D_n + U_m$

References

1. Adelnia, H.; Gavgani, J.N.; Riazi, H.; Bidsorkhi, H.C. Transition behavior, su characteristics and film formation of functionalized poly (methyl methacrylate-co-butyl acrylate) particles. *Prog. Org. Coat.* **2014**, *77*, 1826–1833. [CrossRef]

2. Adelnia, H.; Riazi, H.; Saadat, Y.; Hosseinzadeh, S. Synthesis of monodisperse anionic submicron polystyrene particles by stabilizer-free dispersion polymerization in alcoholic media. *Colloid Polym. Sci.* **2013**, *291*, 1741–1748. [CrossRef]

3. Yasuda, H.; Wang, C. Plasma polymerization investigated by the substrate temperature dependence. *J. Polym. Sci. A1* **1985**, *23*, 87–106. [CrossRef]

4. Price, G.J.; Norris, D.J.; West, P.J. Polymerization of methyl methacrylate initiated by ultrasound. *Macromolecules* **1992**, *25*, 6447–6454. [CrossRef]

5. Decker, C. The use of UV irradiation in polymerization. *Polym. Int.* **1998**, *45*, 133–141. [CrossRef]

6. Marans, N.S. Ionizing Irradiation Polymerization of Trioxane with Copolymerizable Stabilizing Comonomers. Google Patents. U.S. Patent 3,366,561, 30 January 1968.

7. Riazi, H.; Mohammadi, N.; Mohammadi, H. Emulsion copolymerization of methyl methacrylate/butyl acrylate/iodine system to monosize rubbery nanoparticles containing iodine and triiodide mixture. *Ind. Eng. Chem. Res.* **2013**, *52*, 2449–2456. [CrossRef]

8. Riazi, H.; Arabi Shamsabadi, A.; Grady, M.; Rappe, A.; Soroush, M. *Experimental and Macroscopic Mechanistic Modeling Studies of the Methyl Acrylate Self Initiation Reaction*; AIChE Annual Meeting: Minneapolis, MN, USA, 2017.

9. Riazi, H.; Arabi Shamsabadi, A.; Grady, M.; Rappe, A.M.; Soroush, M. Experimental and Theoretical Study of the Self-Initiation Reaction of Methyl Acrylate in Free-Radical Polymerization. *Ind. Eng. Chem. Res.* **2017**. [CrossRef]

10. Alsharaeh, E.H.; Ibrahim, Y.M.; El-Shall, M.S. Direct evidence for the gas phase thermal polymerization of styrene. Determination of the initiation mechanism and structures of the early oligomers by ion mobility. *J. Am. Chem. Soc.* **2005**, *127*, 6164–6165. [PubMed]

11. Pan, G.; Sudol, E.D.; Dimonie, V.L.; El-Aasser, M.S. Thermal self-initiation of styrene in the presence of TEMPO radicals: Bulk and miniemulsion. *J. Polym. Sci. A1* **2004**, *42*, 4921–4932. [CrossRef]

12. Matyjaszewski, K.; Davis, T.P. *Handbook of Radical Polymerization*; John Wiley & Sons: Hoboken, NJ, USA, 2003; p. 878.

13. Vana, P.; Barner-Kowollik, C.; Davis, T.P.; Matyjaszewski, K. *Radical Polymerization*; Encyclopedia of Polymer Science and Technology; John Wiely & Sons: Hoboken, NJ, USA, 2003.

14. Moad, G.; Solomon, D.H. *The Chemistry of Radical Polymerization*, 2rd ed.; Elsevier: Amsterdam, The Netherlands, 2006; pp. 108–109.

15. Wang, W.; Hutchinson, R.A. Free-Radical Acrylic Polymerization Kinetics at Elevated Temperatures. *Chem. Eng. Technol.* **2010**, *33*, 1745–1753. [CrossRef]

16. Srinivasan, S.; Kalfas, G.; Petkovska, V.I.; Bruni, C.; Grady, M.C.; Soroush, M. Experimental study of the spontaneous thermal homopolymerization of methyl and n-butyl acrylate. *J. Appl. Polym. Sci.* **2010**, *118*, 1898–1909. [CrossRef]

17. Liu, S.; Srinivasan, S.; Grady, M.C.; Soroush, M.; Rappe, A.M. Computational study of cyclohexanone-monomer co-initiation mechanism in thermal homo-polymerization of methyl acrylate and methyl methacrylate. *J. Phys. Chem. A* **2012**, *116*, 5337–5348. [CrossRef] [PubMed]

18. Mishra, M.; Yagci, Y. *Handbook of Radical Vinyl Polymerization*; CRC Press: Boca Raton, FL, USA, 1998; p. 48.

19. Mattamal, G.J. US FDA perspective on the regulations of medical-grade polymers: Cyanoacrylate polymer medical device tissue adhesives. *Exp. Rev. Med. Devices* **2008**, *5*, 41–49. [CrossRef] [PubMed]

20. Frommelt, H. Polymers for medical applications. *Macromol. Symp.* **1987**, *12*, 281–301. [CrossRef]

21. Jin, T.; Zhang, H. Biodegradable polylactic acid polymer with nisin for use in antimicrobial food packaging. *J. Food Sci.* **2008**, *73*, 127–134. [CrossRef] [PubMed]

22. Panagiotopoulou, M.; Beyazit, S.; Nestora, S.; Haupt, K.; Bui, B.T.S. Initiator-free synthesis of molecularly imprinted polymers by polymerization of self-initiated monomers. *Polymer* **2015**, *66*, 43–51. [CrossRef]

23. Sun, Y.; Wu, Y.; Chen, L.; Fu, Z.; Shi, Y. Thermal Self-Initiation in Stable Free-Radical Polymerization of Styrene. *Polym. J.* **2009**, *41*, 954–960. [CrossRef]

24. Peck, A.N.; Hutchinson, R.A. Secondary reactions in the high-temperature free radical polymerization of butyl acrylate. *Macromolecules* **2004**, *37*, 5944–5951. [CrossRef]

25. Yu, X.; Pfaendtner, J.; Broadbelt, L.J. Ab initio study of acrylate polymerization reactions: Methyl methacrylate and methyl acrylate propagation. *J. Phys. Chem. A* **2008**, *112*, 6772–6782. [CrossRef] [PubMed]

26. Moghadam, N.; Liu, S.; Srinivasan, S.; Grady, M.C.; Rappe, A.M.; Soroush, M. Theoretical study of intermolecular chain transfer to polymer reactions of alkyl acrylates. *Ind. Eng. Chem. Res.* **2015**, *54*, 4148–4165. [CrossRef]

27. Hui, C.M.; Dang, A.; Chen, B.; Yan, J.; Konkolewicz, D.; He, H.; Ferebee, R.; Bockstaller, M.R.; Matyjaszewski, K. Effect of thermal self-initiation on the synthesis, composition, and properties of particle brush materials. *Macromolecules* **2014**, *47*, 5501–5508. [CrossRef]

28. Shalati, M.D.; Scott, R.M. Thermal polymerization of dimethylaminoethyl methacrylate. *Macromolecules* **1975**, *8*, 127–130. [CrossRef]

29. Zorn, A.M.; Junkers, T.; Barner-Kowollik, C. Synthesis of a Macromonomer Library from High-Temperature Acrylate Polymerization. *Macromol. Rapid Commun.* **2009**, *30*, 2028–2035. [CrossRef] [PubMed]

30. Junkers, T.; Bennet, F.; Koo, S.P.; Barner-Kowollik, C. Self-directed formation of uniform unsaturated macromolecules from acrylate monomers at high temperatures. *J. Polym. Sci. Pol. Chem.* **2008**, *46*, 3433–3437. [CrossRef]

31. Beckert, F.; Rostas, A.M.; Thomann, R.; Weber, S.; Schleicher, E.; Friedrich, C.; Mulhaupt, R. Self-initiated free radical grafting of styrene homo-and copolymers onto functionalized graphene. *Macromolecules* **2013**, *46*, 5488–5496. [CrossRef]

32. Soroush, M.; Grady, M.C.; Kalfas, G.A. Free-radical polymerization at higher temperatures: Systems impacts of secondary reactions. *Comput. Chem. Eng.* **2008**, *32*, 2155–2167. [CrossRef]

33. Meyer, T.; Keurentjes, J. *Handbook of Polymer Reaction Engineering*; Wiley-VCH: Weinheim, Germany, 2005; p. 169.

34. Zhang, C.; Wang, X.; Liu, L.; Wang, Y.; Peng, X. Modeling the spontaneous initiation of the polymerization of methyl methacrylate. *J. Mol. Model.* **2008**, *14*, 1053–1064. [CrossRef] [PubMed]

35. Srinivasan, S.; Lee, M.W.; Grady, M.C.; Soroush, M.; Rappe, A.M. Self-initiation mechanism in spontaneous thermal polymerization of ethyl and *n*-butyl acrylate: A theoretical study. *J. Phys. Chem. A* **2010**, *114*, 975–7983. [CrossRef] [PubMed]

36. Liu, S.; Srinivasan, S.; Tao, J.; Grady, M.C.; Soroush, M.; Rappe, A.M. Modeling spin-forbidden monomer self-initiation reactions in spontaneous free-radical polymerization of acrylates and methacrylates. *J. Phys. Chem. A* **2014**, *118*, 9310–9318. [CrossRef] [PubMed]

37. Srinivasan, S.; Lee, M.W.; Grady, M.C.; Soroush, M.; Rappe, A.M. Computational study of the self-initiation mechanism in thermal polymerization of methyl acrylate. *J. Phys. Chem. A* **2009**, *113*, 10787–10794. [CrossRef] [PubMed]

38. Liu, S.; Srinivasan, S.; Grady, M.C.; Soroush, M.; Rappe, A.M. Backbiting and β-scission reactions in free-radical polymerization of methyl acrylate. *Int. J. Quantum Chem.* **2014**, *114*, 345–360. [CrossRef]

39. Shamsabadi, A.A.; Moghadam, N.; Srinivasan, S.; Corcoran, P.; Grady, M.C.; Rappe, A.M.; Soroush, M. Study of *n*-Butyl Acrylate Self-Initiation Reaction Experimentally and via Macroscopic Mechanistic Modeling. *Processes* **2016**, *4*, 15. [CrossRef]
40. Brandrup, J.; Immergut, E.H.; Grulke, E.A.; Abe, A.; Bloch, D.R. *Polymer Handbook*; Wiley: New York, NY, USA, 1989; p. II/120.
41. Moghadam, N.; Srinivasan, S.; Grady, M.C.; Rappe, A.M.; Soroush, M. Theoretical study of chain transfer to solvent reactions of alkyl acrylates. *J. Phys. Chem. A* **2014**, *118*, 5474–5487. [CrossRef] [PubMed]

Article

Computational Package for Copolymerization Reactivity Ratio Estimation: Improved Access to the Error-in-Variables-Model

Alison J. Scott and Alexander Penlidis *

Institute for Polymer Research (IPR), Department of Chemical Engineering, University of Waterloo, Waterloo, ON N2L 3G1, Canada; ajscott@uwaterloo.ca
* Correspondence: penlidis@uwaterloo.ca; Tel.: +1-519-888-4567 (ext. 36634)

Received: 12 December 2017; Accepted: 13 January 2018; Published: 20 January 2018

Abstract: The error-in-variables-model (EVM) is the most statistically correct non-linear parameter estimation technique for reactivity ratio estimation. However, many polymer researchers are unaware of the advantages of EVM and therefore still choose to use rather erroneous or approximate methods. The procedure is straightforward but it is often avoided because it is seen as mathematically and computationally intensive. Therefore, the goal of this work is to make EVM more accessible to all researchers through a series of focused case studies. All analyses employ a MATLAB-based computational package for copolymerization reactivity ratio estimation. The basis of the package is previous work in our group over many years. This version is an improvement, as it ensures wider compatibility and enhanced flexibility with respect to copolymerization parameter estimation scenarios that can be considered.

Keywords: copolymerization kinetics; copolymer composition; design of experiments; error-in-variables-model (EVM); parameter estimation; polymer reaction engineering; reactivity ratios

1. Introduction

In copolymerization kinetics, reactivity ratios are important parameters. Not only do reactivity ratio estimates specify the degree of incorporation of each comonomer into the copolymer (i.e., average copolymer composition) but they also provide information about other copolymer microstructural indicators (namely azeotropic point, sequence length distribution, triad fractions and so on). This knowledge of kinetics and microstructure can be useful in synthesizing copolymers with specific desirable properties for specific applications. Thus, polymer chemists and polymer reaction engineers require reliable reactivity ratio estimates.

Over the years, many different (and incorrect) methods have been implemented for reactivity ratio estimation. Linear parameter estimation techniques (such as the Mayo-Lewis method (method of intersections), the Fineman-Ross method and the Kelen-Tüdős method) were used previously due to lack of computational power. However, these techniques should not be used, as linear estimation techniques applied to non-linear models result in faulty parameter estimates and a distorted error structure [1–3]. Other common sources of error in parameter estimation include poorly designed experiments—too few (usually unreplicated) data points, chosen at random—inherent experimental difficulties (especially at low conversion levels) and inappropriate kinetic models. Ultimately, this has created a wide variety of reactivity ratios in the literature, even for similar copolymer systems (see, for example, reactivity ratios associated with the copolymer of 2-acrylamido-2-methylpropane sulfonic acid and acrylamide, as summarized by Scott et al. [4]).

The most statistically correct technique for reactivity ratio estimation is the error-in-variables-model (EVM). EVM is a non-linear parameter estimation technique that considers the error present in all variables. The procedure is fairly straightforward but somewhat computationally intensive. As a result, researchers often revert back to "historical" (and incorrect) linear parameter estimation techniques. It is speculated that researchers choose not to use EVM for two main reasons: (1) They are unaware of EVM and its advantages, and/or (2) They are intimidated by the complexity of the background mathematics required to use EVM. Therefore, the goal of the current work (based on past work as described in references [1–3]) is to make EVM more accessible to all researchers by analyzing a variety of copolymerization case studies using a ready-to-use computational package.

A series of five case studies presented herein revisits copolymerization data from the literature; each analysis has a specific goal in mind. Initially, we will look at current "best practices" and their shortcomings (Section 4.1) by exploring linear parameter estimation techniques (Exhibit A) and the limitations associated with low conversion data sets (Exhibit B). Next, we will demonstrate how to maximize and exploit information content from experimental data (Section 4.2). More specifically, case studies will exhibit the benefits of using cumulative copolymerization data (Exhibit C), the need for replicated experiments (Exhibit D) and the advantages of sequential design of experiments (Exhibit E).

2. A Brief Overview of EVM for Reactivity Ratio Estimation

2.1. Copolymerization Models

Monomer reactivity ratios (r_1 and r_2) are parameters used to describe the potential for homopropagation relative to cross-propagation. Reactivity ratios can be estimated using experimental data and a copolymerization model, if the unreacted monomer composition in the polymerizing mixture and the cumulative copolymer composition are known [1–5].

The Mayo-Lewis equation (see Equation (1)), also called the instantaneous copolymer composition (ICC) equation, is the most widely used copolymerization model. Equation (1) can be used to determine the instantaneous mole fraction of monomer 1 incorporated into the copolymer (F_1) given the comonomer composition in the polymerizing mixture (as mole fractions of unbound monomer, f_i). It is important to note that the Mayo-Lewis equation provides the instantaneous copolymer composition, which means that the model is only applicable for low conversion data (typically <10%, where composition drift is minimal).

$$F_1 = \frac{r_1 f_1^2 + f_1 f_2}{r_1 f_1^2 + 2 f_1 f_2 + r_2 f_2^2} \tag{1}$$

where $r_1 = \frac{k_{11}}{k_{12}}$ and $r_2 = \frac{k_{22}}{k_{21}}$ (k_{ij} is the rate constant for each of the four possible propagation reactions, with active center i adding monomer j).

In order to analyze copolymerization data for medium or high conversion levels, the cumulative form of the copolymer composition model becomes necessary. Direct numerical integration (DNI) requires combining and solving (simultaneously) an instantaneous mole balance and a cumulative mole balance (after reaching a certain molar conversion level, X_n). The instantaneous mole balance (Equation (2)) is an ordinary differential equation, from which f_i can be found at any conversion level (initial conditions $f_1 = f_{1,0}$ at $X_n = 0$). The cumulative mole fraction corresponding to X_n is given by the well-known Skeist equation (Equation (3)). DNI is a direct numerical approach and does not rely on model transformations or other potentially restrictive assumptions. This is a significant advantage over other estimation approaches with copolymerization models [3,6].

$$\frac{df_1}{dX_n} = \frac{f_1 - F_1}{1 - X_n} \tag{2}$$

$$\overline{F}_1 = \frac{f_{1,0} - f_1(1 - X_n)}{X_n} \tag{3}$$

2.2. Error-in-Variables-Model (EVM)

A full statistical explanation of the error-in-variables-model (and enumeration of its benefits) has been presented previously; the interested reader should refer to Reilly and Patino-Leal [7] or Kazemi et al. [3,6,8]. Only the basics are presented herein, for the reader to have a brief overview before we tackle the case studies.

As mentioned previously, EVM forces the researcher to consider all sources of error, including the error associated with independent variables (such as feed composition) and (measured) cumulative copolymer composition. To obtain estimates of the "true" values of both the independent variables and the parameters, the EVM program uses a nested-iterative loop (this is represented schematically in Figure 1, with variables defined in the discussion below). The inner loop searches for "true" values of the independent variables, since there is inevitably some error associated with the measured values. Mathematically, we can relate the vector of measurements ($\underline{x}_i$) to the vector of their unknown "true" values ($\underline{\xi}_i$) and an error term ($k\varepsilon_i$), according to Equation (4). In the error term, k is a constant that represents the magnitude of the error and $\underline{\varepsilon}$ (error) is a random variable that is typically uniformly distributed on the interval [−1, 1] (an additional explanation is included in Appendix A, for the interested reader). At the same time, the outer loop uses a copolymerization model (such as the ICC model, Equation (1)) to relate the "true" variables and the parameter (reactivity ratio) estimates, as shown in Equation (5).

$$\underline{x}_i = \underline{\xi}_i(1 + k\underline{\varepsilon}_i) \tag{4}$$

$$\underline{g}\left(\underline{\xi}_i, \underline{\theta}\right) = 0 \tag{5}$$

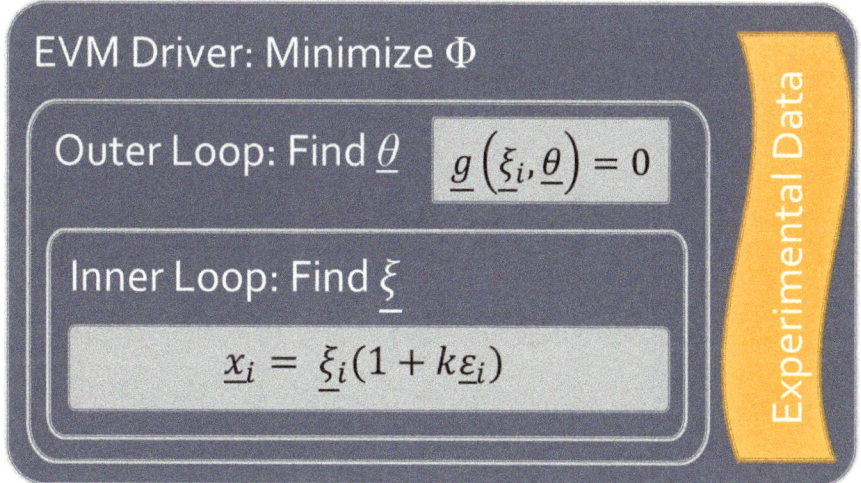

Figure 1. Nested-iterative algorithm for the error-in-variables-model (EVM).

From a statistical perspective, the program uses this nested-iterative approach to minimize the sum of squares between the observed and predicted values, both in terms of the error in the independent variables and in terms of the parameter estimates. When the objective function (Equation (6)) is minimized, the program has found the best estimates for both the independent variables and the parameters (reactivity ratios).

$$\Phi = \frac{1}{2}\sum_{i=1}^{n} r_i(\underline{\overline{x}}_i - \underline{\hat{\xi}}_i)'\underline{V}^{-1}\left(\underline{\overline{x}}_i - \underline{\hat{\xi}}_i\right) \tag{6}$$

where n is the number of experimental trials (runs), r_i is the number of replicates for the ith trial, $\bar{x}_i$ is the average of the r_i measurements (x_i), $\hat{\underline{\zeta}}_i$ is an estimate of the true values of the variables ($\underline{\zeta}_i$) and $\underline{V}$ is the variance-covariance matrix of the variables (which provides information about measurement error of the variables involved).

Alternatively, minimizing the objective function can be considered graphically, as in Figure 2. Given a model and some measured (independent) data, the inner loop minimizes the horizontal distances between the data points and the model (curve). At the same time, the outer loop minimizes the vertical distances between the data points and the model (that is, the outer loop attempts to reconcile model predictions and measurements).

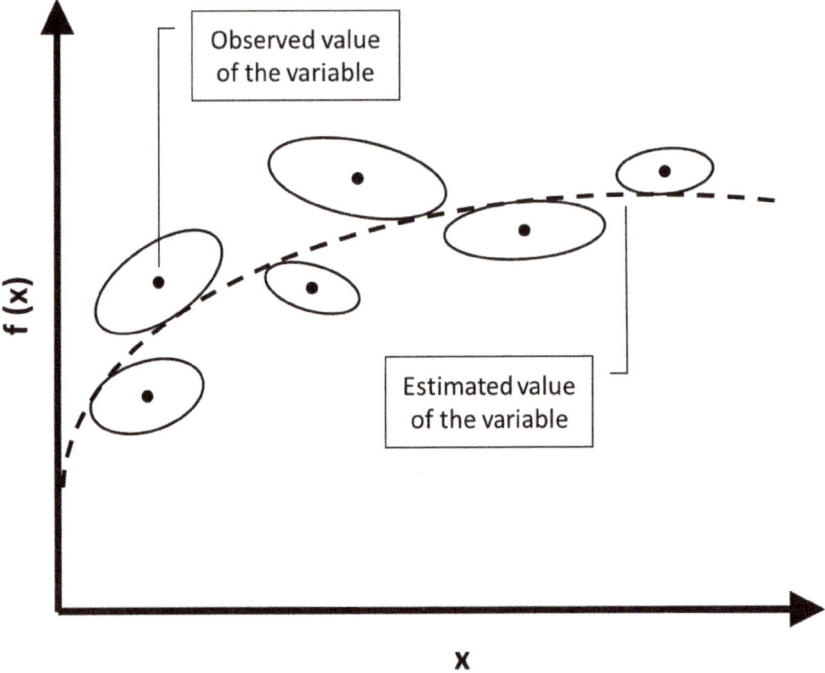

Figure 2. Graphical representation of EVM (inspired by [2]).

The computational package described herein that employs EVM for reactivity ratio estimation is based on the RREVM program created by Dubé et al. [1] (in Fortran 77), which was later updated by Polic et al. [2]. The program version was further updated and converted to MATLAB by Kazemi et al. [3,6,8]. The new and improved software version in MATLAB ensures wider compatibility and allows for possible extensions to other multi-component systems [9]. Also, using MATLAB as the program platform allows for open-source programming, which gives researchers the option of modifying and tailoring the program as needed.

3. Program Description

3.1. Overview

Although the technical aspects of EVM were kept to a minimum in Section 2 (since more details can be found in the references), several comments are now in order about the modifications to the program, in order to make it more user-friendly. The program has been equipped with a graphical user interface (GUI), so that very little knowledge of MATLAB (or programming, in general) is required to invoke the EVM algorithm. Once users open the QuickStart file and execute the program, they are presented with a series of instructions and user prompts. Details and program screenshots are presented in Appendix B for the interested reader, whereas a brief overview is presented in this section.

First, the user must choose their preferred method of data entry (Figure A1). Data input may be manual (with step by step prompts) or may employ a user-prepared data file. Next, the user must indicate whether the copolymerization information (data) for analysis is instantaneous (below 10% conversion) or cumulative (medium-high conversion).

Once these preliminary decisions have been made, the user is prompted to provide the copolymerization data required for analysis (Section 3.2). Finally, the program evaluates the data and presents the results (Section 3.3). The program typically converges within seconds for the instantaneous analysis and in under one minute for the cumulative analysis (usually within less than 10 to 20 iterations in both cases). The time (or number of iterations) required for convergence is a consequence of the location of the initial estimates and the precision required in the estimates; better initial estimates will result in faster program convergence. The EVM program is also equipped with a "time-out" option, which occurs if no solution has been found after a pre-determined number of iterations.

3.2. Program Requirements

3.2.1. Instantaneous Model

Preliminary estimates of r_1 and r_2 act as "starting points" for EVM (Figure A2). Depending on how much is known about the copolymer system, this information may be acquired from either the literature or preliminary experiments. Literature values can be good starting guesses for reactivity ratios, either from a prior study on the same copolymer system (or even from existing values from a similar system) or from simple trending analysis based on preliminary/screening experiments over a limited range of conditions. Prior knowledge can provide valuable information for both the design and estimation steps. If these preliminary estimates are far from the real ones, convergence may simply take slightly longer than typical orders of magnitude given at the end of Section 3.1.

For the instantaneous case, the copolymerization data required are the feed composition ($f_{1,0}$) and cumulative copolymer composition ($\bar{F}_1$), both in terms of monomer 1 (see sample program prompt in Figure A3). ($\bar{F}_1$ is approximated by F_1 for low conversion experiments treated by the instantaneous case). As mentioned previously, the ICC model assumes that composition drift does not occur. Therefore, it is recommended that only low conversion data (below 10% conversion) be included in this analysis.

The final prompt prior to parameter estimation is a review of the default settings (Figure A4). This window gives users the opportunity to check settings such as the error type (additive vs. multiplicative error), the error tolerance level and the variance-covariance matrix for the copolymerization system. Details regarding these input values are presented in the *Default Settings* section (Section 3.2.3).

If the user prefers to use a pre-made data file for program input (see again Figure A1), the same information is required: preliminary reactivity ratio estimates, experimental data and program settings. However, all of the data input is presented in a single '.txt' file, which can be saved, modified (as necessary) and re-analyzed. This is particularly advantageous if a data set is being altered slightly between analyses, as in some of the case studies presented in Section 4. For the interested reader, a sample data file is presented in Appendix B (Figure A5).

3.2.2. Cumulative Model

The analysis with the cumulative model uses many of the same inputs as the instantaneous analysis but the direct numerical integration (DNI) requires additional information. After the user provides preliminary reactivity ratio estimates, one is prompted to input the molecular weights (MW_i)

of both comonomers. This information is required to relate weight conversion data (X_w, which can be experimentally determined using gravimetry) to molar conversion data (X_n, which is used in the DNI as per Equations (2) and (3)). The program converts X_w to X_n according to Equation (7).

$$X_n = X_w \frac{MW_1 f_{1,0} + MW_2 f_{2,0}}{MW_1 \overline{F}_1 + MW_2 \overline{F}_2} \tag{7}$$

The next program requirement is the input of the copolymerization data. In this case, since medium or even high conversion data may be analyzed, the conversion values must be included for each point. Therefore, the user enters three arrays of data: X_w (measured mass conversion), $f_{1,0}$ (known initial feed composition) and $\overline{F}_1$ (measured cumulative copolymer composition). Again, initial reactivity ratio estimates, monomer molecular weights and experimental data may be provided through this series of prompts or in a single data file.

3.2.3. Default Settings

A number of default settings are included in the program to ensure ease of implementation (see again Figure A4). However, settings such as the number of parameters, equations and/or variables involved should not be modified. Changing these settings (and modifying the associated source code) allows for expansion of the program to other applications, such as reactivity ratio estimation for multi-component polymerizations (a program for ternary reactivity ratio estimation has already been developed and applied to experimental terpolymerization data) [9].

The only change that might be made to the default settings is to the variance-covariance matrix, $\underline{V}$. The matrix dimension must not be modified but individual entries may be changed to incorporate prior knowledge. For example, since the instantaneous model uses two variables ($f_{1,0}$ and $\overline{F}_1$), the variance-covariance matrix is a 2×2 matrix. Additional information about default entries in the $\underline{V}$ matrix can be found in Appendix A.

3.3. Results & Diagnostics

Once the program has all necessary (input) data, EVM acquires the best possible estimates of the reactivity ratios ($\underline{\theta}$) and the independent variables ($\underline{\xi}_i$).

Additional outputs are the objective function value (Φ, minimized as per Equation (6)) and $\underline{G}$, which is the expected value of the second derivative of Φ with respect to the parameters. This is expressed mathematically in Equation (8).

$$\underline{G} = E\left[\frac{d^2\Phi}{d\theta_i d\theta_j}\right] = \sum_{i=1}^{n} r_i \underline{Z}'_i \left(\underline{B}_i \underline{V} \underline{B}'_i\right)^{-1} \underline{Z}_i \tag{8}$$

For the interested reader, more information about Equation (8) (and relevant variables) is given in Appendix A. However, since the program calculates the $\underline{G}$ matrix "behind the scenes", the average user should focus on the fact that the $\underline{G}$ matrix gives valuable information about the parameters ($\underline{\theta}$, specifically r_1 and r_2). In fact, the inverse of the $\underline{G}$ matrix provides an approximation of the variance-covariance matrix for the parameters. With this information, the MATLAB program can plot joint confidence regions (JCRs), which are discussed in what follows.

Joint Confidence Regions

JCRs are typically elliptical contours that quantify the level of uncertainty in the parameter estimates; smaller JCRs indicate higher precision and therefore more confidence in the estimation results.

In this program, the joint confidence region for parameter estimates can be visualized using an "error ellipse" (Equation (9)). This assumes that the error be normally distributed and that the variance be known.

$$(\underline{\theta} - \underline{\hat{\theta}})' \underline{G}(\underline{\theta} - \underline{\hat{\theta}}) \leq \chi^2_{p,\alpha} \tag{9}$$

where $\chi^2_{p,\alpha}$ represents the chi-squared distribution for p parameters and a confidence level of $(1 - \alpha)$. The program uses $\chi^2_{2,0.05} = 5.991$ to plot JCRs at the 95% confidence level.

Perhaps the most useful information from the calculation of JCRs is the degree of precision (that is, the size and shape of the error ellipse). Ideally, the JCR will be small and round (as in Figure 3, ellipse "A"). A small JCR confirms that the parameter estimates are close to the "true" values and a round JCR indicates that the two parameter estimates have approximately the same amount of associated uncertainty. If, on the other hand, the JCR is long and narrow, it suggests that one parameter may be well-defined, whereas the other parameter may have a significant amount of associated uncertainty (see, for example, Figure 3, ellipse "B").

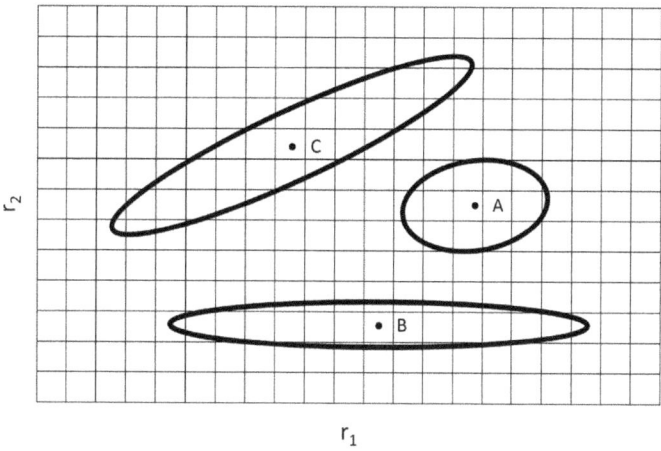

Figure 3. Sample Joint Confidence Regions (JCRs).

Another important piece of information is the degree of parameter correlation, which can be evaluated according to the slope of the JCR. Parameter correlation is something that should be avoided as much as possible and can be minimized by using designed experiments. If there is a high degree of parameter correlation, the elliptical JCR will be at an angle, as in Figure 3, ellipse "C". Well-behaved copolymerization systems should have reactivity ratios with similar degrees of uncertainty and minimal correlation.

4. Case Studies

4.1. Current "Best Practices" and Their Shortcomings

4.1.1. Exhibit A: Why Do Researchers Continue to Use Linear Parameter Estimation Techniques?

As mentioned in the introduction, linear parameter estimation techniques were originally used for reactivity ratio estimation (RRE) due to lack of computational power. However, since the required technology is now readily available, linear parameter estimation techniques should no longer be used for RRE. Linearizing or transforming the model distorts the error structure and may result in faulty parameter estimates. The statement may seem obvious but it is still worth emphasizing:

linear parameter estimation techniques should not be used for the estimation of parameters in non-linear models!

Although most polymer researchers know that linear techniques are inaccurate, they are still taught in both introductory and graduate level polymer chemistry/science courses. Additionally, in perhaps the most commonly perused non-technical "reference," Wikipedia, only the outdated linear techniques are mentioned. It is no wonder, then, that researchers continue to use incorrect parameter estimation techniques.

Exhibit A presents an overview of recent literature [10–13] regarding the copolymerization of 2-methylene-1,3-dioxepane (MDO; monomer 1) and vinyl acetate (VAc; monomer 2) (see also Table 1; RR stands for reactivity ratio). This copolymer has gained considerable attention in the past decade, largely due to its degradable properties. Researchers are especially interested in the reactivity ratios for the system, as reactivity ratios provide information about the copolymer microstructure. However, the reactivity ratio estimation (RRE) techniques used in this field are often incorrect. This case study will focus primarily on the issue of linear parameter estimation techniques, but invalid low conversion assumptions (that is, inappropriate use of the instantaneous copolymerization model) and error-prone data cannot be overlooked. Therefore, to demonstrate the advantages of EVM, select data from the literature will be re-evaluated (properly) and comparisons will be conducted.

Table 1. Summary of reactivity ratio estimation (RRE) studies for 2-methylene-1,3-dioxepane (MDO; monomer 1)/vinyl acetate (VAc; monomer 2) copolymerization.

Ref.	RRE Technique	RRE Results		Comments
		r_1	r_2	
[10]	Kelen-Tüdős (K-T)	0.47	1.56	• Linear RRE technique used • Inappropriate low conversion assumption (reactivity ratios are "different enough" that composition drift is possible) • Low conversion (<20%) data not presented; cannot be re-evaluated with EVM
[11]	Fineman-Ross (F-R)	0.93	1.71	• Linear RRE technique used • Unequal weighting of experimental data • Instantaneous model applied to high conversion data (58–78% conversion reported) • F-R plot axes unintentionally flipped in original work (which changes RR estimates) • More comments in what follows
[12]	Non-Linear Least Squares (NLLS)	1.03	1.22	• Non-linear RRE technique used (good!) • Controlled radical polymerization (RAFT) data used for RRE, therefore parameter estimates are "apparent" reactivity ratios (as per Feldermann et al. [14])
[13]	Fineman-Ross (F-R)	0.14	1.89	• Linear RRE technique used • Inappropriate low conversion assumption (as with [10], reactivity ratios are "different enough" that composition drift is possible) • Suggest that low MDO reactivity (compared to other MDO/VAc RRE results in the literature) a result of low temperature; however, effect of temperature on RRs is usually weak

Evaluation of MDO/VAc Copolymerization Data: Fineman-Ross vs. EVM

In a recent study by Undin et al. [11], experimental data from six distinct feed compositions were used to estimate reactivity ratios for the MDO/VAc copolymerization. These six (batch) runs were allowed to continue until conversion did not change and the final conversion and composition measurements were reported. Finally, the reactivity ratios for the system were calculated using the Fineman-Ross (F-R) method. However, as mentioned briefly in Table 1, the data on the x and y axes were unintentionally flipped in the analysis; thus, the reactivity ratio estimates originally reported are not representative of the experimental data collected.

Besides this unintended error, there are several other problems with the analysis, including (1) the use of undesigned data (that is, no design of experiments used for the selection of feed compositions); (2) the lack of composition drift considerations (RRE experiments should be performed at low conversion, or a cumulative model should be used); (3) the use of an outdated (and linear!) RRE technique. For the purposes of this discussion, we will focus on the use of the F-R method for RRE but the other important points should also be noted and kept in mind.

As discussed by Hagiopol [15], the F-R method is often justified by its simplicity. However, it has many shortcomings, including unequal weighting of experimental data and symmetry issues (i.e., calculation results depend on which monomer is selected as M_1). The data set presented in [11] is especially vulnerable to these shortcomings, largely due to the undesigned initial feed compositions (collection and use of undesigned data for parameter estimation also induce considerable correlation between the parameters, which is highly undesirable). As shown in Table 2, some of the data are obtained under fairly low M_1 comonomer feed fraction; these conditions tend to have the greatest influence on the slope of a line, which ultimately affects reactivity ratio estimates obtained using the F-R method [15].

Table 2. RRE data for the copolymerization of MDO (monomer 1)/VAc (monomer 2) [11].

Sample	Monomer Feed		Copolymer Composition	
	$f_{1,0}$	$f_{2,0}$	$\bar{F}_1$	$\bar{F}_2$
MDO70	0.70	0.30	0.66	0.34
MDO50	0.50	0.50	0.42	0.58
MDO30	0.30	0.70	0.23	0.77
MDO10	0.10	0.90	0.06	0.94
MDO5	0.05	0.95	0.03	0.97
MDO1	0.01	0.99	0.005	0.995

The more pressing concern with the F-R method (also described by Hagiopol [15]) is the lack of symmetry. Thus, values of r_1 and r_2 depend on which monomer is selected as M_1. To demonstrate this point, the data collected by Undin et al. [11] are evaluated with M_1 = MDO (which was performed incorrectly in the original work; see Figure 4a) and with M_1 = VAc (performed herein for the demonstration; see Figure 4b).

It is clear from Figure 4 that the reactivity ratio estimates depend on which comonomer is selected as M_1; the fact that two reactivity ratio pairs can be obtained from a single estimation technique is problematic. It is also interesting to note that both analyses give $r_1 > 1$ and $r_2 > 1$. While this is physically impossible, it is a side-effect of experimental (and estimation) error. In reality, these results suggest that both reactivity ratios should be close to unity (which agrees with the findings of Undin et al. [11] and Hedir et al. [12]) but that at least one reactivity ratio is <1.

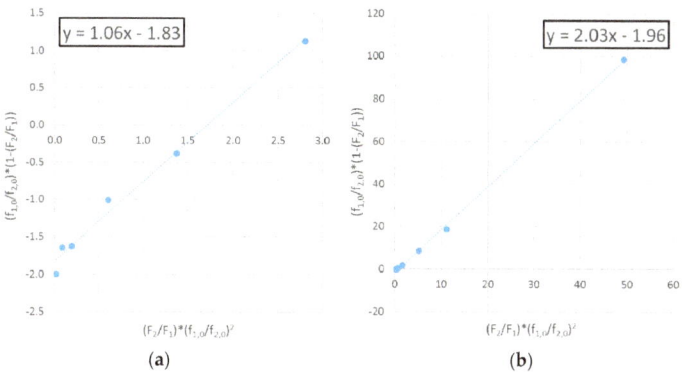

(a)   (b)

Figure 4. Fineman-Ross plots for the copolymerization of MDO/VAc with (**a**) M_1 = MDO (r_{MDO} = 1.06; r_{VAc} = 1.83 and (**b**) M_1 = VAc (r_{MDO} = 1.96; r_{VAc} = 2.03).

The issue of symmetry (combined with the statistical inaccuracy of using linear parameter estimation to evaluate non-linear models) highlights the need for a non-linear parameter estimation technique like EVM. When using EVM for reactivity ratio estimation, the influence of which comonomer is defined as M_1 has no impact on the parameter estimates. (If RR estimates are slightly different based on the choice of M_1, this is due to experimental error in the data). As shown in Figure 5, reactivity ratio estimates are within the JCR, regardless of which monomer is identified as M_1. That is, slight discrepancies between EVM-obtained reactivity ratio estimates are well within the expected error (1% error in $f_{i,0}$ and 10% error in $\overline{F}_i$; more on typical error levels in Appendix A). As expected, using measured/reported values as program inputs (in this case, $f_{MDO,0}$ and $\overline{F}_{MDO}$) provides us with a greater degree of confidence in our results; note that the JCR in Figure 5a is smaller than that in Figure 5b. There is also significantly more parameter correlation visible in Figure 5b, as evidenced by the diagonal nature of the (more elongated) JCR. This, again, is as expected; the VAc data set was calculated from the measured MDO composition data, so correlation is inevitable here. For the interested reader, data files used for this analysis are provided in Appendix C (Section C.1).

In using EVM to re-analyze the data, $r_1 > 1$ and $r_2 > 1$ is still observed (see again Figure 5). This outcome is likely a result of using cumulative composition data in an instantaneous model, since composition drift was not taken into account for this data set and conversion levels up to 80% are reported. Even the most statistically correct technique cannot reconcile cumulative experimental data with an instantaneous model (and, in this case, appropriate conversion data are unavailable for reanalysis with the cumulative EVM program).

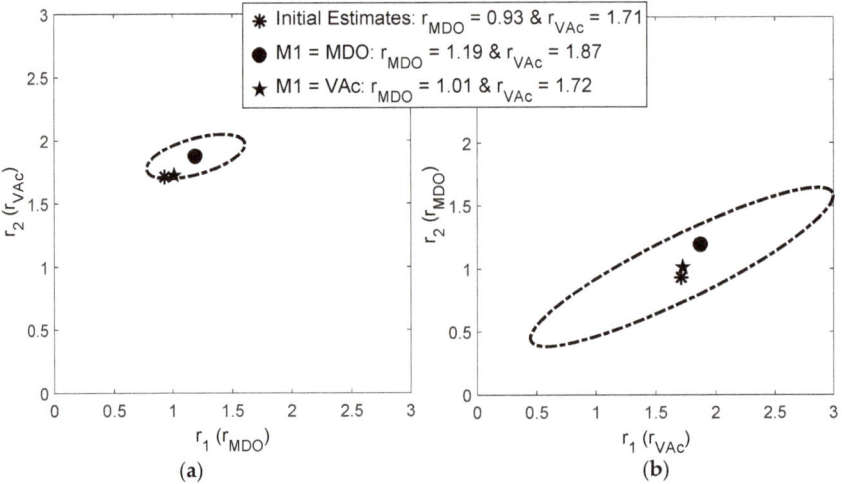

Figure 5. EVM-obtained RR estimates and JCRs for the copolymerization of MDO/VAc with (a) M_1 = MDO (r_{MDO} = 1.19; r_{VAc} = 1.87) and (b) M_1 = VAc (r_{MDO} = 1.01; r_{VAc} = 1.72).

Finally, we can visually evaluate the prediction performance of the reactivity ratio estimates, which involves comparing the experimental values to those predicted by the ICC equation (Equation (1)). As shown in Figure 6, the symmetry issues associated with the Fineman-Ross (F-R) technique have a significant impact on the prediction performance (red curves, Figure 6a). In contrast, both of the predictions using EVM-obtained RR estimates (blue curves, Figure 6b) are in agreement with each other and with the experimental data. This is compelling evidence to choose non-linear parameter estimation techniques like EVM over the statistically incorrect linear parameter estimation techniques.

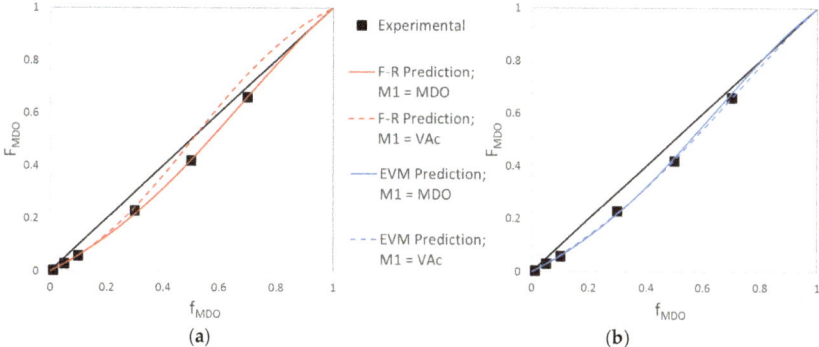

Figure 6. Comparison of prediction performance for RR estimates obtained by (**a**) F-R and (**b**) EVM.

4.1.2. Exhibit B: Are Researchers Addressing the Limitations of Low Conversion Data Analysis?

As described in Section 2.1, the instantaneous copolymer composition equation (Equation (1)) is only valid at low conversion levels (<10%). This limitation, though sometimes difficult to achieve experimentally, allows researchers to assume that composition drift does not occur in the samples being analyzed. Hence, the cumulative and instantaneous mole fractions in the copolymer are about the same. This experimental fact (and subsequent analysis) is considered "best practice," and is used throughout the reactivity ratio estimation literature (see, for example, [10,12–14,16,17]).

However, limiting kinetic investigations to low conversion levels presents some fundamental challenges. In spite of our best efforts to validate the "lack of composition drift" assumption, there is almost inevitably some change in feed composition with increasing conversion. From a more practical perspective, collecting low conversion data presents experimental challenges and the collected data are extremely prone to error.

Researchers should be aware of these limitations and should act accordingly. One might choose to include conversion data in the analysis (using a cumulative model and DNI, as per Section 2.1) to account for composition drift. Alternatively (rather, in addition), researchers might use design of experiments and experimental replication to address the inevitable error associated with the data collected. If nothing else, parameter estimation using EVM considers the error present in all variables, which can account for some of the experimental error. Ultimately, though, even the most statistically correct technique cannot compensate for bad data collection!

Evaluation of HEA/DCP Copolymerization Data

Recent work by Suresh et al. [16] describes the synthesis and reactivity ratio estimation of photosensitive copolymers based on 4-(3-(2,4-dichorophenyl)-3-oxoprop-1-enyl) phenylacrylate (DCP; monomer 2). In the study, DCP was copolymerized with hydroxyethyl acrylate (HEA; monomer 1) and with styrene and reactivity ratios were determined to better understand copolymerization behavior. However, as established in Exhibit A (Section 4.1.1), researchers often revert back to linear parameter estimation techniques and the authors (incorrectly) used the Fineman-Ross (F-R) and Kelen-Tüdös (K-T) methods for parameter estimation (see Table 3).

Since the virtues of EVM over linear parameter estimation techniques have already been established, the goal of the current case study is to emphasize the limitations of low conversion data and demonstrate how they can be addressed using EVM. All experimental data that Suresh et al. [16] used for reactivity ratio estimation were kept below 15% conversion, so that the instantaneous copolymerization equation could be used for parameter estimation. But, is "below 15% conversion" enough? As mentioned previously, this is largely considered "best practice," but does not account for composition drift even at low conversions nor for experimental error. As shown in Figure 7, only five (seemingly unreplicated) data points were collected for reactivity ratio estimation, with obvious discrepancies between the experimental data and the model predictions.

Table 3. Summary of RRE results for hydroxyethyl acrylate (HEA; monomer 1)/4-(3-(2,4-dichorophenyl)-3-oxoprop-1-enyl) phenylacrylate (DCP; monomer 2) copolymerization.

Ref.	RRE Technique	RRE Results	
		r_1	r_2
[16]	Fineman-Ross (F-R)	1.53 ± 0.10	0.76 ± 0.16
[16]	Kelen-Tüdös (K-T)	1.67 ± 0.13	0.58 ± 0.05
[16]	Extended K-T	1.65 ± 0.13	0.60 ± 0.08
Current Work	Instantaneous EVM	1.28 *	0.56 *
Current Work	Cumulative EVM	1.32 *	0.55 *

* Note: For EVM-obtained reactivity ratio estimates, statistically correct JCRs are presented instead of approximate confidence intervals (derived on a linear hypothesis); see Figure 8.

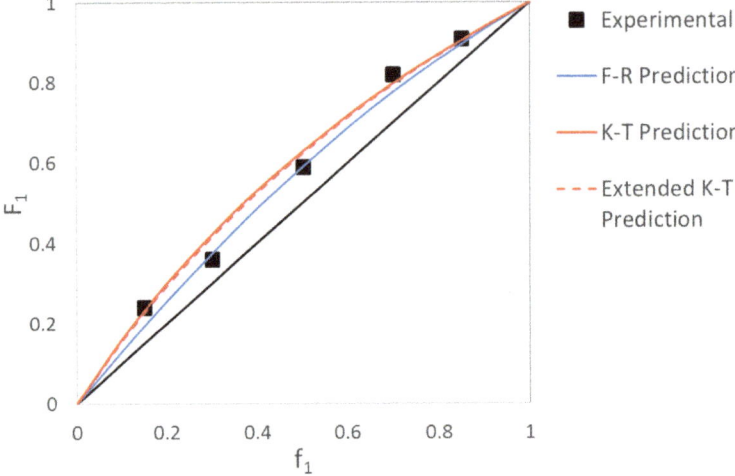

Figure 7. Prediction performance of RR estimates obtained by linear RRE techniques.

It is likely that these discrepancies are due to experimental error; this type of behavior is observed very often, especially for low conversion data. In order to address this, researchers should review potential sources of error; they are likely (1) unidentified composition drift and/or (2) experimental difficulties. This case study will demonstrate both of these sources of error (and how to handle them). This is another very important, yet implicit, contribution of EVM. EVM, if nothing else, forces one to think about the possible sources of variation (and quantify them). Relevant data, program screenshots and results are available in Appendix C, Section C.2.

To account for composition drift (even at low conversions), the cumulative copolymerization model (Equations (2) and (3)) should be used. Using direct numerical integration to solve this system of equations ensures that the feed composition (f_1) is considered as a function of conversion, thus taking any composition drift into account mathematically. To establish whether unidentified composition drift is the culprit in the current experimental data set, we can evaluate the data using both the instantaneous and cumulative models and compare the results (Figure 8, discussed below). In reality, using the cumulative model would also increase the amount of data available for analysis, as the copolymerization would be allowed to go to higher conversion levels (and data would continue to be collected); in addition, the experimental information is enhanced, anyway, since both conversion and copolymer composition data are included; see also Section 4.2.1. Ultimately, this would increase the degree of confidence in the reactivity ratio estimates and decrease the size of the JCR.

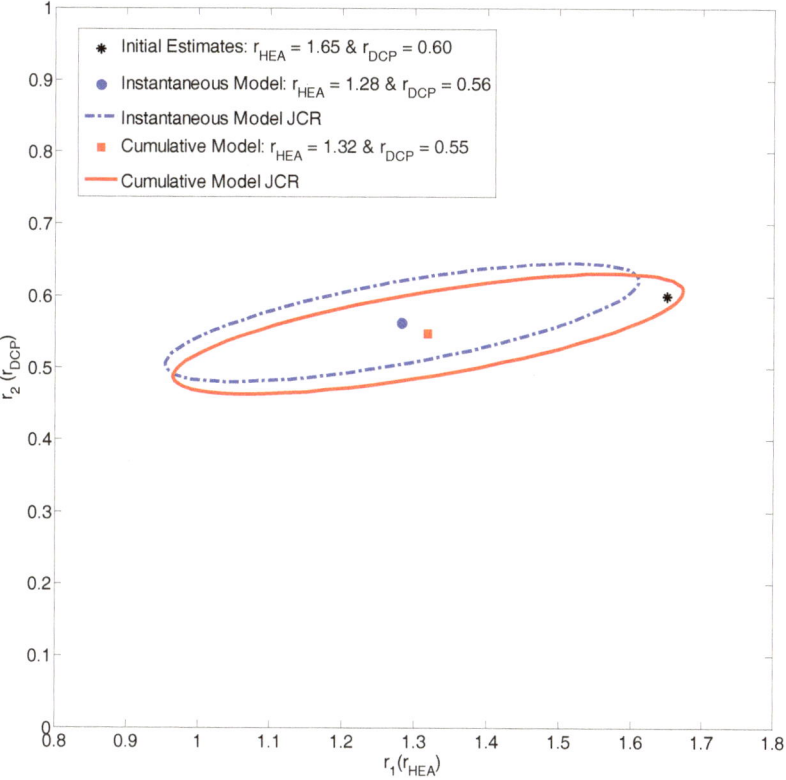

Figure 8. EVM-obtained RR estimates and JCRs for the copolymerization of HEA/DCP using the instantaneous model (r_{HEA} = 1.28; r_{DCP} = 0.56) and the cumulative model (r_{HEA} = 1.32; r_{DCP} = 0.55).

Figure 8 indicates that the two EVM-obtained reactivity ratio estimates are in good agreement and the JCR sizes and orientations are similar. Thus, in this case, the effect of composition drift is likely minimal. Therefore, we will continue our troubleshooting by investigating the second source of error: experimental difficulties. Since no replicate data are available, we cannot calculate the error associated with the composition measurements shown in Figure 7. However, as discussed in Section 2.2, EVM considers the error present in all variables throughout the parameter estimation process. The program default values of 1% error (associated with $f_{1,0}$) and 5% error (associated with $\overline{F}_1$) were therefore used in the analysis (see Appendix A for additional information).

The reactivity ratios calculated using the EVM program are as described in Table 3 and Figure 8. The converged program also provides the best possible estimates of "true" values of the variables, a feature which was described in Section 3.3 (see also Appendix C, Section C.2. Thus, in Figure 9, we can compare the experimental (measured) values to the "true" experimental values and the EVM model predictions (both instantaneous and cumulative).

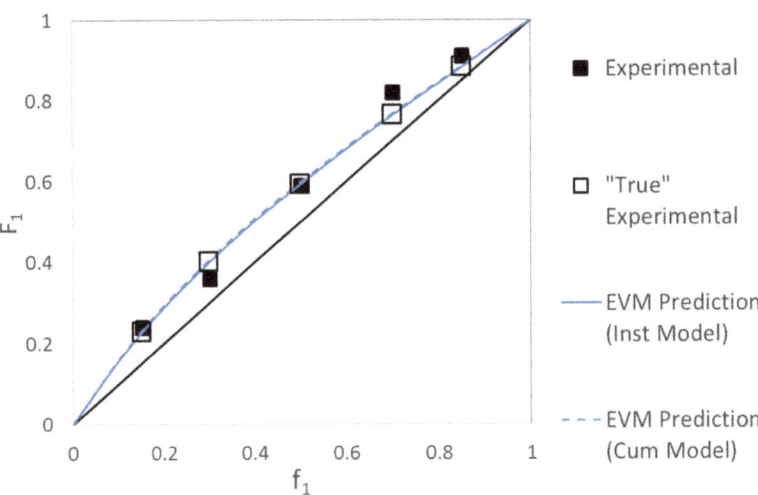

Figure 9. Prediction performance of RR estimates obtained using EVM.

Figure 9 indicates that the experimental data points were subject to some degree of error, especially for $f_{1,0} = 0.3$ and $f_{1,0} = 0.7$. Thus, experimental difficulties are likely the culprit here. The best way to mitigate this type of problem is to design experiments, replicate copolymerization runs and use EVM for parameter estimation. Additionally, using full conversion data with the cumulative model gives a more complete picture of copolymerization kinetics; researchers do not have to struggle with the experimental challenges of collecting low conversion data.

4.2. Maximizing and Exploiting Information Content

4.2.1. Exhibit C: What Happens when We Take Advantage of ALL Copolymerization Data?

Collecting low conversion data presents some challenges (as described in Exhibit B) but the instantaneous model is statistically valid if managed properly. That is, good control over conversion levels (typically < 10%), statistical design of experiments, experimental replication, and/or a non-linear parameter estimation technique should be used.

Evaluation of BMA/BA Copolymerization Data: Low Conversion Analysis

Ren et al. [18] recently investigated the copolymerization kinetics of *n*-butyl methacrylate (BMA; monomer 1) and *n*-butyl acrylate (BA; monomer 2). Originally, the group collected copolymerization data at low conversion levels (<10%) so that the instantaneous model (Equation (1)) could be used for analysis. After estimating preliminary reactivity ratios using the RREVM program [1], additional (replicated) designed experiments were completed to improve the quality of the data.

This low conversion data set has been re-analyzed using the new MATLAB-based EVM program. As with the original investigation, analysis was first performed with the preliminary data only (9 equidistant feed compositions). Then, the analysis was repeated with preliminary data supplemented by the designed replicates (feed compositions selected using the Tidwell-Mortimer criterion [4,19]). A comparison of reactivity ratio estimates is shown in Table 4 and the original data and program output are provided in Appendix C, Section C.3.

Table 4. Summary of RRE results for *n*-butyl methacrylate (BMA; monomer 1) and *n*-butyl acrylate (BA; monomer 2) copolymerization.

Ref.	RRE Technique	RRE Results	
		r_1	r_2
[18]	Preliminary Estimates (RREVM) [1]	2.100	0.489
[18]	Estimates from Tidwell-Mortimer Designed Experiments (RREVM) [1,2]	2.008	0.460
Current Work	Instantaneous EVM (preliminary data)	2.109	0.492
Current Work	Instantaneous EVM (preliminary data & designed replicates)	2.012	0.462
Current Work	Cumulative EVM	2.114	0.500

Good agreement is observed between reactivity ratio estimates, no matter what the amount (or type) of data used. This indicates well-behaved data; the low conversion analysis was done in a methodical and statistically correct manner.

Evaluation of BMA/BA Copolymerization Data: Medium-High Conversion Analysis

This type of low conversion analysis (as performed by Ren et al. [18]) would be sufficient for reactivity ratio estimation, especially since design of experiments was included in the investigation. However, in this case, what if one had also included additional (medium-high conversion) experimental data? Ren et al. [18] chose to run three feed compositions up to high conversion values; the full conversion experimental data was used to evaluate the prediction performance of the reactivity ratio estimates. The analysis showed good agreement between model predictions and experimental data, thus confirming the reactivity ratio estimates.

Let's now take this a step further for illustration purposes. We know that using the cumulative model and direct numerical integration provides us with the opportunity to "repurpose" this cumulative (medium-high conversion) data for improved reactivity ratio estimation. With a cumulative model, there is potential to obtain significantly more information (that is, more data points) from each experiment. Since researchers are not limited to low conversion, less experimental tedium is required to obtain the same degree of accuracy, as long as the experiments are well-designed.

A direct comparison of the preliminary analysis (9 feed compositions) and the cumulative analysis (3 feed compositions) results is provided in Figure 10. The same initial estimates were used in both cases to ensure that both of the RRE techniques had the same starting point. It is interesting to note that both the reactivity ratio estimates and the JCR areas are almost identical for the two data sets. This provides us with two main conclusions: that the parameter estimation results using the instantaneous and cumulative models are in agreement and that the degree of confidence in our results is approximately the same (regardless of which data set and/or model is being used). Therefore, in this case, 3 full conversion runs have approximately the same information content as 9 runs that are limited to low conversion levels.

This result should motivate researchers to think carefully about their preliminary experimental work. By strategically selecting feed compositions (using design of experiments techniques like Tidwell-Mortimer [4,19] or EVM [8,9]) and collecting copolymerization data up to medium or high conversion levels, it is possible to obtain sufficient information about a new system. The results shown herein suggest that preliminary experimental work can almost be reduced to 1/3 of the original load, without any loss of information content. Therefore, researchers should be encouraged to make use of all copolymerization data by employing the cumulative copolymerization model.

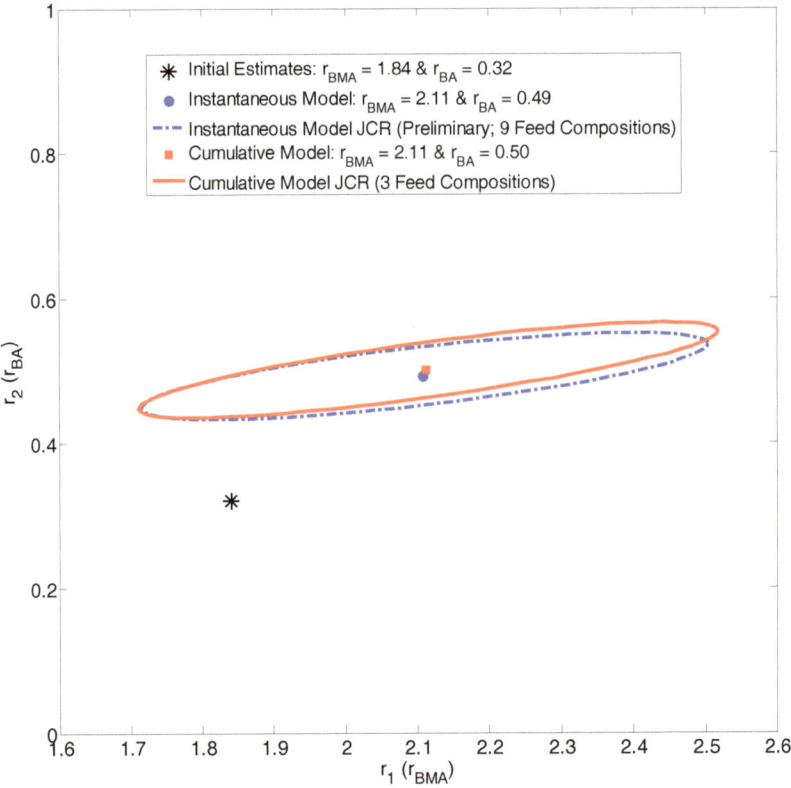

Figure 10. Comparison of results for the copolymerization of BMA/BA using the instantaneous model ($r_{BMA} = 2.11$; $r_{BA} = 0.49$) and the cumulative model ($r_{BMA} = 2.11$; $r_{BA} = 0.50$).

4.2.2. Exhibit D: How Many Replicates Do We Really Need?

As demonstrated in Exhibit C, experimental replication is an important aspect of reactivity ratio estimation. Collection of data for reactivity ratio estimation, especially at low conversion, is subject to experimental error; replicating experiments helps researchers account for experimental error and potential lurking variables. In deciding on the number of required replicates, a number of aspects must be considered [20]. The goal of the current case study is not to do a statistical evaluation of the number of replicates required but rather to look at pre-existing data and demonstrate once more the importance of experimental replicates.

Many reactivity ratio estimation studies (including [18,21–23]) have used the Tidwell-Mortimer (T-M) criterion for design of experiments to select the feed compositions at which to run copolymerizations for reactivity ratio estimation. In most cases, two optimal feed compositions (according to T-M) are used and replicated four times each. To establish the importance of replication and symmetry, this case study will analyze subsets of an original data set from the copolymerization of styrene (Sty; monomer 1) and ethyl acrylate (EA; monomer 2) [21].

When experiments are statistically designed, it is possible to obtain accurate reactivity ratio estimates from reduced experimental effort. Replication ensures that experimental results are repeatable, gives us an estimate of the experimental error and increases the degree of confidence in the resulting reactivity ratio estimates. The data reported by McManus and Penlidis [21] included an initial run and 3 replicates at two feed compositions (8 data points total); the data set is available in Appendix C, Section C.4.

To create this collection of reactivity ratio estimates and JCRs (Figure 11), the original data set [21] was revisited with the MATLAB-based RREVM program (blue JCR and circle point estimate in Figure 11; 3 replicates). Then, one randomly selected replicate at each feed composition was removed and the data was re-evaluated (red JCR and square point estimate; 2 replicates). This process was repeated, thus leaving half of the original data set (green JCR and triangle point estimate; 1 replicate). Finally, only one data point at each feed composition was used in the analysis (black JCR and diamond point estimate; no replicates).

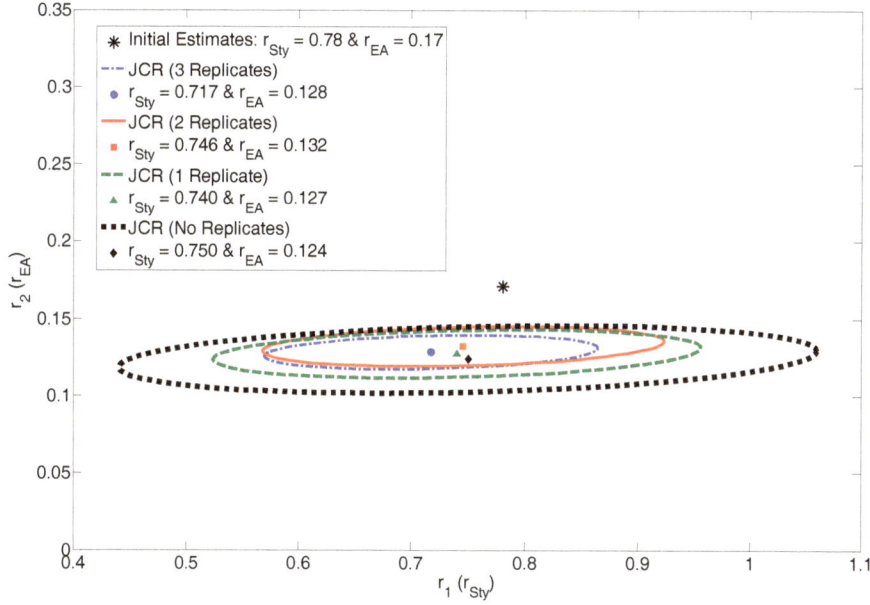

Figure 11. Importance of replication for the copolymerization of Sty/EA.

As shown in Figure 11, the reactivity ratio estimates are all similar for the copolymerization of styrene and ethyl acrylate, regardless of how many replicates are used (in this specific case, for this specific data set). The JCRs form a series of concentric ellipses and the slight skew (off-centeredness) depends on which of the replicates are randomly removed. Clearly, using the full data set (3 replicates) gives a much smaller JCR (and a much higher degree of confidence) compared to the other estimates. In general, the JCR size increases (that is, the uncertainty becomes greater) as the number of replicates decreases.

At first glance, it seems as though the uncertainty in r_1 is much greater than that in r_2 (as demonstrated by the horizontal growth in the JCRs as the number of replicates decreases). However, this is partially due to the fact that r_1 is approximately 6 times larger than r_2. Thus, the same relative error will have a larger absolute value in r_1 compared to r_2. Another factor that should be considered is the feed compositions used to collect the experimental data ($f_{1,0}$). The T-M criterion suggested $f_{1,0} = 0.0788$ and $f_{1,0} = 0.7193$, where monomer 1 is styrene [21]. While this is the statistically correct approach (and an excellent starting point), this means that the experimental data contain information only about a "monomer 2"-rich system (given $f_{1,0} = 0.0788$, $f_{2,0} = 0.9212$). While $f_{1,0} = 0.7193$ contains more styrene than ethyl acrylate, an even higher $f_{1,0}$ would further improve the degree of confidence in

r_1. This will be discussed further (and proven through a case study) in Exhibit E, which demonstrates the effectiveness of a sequential design of experiments.

In looking at the importance of replication, it is also worth examining the effect of symmetry on reactivity ratio estimates and JCRs. Specifically, the next step in the investigation looks at how estimation results are affected when replicates are only available for one of the two recipes. Again, the styrene/ethyl acrylate data set from McManus and Penlidis [21] was employed.

The reactivity ratio estimates and JCRs presented in Figure 12 tell an interesting story. When replicates are only included from the high $f_{1,0}$ runs, the error in r_2 (vertical error, in this case) increases (red JCR and square). The reverse is true when all of the replicates are from the low $f_{1,0}$ (high $f_{2,0}$) runs; uncertainty in r_1 becomes much more substantial (green JCR and triangle). This observation is in agreement with the results of Figure 11, as the inclusion of "monomer 1"-rich data improves the degree of confidence in r_1.

Both asymmetrical data sets give reasonably good estimates of reactivity ratios for the styrene and ethyl acrylate copolymerization. Ultimately, limited replication is better than no replication at all. However, it is no coincidence that the JCR from the fully replicated data set is located almost in the intersection of the other two curves (between the red and green curves in Figure 12). Including all of the experimental replicates in the analysis ensures that we have the highest degree of confidence in both r_1 and r_2, thus decreasing the JCR area as much as possible.

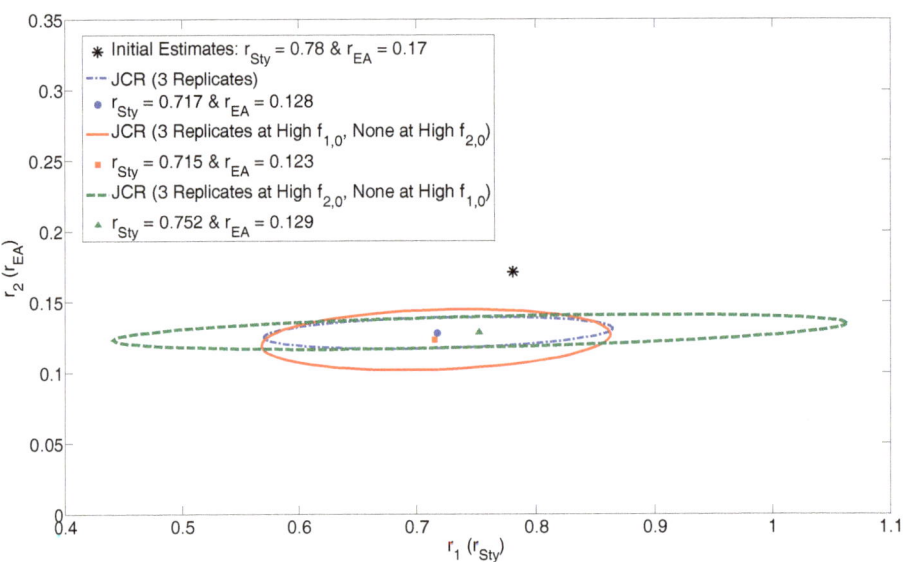

Figure 12. Effect of symmetry in replication for the copolymerization of Sty/EA.

4.2.3. Exhibit E: Can We Use Design of Experiments to Increase Confidence in Our Results?

In both Exhibit C and Exhibit D, we observed that design of experiments (specifically, using the Tidwell-Mortimer criterion) is an important aspect of reactivity ratio estimation. Reactivity ratio estimates obtained using designed data are more accurate and more precise than preliminary experiments, because they use a combination of prior knowledge and statistical principles to increase confidence in the final estimates.

In the current case study, we will look at experimental data for the copolymerization of butyl acrylate (BA; monomer 1) and methyl methacrylate (MMA; monomer 2). The original investigation by

Dubé and Penlidis [22] was a detailed, multi-step analysis but only data from the first step are used in the current exhibit.

In investigating the effect of experimental design on the confidence in our estimation results, there are two important pieces of information to consider. If the preliminary estimates of r_1 are smaller than r_2, then (1) uncertainty in r_1 will seem much lower than uncertainty in r_2 (as explained previously in Section 4.2.2, the same relative error will have a larger absolute value in r_2 compared to r_1) and (2) the Tidwell-Mortimer design will suggest recipes rich in monomer 1. As shown in Figure 13 (black JCR; 2 feed compositions), $r_1 < r_2$ for the BA/MMA system and there is more uncertainty in r_2 (that is, in the vertical direction). Generally speaking, we find that the JCR is "stretched" along the axis of the larger reactivity ratio estimate, which is due to both the absolute error and the selected feed compositions.

Since the absolute error is experiment-dependent, it is a fact of life that one has to live with. Thus, this case study focuses on item (2) described above: how do the feed compositions (selected randomly or via design of experiments) affect the reactivity ratio estimates and associated JCRs? In the case of BA/MMA copolymerization (given preliminary estimates $r_1 = 0.51$ and $r_2 = 2.38$ from Grassie et al. [24]), the Tidwell-Mortimer criterion suggests the following feed compositions: $f_{1,0} = 0.543$ and $f_{1,0} = 0.798$, where monomer 1 is butyl acrylate [22]. Based on this criterion, all of the experimental data collected are rich in monomer 1, which provides us with more certainty in r_1 (see again Figure 13; black JCR; 2 feed compositions).

At the next step, we can use EVM-based sequential design of experiments [8]. This allows for further refinement of the reactivity ratio estimates, a higher degree of certainty and therefore smaller JCRs. The procedure is described below:

(1) EVM is applied to instantaneous (low conversion) data (from [22]) to estimate reactivity ratios. Feed compositions are selected according to Tidwell-Mortimer design and four runs are done at each level: $f_{1,0} = 0.543$ and $f_{1,0} = 0.798$.

(2) Parameter estimation results from EVM are recorded (see Appendix C, Section C.5). Specifically, reactivity ratio estimates (r_1 and r_2) and the $\underline{G}$ matrix (Appendix A) are required for sequential design of experiments.

(3) The EVM-based sequential design of experiments program (using data from step (2), as well as the preliminary feed compositions from step (1)) is employed. Details on the design have been reported by Kazemi et al. [8].

From the sequential design of experiments, we find that the "next best" feed composition for analysis of the BA/MMA copolymerization is $f_{1,0} = 0.100$. This indication that more monomer 2-rich data is required is very reasonable, since all data collected to this point has been rich in monomer 1. By introducing experimental data rich in monomer 2, the uncertainty in r_2 should decrease.

In the absence of experimental data for $f_{1,0} = 0.100$, data were simulated using the instantaneous copolymerization model and random error was added (based on the variance reported in the original study [21]). As was the case for the other feed compositions, four data points at $f_{1,0} = 0.100$ were added to the analysis. These new data points, along with the original data (shown in Appendix C, Section C.5) were then used to re-estimate the reactivity ratios with EVM. The results are shown alongside the original analysis in Figure 13 (red JCR; 3 feed compositions).

The inclusion of data rich in monomer 2 drastically improves the degree of certainty in our reactivity ratio estimates. While the point estimates are unaffected, the error in r_2 is significantly reduced. This is as expected: when data rich in monomer 2 are available, we can have greater confidence in r_2.

This result demonstrates the importance of design of experiments for reactivity ratio estimation. We are able to maximize the information content from a minimal number of runs and we are able to decrease the degree of uncertainty in our parameter estimates. Sequential designs are extremely useful and revealing and minimize the overall experimental effort.

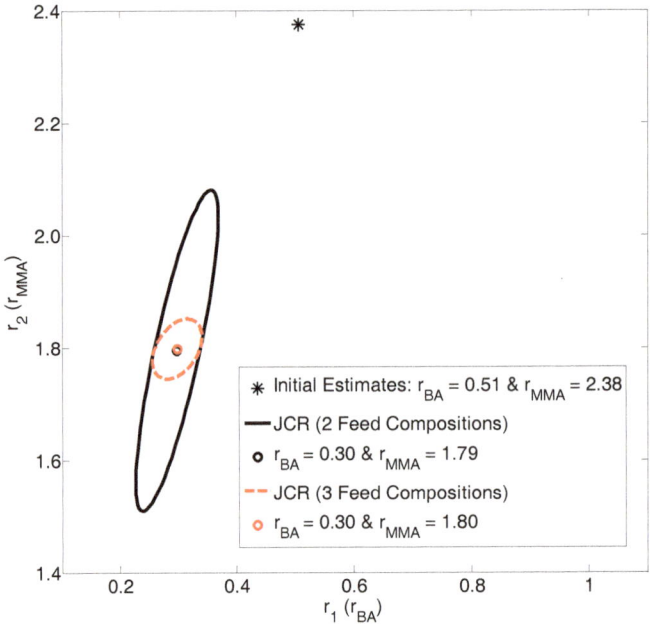

Figure 13. Effect of sequentially designed experiments for the copolymerization of BA/MMA.

5. Conclusions

The MATLAB-based program for reactivity ratio estimation (using the error-in-variables-model) is statistically correct, accurate and user-friendly. This ready-to-use software can easily be employed to evaluate conversion and composition data, providing researchers with precise reactivity ratio estimates within seconds. Using EVM for reactivity ratio estimation should provide polymer researchers with confidence, especially as they continue to use reactivity ratios to predict copolymer properties and microstructure for specific applications.

We hope that the case studies presented herein will motivate polymer chemists and engineers to think critically about which technique(s) they use for parameter estimation. Current "best practices" have serious shortcomings, especially as linear parameter estimation techniques continue to be used. Ignoring the limitations of low conversion data is imprudent; inappropriate statistical approaches can undermine even carefully collected data. The case studies have also shown that the error-in-variables-model can maximize information content from copolymerization data. Researchers can ensure the accuracy of their results (while minimizing required time and resources) by collecting and analyzing full conversion data. Also, as demonstrated by the case studies, replication and design of experiments are key to a high degree of confidence.

Ultimately, the computer program described and demonstrated herein provides both the ease-of-use required for introductory studies and the flexibility needed for extensions to complex and/or multi-component systems. Therefore, it should find use in both industry and academia. The program can be obtained by contacting the authors.

Acknowledgments: The authors wish to acknowledge financial support from the Natural Sciences and Engineering Research Council (NSERC) of Canada and the Canada Research Chair (CRC) program. In addition, thanks go to UWW/OMNOVA Solutions, Akron, OH, USA, for special support to A.J.S.

Author Contributions: A.J.S. made the EVM program "user-friendly" within MATLAB, researched and analyzed the case studies and wrote the paper. A.P. supervised the work and corrected several drafts of the paper.

Conflicts of Interest: The authors declare no conflict of interest.

An Overview of the Appendices

In Appendix A, information about relevant statistical principles is presented. This is intended to provide the interested reader with additional insight about "behind-the-scenes" mathematical details. Appendix B presents a general program description, including screenshots with default program values. Finally, Appendix C contains details and data for each of the specific case studies presented in the main text (Section 4).

Appendix A. Relevant Statistical Principles

The user need not have a detailed understanding of the statistical principles used in the error-in-variables-model. However, for the interested reader, some additional information is included in what follows.

Appendix A.1. Additive and Multiplicative Error

The magnitude (and type) of error associated with the variables can be determined through independent replication. Typically, the relationship between a variable and its error is either additive (absolute) or multiplicative (relative). Multiplicative error is typically assumed because error is presented as a percentage of the measurement (and is, therefore, relative in nature). However, if a user has insight about a system that indicates additive error, it is possible to modify the program accordingly.

The relationships between the "true" value of the variable ($\underline{\zeta}_i$) and the measured/recorded values ($\underline{x}_i$) are shown in Equations (A1) and (A2) for additive and multiplicative error, respectively. Note that Equation (A2) has been shown previously as Equation (4) but is repeated here (more generally) for reference.

$$x = \zeta + k\varepsilon \tag{A1}$$

$$x = \zeta(1 + k\varepsilon) \tag{A2}$$

As explained before, k is a constant that reflects the uncertainty of the variables (for example, if 5% error is assumed for the multiplicative case, $k = 0.05$). Error, ε, is a random variable that is typically uniformly distributed between -1 and 1 [25].

When multiplicative error is assumed, it becomes necessary to transform Equation (A2) so that the error term is additive. Taking the natural logarithm of both sides gives Equation (A3). Note that $\ln(1 + k\varepsilon)$ can be replaced by $k\varepsilon$, as long as the magnitude of the error does not exceed 10% ($k \leq 0.10$).

$$\ln(x) = \ln(\zeta) + k\varepsilon \tag{A3}$$

Regardless of error structure, the value of k (the degree of uncertainty) manifests itself in the same way in the variance-covariance matrix. This is shown in Equations (A4) through (A7). Equation (A4) gives the variance of x (for the additive case), whereas Equation (A5) gives the variance of $\ln(x)$ (for the multiplicative case).

$$V(x) = V(\zeta + k\varepsilon) = k^2 V(\varepsilon) \tag{A4}$$

$$V(\ln(x)) = V(\ln(\zeta) + k\varepsilon) = k^2 V(\varepsilon) \tag{A5}$$

Equations (A6) and (A7) are relevant to both error structures since $V(x) = V(\ln(x))$, as shown above.

$$V(\varepsilon) = E\left(\varepsilon^2\right) - [E(\varepsilon)]^2 = \int_{-1}^{1} \frac{\varepsilon^2}{2} d\varepsilon = \frac{1}{3} \tag{A6}$$

$$V(x) = V(\ln(x)) = \frac{k^2}{3} \tag{A7}$$

The variance estimate shown in Equation (A7) is applied to different variables, which populate the variance-covariance matrix for the EVM program. The program's default settings assume 1% error associated with feed composition ($x_1 = f_{1,0}$ and $k_1 = 0.01$) and 5% error associated with cumulative copolymer composition ($x_2 = \bar{F}_i$ and $k_2 = 0.05$). Therefore, the variance-covariance matrix, $\underline{V}$, for the instantaneous (low conversion) case is shown in Equation (A8).

$$\underline{V} = \begin{bmatrix} V(x_1) & 0 \\ 0 & V(x_2) \end{bmatrix} = \begin{bmatrix} \frac{k_1^2}{3} & 0 \\ 0 & \frac{k_2^2}{3} \end{bmatrix} = \begin{bmatrix} \frac{0.01^2}{3} & 0 \\ 0 & \frac{0.05^2}{3} \end{bmatrix} = \begin{bmatrix} 0.0000\bar{3} & 0 \\ 0 & 0.0008\bar{3} \end{bmatrix} \tag{A8}$$

This can be cross-referenced with the input data shown in Appendix B (Figures A4 and A5).

Appendix A.2. Calculation of $\underline{G}$ as an EVM Program Output

Only the very basics are presented in what follows. A detailed discussion of the nested-iterative EVM algorithm has been presented by Reilly and Patino-Leal [7] and has more recently been described by Kazemi et al. [3,6].

Equation (A9) below is the definition of $\underline{G}$, the second derivative of Φ with respect to the parameters, given earlier by Equation (8) (Section 3.3).

$$\underline{G} = E \left[\frac{d^2\Phi}{d\theta_i d\theta_j} \right] = \sum_{i=1}^{n} r_i \underline{Z}_i' \left(\underline{B}_i \underline{V} \underline{B}_i' \right)^{-1} \underline{Z}_i \tag{A9}$$

r_i and $\underline{V}$ are as defined in Equation 6 (recall that r_i is the number of replicates for the ith trial and $\underline{V}$ is the variance-covariance matrix of the variables). $\underline{Z}_i$ is the vector of partial derivatives of the function $\underline{g}\left(\underline{\xi}_i, \underline{\theta}\right)$ (that is, the model, e.g., see Equation (5)) with respect to the parameters for the mth element (see Equation (A10)) and $\underline{B}_i$ is the vector of partial derivatives of the function $\underline{g}\left(\underline{\xi}_i, \underline{\theta}\right)$ with respect to the variables.

$$\underline{Z}_i = \left[\frac{\partial \underline{g}\left(\underline{\xi}_i, \underline{\theta}\right)}{\partial \theta_m} \right] \tag{A10}$$

$$\underline{B}_i = \left[\frac{\partial \underline{g}\left(\underline{\xi}_i, \underline{\theta}\right)}{\partial \left(\underline{\xi}_i\right)} \right] \tag{A11}$$

Appendix B. General Program Description

The following screenshots from the MATLAB-based EVM program are meant to supplement descriptions in the main text (especially Section 3.2). The analysis of instantaneous copolymerization data is presented herein but the same general information is relevant to the analysis of cumulative data.

We will start by demonstrating the manual data input option, then we will show the same information using the "data file" option. The prompts contain sample data from McManus and Penlidis [21]. If the user decides to input data using a pre-made data file, the same inputs are required (with slightly different formatting).

Running the program brings up the "QuickStart" menu (Figure A1), which was described in Section 3.1. Figures A2–A4 show the pop-up menus (that is, the required data) for manual input of instantaneous composition data. Typically, only the preliminary reactivity ratio estimates (Figure A2) and the copolymerization (composition) data (Figure A3) need to be modified by the user; the "form" of the data entry can be observed in the screenshots below.

As for Figure A4, the only "default" value that may require updating (at the user's discretion) is the variance-covariance matrix. The default values were explained in Section A.1. If necessary, the magnitude of the matrix entries may be modified but the size of the matrix itself should not be changed (that is, it should remain a 2×2 matrix for the instantaneous case).

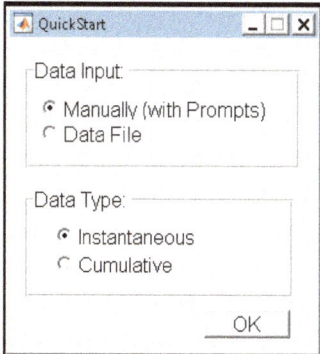

Figure A1. "QuickStart" Menu.

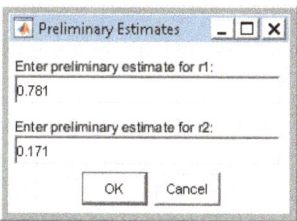

Figure A2. "Preliminary Estimates" prompt.

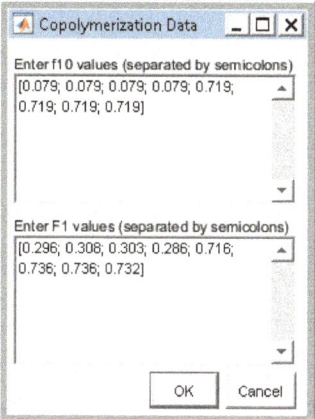

Figure A3. "Copolymerization Data" prompt.

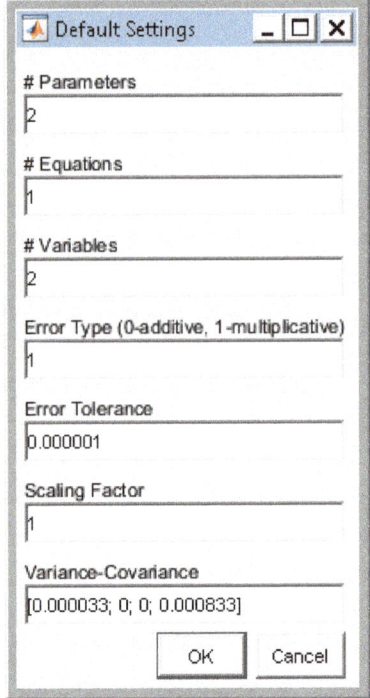

Figure A4. "Default Settings" prompt.

An alternative to the step-by-step prompts is to include all of the required estimation data in a single ".txt" data file. When a user chooses the data file input option, a pop-up window containing their files appears (that is, any files in the same folder as the "QuickStart" file). Once an appropriate file is selected, the program will access the data and run automatically.

As mentioned previously, a data file includes all of the same data as the prompts but it can be saved, modified and reused. It can either be created in Notepad or in MATLAB but should have the extension ".txt". The data file must be prepared prior to running the EVM program, since the program will access the data file "behind the scenes" to obtain the required information. A sample data file (here for the McManus and Penlidis [21] data set) is shown in Figure A5.

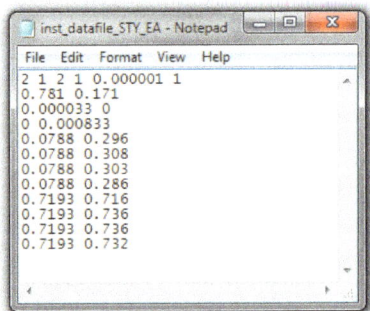

Figure A5. Sample data file.

In Figure A5, line 1 provides the "Default Settings" for the program, such as the number of parameters, equations and variables (the reader will notice that these values are the same as in Figure A4, with slightly different formatting). Line 2 contains preliminary reactivity ratio estimates (recall Figure A2), while lines 3 and 4 can be combined to form the variance-covariance matrix for the variables. Finally, lines 5 through 12 are the experimental copolymerization data (recall Figure A3). The first column represents the initial feed composition ($f_{1,0}$) and the second column represents the corresponding measured cumulative copolymer composition ($\bar{F}_1$).

Appendix C. Data & Screenshots from Case Studies

In this section, additional data and program screenshots are presented for the interested reader (or potential program user). Data files are typically shown (rather than step-by-step prompts), which allows for easy reference and modification. However, the same data could be fed to the EVM program using the prompts described in Section 3 and Appendix B. The time required for parameter estimation (that is, parameter convergence) is listed for each case study below (as run on an Intel(R) Core™ i7-860 processor). However, one should note that execution time is dependent on a number of factors including preliminary parameter estimates, data sets (size and associated error), computer processor, background tasks, etc.

Appendix C.1. Screenshots from MDO/VAc Copolymerization Analysis

The additional information shown herein is relevant to Exhibit A (Section 4.1.1), the copolymerization of 2-methylene-1,3-dioxepane (MDO) and vinyl acetate (VAc). In spite of the fact that cumulative copolymer composition was measured and reported by Undin et al. [11], information from relevant conversion data was not included. Thus, the analysis conducted was based on the instantaneous model. Regardless of which monomer was selected as M_1, all calculations converged in less than five seconds.

Two data files (used to obtain the results shown in Figure 5) are presented in Figure A6. Both contain the same data set (for the MDO/VAc copolymerization from Undin et al. [11]) but differ in terms of which comonomer is identified as M_1. Here, Figure A6a is as reported (M_1 = MDO) and Figure A6b is the reverse (M_1 = VAc).

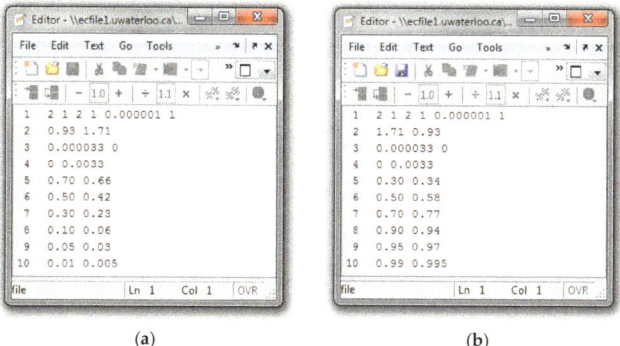

(a)   (b)

Figure A6. Data files for the analysis of MDO/VAc using EVM assuming (a) M_1 = MDO and (b) M_1 = VAc.

Text output is shown in Figure A7 for M_1 = VAc, as a sample program output (described in Section 3.3). THETA gives the reactivity ratio estimates (r_1 and r_2), XI gives the best estimates of the "true" values of the variables (compare to lines 5 through 10 in Figure A6b), Phi is the objective function value (recall Equation (6)) and G is the expected value of the second derivative of Φ with respect to the parameters (recall Equation (8)).

Finally, the user is asked "Do you wish to continue with Joint Confidence Calculations?" By typing "Y," the JCR shown in Figure 5b is obtained.

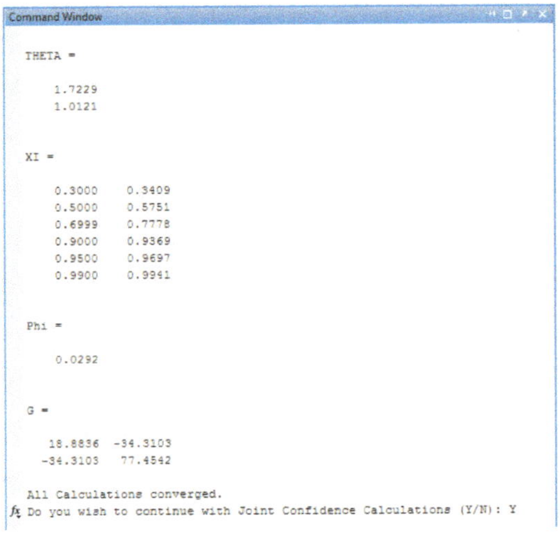

```
Command Window

    THETA =

        1.7229
        1.0121


    XI =

        0.3000    0.3409
        0.5000    0.5751
        0.6999    0.7778
        0.9000    0.9369
        0.9500    0.9697
        0.9900    0.9941


    Phi =

        0.0292


    G =

        18.8836   -34.3103
       -34.3103    77.4542

    All Calculations converged.
    fx Do you wish to continue with Joint Confidence Calculations (Y/N): Y
```

Figure A7. Program output for the analysis of MDO/VAc.

Appendix C.2. Screenshots from HEA/DCP Copolymerization Analysis

The additional information shown herein is relevant to Exhibit B (Section 4.1.2), the copolymerization of hydroxyethyl acrylate (HEA) and 4-(3-(2,4-dichlorophenyl)-3-oxoprop-1-enyl) phenylacrylate (DCP) (original data from Suresh et al. [16]).

The two data files (Figure A8) contain similar information, as discussed in Section 3.2.2. The additional data required by the cumulative model are the monomer molecular weights and the conversion values. The sample output shown below (Figure A9) is for the cumulative analysis; XI here is the best estimate for the cumulative copolymer composition $(\overline{F}_1)$. JCRs for the two analyses are shown in Figure 8. The instantaneous analysis required less than five seconds to estimate reactivity ratios, whereas the cumulative model converged in less than twenty seconds.

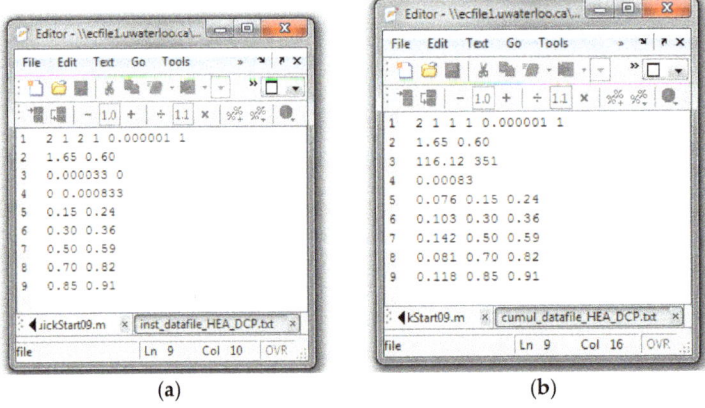

Figure A8. Data files for the (**a**) instantaneous and (**b**) cumulative analysis of HEA/DCP.

Figure A9. Program output for the analysis of HEA/DCP using the cumulative model.

Appendix C.3. Screenshots from BMA/BA Copolymerization Analysis

The additional information shown herein is relevant to Exhibit C (Section 4.2.1), the copolymerization of *n*-butyl methacrylate (BMA) and *n*-butyl acrylate (BA) (original data from Ren et al. [18]). The analysis of low conversion data (using the instantaneous model) is shown first, followed by the analysis of full conversion data (using the cumulative model). Only the preliminary (instantaneous) data analysis and the cumulative data analysis (rows 3 and 5 in Table 4; featured in Figure 10) are shown herein. However, the same general procedure was applied to the "Instantaneous EVM (preliminary data & designed replicates)" analysis (refer to row 4 in Table 4). Instantaneous analyses converged in under five seconds, whereas the cumulative analysis took less than twenty seconds.

Figure A10a shows the preliminary (low conversion) data set reported by Ren et al. [18]. Figure A10b shows the full conversion data (again reported by Ren et al. [18]) which was originally used for reactivity ratio prediction performance (see original work, Figure 5). Non-truncated data points are a result of plot digitization. Next, Figure A11 shows the program output for the preliminary analysis (data from Figure A10a) and Figure A12 shows the same for the analysis of full conversion data (data from Figure A10b). Notice here that both the reactivity ratio estimates (THETA) and the $\underline{G}$ matrices (G) are similar, which translates into similar JCRs, as observed in Figure 10.

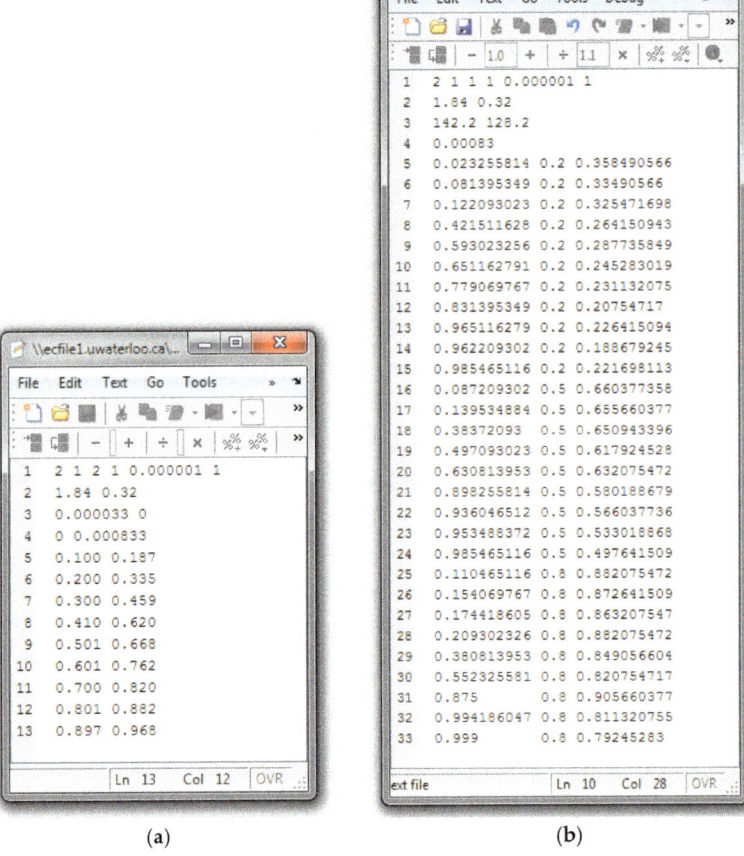

Figure A10. Data files for the (**a**) instantaneous and (**b**) cumulative analysis of BMA/BA.

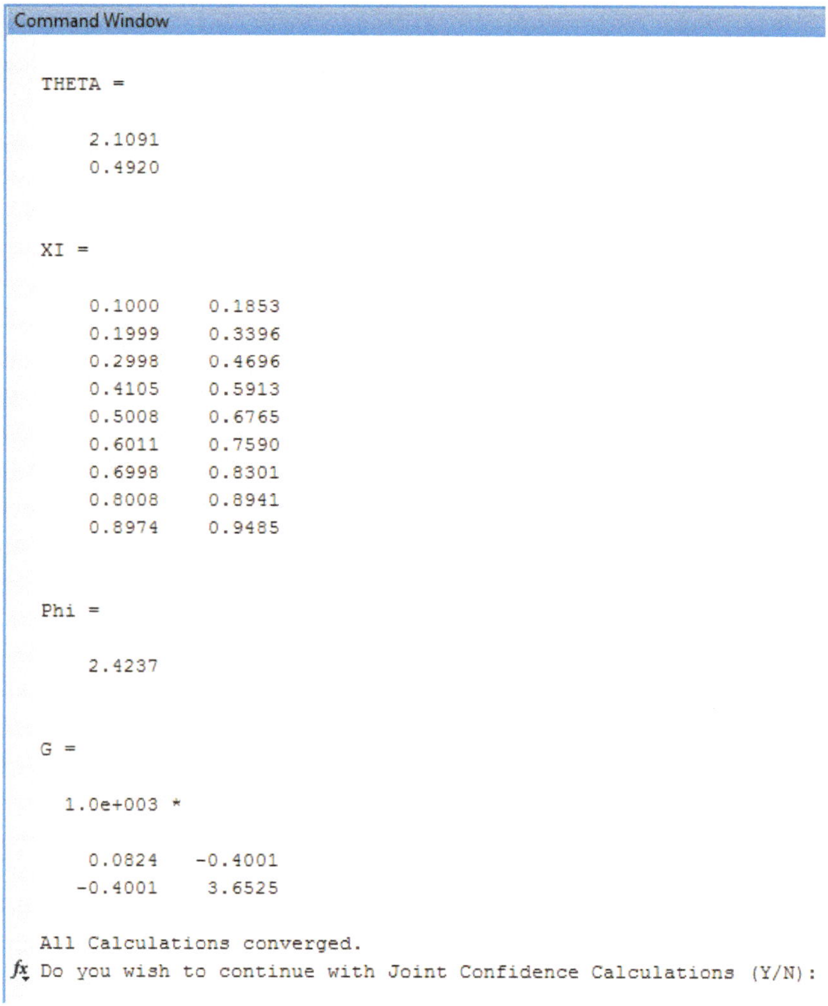

```
Command Window

  THETA =

       2.1091
       0.4920


  XI =

       0.1000    0.1853
       0.1999    0.3396
       0.2998    0.4696
       0.4105    0.5913
       0.5008    0.6765
       0.6011    0.7590
       0.6998    0.8301
       0.8008    0.8941
       0.8974    0.9485


  Phi =

       2.4237


  G =

     1.0e+003 *

       0.0824   -0.4001
      -0.4001    3.6525

  All Calculations converged.
fx Do you wish to continue with Joint Confidence Calculations (Y/N):
```

Figure A11. Program output for the preliminary analysis of BMA/BA using the instantaneous model.

```
Command Window

  THETA =

      2.1136
      0.5004


  XI =

      0.3352
      0.3294
      0.3253
      0.2913
      0.2689
      0.2602
      0.2403
      0.2313
      0.2074
      0.2072
      0.2033
      0.6679
      0.6634
      0.6390
      0.6248
      0.6049
      0.5434
      0.5300
      0.5222
      0.5069
      0.8904
      0.8890
      0.8883
      0.8871
      0.8804
      0.8715
      0.8417
      0.8049
      0.8002


  Phi =

      50.2464


  G =

     1.0e+003 *

      0.1065   -0.5339
     -0.5339    4.0878

 All Calculations converged.|
fx Do you wish to continue with Joint Confidence Calculations (Y/N):
```

Figure A12. Program output for the analysis of BMA/BA using the cumulative model.

Appendix C.4. Screenshots from Sty/EA Copolymerization Analysis

The additional information shown herein is relevant to Exhibit D (Section 4.2.2), the copolymerization of styrene (Sty) and ethyl acrylate (EA). Since all of the recorded data was for low conversion experiments, only the instantaneous model was used (and always converged in under five seconds). Several examples of low conversion data files have already been shown throughout the appendix (see, for example, Figures A6, A8a and A10a), so the information is not repeated herein.

However, as explained in Section 4.2.2, subsets of the original data (reported by McManus and Penlidis [21]) were analyzed to demonstrate the importance of experimental replicates. Specific data used at each stage of the analysis (and resulting reactivity ratio estimates) are shown in Tables A1 and A2. As stated in the main text, runs were randomly selected for removal during this exercise. The investigation could be repeated with different runs removed (or, the same runs removed in a different order) and the general observations would be the same.

Table A1. Experimental data for investigating the importance of replication (see Figure 11).

3 Replicates		2 Replicates		1 Replicate		No Replicates	
$f_{1,0}$	F_1	$f_{1,0}$	F_1	$f_{1,0}$	F_1	$f_{1,0}$	F_1
0.079	0.296	0.079	0.296	0.079	0.296	0.079	0.296
0.079	0.308	0.079	0.308	0.079	0.308	0.079	0.308
0.079	0.303	0.079	0.303	0.079	0.303	0.079	0.303
0.079	0.286	0.079	0.286	0.079	0.286	0.079	0.286
0.719	0.716	0.719	0.716	0.719	0.716	0.719	0.716
0.719	0.736	0.719	0.736	0.719	0.736	0.719	0.736
0.719	0.736	0.719	0.736	0.719	0.736	0.719	0.736
0.719	0.732	0.719	0.732	0.719	0.732	0.719	0.732
$r_1 = 0.717$		$r_1 = 0.746$		$r_1 = 0.740$		$r_1 = 0.750$	
$r_2 = 0.128$		$r_2 = 0.132$		$r_2 = 0.127$		$r_2 = 0.124$	

Table A2. Experimental data for investigating the effect of replicate symmetry (see Figure 12).

3 Replicates		High $f_{1,0}$ Replicates		High $f_{2,0}$ Replicates	
$f_{1,0}$	F_1	$f_{1,0}$	F_1	$f_{1,0}$	F_1
0.079	0.296	0.079	0.296	0.079	0.296
0.079	0.308	0.079	0.308	0.079	0.308
0.079	0.303	0.079	0.303	0.079	0.303
0.079	0.286	0.079	0.286	0.079	0.286
0.719	0.716	0.719	0.716	0.719	0.716
0.719	0.736	0.719	0.736	0.719	0.736
0.719	0.736	0.719	0.736	0.719	0.736
0.719	0.732	0.719	0.732	0.719	0.732
$r_1 = 0.717$		$r_1 = 0.715$		$r_1 = 0.752$	
$r_2 = 0.128$		$r_2 = 0.123$		$r_2 = 0.129$	

Appendix C.5. Screenshots from BA/MMA Copolymerization Analysis

The additional information shown herein is relevant to Exhibit E (Section 4.2.3), the copolymerization of butyl acrylate (BA) and methyl methacrylate (MMA). The original data set from Dubé and Penlidis [22] employed the Tidwell-Mortimer design of experiments; the data file is shown in Figure A13 and the program output (obtained in less than five seconds) is shown in Figure A14. The associated JCR was shown previously (recall Figure 13).

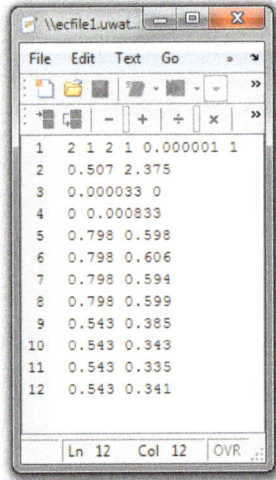

Figure A13. Data file for the preliminary analysis of BA/MMA.

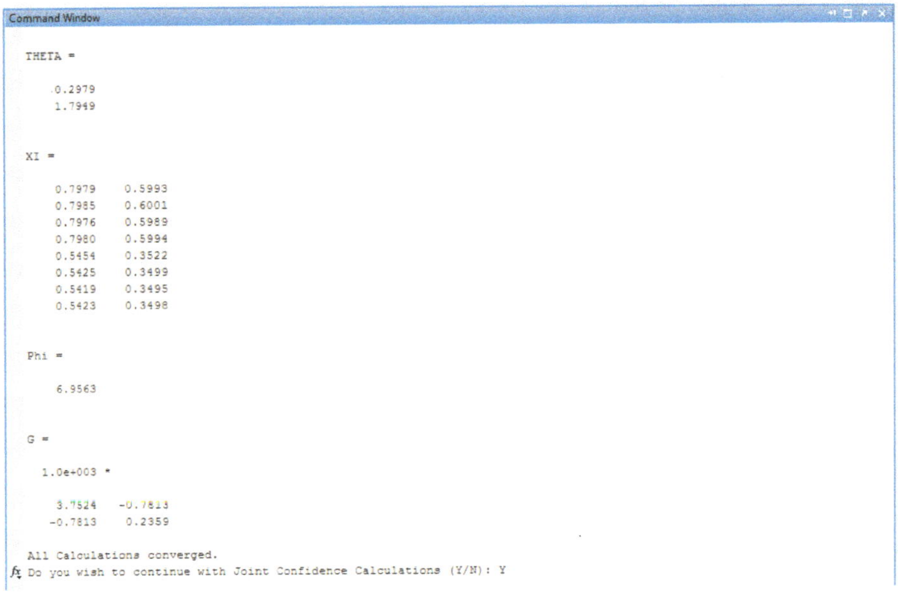

```
Command Window

THETA =

    .0.2979
     1.7949


XI =

    0.7979    0.5993
    0.7985    0.6001
    0.7976    0.5989
    0.7980    0.5994
    0.5454    0.3522
    0.5425    0.3499
    0.5419    0.3495
    0.5423    0.3498


Phi =

     6.9563


G =

    1.0e+003 *

     3.7524    -0.7813
    -0.7813     0.2359

All Calculations converged.
fx Do you wish to continue with Joint Confidence Calculations (Y/N): Y
```

Figure A14. Program output for the preliminary analysis of BMA/BA.

Given the output data from Figure A14, the EVM-based sequential design of experiments can be used to select the next optimal feed composition. Using EVM for sequential design of experiments has been discussed by Kazemi et al. [8]. From the design of experiments, the next logical feed composition is $f_{1,0} = 0.1$. Simulated data (shown in addition to the preliminary data set) used for subsequent analysis are shown in Figure A15, with the results shown in Figure A16. Again, the program converged in under five seconds.

\\ecfile1.uwaterloo.ca\...
File Edit Text Go Tools
1 2 1 2 1 0.000001 1
2 0.507 2.375
3 0.000033 0
4 0 0.000833
5 0.798 0.598
6 0.798 0.606
7 0.798 0.594
8 0.798 0.599
9 0.543 0.385
10 0.543 0.343
11 0.543 0.335
12 0.543 0.341
13 0.100 0.0585
14 0.100 0.0574
15 0.100 0.0561
16 0.100 0.0550
Ln 16 Col 6 OVR

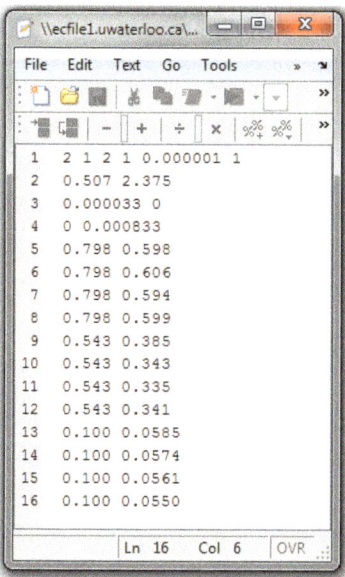

Figure A15. Data file for the sequential (DOE) analysis of BA/MMA.

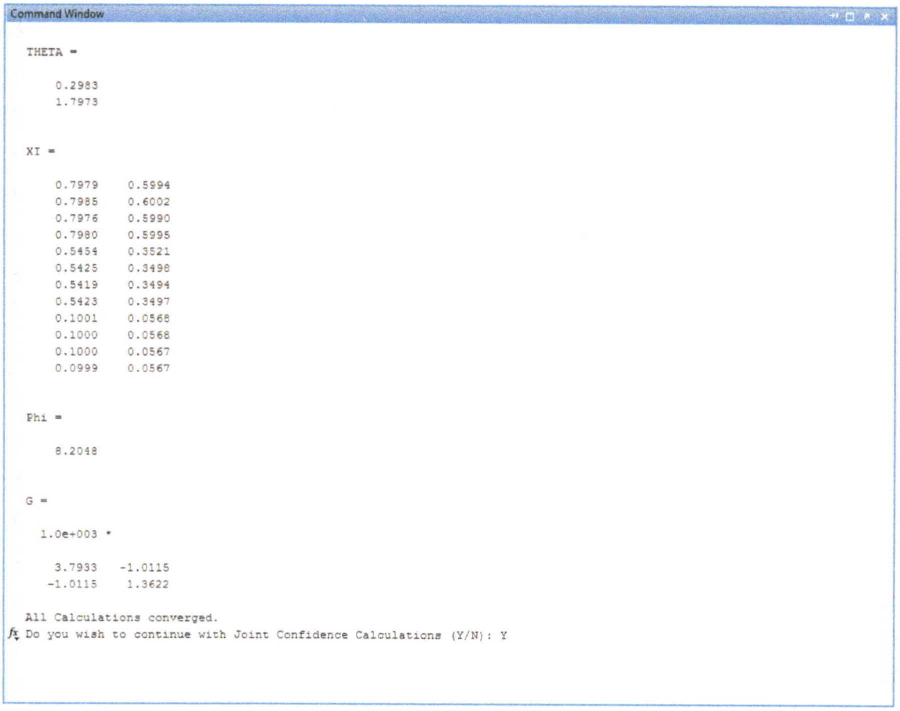

Command Window
THETA =
0.2983
1.7973
XI =
0.7979 0.5994
0.7985 0.6002
0.7976 0.5990
0.7980 0.5995
0.5454 0.3521
0.5425 0.3498
0.5419 0.3494
0.5423 0.3497
0.1001 0.0568
0.1000 0.0568
0.1000 0.0567
0.0999 0.0567
Phi =
8.2048
G =
1.0e+003 *
3.7933 -1.0115
-1.0115 1.3622
All Calculations converged.
Do you wish to continue with Joint Confidence Calculations (Y/N): Y

Figure A16. Program output for the sequential (DOE) analysis of BMA/BA.

References

1. Dubé, M.A.; Amin Sanayei, R.; Penlidis, A.; O'Driscoll, K.F.; Reilly, P.M. A microcomputer program for estimation of copolymerization reactivity ratios. *J. Polym. Sci. Part A Polym. Chem.* **1991**, *29*, 703–708. [CrossRef]

2. Polic, A.L.; Duever, T.A.; Penlidis, A. Case studies and literature review on the estimation of copolymerization reactivity ratios. *J. Polym. Sci. Part A Polym. Chem.* **1998**, *36*, 813–822. [CrossRef]

3. Kazemi, N.; Duever, T.A.; Penlidis, A. A powerful estimation scheme with the error-in-variables-model for nonlinear cases: Reactivity ratio estimation examples. *Comput. Chem. Eng.* **2013**, *48*, 200–208. [CrossRef]

4. Scott, A.J.; Riahinezhad, M.; Penlidis, A. Optimal design for reactivity ratio estimation: A comparison of techniques for AMPS/acrylamide and AMPS/acrylic acid copolymerizations. *Processes* **2015**, *3*, 749–768. [CrossRef]

5. Scott, A.J.; Penlidis, A. Copolymerization. In *Elsevier Reference Module in Chemistry, Molecular Sciences and Chemical Engineering*; Reedijk, J., Ed.; Elsevier: Waltham, MA, USA, 2017.

6. Kazemi, N.; Duever, T.A.; Penlidis, A. Reactivity ratio estimation from cumulative copolymer composition data. *Macromol. React. Eng.* **2012**, *5*, 385–403. [CrossRef]

7. Reilly, P.M.; Patino-Leal, H. A Bayesian study of the error-in-variables model. *Technometrics* **1981**, *23*, 221–231. [CrossRef]

8. Kazemi, N.; Duever, T.A.; Penlidis, A. Design of experiments for reactivity ratio estimation in multicomponent polymerizations using the error-in-variables approach. *Macromol. Theory Simul.* **2013**, *22*, 261–272. [CrossRef]

9. Scott, A.J.; Kazemi, N.; Penlidis, A. AMPS/AAm/AAc terpolymerization: Experimental verification of the EVM framework for ternary reactivity ratio estimation. *Processes* **2017**, *5*, 9. [CrossRef]

10. Agarwal, S.; Kumar, R.; Kissel, T.; Reul, R. Synthesis of degradable materials based on caprolactone and vinyl acetate units using radical chemistry. *Polym. J.* **2009**, *42*, 650–660. [CrossRef]

11. Undin, J.; Illanes, T.; Finne-Wistrand, A.; Albertsson, A. Random introduction of degradable linkages into functional vinyl polymers by radical ring-opening polymerization, tailored for soft tissue engineering. *Polym. Chem.* **2012**, *3*, 1260–1266. [CrossRef]

12. Hedir, G.; Bell, C.; Ieong, N.; Chapman, E.; Collins, I.; O'Reilly, R.; Dove, A. Functional degradable polymers by xanthate-mediated polymerization. *Macromolecules* **2014**, *47*, 2847–2852. [CrossRef]

13. Ding, D.; Pan, X.; Zhang, Z.; Li, N.; Zhu, J.; Zhu, X. A degradable copolymer of 2-methylene-1,3-dioxepane and vinyl acetate by photo-induced cobalt-mediated radical polymerization. *Polym. Chem.* **2016**, *7*, 5258–5264. [CrossRef]

14. Feldermann, A.; Toy, A.; Phan, H.; Stenzel, M.; Davis, T.; Barner-Kowollik, C. Reversible addition fragmentation chain transfer copolymerization: Influence of the RAFT process on the copolymer composition. *Polymer* **2004**, *45*, 3997–4007. [CrossRef]

15. Hagiopol, C. *Copolymerization: Toward a Systematic Approach*; Plenum Publishers: New York, NY, USA, 1999.

16. Suresh, J.; Karthik, S.; Arun, A. Photocrosslinkable polymer based on 4-3-(2,4-dichlorophenyl)-3-oxoprop-1-enyl) phenylacrylate: Synthesis, reactivity ratio, and crosslinking studies. *Mater. Sci. Pol.* **2016**, *34*, 834–844. [CrossRef]

17. Zhang, G.; Zhang, L.; Gao, H.; Konstantinov, I.; Arturo, S.; Yu, D.; Torkelson, J.; Broadbelt, L. A combined computational and experimental study of copolymerization propagation kinetics for 1-ethylcyclopentyl methacrylate and methyl methacrylate. *Macromol. Theory Simul.* **2016**, *25*, 263–273. [CrossRef]

18. Ren, S.; Hinojosa-Castellanos, L.; Zhang, L.; Dubé, M.A. Bulk free-radical copolymerization of *n*-butyl acrylate and *n*-butyl methacrylate: Reactivity ratio estimation. *Macromol. React. Eng.* **2017**, *11*, 1600050. [CrossRef]

19. Tidwell, P.W.; Mortimer, G.A. An improved method of calculating copolymerization reactivity ratios. *J. Polym. Sci. Part A Polym. Chem.* **1965**, *3*, 369–387. [CrossRef]

20. Cochran, W.G.; Cox, G.M. *Experimental Designs*; John Wiley & Sons, Inc.: New York, NY, USA, 1957.

21. McManus, N.; Penlidis, A. A kinetic investigation of styrene/ethyl acrylate copolymerization. *J. Polym. Sci. Part A Polym. Chem.* **1996**, *34*, 237–248. [CrossRef]

22. Dubé, M.A.; Penlidis, A. A systematic approach to the study of multicomponent polymerization kinetics—The butyl acrylate/methyl methacrylate/vinyl acetate example: 1. Bulk copolymerization. *Polymer* **1995**, *36*, 587–598. [CrossRef]

23. Zhang, Y.; Dubé, M.A. Copolymerization of *n*-butyl methacrylate and D-limonene. *Macromol. React. Eng.* **2014**, *8*, 805–812. [CrossRef]

24. Grassie, N.; Torrance, B.; Fortune, J.; Gemell, J. Reactivity ratios for the copolymerization of acrylates and methacrylates by nuclear magnetic resonance spectroscopy. *Polymer* **1965**, *6*, 653–658. [CrossRef]

25. Rossignoli, P.J.; Duever, T.A. The estimation of copolymer reactivity ratios: A review and case studies using the error-in-variables model and nonlinear least squares. *Polym. React. Eng.* **1995**, *3*, 361–395.

Article

Solving Materials' Small Data Problem with Dynamic Experimental Databases

Michael McBride, Nils Persson, Elsa Reichmanis and Martha A. Grover *

School of Chemical and Biomolecular Engineering, Georgia Institute of Technology, Atlanta, GA 30332, USA; mmcbride6@gatech.edu (M.M.); nils.persson@chbe.gatech.edu (N.P.); elsa.reichmanis@chbe.gatech.edu (E.R.)
* Correspondence: martha.grover@chbe.gatech.edu; Tel.: +1-404-894-2878

Received: 7 June 2018; Accepted: 25 June 2018; Published: 27 June 2018

Abstract: Materials processing is challenging because the final structure and properties often depend on the process conditions as well as the composition. Past research reported in the archival literature provides a valuable source of information for designing a process to optimize material properties. Typically, the issue is not having too much data (i.e., big data), but rather having a limited amount of data that is sparse, relative to a large number of design variables. The full utilization of this information via a structured database can be challenging, because of inconsistent and incorrect reporting of information. Here, we present a classification approach specifically tailored to the task of identifying a promising design region from a literature database. This design region includes all high performing points, as well as some points having poor performance, for the purpose of focusing future experiments. The classification method is demonstrated on two case studies in polymeric materials, namely: poly(3-hexylthiophene) for flexible electronic devices and polypropylene–talc composite materials for structural applications.

Keywords: materials; processing; polymers; database; classification; informatics

---

1. Introduction

Research institutions and their funding sources have made a strong push in recent years to leverage the tools of modern data analytics for scientific research. This effort has fostered the creation of large centralized data repositories, the proliferation of more open-source software and code libraries, and a general move toward 'open science' [1]. The tools of data analytics and informatics were developed to analyze truly massive databases containing millions to billions of entries, such as the collective Google search history of a country's population, or the Amazon shopping history of the same. As such, early efforts in the sciences have focused on the assembly of the largest possible collections of data, namely: thermodynamic constants for every known chemical compound, density functional theory (DFT) simulations of every known crystalline material, or characterization data from every known biological protein [2–6].

While the scope of these centralized databases is vast, they still do not capture the full extent of the scientific data collected and published on a daily basis. Day-to-day research is not driven by a desire to fill in the billionth row of a central repository, but rather to solve problems and produce knowledge in specialized fields with moderately-sized communities and dynamic trends. The design of a new research study is most often based upon the results of a few dozen previous studies at most. This would not be considered 'big data' by any measure, yet these small sets of experimental results play a crucial role in daily decision-making. In order to use data analytics to guide this decision-making process, it would be beneficial if the datasets from related publications were assembled into interim databases with a common schema. It is likely that many researchers already aggregate data in this way, especially for a formal literature review. This is all the more reason to develop tools and strategies to work quantitatively with these small, sometimes sparse datasets.

Experimental research in materials processing depends crucially on the selection of a promising initial design region within the available process parameters. When processing a material, there are many potential design variables by which to optimize its properties and performance, including formulation, equipment selection, and settings. The challenge of process design for materials differs from chemical production, because the process influences the material's structure, which further influences its properties. Some crystalline solids and disordered liquids are processed to reach an equilibrium structure, but many materials, such as the soft materials presented here, are guided to process-dependent non-equilibrium states, sometimes with multiple phases. Furthermore, interfacial structures can often influence or even dictate the final performance beyond the bulk material properties. Selecting a process variable design space for a new experiment should thus be approached using as much quantitative knowledge as possible [7].

Data mining approaches to experimental design have experienced a recent uptick in interest. For example, Kim et al. performed an automated review of the synthesis conditions of metal oxides across over 12,000 manuscripts. Their data extraction pipeline consisted of a combination of trained machine learning models and software such as ChemDataExtractor. A decision tree classifier considering 27 synthesis variables was trained to predict whether or not a titania synthesis route would produce nanotubes, achieving 82% accuracy [8]. Agrawal et al. compared a range of machine learning approaches, including multivariate polynomial regression ($R^2 = 0.9801$), support vector machines ($R^2 = 0.9594$), and artificial neural networks ($R^2 = 0.9724$), to predict the fatigue strength of steels [9]. Input variables describing chemical composition, processing temperatures and times, and upstream processing details were taken from the National Institute of Materials Science (NIMS) MatNavi [10]. Ren et al. combined literature data and high-throughput experiments to predict the likelihood of finding metallic glasses in the Co–V–Zr ternary [11].

The use of the design of experiment methodologies to generate ideal process–property databases is another approach to enable subsequent data mining of process–property relationships. This is exemplified by AbuOmar et al. and Zhang et al., where polymer nanocomposite databases were generated via carefully designed experiments [12,13]. The databases generated using uniform design approaches typically only survey low-dimensional design spaces with a priori domain knowledge to appropriately grid experimental values [14]. The common element among these previous studies is the accessibility of fully informed databases with all descriptors fully characterized and reported.

The scientific literature can be conceived as an unstructured materials database, rich with process–property data points. However, the nature of exploratory materials discovery suggests a high-dimensional design space, with individual publications providing a limited number of data points. There are several challenges in constructing and analyzing a literature database, even with manual construction by domain experts, namely: (1) literature reports do not have standardized fields for frequently-reported quantities; (2) authors report non-overlapping sets of information, leading to data sparsity; and (3) reporting of design variables and property measurements suffer from variability due to measurement equipment, inconsistent application of mathematical models, and human error [15]. Persson et al. provides an illustrative example of these challenges by mining the literature on poly(3-hexylthiophene) organic field effect transistors [16]. Over 200 data points from 19 publications describing the role of 28 processing variables were curated to predict charge carrier mobility. Identifying a subset of the five most similar devices reduced the range of charge carrier mobility values from over six orders of magnitude to three. The presence of unreported, missing data limited the applicability of the standard classification and regression materials informatics techniques.

Herein, we demonstrate how the construction of a structured database containing relevant literature results can be used to guide experimental design for materials processing. This tailored approach to classification is proposed to specifically handle datasets with missing data, but is also applicable to fully characterized databases. The set of best performing points indicates a promising design region for future experiments. Two case studies are presented to demonstrate the approach, with each database containing 100–200 data points. The number of data points is largely governed by the available relevant publications. The analysis is primarily intended to guide future experiments into regions likely to have the desired properties. Further local optimization can then be performed using response surface methodology [17,18]. Once the optimal point is identified, then additional design variables could be added to enable further improvement, or previously unexplored regions could be explored.

2. Materials and Methods

2.1. Database Construction

The databases are constructed manually using a Microsoft Excel spreadsheet. The data extracted from relevant publications is found in text, tables, and graphical format. Each row in the spreadsheet corresponds to a particular data point. The columns represent the process design variables, which may be numerical (continuous or integer) or categorical. In some cases, the categorical variables (e.g., solvent type) are augmented with numerical descriptors to quantify their properties (e.g., solvent boiling point). In other cases, the categorical variables are hierarchical (e.g., deposition method: spin-coat) with associated numerical values quantifying that category (e.g., spin rate and spin time). Each property measurement is also associated with a column. Additional columns are included so as to document the source of each datapoint, including its digital object indicator (DOI). Further metrics of the journal impact factor and the number of citations of the paper can also be included, but present additional complications as they change over time.

While this manual, ad hoc approach is not ideal from a scalability standpoint, the cost of automating this step can be prohibitive in terms of time and expertise. Furthermore, the automation of data extraction requires a significant number of labeled training examples, and the construction of the training set frequently accomplishes so much of the labeling that it makes more sense to simply finish the job manually. This trade-off should be evaluated on a case-by-case basis.

2.2. Classification Approach

A common approach to the classification of data is the support vector machine (SVM), which, in its simplest form, constructs a hyperplane to divide two prelabeled classes. This approach is exemplified by Kim et al., who separated metal oxide structures into nanotube-forming and non-nanotube-forming by their processing conditions [8]. SVMs are optimized to maximize the distance between the decision boundary and the nearest point from each class. In other words, SVMs try to minimize the occurrence of misclassification on both sides of the boundary. In exploratory research, however, it is much worse to exclude a potential positive result from consideration than it is to run an experiment that turns out negative. Here, we apply a different approach to classification, in which the objective is to retain all datapoints with good properties in the 'promising region'.

A property of interest is selected, then a critical value of that property is specified by the user. All datapoints in the database are thus labeled as 'high' or 'low', depending on whether they lie above or below this critical value. The region of the design space that contains all the 'high' points is then constructed. Figure 1 illustrates this classification approach. This approach helps to ensure that potential positive results are never excluded from the promising region. It is, however, inherently sensitive to outlier points, in which a 'high' property value has been incorrectly reported. These outliers should be uncommon, because such results would most likely have received great interest and scrutiny in the review process. Nonetheless, points with a reported 'high' performance should receive further investigation in subsequent experiments, that is, they should be repeated to validate (or invalidate) their good performance.

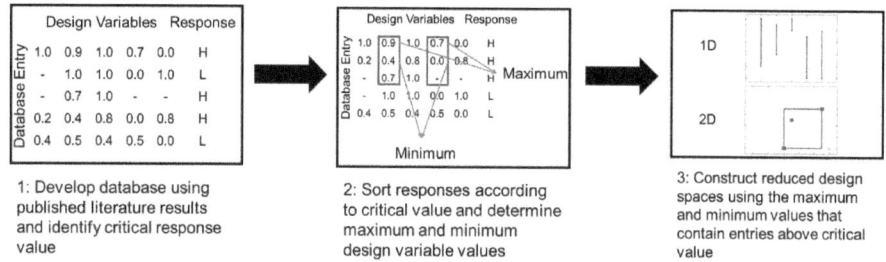

1: Develop database using published literature results and identify critical response value

2: Sort responses according to critical value and determine maximum and minimum design variable values

3: Construct reduced design spaces using the maximum and minimum values that contain entries above critical value

Figure 1. Flow chart of classification approach to constructed reduced design regions.

One possible implementation for quantifying the promising region is to construct a convex hull containing all of the 'high' points in the literature database. Here, we take a simpler approach, which is to calculate the upper and lower bounds of each design variable within which all of the 'high' points are contained, defining a box. The rationale is to provide better intuition and visualization for the user. For each design variable, one compares the minimum and maximum values for 'high' performance to the minimum and maximum values in the entire database, quantifying the percent reduction of the design space, which can be achieved by focusing future experiments on past 'high' performing regions. This reduction r_s can expressed mathematically, as follows:

$$r_s = \frac{\prod_1^{n_s} l_{s_i}}{\prod_1^{n_s} L_{s_i}} \tag{1}$$

where S is the set of indices for a subset of the design variables, n_s is the dimensionality of the design space (equivalent to the size of S), l_i is the span of variable i that contains 'high' points, and L_i is the span of variable i over all points in the database. In the one-dimensional (1-D) case, r_s is the relative length of the line segment that contains the 'high' points, while in the two-dimensional (2-D) case, r_s is the relative area that contains the 'high' points. Each design variable, or combination of design variables, is then represented with a value of r_s between 0 and 1, where it is hypothesized that a smaller value signifies that the associated design variable is a better indicator of a 'high' performance.

While r_s expresses the volumetric reduction of the design space, it is also useful to understand the fraction of the original data contained in the promising region, as data density is likely non-uniform. This is expressed as follows:

$$F_r = \frac{|d|}{|D|} \tag{2}$$

where d is the number of points contained within the box and D is the total number of observations in the database. Each observation corresponds to a row in the Excel database spreadsheet.

It is certainly possible that the box excludes potential regions of high performance that were not investigated in previous studies. One could apply a padding parameter to the box bounds to reduce this risk. Alternatively, future experiments can explore additional regions, locally, through response surface methodology, or more globally, with random experimental settings or a particular focus on unexplored regions. Certainly, physical models and domain expertise will also guide future experiments, but the details of the experimental design are beyond the scope of this article [17–19].

2.3. Case Studies

2.3.1. Poly(3-Hexylthiophene) (P3HT)

Poly(3-hexylthiophene) is a semiconducting polymer that has been widely studied for application in large-area flexible electronics [20]. Such systems could access new markets beyond silicon transistors if they can be printed economically in a roll-to-roll process. Thin films of P3HT exhibit hole mobilities that vary by orders of magnitude, depending on how they are processed [21]. Distinct fibrillar morphologies are observed in the atomic force microscopy, and the morphology can be influenced by the process, and correlated with the hole mobility [22,23]. Design variables that are reported in the literature include polymer molecular weight and regioregularity, solvent and concentration, deposition method and film thickness, and annealing time [16]. This system is depicted in Figure 2.

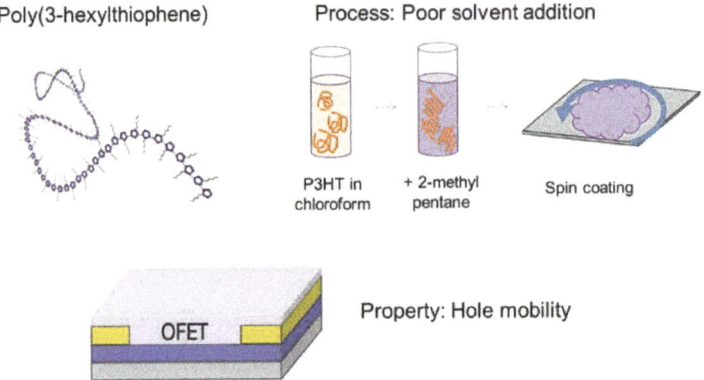

Figure 2. One possible processing route for poly-3-hexylthiophene (P3HT), and a key property of interest in organic field-effect transistor (OFET) performance.

The analyzed database, which was presented in a previous publication, is comprised of 218 datapoints from 19 publications. To model process–property relationships to obtain high mobility devices, 29 design variables were identified that are either numerical (20 design variables) or categorical (9 design variables). The design variables were removed from consideration if all relevant entries were identical, exceptionally sparse, or deemed irrelevant according to expert opinion (e.g., dip rate, dip time, process environment, and electrode material). The mobility values in this database range from 1.0×10^{-6} to 2.8×10^{-1} cm^2/V·s.

This database is included in the supplementary information and is also publicly accessible at: http://www.github.com/Imperssonator/OFET-Database.

2.3.2. Polypropylene–Talc Composite

The properties of polymeric materials can be altered and improved by mixing them with fillers to create a composite material [24–26]. In addition, the processing method impacts the properties [27]. Polypropylene (PP) has relatively strong mechanical properties for a polymer and it is widely available. If its properties could be further enhanced, polypropylene might be a candidate for replacing metals, for example in the additive manufacturing of automotive parts. Here, we consider talc as our filler. Talc is a clay material composed of magnesium silicate. It is a good filler candidate for our study because it has been reported to enhance polypropylene performance in many literature studies and because it is relatively inexpensive, enhancing commercial viability [28,29]. This system is illustrated in Figure 3.

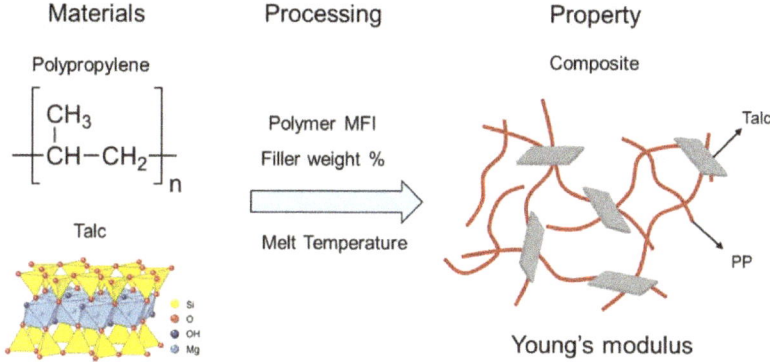

Figure 3. Proposed polymer and filler to develop high strength composites and a key property of interest: Young's modulus.

The polypropylene-talc database is comprised of 140 datapoints from 22 publications with the goal of improving the strength of polypropylene. The Young's modulus was selected as the relevant mechanical parameter representing the materials' strength. Fourteen design variables were selected for this analysis, with eleven being numerical and three being categorical. Expert opinion was leveraged to characterize differing melt mixer equipment by recording the highest temperature in a multistate extruder. The Young's modulus values for this database vary from 0.38 to 6.94 GPa.

This database is included in the supplementary information and is also publicly accessible at: https://github.com/DocMike/TALC-Database.

3. Results

3.1. Case Study 1: Poly-3-Hexylthiophene

The breadth of processing conditions recorded in the P3HT database illustrates the utility of the proposed classification approach. In this database, a mobility cutoff value of 0.1 cm^2/V·s was selected to differentiate between 'high' and 'low' devices. The design variables extracted from the literature can be grouped into five main categories, namely: (1) polymer characteristics, (2) solvent environment, (3) deposition conditions, (4) post-deposition processing, and (5) transistor configuration. All of these variables have been hypothesized to influence the morphology of the final semiconducting film and thus the charge carrier mobility [30,31].

Figure 4 illustrates the classification approach applied to the P3HT database for 13 of the continuous design variables contained in the database, out of a total of 29 design variables of numerical and categorical type. The one-dimensional analysis is shown in Figure 4a for these 13 design variables, with results scaled to the full range of values reported in the database. For each of these design variables, the difference between the maximum and minimum values associated with a 'high' performance l_i, is divided by L_i, the difference between the maximum and minimum values associated with all entries in the database, according to Equation (1). In some cases, a 'high' performance is observed only for a small fraction of the full reported range, such as for regioregularity, spin rate, and channel width, such that $r_s < 0.1$. In other cases, a 'high' performance can be observed at most of the reported values (molecular weight, initial concentration, boiling point, annealing temperature, and time), such that $r_s > 0.9$. In other intermediate cases, a restricted range can be observed, namely: polydispersity, Hansen radius, film thickness, and channel length.

One might first focus on regioregularity, spin rate, and channel width to reduce the design space for future experiments. However, these are not the best candidates. Regioregularity must be high for a 'high' performance, which is well known; only a small number of the earlier papers used polymers with lower values of regioregularity [32,33]. Moreover, precision synthesis and characterization of a specific regioregularity is infeasible [34]. Of the 37 'high' devices, only 8 report a spin rate and all 8 arise from the same publication. Thus spin rate is not an ideal candidate. The channel width is also shown to impact the observed mobility, but device physics indicate that it should not influence the transistor performance [35]. The standardization of test conditions is necessary to improve the quantification of results in polymer organic electronics, and future experiments should conform to this common standard [36].

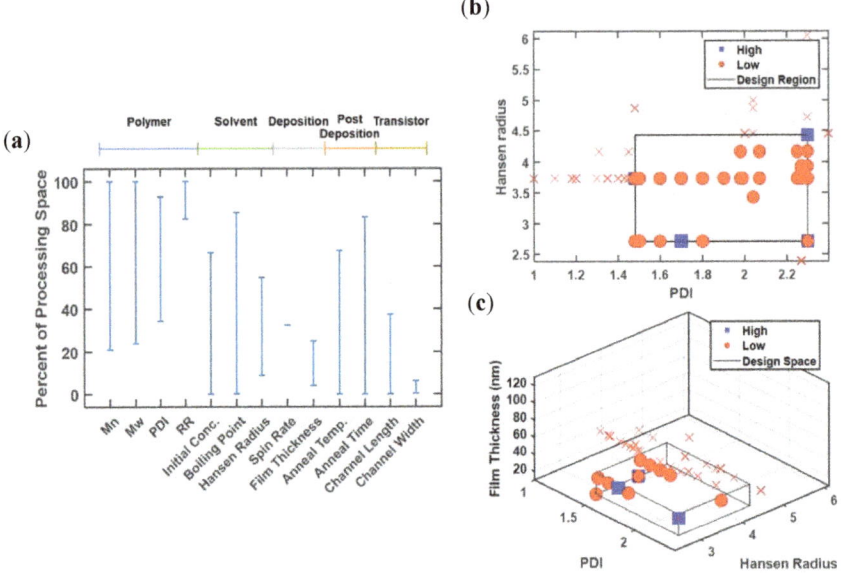

Figure 4. Representative analysis plots of (a) one-, (b) two-, and (c) three-dimensional classifying design variables, resulting in charge mobility values exceeding 0.1 cm^2/V·s. Blue squares are data points above the cutoff, red dots represent points below the cutoff but within the target design region, and red x markers indicate all other data points below the cutoff but not within the target region. All axis ranges denote the full range of values present in the database.

The more revealing design variables are those with a moderate reduction of the design space, namely: polydispersity, Hansen radius, film thickness, and channel length. Like channel width, channel length affects performance through device physics rather than material properties, and can be excluded. This leaves three key design variables, which can be visualized in two or three dimensions, as shown in Figure 4b,c. A high value of polydispersity is required to achieve a good performance, but low values of the Hansen radius and film thickness are also needed.

The two-dimensional plots can be generated for each pairwise combination of design variables, or on the selected promising design variables identified in the one-dimensional analysis. Figure 4b delineates the promising design region using the polymer polydispersity index (PDI) and Hansen radius. The area of the identified design region relative to the full area spanned by the design variables is quantified by $r_s = 0.27$, for $S = \{3,7\}$. Thus, future experimental efforts can be focused on only 27% of the original design region. When film thickness is additionally added as a classifier, $r_s = 0.056$, for $S = \{3,7,9\}$.

However, quantifying the reduction using r_s without visualizing results or utilizing additional metrics can lead to misleading conclusions, depending on the quality of the information in the database.

Issues with only characterizing the relative reduction can be grouped into two main categories, as follows: (1) limited coverage of the entire design region (sparsity), and (2) inconsistent reporting and characterization. These cases are illustrated in Figure 5. Note that in Figure 5b, there are no datapoints shown with regioregularity (RR) values below 90%, despite their presence in the database, because film thickness was not reported for any of those datapoints.

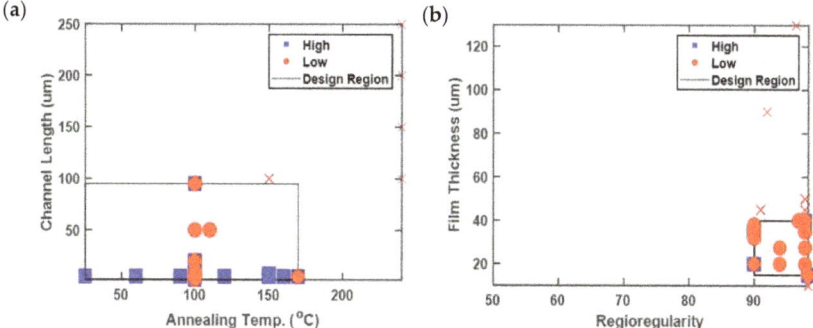

Figure 5. Illustrative two-dimensional plots indicating (**a**) limited collective screening of the entire design region (sparsity), and (**b**) inconsistent reporting and characterization. Axis values denote the full range of values present in the database.

The fraction of points contained within the reduced designed region (F_r) serves as a simple indicator for the distribution and sparsity of points throughout the entire design region. Table 1 highlights both the reduced area and the fraction of points contained in this region. In general, r_s is smaller than F_r, indicating that the volumetric reduction of the design space is greater than the fraction of database points excluded. This can be rationalized in the context of Schrier's Dark Reaction Project; negative results are infrequently reported in the literature [37]. It makes sense that data density would be concentrated in the promising region, because other experimentalists intuitively concentrate their effort in a similar region, and because they omit negative results from outside that region. This is a source of significant bias in the training set, but it can also serve as a sanity check on the selection of the promising region. Nonetheless, the importance of reporting negative results cannot be overstated.

Table 1. Metrics for the two-dimensional analysis of the poly-3-hexylthiophene (P3HT) system. PDI—polymer polydispersity index, RR—polymer regioregularity.

	PDI/Hansen Radius	PDI/Film Thickness	Hansen Radius/Film Thickness	Channel Length/Annealing Temp	RR/Film Thickness
r_s (%)	27	12	10	25	4
F_r (%)	45	30	53	45	63

This classification approach can theoretically be scaled to any dimension. However, the reliability of the identified target region containing high performing devices decreases as the dimensions increase. This reliability issue is illustrated in Figure 4c with a box denoting the target design region. The data is distributed throughout the x and y plane (PDI and Hansen radius), indicating that these two variables are highly reported. In contrast, no results are presented with a thickness greater than ~40 nm, despite thickness values as high as 130 nm being reported in the database. This problem arises from a lack of standardized reporting. Despite the fact that 146 of the entries report PDI, 190 report Hansen radius values, and 149 report film thickness, only 105 entries have all three of these variables specified. This issue can also impact the relative volume of the target design region. While the two-dimensional analysis (Figure 4b) points to a 'high' device with a PDI value of ~2.3 and a Hansen radius of ~4.5, this database entry does not contain a corresponding film thickness and as a result does not appear in Figure 4c. Thus, the analysis of higher-dimensional design spaces to simultaneously optimize numerous processing conditions relies heavily upon consistent and standardized reporting.

In addition to the 13 numerical design variables, three relevant categorical variables were identified. The choice of solvent influences the ability of the polymer chains to self-assemble in the solution and the solvent evaporation rate during the film deposition phase. The analysis performed in this study indicated that the use of chloroform and trichlorobenzene result in 'high' performing devices, while toluene, thiophene, benzene, chlorobenzene, and styrene do not. Section 3.1.2 will discuss how solubility parameters (Hansen radius) were used to represent a solvent on a continuous scale. Of the 37 'high' performing devices, 28 films were formed through drop casting, 8 via spin coating, and 1 film using dip casting, spanning all deposition methods in the database. It is likely that the interaction of these deposition methods with other process variables was optimized to produce favorable thin film microstructures. Finally, the treatment of the silicon dioxide capacitance layer to modify the wettability and polymer-substrate interface using hexamethyldisilane (HMDS) or no treatment at all can both result in 'high' devices. In contrast, perfluorodecyltrichlorosilane (FDTS) and octadecyltrichlorosilane (OTS) surface treatments produced unfavorable results. Ideally, surface free energy would be reported as a way to quantify the surface treatment, similar to the Hansen radius for the solvent.

Once the target design region has been identified, hypothesis generation and future experiment planning can occur. Here, we briefly discuss the physical intuition that can guide experimental design beyond the selection of the promising design region.

3.1.1. Polydispersity Index

Long polymer chains are required to extend across grain boundaries and thus provide high-mobility pathways through otherwise amorphous regions of the film [33,38,39]. The one-dimensional analysis of polymer characteristics (Figure 4a) suggests that 'high' performance devices can be obtained with number average molecular weights ranging from 26 kDa to 117 kDa. The PDI indicates the spread of the molecular weight distribution in a polymer sample, so a sample with a low average molecular weight could still contain many long chains if it has a high PDI. Our analysis indicates that a PDI > 1.5 is required for a high-performance device. However, there is experimental difficulty in synthesizing polymers with tailored PDIs for systematic studies [34]. Instead, techniques to either fractionate the P3HT samples or blend multiple polymers samples to target a specified PDI have been employed [40,41].

3.1.2. Hansen Radius

Thin film formation is a complex and dynamic process in which solvent evaporation governs structural organization mechanisms such as polymer aggregation and phase separation [42]. These processes are captured, albeit incompletely, through the initial polymer concentration, solvent boiling point, and Hansen radius, in addition to the equipment deposition parameters (e.g., spin rate, spin time). According to the one-dimensional analysis, the Hansen radius provides better discrimination between high and low performance devices compared with both the initial polymer concentration and the boiling point. The Hansen radius is a numerical descriptor of solvent–polymer interaction energy and can be used to reduce the number of design variables when solvent mixtures are utilized [11,43–45]. As an example, the dissolution of P3HT in a good solvent, followed by the addition of a poor solvent, was utilized as a processing method by 3 of the 19 papers in the database [43,45,46]. To fully characterize this, two categorical variables (good solvent, poor solvent) and one numerical variable (volume fraction of poor solvent) need to be specified. By applying the Hansen solubility model, these three variables can instead be reported as a single numerical value, the Hansen radius, simplifying the interpretation of the database and extraction of process–property relationships.

3.1.3. Film Thickness

An analysis of the database suggests that film thicknesses ranging from 25 to 40 nm are desired in order to obtain high-performing devices. This observation is in general agreement with work by Joshi et al., which explored the thickness dependence of the charge mobility of low molecular weight P3HT [47]. In their study, mobility plateaued after a thickness of 15 nm, due to the orientation of crystalline grains at the transistor oxide-P3HT interfaces. Varying the thickness of the P3HT films is usually the result of changing another design variable, such as polymer concentration or spin rate, and is highly dependent on the kinetics of the solution-to-film phase transition [47,48]. Ideally, an influential design variable should be able to be varied independently from other process conditions. Recent advances in blade and shear coating techniques, currently not captured by the database, offer a promising approach to independently control the film thickness to explore its impact on charge carrier mobility [49,50].

3.2. Case Study 2: Polypropylene–Talc Composite

The polypropylene–talc database was constructed from 22 publications, with a total of 140 data points. The property of interest that was selected was Young's modulus. While there are many ways to quantify mechanical properties (fracture strength, toughness, etc.), Young's modulus was selected here, in part because of the importance of stiffness in a mechanical component such as an automotive part. It is also widely reported in the literature because it is relatively easy to measure. A cutoff value of 5 GPa was selected for the Young's modulus, with 11 points (8%) having 'high' values above the cutoff value.

Similar to the P3HT study, the minimum and maximum reported values of 11 continuous design variables were identified in the full database and the high-performing subset to calculate r_s. The results of this one-dimensional analysis are shown in Figure 6a, using variables scaled to the minimum and maximum values in the database. Several design variables have an extremely limited range of values associated with high performance, namely, polymer weight-average molecular weight, polymer polydispersity, filler surface area, and composite density. In all four cases, this large reduction in range is due to very low rates of reporting. For example, only 30 data points have the molecular weight reported, and only two of those points have a 'high' performance. Both data points are from the same publication and have the same molecular weight. Other design variables have no range reported at all, including mold temperature and composite melt flow index (MFI), again due to underreporting—none of the high performing points have values for mold temperature or composite MFI.

In contrast, filler density and melt temperature are reported for most entries. Since the only filler in the database is talc, there is little actual variation in its reported density, except for one outlier entry that may be misreported [28]. The melt temperature also extends across most of the range in the database, although low values (lowest 20% of the reported range) seem to be undesirable since they are never reported together with a 'high' performance.

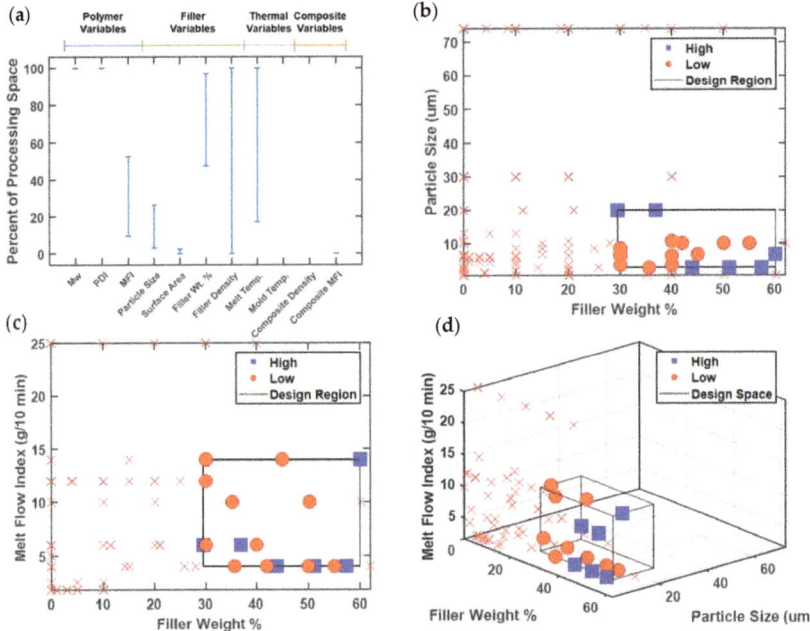

Figure 6. Representative analysis plots of (**a**) one-, (**b**,**c**) two-, and (**d**) three-dimensional analysis, classifying design variables that result in Young's modulus exceeding 5 GPa. All axis values denote the full range of values present in the database. Blue squares are data points above the cutoff, red dots represent points below the cutoff but within the target design region, and the red x markers indicate all other data points below the cutoff but not within the target design region.

The design variables with intermediate ranges in Figure 6a are polymer MFI, filler particle size, and filler weight percent. These three design variables contribute the most information to the future experimental design. It is notable that all three are related to material composition, rather than processing conditions. The melt flow index correlates inversely with the molecular weight, but is much more commonly reported in the database, because of the difficulty of dissolving polypropylene for the characterization of molecular weight by size exclusion chromatography [51]. The database analysis suggests that high stiffness requires a low polymer MFI, or equivalently, a high molecular weight. The filler particle size must fall in the lowest third of the reported values, while the filler weight percent must fall within the upper half of the values reported in the database.

The data can also be viewed in higher-dimensional spaces, as follows: Figure 6b illustrates the promising design region associated with filler size and weight percent. Note here that original variable scales are used, rather than the scaled variable ranges of Figure 6a. Future experiments can focus on this smaller box having low (but not too low) filler size and high filler weight %. The unexplored region of particle size in the range 35–70 μm could also be tested, depending on the experimental budget and availability of larger talc particles. Figure 6c again shows a two-dimensional representation, with polymer MFI and filler weight %. The promising design region contains low values of MFI, although there is little data available at higher MFIs. Future experiments could attempt to expand this region into higher MFI values (lower molecular weight), which would also lower viscosity and thus ease processing.

The calculated metrics, including the relative area of the new design region, are shown in Table 2. Similar volumetric reductions (r_s) in the identified design region are observed for the PP–talc database compared to the P3HT study. However, similar values of data reduction in the design region (F_s) for the PP–talc database suggests that the community has more thoroughly sampled the entire design region compared with the P3HT community. This could arise from the reduced number of considered design variables for the polymer composites (14 variables) compared with P3HT for organic electronics (29 variables).

Table 2. Quantified metrics of two-dimensional analysis for the PP–talc system. MFI—melt flow index.

	MFI/Particle Size	MFI/Filler wt.%	Particle Size/Filler wt.%
r_s (%)	10	21	12
F_r (%)	28	14	15

Finally, a three-dimensional representation is shown in Figure 6d. The promising design region is about 5% of the volume of the full range of design variables reported in the database. Also notable in Figure 6d is the lack of data points at the high particle size of 74 μm, which were previously plotted in Figure 6b [52]. Since no value of polymer MFI was reported for those points, they cannot be visualized in Figure 6d, and more importantly, they cannot be reproduced since the value of the polymer MFI that was used is unknown. Since Gafur et al. did report the weight-average molecular weight, a model could potentially be used to estimate the unreported MFI [51,52]. Physically-based correlations provide a potential route to filling in missing data, which may be more effective than linear imputation when the relationships between design variables are known and are nonlinear, such as the inverse correlation between MFI and weight-average molecular weight (Mw). However, since the Gafur et al. data points have a 'low' performance, such modeling would not change the promising region.

The polypropylene–talc database contains 11 continuous design variables, but also incorporates three categorical variables, namely compatibilizer, talc surface treatment, and the film formation method. Only 23 data points report a compatibilizer with none of these data points resulting in 'high' performance. Twenty-two data points report a surface treatment, but only aminopropyl-trimethoxysilane (4 data points) resulted in Young's modulus values above 5 GPa. This categorical variable presents a seldom-used processing condition to further explore. Films are formed either through extrusion or compression molding, but only compression molding produced a 'high' performance. Better understanding of the mechanistic reasons for improved performance via compression molding could help identify new processing conditions, but in the short term, compression molding is the preferred technique.

4. Discussion

Significant advances have been made in applying materials informatics and machine learning techniques to leverage the combined knowledge of research communities. The advent of large centralized materials data repositories that are publicly accessible has been a tremendous boon to accelerated materials discovery and process optimization. However, the curation and use of materials data that are generated on a day-to-day basis, within a given material system under a formalized materials informatics lens, is still in its infancy. Smaller, more specific databases relevant to a particular application will be required to rapidly provide chemical compositions and optimized processing conditions.

A key challenge seldom addressed in 'small data' materials research is how to extract meaningful process–structure relationships to target desired properties. Small data can be curated in two main approaches, namely (1) controlled experiments via the design of experiments or high-throughput methods within a single laboratory, and/or (2) mining the literature to leverage experiments conducted by the community. The former datasets are often well-structured, allowing process–property information extraction via material informatics and/or machine learning methodologies [9,11]. The latter approach involves a significantly wider scope of potential design variables, resulting in an unstructured database rife with missing and noisy data. Missing data limits the applicability of materials informatics approaches, including decisions tress, neural networks, and support vector machines. Instead, we have proposed an approach that uses all available data in a small material database to identify promising design regions for future experimentation. This approach aims to provide quantitative guidance to focus experimental work based on the collective work on the community. Ideally, these future experiments will fully characterize all design variables to enable regression and machine learning approaches. The metrics, r_s and F_r have been proposed to describe the relative importance of process variable combinations in determining material performance. Design regions with smaller r_s and F_r values should be prioritized for future experiments. Overall, the analysis presented is general and can be expanded to examine any subset of relevant design variables.

Several opportunities and obstacles still await the widespread use of literature-guided materials databases. Once the database has been reduced to a targeted design region that contains the 'high' performing data points, justifications for the presence of 'low' data points in this region can be suggested. In the polypropylene-talc case study, the three-dimensional promising region contains six high performing points, but also nine points with a low performance. Attempts to classify based on additional design parameters were unsuccessful. The impact factor of the journal was added as an additional classifier, which did separate the three studies containing high performance from the three studies containing low performance. With a more complex database, the incorporation of additional variables has the potential to distinguish and inform more reliable and robust datasets. A potentially problematic subset of data, termed 'hierarchical data', is prominent amongst materials literature. Hierarchical data is defined here as paired categorial and numerical data, in which the numerical data is only relevant for that category. An example from the P3HT case study is the spin rate that is only relevant to spin cast P3HT devices, rather than dip coated devices, which are quantified instead by the dip rate. Hierarchical data requires special attention to ensure that the numerical data is not used as a predictor without the associated categorical variables.

The most significant of challenges is the handling of missing data to ensure quality predictors that reflect all of the relevant publications and not just the most-well characterized. As discussed, mining data from highly exploratory work can result in unreported data from a lack of standardized reporting templates, the discovery of new important promising conditions and/or access to various equipment and characterization techniques. Data imputation techniques have been recently developed to fill in missing values. These techniques range in complexity from imputing the population mean value to modified nonlinear iterative partial least squares regression. The latter approach has been shown to be an effective technique when the missing data is randomly distributed, rather than in structured blocks [53]. Research by Nelson et al. and Ferrer et al. has shown that principal component analysis approaches are better suited to structured blocks of missing data that may be more prevalent when developing literature databases [54,55]. Data imputation has been applied to materials datasets but, to our knowledge, not been used on literature databases. For example, Verpoort et al. trained an artificial neural network to identify erroneous values and impute missing data based on polymer composite properties provided by manufacturers' datasheets [56]. However, only eight out of thousands of data points were missing and imputed, a marked difference to the amount of missing data found in the literature databases. Secondary approaches to handle missing data involve physical models, where known data serves as the input to an established model to predict the missing values. In the talc case study, experimental data relating polymer molecular weight to MFI was highlighted as an example [51]. Relating the spin rate to predict missing film thickness values could serve as an example for the P3HT case study. However, accurate physical based models may require additional experiments for validation.

The development of a material database from literature data is still largely a manual process, with numerous decisions made by the curator [8]. Data must be extracted from text, tables, and graphs, and must be organized in a meaningful manner. At present, automated data extraction tools are in their infancy and their implementation presents a major technological hurdle to researchers not well-versed in machine learning. Manually extracted data will be instrumental in providing training sets for the development of such tools. In pursuit of this goal, it is essential that ontologies and schema remain flexible. Research data does not necessarily cluster into fields with hard boundaries, and as such, the entries from one field's database should be adaptable to other fields' databases.

Finally, experimental databases are dynamic, living documents that should be easily accessible to the community. To fully utilize an experimental database, new experiments must be added, and the analysis needs to be updated on a continual basis. The question of who curates and maintains material databases is still an ongoing question in the materials community. Kalidindi et al. suggested the National Laboratories as a central entity to create a standardized strategy for materials data analysis, but this may only be appropriate in the case of big data [1]. The Protein Databank demonstrates a different model with an independent consortium as the curating body [57]. As small dynamic experimental databases are essentially quantified in literature reviews, the responsibility may rest on individual research groups to compile data for review articles, and on publishers for long term curation.

Supplementary Materials: The following are available online at http://www.mdpi.com/2227-9717/6/7/79/s1, Talc-Database.xlsx and P3HT-Database.xlsx.

Author Contributions: Conceptualization, M.M. and M.A.G.; methodology, M.M. and M.A.G.; formal analysis, M.M.; data curation, M.M. and N.P.; writing (original draft preparation), M.M. and M.A.G.; writing (review and editing), N.P. and E.R.; funding acquisition, E.R. and M.A.G.

Funding: This research was funded by financial support from Konica Minolta.

Acknowledgments: The authors thank Carson Meredith, Guoyan Zhang, Zihao Li, Jun Amano, and Leiming Wang for helpful discussions on the polypropylene–talc case study.

Conflicts of Interest: The authors declare no conflict of interest.

References

1. Kalidindi, S.R.; De Graef, M. Materials data science: Current status and future outlook. *Annu. Rev. Mater. Res.* **2015**, *45*, 171–193. [CrossRef]
2. Citrine Informatics. Available online: http://www.citrination.com (accessed on 7 June 2018).
3. Calphad (Computer Coupling of Phase Diagrams and Thermochemistry). Available online: http://www.calphad.org (accessed on 7 June 2018).
4. The Materials Project. Available online: http://www.materialsproject.org (accessed on 7 June 2018).
5. Open Quantum Materials Database. Available online: http://oqmd.org (accessed on 7 June 2018).
6. Nist (National Institute of Standards and Technology) Data Gateway. Available online: http://srdata.nist.gov/gateway/gateway?dblist=1 (accessed on 7 June 2018).
7. Casciato, M.J.; Vastola, J.T.; Lu, J.C.; Hess, D.W.; Grover, M.A. Initial experimental design methodology incorporating expert conjecture, prior data, and engineering models for deposition of iridium nanoparticles in supercritical carbon dioxide. *Ind. Eng. Chem. Res.* **2013**, *52*, 9645–9653. [CrossRef]
8. Kim, E.; Huang, K.; Saunders, A.; McCallum, A.; Ceder, G.; Olivetti, E. Materials synthesis insights from scientific literature via text extraction and machine learning. *Chem. Mater.* **2017**, *29*, 9436–9444. [CrossRef]
9. Agrawal, A.; Deshpande, P.; Cecen, A.; Basavarsu, G.; Choudhary, A.; Kalidindi, S. Exploration of data science techniques to predict fatigue strength of steel from composition and processing parameters. *Integr. Mater. Manuf. Innov.* **2014**, *3*, 1–19. [CrossRef]
10. Matnavi Nims Materials Database. Available online: http://mits.nims.go.jp/index_en.html (accessed on 7 June 2018).
11. Ren, F.; Ward, L.; Williams, T.; Laws, K.J.; Wolverton, C.; Hattrick-Simpers, J.; Mehta, A. Accelerated discovery of metallic glasses through iteration of machine learning and high-throughput experiments. *Sci. Adv.* **2018**, *4*, 1–11. [CrossRef] [PubMed]
12. AbuOmar, O.; Nouranian, S.; King, R.; Bouvard, J.L.; Toghiani, H.; Lacy, T.E.; Pittman, C.U. Data mining and knowledge discovery in materials science and engineering: A polymer nanocomposites case study. *Adv. Eng. Inform.* **2013**, *27*, 615–624. [CrossRef]

13. Zhang, S.L.; Zhang, Z.X.; Xin, Z.X.; Pal, K.; Kim, J.K. Prediction of mechanical properties of polypropylene/waste ground rubber tire powder treated by bitumen composites via uniform design and artificial neural networks. *Mater. Des.* **2010**, *31*, 1900–1905. [CrossRef]

14. Ling, J.; Hutchinson, M.; Antono, E.; Paradiso, S.; Meredig, B. High-dimensional materials and process optimization using data-driven experimental design with well-calibrated uncertainty estimates. *Integr. Mater. Manuf. Innov.* **2017**, *6*, 207–217. [CrossRef]

15. Park, J.; Howe, J.D.; Sholl, D.S. How reproducible are isotherm measurements in metal–organic frameworks? *Chem. Mater.* **2017**, *29*, 10487–10495. [CrossRef]

16. Persson, N.; McBride, M.; Grover, M.; Reichmanis, E. Silicon valley meets the ivory tower: Searchable data repositories for experimental nanomaterials research. *Curr. Opin. Solid State Mater. Sci.* **2016**, *20*, 338–343. [CrossRef]

17. Box, G.E.; Wilson, K.B. On the experimental attainment of optimum conditions. *J. R. Stat. Soc.* **1951**, *13*, 1–45.

18. Montgomery, D.C. *Design and Analysis of Experiments*, 7th ed.; Wiley: New York, NY, USA, 2009.

19. Kim, S.; Kim, H.; Lu, J.-C.; Casciato, M.J.; Grover, M.A.; Hess, D.W.; Lu, R.W.; Wang, X. Layers of experiments with adaptive combined design. *Nav. Res. Logist.* **2015**, *62*, 127–142. [CrossRef]

20. Dimitrakopoulos, C.D.; Malenfant, P.R.L. Organic thin film transistors for large area electronics. *Adv. Mater.* **2002**, *14*, 99–117. [CrossRef]

21. Persson, N.E.; Chu, P.H.; McBride, M.; Grover, M.; Reichmanis, E. Nucleation, growth, and alignment of poly(3-hexylthiophene) nanofibers for high-performance ofets. *Acc. Chem. Res.* **2017**, *50*, 932–942. [CrossRef] [PubMed]

22. Persson, N.; McBride, M.; Grover, M.; Reichmanis, E. Automated analysis of orientational order in images of fibrillar materials. *Chem. Mater.* **2016**, *29*, 3–14. [CrossRef]

23. Persson, N.E.; Rafshoon, J.; Naghshpour, K.; Fast, T.; Chu, P.H.; McBride, M.; Risteen, B.; Grover, M.; Reichmanis, E. High-throughput image analysis of fibrillar materials: A case study on polymer nanofiber packing, alignment, and defects in organic field effect transistors. *ACS Appl. Mater. Interfaces* **2017**, *9*, 36090–36102. [CrossRef] [PubMed]

24. Shubhra, Q.T.H.; Alam, A.; Quaiyyum, M.A. Mechanical properties of polypropylene composites. *J. Thermoplast. Compos. Mater.* **2011**, *26*, 362–391. [CrossRef]

25. Ahmed, S.; Jones, F.R. A review of particulate reinforcement theories for polymer composites. *J. Mater. Sci.* **1990**, *25*, 4933–4942. [CrossRef]

26. Paul, D.R.; Robeson, L.M. Polymer nanotechnology: Nanocomposites. *Polymer* **2008**, *49*, 3187–3204. [CrossRef]

27. Samuels, R.J. Polymer structure: The key to process-property control. *Polym. Eng. Sci.* **1985**, *25*, 864–874. [CrossRef]

28. Premalal, H.; Ismail, H.; Baharin, A. Comparison of the mechanical properties of rice husk powder filled polypropylene composites with talc filled polypropylene composites. *Polym. Test.* **2002**, *21*, 833–839. [CrossRef]

29. Pukanszky, B.; Belina, K.; Rockenbauer, A.; Maurer, R.H.J. Effect of nucleation, filler anisotropy and orientation on the properties of pp composites. *Composites* **1993**, *3*, 205–214.

30. Rivnay, J.; Mannsfeld, S.C.; Miller, C.E.; Salleo, A.; Toney, M.F. Quantitative determination of organic semiconductor microstructure from the molecular to device scale. *Chem. Rev.* **2012**, *112*, 5488–5519. [CrossRef] [PubMed]

31. Arias, A.C.; MacKenziew, J.D.; McCulloch, I.; Rivnay, J.; Salleo, A. Materials and applications for large area electronics: Solution-based approaches. *Chem. Rev.* **2010**, *110*, 3–24. [CrossRef] [PubMed]

32. Sirringhaus, H.; Brown, P.J.; Friend, R.H.; Nielsen, M.; Bechgaard, K.; Langeveld-Voss, B.; Spiering, A.; Janssen, R.; Meijer, E.; Herwig, P.; et al. Two-dimensional charge transport in self-organized, high-mobility conjugated polymers. *Nature* **1999**, *401*, 685–688. [CrossRef]

33. Kline, R.; McGehee, M.; Kadnikova, E.; Liu, J.; Frechet, J.; Toney, M.F. Dependence of regioregular poly(3-hexylthiophene) film morphology and field-effect mobility on molecular weight. *Macromolecules* **2005**, *38*, 3312–3319. [CrossRef]

34. Bronstein, H.A.; Luscombe, C.K. Externally initiated regioregular p3ht with controlled molecular weight and narrow polydispersity. *J. Am. Chem. Soc.* **2009**, *131*, 12894–12895. [CrossRef] [PubMed]

35. Horowitz, G. Organic field effect transistors. *Adv. Mater.* **1998**, *10*, 365–377. [CrossRef]

36. Choi, D.; Chu, P.-H.; McBride, M.; Reichmanis, E. Best practices for reporting organic field effect transistor device performance. *Chem. Mater.* **2015**, *27*, 4167–4168. [CrossRef]
37. Raccuglia, P.; Elbert, K.C.; Adler, P.D.; Falk, C.; Wenny, M.B.; Mollo, A.; Zeller, M.; Friedler, S.A.; Schrier, J.; Norquist, A.J. Machine-learning-assisted materials discovery using failed experiments. *Nature* **2016**, *533*, 73–76. [CrossRef] [PubMed]
38. Kline, R.; McGehee, M.; Kadnikova, E.; Liu, J.; Frechet, J. Controlling the field-effect mobility of regioregular polythiophene by changing the molecular weight. *Adv. Mater.* **2003**, *15*, 1519–1522. [CrossRef]
39. Zen, A.; Pfaum, J.; Hirschmann, S.; Zhuang, W.; Jaiser, F.; Asawapirom, U.; Rabe, J.; Scherf, U.; Neher, D. Effect of molecular weight and annealing of poly(3-hexylthiophene)s on the performance or organic field-effect transistors. *Adv. Funct. Mater.* **2004**, *14*. [CrossRef]
40. Himmelberger, S.; Vandewal, K.; Fei, Z.; Heeney, M.; Salleo, A. Role of molecular weight distribution on charge transport in semiconducting polymers. *Macromolecules* **2014**, *47*, 7151–7157. [CrossRef]
41. Scharsich, C.; Lohwasser, R.; Sommer, M.; Asawapirom, U.; Scherf, U.; Thelakkat, M.; Neher, D.; Köhler, A. Control of aggregate formation in poly(3-hexylthiophene) by solvent, molecular weight, and synthetic method. *J. Polym. Sci. Part B Polym. Phys.* **2012**, *50*, 442–453. [CrossRef]
42. Chang, M.; Lim, G.; Park, B.; Reichmanis, E. Control of molecular ordering, alignment, and charge transport in solution-processed conjugated polymer thin films. *Polymers* **2017**, *9*, 212. [CrossRef]
43. Chang, M.; Choi, D.; Fu, B.; Reichmanis, E. Solvent based hydrogen bonding: Impact on poly(3-hexylthiophene) nanoscale morphology and charge transport characteristics. *ACS Nano* **2013**, *7*, 5402–5413. [CrossRef] [PubMed]
44. Roesing, M.; Howell, J.; Boucher, D. Solubility characteristics of poly(3-hexylthiophene). *J. Polym. Sci. Part B Polym. Phys.* **2017**. [CrossRef]
45. Choi, D.; Chang, M.; Reichmanis, E. Controlled assembly of poly(3-hexylthiophene): Managing the disorder to order transition on the nano- through meso-scales. *Adv. Funct. Mater.* **2015**, *25*, 920–927. [CrossRef]
46. Verilhac, J.; LeBlevennec, G.; Djurado, D.; Rieutord, F.; Chouiki, M.; Travers, J.; Pron, A. Effect of macromolecular parameters and processing conditions on supramolecular organisation, morphology and electrical transport properties in thin layers of regioregular poly(3-hexylthiophene). *Synth. Met.* **2006**, *156*, 815–823. [CrossRef]
47. Joshi, S.; Grigorian, S.; Pietsch, U.; Pingel, P.; Zen, A.; Neher, D.; Scherf, U. Thickness dependence of the crystalline structure and hole mobility in thin films of low molecular weight poly(3-hexylthiophene). *Macromolecules* **2008**, *41*, 6800–6808. [CrossRef]
48. Na, J.Y.; Kang, B.; Sin, D.H.; Cho, K.; Park, Y.D. Understanding solidification of polythiophene thin films during spin-coating: Effects of spin-coating time and processing additives. *Sci. Rep.* **2015**, *5*, 13288. [CrossRef] [PubMed]
49. Chu, P.H.; Kleinhenz, N.; Persson, N.; McBride, M.; Hernandez, J.; Fu, B.; Zhang, G.; Reichmanis, E. Toward precision control of nanofiber orientation in conjugated polymer thin films: Impact on charge transport. *Chem. Mater.* **2016**, *28*, 9099–9109. [CrossRef]
50. Chang, M.; Choi, D.; Egap, E. Macroscopic alignment of one-dimensional conjugated polymer nanocrystallites for high-mobility organic field-effect transistors. *ACS Appl. Mater. Interfaces* **2016**, *8*, 13484–13491. [CrossRef] [PubMed]
51. Bermner, T.; Rudin, A. Melt flow index values and molecular weight distributions of commercial thermoplastics. *J. Appl. Polym. Sci.* **1990**, *41*, 1617–1627. [CrossRef]
52. Gafur, M.A.; Nasrin, R.; Mina, M.F.; Bhuiyan, M.A.H.; Tamba, Y.; Asano, T. Structures and properties of the compression-molded istactic-polypropylene/talc composites: Effect of cooling and rolling. *Polym. Degrad. Stab.* **2010**, *95*, 1818–1825. [CrossRef]
53. Nelson, P.R.C. Treatment of Missing Measurements in PCA and PLS Models. Ph.D. Thesis, Master University, Hamilton, ON, Canada, 2002.
54. Nelson, P.R.C.; Taylor, P.A.; MacGregor, J.F. Missing data methods in pca and pls: Score calculations with incomplete observations. *Chemom. Intell. Lab. Syst.* **1996**, *35*, 45–65. [CrossRef]
55. Folch-Fortuny, A.; Arteaga, F.; Ferrer, A. Pca model building with missing data: New proposals and a comparative study. *Chemom. Intell. Lab. Syst.* **2015**, *146*, 77–88. [CrossRef]

56. Verpoort, P.C.; MacDonald, P.; Conduit, G.J. Materials data validation and imputation with an artificial neural network. *Comput. Mater. Sci.* **2018**, *147*, 176–185. [CrossRef]

57. Berman, H.M.; Westbrook, J.; Feng, Z.; Gilliland, G.; Bhat, T.N.; Weissig, H.; Shindyalov, I.N.; Bourne, P.E. The Protein Data Bank. Available online: https://www.rcsb.org/ (accessed on 7 June 2018).

Article

The Impact of Global Sensitivities and Design Measures in Model-Based Optimal Experimental Design

René Schenkendorf [1,2,*,†], Xiangzhong Xie [1,2,3], Moritz Rehbein [2,4], Stephan Scholl [2,4] and Ulrike Krewer [1,2]

1   Institute of Energy and Process Systems Engineering, TU Braunschweig, Franz-Liszt-Straße 35, 38106 Braunschweig, Germany; x.xie@tu-braunschweig.de (X.X.); u.krewer@tu-braunschweig.de (U.K.)
2   Center of Pharmaceutical Engineering (PVZ), TU Braunschweig, Franz-Liszt-Straße 35a, 38106 Braunschweig, Germany
3   International Max Planck Research School (IMPRS) for Advanced Methods in Process and Systems Engineering, Sandtorstraße 1, 39106 Magdeburg, Germany
4   Institute for Chemical and Thermal Process Engineering, TU Braunschweig, Langer Kamp 7, 38106 Braunschweig, Germany; moritz.rehbein@tu-braunschweig.de (M.R.); s.scholl@tu-braunschweig.de (S.S.)
*   Correspondence: r.schenkendorf@tu-braunschweig.de; Tel.: +49-531-391-65601
†   Current address: Institute of Energy and Process Systems Engineering, TU Braunschweig, Franz-Liszt-Straße 35, 38106 Braunschweig, Germany.

Received: 14 February 2018; Accepted: 19 March 2018; Published: 21 March 2018

Abstract: In the field of chemical engineering, mathematical models have been proven to be an indispensable tool for process analysis, process design, and condition monitoring. To gain the most benefit from model-based approaches, the implemented mathematical models have to be based on sound principles, and they need to be calibrated to the process under study with suitable model parameter estimates. Often, the model parameters identified by experimental data, however, pose severe uncertainties leading to incorrect or biased inferences. This applies in particular in the field of pharmaceutical manufacturing, where usually the measurement data are limited in quantity and quality when analyzing novel active pharmaceutical ingredients. Optimally designed experiments, in turn, aim to increase the quality of the gathered data in the most efficient way. Any improvement in data quality results in more precise parameter estimates and more reliable model candidates. The applied methods for parameter sensitivity analyses and design criteria are crucial for the effectiveness of the optimal experimental design. In this work, different design measures based on global parameter sensitivities are critically compared with state-of-the-art concepts that follow simplifying linearization principles. The efficient implementation of the proposed sensitivity measures is explicitly addressed to be applicable to complex chemical engineering problems of practical relevance. As a case study, the homogeneous synthesis of 3,4-dihydro-1*H*-1-benzazepine-2,5-dione, a scaffold for the preparation of various protein kinase inhibitors, is analyzed followed by a more complex model of biochemical reactions. In both studies, the model-based optimal experimental design benefits from global parameter sensitivities combined with proper design measures.

Keywords: optimal experimental design; global parameter sensitivities; optimal design measures; robustification; point estimate method

1. Introduction

Mathematical models have been proven to be an indispensable tool in chemical engineering research and design. For instance, model-based and computer-aided concepts are the standard in process analysis [1–3], process design [4–6], and condition monitoring [7–9]. To gain the most benefit of model-based approaches, the model-building process itself becomes a crucial step in computer-aided process analysis and design. When it comes to mathematical modeling and model-based design, model developers are challenged by the problem of providing meaningful results. The situation is the same for the computer-aided process design of pharmaceutical processes and active pharmaceutical ingredients (API) syntheses [10,11]. In addition to a proper model structure determining the interconnection of involved quantities and species, the most critical part is to identify related model parameters [12,13]. When assuming a correct model structure, any mismatch of measurement data and simulation results is mainly attributed to biased model parameters. Optimal tuning of those model parameters according to the experimental data is an essential step in the process of model development where model parameters are iteratively adapted until simulation results fit the given measurement data best [12,14]. Thus, the quality of model-based results depends significantly on the quality of the parameter estimates. The parameter quality itself, however, depends on the quality of the measurement data and the conducted experiments, respectively. The operating conditions of the experiment indirectly determine the sensitivity of the model parameters to the simulation results. The parameters with a high sensitivity are likely to be reliably identified. Model parameters with low sensitivity, in turn, can be changed by an order of magnitude with little to no effect. Any attempt at a manual or algorithmic tuning of such insensitive parameters is error prone by definition. To avoid this kind of ill-posed parameter identification problem, model-based optimal experimental design (MB-OED) comes into play [14–17]. The aim of MB-OED is to redesign the experimental setup ensuring high sensitivities of all parameters over the course of the experimental run. To this end, a dynamic optimization problem has to be solved, which typically includes the following generic steps; see also Figure 1 for their interaction. First, the parameter sensitivities of a given model candidate have to be quantified. Based on the parameter sensitivities, a suitable design measure is defined and evaluated, which translates parameter sensitivities into algorithmically evaluable quantities. After specifying potential design variables of the experimental setup, which are practically feasible and allow the control of the main process states, an optimizer routine adapts the experimental setup iteratively. The resulting optimal design variables determine the setup of the next experimental run, which is expected to produce new informative measurement data. These data, in turn, represent operating conditions at which, in general, the model parameters show an increased sensitivity leading to more precise parameter estimates. This MB-OED loop is reiterated until the parameter estimates fulfill the needs of particular modeling tasks. While following the proposed work flow, i.e., (1) sensitivity quantification, (2) design measure evaluation, (3) dynamic optimization, and (4) running an optimally designed experiment, the final result depends critically on the sensitivity measure and its proper implementation. Any improper choice at this point will lead to sub-optimal experimental designs and non-informative data. Thus, reliable sensitivity measures are mandatory for MB-OED to provide the most informative data possible. This aspect is particularly relevant for syntheses of active pharmaceutical ingredients, where typically the number of novel drugs and the number of experimental runs are limited.

In the literature, most implementations of MB-OED are based on local sensitivities as part of the Fisher information matrix (FIM) [12,14,16,18–20]. Technically, the inverse of the FIM is used to approximate the covariance matrix of the parameter estimates [12,14,18]. This local approach assumes a linear relation between parameter perturbations and model responses—at least locally in the neighborhood of the correct reference parameter values. As the actual parameter values, however, are typically unknown, their current best estimates have to be inserted instead. Thus, the calculated local sensitivities may fall short as they do not describe the underlying nonlinearities and because of biased reference parameters [16,21]. To compensate for biased references, sensitivity gradients can be evaluated at various but representative parameter values within the given parameter domain following a Bayesian experimental design principle [22–24]. Here, the basic idea is to get an improved approximation of the local sensitivities based on an averaged measure [25]. Nonlinear effects, however, still remain unaddressed. Most parameter identification problems of practical relevance in the field of pharmaceutical processes are characterized by nonlinear impacts of model parameters on model responses [19]. Thus, local sensitivities, in general, might lead to oversimplified inferences, less robust experimental designs, and an unnecessarily high number of experimental reiterations [21]. This last point, as mentioned previously, is of critical relevance when analyzing novel but exceedingly rare drug candidates.

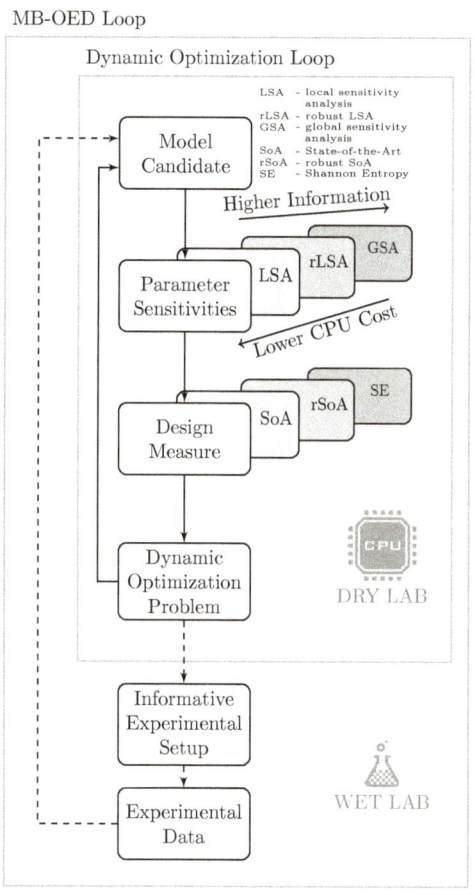

Figure 1. Key elements of the model-based optimal experimental design (MB-OED): the parameter sensitivity approach and the implemented design measure determine the quality of the designed experiment significantly.

Alternatively, in the last decade, the implementation of global sensitivities for MB-OED has been introduced in the literature [25–27]. Global sensitivities by definition take into account nonlinear effects and multivariate parameter dependencies. To the best of the authors' knowledge, the usefulness of GSA and its practical implementation for pharmaceutical and (bio)chemical processes have not been analyzed thus far. This manuscript shows how global sensitivities can be determined efficiently by a deterministic sampling rule even for complex models with a large number of model parameters, which makes GSA available for advanced MD-OED strategies.

The focus of this study is to compare and assess different parameter sensitivity measures, as well as an analysis of the need for novel design criteria for MB-OED. In particular, local sensitivity analysis (LSA), robust local sensitivity analysis (rLSA), and global sensitivity analysis (GSA) are introduced. In addition to state-of-the-art (SoA) design criteria, further alternatives are discussed in the following. Thus, the paper unfolds as follows. First, the basics of parameter sensitivity analysis as a key element of optimal experimental design are presented in Section 2. Here, the focus is also on the efficient implementation of global parameter sensitivities. Based on the derived sensitivity measures, optimal design metrics are derived in Section 3. In Section 4, the proposed concept is illustrated by two pharmaceutically relevant processes, the homogeneous synthesis of 3,4-dihydro-1*H*-1-benzazepine-2,5-dione, a scaffold for the preparation of various protein kinase inhibitors, and a more complex biochemical synthesis example. The conclusions are given in Section 5.

2. Sensitivity Measures

In the literature, different methods and concepts for parameter sensitivity analysis (SA) exist, ranging from local to global approaches [28–30]. In the context of MD-OED, sensitivity analysis is an inherent part of the optimization problem aiming to provide an optimal range of sensitivities over the course of the experimental run. In the following, we consider nonlinear state-space models of the form

$$\frac{dx(t)}{dt} = f(x(t), u(t), \theta),$$ (1)
$$y(t) = h(x(t), u(t), \theta),$$ (2)

where $x(t)$ is the vector of model states ($x \in \mathbb{R}^n$), $u(t)$ is the vector of system inputs ($u \in \mathbb{R}^r$), θ is the vector of parameters ($\theta \in \mathbb{R}^p$), and $y(t)$ is the vector of model outputs ($y \in \mathbb{R}^m$). Typically, only a minor subset of the states $x(t)$ can be measured directly or indirectly; i.e., $m < n$. The proposed concepts can be extended to more general model types, e.g., differential algebraic or partial differential equation systems, but are discussed here solely for nonlinear state space models for the sake of clarity. As the model parameters θ are typically unknown, they have to be identified by minimizing the error between simulation results, $y(t)$, and measurement data, $y^{\text{data}}(t)$. By following a maximum likelihood estimation procedure and assuming additive white measurement noise with a standard deviation of one [14], the parameter identification problem simplifies to

$$\hat{\theta} = \arg\min_{\theta} \sum_{i=1}^{K} ||y^{\text{data}}(t_i) - y(t_i)||^2,$$ (3)

where K is the number of measurement time points. According to the measurement noise and the Doob–Dynkin lemma [31], the identified model parameters $\hat{\theta}$ are random variables as well; for example, measurement uncertainties induce stochastic parameter uncertainties. The resulting parameter uncertainties, in turn, are determined by the data quality and parameter sensitivities alike.

2.1. Local Parameter Sensitivities

In practice, local sensitivities are the standard for analyzing the effect of parameter variations on model-based results [28,29,32]. Local sensitivities S^{LSA} are typically expressed by partial derivatives of the j-th model output, $y[j](t)$, with respect to the i-th model parameters, $\theta[i]$, as

$$S^{\text{LSA}}[i, j](t) = \frac{\partial y[j](t)}{\partial \theta[i]}$$ (4)

and summarized by the sensitivity matrix $S^{\text{LSA}}(t) \in \mathbb{R}^{m \times p}$ while $j = 1, \dots, m$ and $i = 1, \dots, p$. In the field of dynamical systems, the governing equations of the original model (Equation (2)) have to be extended by sensitivity terms of the states, S_x^{LSA}, which are solved in parallel according to

$$\dot{S}_x^{\text{LSA}} = \frac{\partial f}{\partial x} \cdot S_x^{\text{LSA}} + \frac{\partial f}{\partial \theta}; \quad S_x^{\text{LSA}}(t=0) = 0_{n \times p}. \tag{5}$$

Thus, the output sensitivities can be expressed as

$$S^{\text{LSA}} = \frac{\partial h}{\partial x} \cdot S_x^{\text{LSA}} + \frac{\partial h}{\partial \theta}. \tag{6}$$

Alternatively, the gradients (Equation (4)) can be calculated with various numerical methods, e.g., automatic differentiation or finite differences. Here, the interested reader is referred to [32,33] and references therein.

Technically, local sensitivities provide information about the relevance of model parameters in the neighborhood of a reference point within the defined parameter space, a reference which is typically unknown for most practical problems. In an ideal world, hypothetically, the reference point corresponds to the exact model parameters. Moreover, the general idea of local SA measures is based on linearization principles; i.e., the original, nonlinear model (Equation (2)) is linearized at the (unknown) reference point. Thus, nonlinear effects are ignored although the analyzed models may have strong nonlinearities involved [21]. The resulting approximation errors may render the outcome of such a sensitivity analysis meaningless and difficult to interpret in general. To compensate for the dependency of unknown reference points, a multi-point averaging approach can be applied as described in the following sections.

2.2. Robustification: Multi-Point Averaging Approach

As the reference points for the sensitivity gradients are typically unknown, local sensitivities may provide artifacts; i.e., insensitive parameters are identified as sensitive or vice versa. To avoid this kind of mis-classification, an averaged version of the sensitivity matrix (Equation (4)) seems to be a more robust measure [25,34] following a Bayesian experimental design principle [22–24] and is defined as

$$S^{\text{rLSA}}[i,j](t) = E[S^{\text{LSA}}[i,j](t)] = \int_{\Omega} S^{\text{LSA}}[i,j](t) pdf_{\theta[i]} \, d\theta[i], \tag{7}$$

where $E[\cdot]$ is the expected value, Ω is the support domain, and the probability density function (PDF), $pdf_{\theta[i]}$, represents the assumed parameter variation of the i-th parameter. According to Equation (7), for linear problems, the proposed concept simplifies to the local SA expression [25]. In most practical scenarios, there exists no analytical solution of Equation (7). Thus, Monte Carlo (MC) simulations are evaluated instead to solve a discretized approximation resulting in a multi-point averaging approach equal to

$$S^{\text{rLSA}}[i,j](t) = E[S^{\text{LSA}}[i,j](t)] \approx \sum_{k=1}^{N} S^{\text{LSA}}[i,j](\theta_k[i],t), \tag{8}$$

where $\theta_k[i]$ represents the k-th realization of the i-th parameter according to the given PDF. The sample number, N, determines the total number of function calls of Equation (5) needed to approximate the averaged sensitivity measure, i.e., for each parameter realization, θ_k, Equation (5) has to be vectorized and solved in parallel with the dynamic model (Equation (1)) that is also evaluated for θ_k. By using S^{rLSA} instead of the standard local quantities S^{LSA}, the resulting parameter sensitivities are more representative of nonlinear problems [34–36] and may provide a reliable base for subsequent MB-OED studies. As standard MC simulations might become prohibitive in terms of computational load, efficient sample-based approaches are mandatory and will be discussed in more detail in Section 2.4.

The multivariate interaction, however, of the analyzed model parameters is still not taken into account appropriately by the multi-point averaging concept. Here, the framework of global sensitivities seems to be more suitable to rank vaguely known parameters and to quantify their interactions.

2.3. Global Parameter Sensitivities

Global sensitivity measures treat parameters, θ, and model outcomes, $y(t)$, as random variables and aim to quantify the amount of variance that each parameter, $\theta[i]$, adds to the total variance of the output, $\sigma^2(y(t))$ [29,37,38]. In detail, the parameter ranking is done by the amount of output variance that disappears when the i-th parameter $\theta[i]$ is assumed to be known, $\sigma^2(\theta[i]) = 0$. For this particular parameter, a conditional variance, $\sigma_{-i}^2(y(t)|\theta[i])$, can be derived. Here, the subscript $-i$ indicates that the variance is taken over all parameters other than $\theta[i]$. As $\theta[i]$, in reality, is a random variable, the expected value of the resulting conditional variance, $E_i\left[\sigma_{-i}^2(y(t)|\theta[i])\right]$, has to be analyzed. The subscript notation of E_i indicates that the expected value is taken only over the parameter $\theta[i]$ itself. Finally, the total output variance, $\sigma^2(y(t))$, can be split into two additive terms [29] equal to

$$\sigma^2(y(t)) = \sigma_i^2(E_{-i}[y(t)|\theta[i]]) + E_i[\sigma_{-i}^2(y(t)|\theta[i])]. \tag{9}$$

Here, the variance of the conditional expectation, $\sigma_i^2(E_{-i}[y(t)|\theta[i]])$, represents the contribution of parameter $\theta[i]$ to the total variance $\sigma^2(y(t))$. The normalized expression given in Equation (10) is known as the first-order Sobol sensitivity index [29] and is used in the following to analyze parameter sensitivities and multivariate parameter interactions, respectively:

$$S^{GSA}[i](t) = \frac{\sigma_i^2(E_{-i}[y(t)|\theta[i]])}{\sigma^2(y(t))}. \tag{10}$$

Similar to the multi-point averaging approach (Section 2.2), the integral terms associated with $\sigma^2(y(t))$, $E_{-i}[y(t)|\theta[i]]$, and $\sigma^2(y(t)|\theta[i])$ are commonly evaluated with MC simulations [39]. For global sensitivity studies of complex models, MC simulations have prohibitive computational costs. Thus, while implementing an MB-OED strategy based on global sensitivities, highly efficient methods have to be used. The most relevant concepts in this direction are outlined in the next section.

2.4. Implementation Aspects

Over the last two decades, various methods have been developed to calculate Sobol indices efficiently. The most frequently used approaches are based on advanced MC simulations and polynomial chaos expansion (PCE) principles [40]. In comparison to ordinary MC simulations, advanced sampling concepts aim for a better convergence rate by an improved space-filling sampling strategy, i.e., to avoid clustering effects, as well as gaps in the p-dimensional sampling space [41]. Please note that p represents the dimension of the random vector, i.e., the number of not perfectly known model parameters. For instance, so-called quasi-MC methods have a convergence rate of $\mathcal{O}((\log N)^p/N)$, which for low-dimensional problems can lead to a significant reduction in computational costs in comparison to $\mathcal{O}(1/\sqrt{N})$ for ordinary MC [41]. On the other side, PCE-based approaches aim to replace the original CPU-intensive model with a handy and easy-to-evaluate surrogate model first. Then, some characteristics of the derived PCE models are used to express global sensitivities analytically by using the PCE coefficients directly or to run additional MC simulations at low computational costs [42]. The parameterization of the PCE model, however, requires a significant number of reference simulations of the original model. Both concepts, advanced MC methods and PCE, might be prohibitive for MB-OED problems because of their computational costs. Highly efficient uncertainty handling methods are needed instead. Here, the point estimate method (PEM) provides a fair compromise in terms of CPU load and accuracy [43,44].

The fundamental idea of the PEM is to choose sample points, θ_k, and associated weights, w_k, in accordance with the first central moments of given random variables [43]. However, in comparison to MC simulations, these sample points are generated deterministically and not randomly. Assuming a three-dimensional parameter problem, for instance, the resulting sample set reads as

$$GF(\pm\vartheta) = \begin{bmatrix} \theta[1] \\ \theta[2] \\ \theta[3] \end{bmatrix} + \left\{ \begin{bmatrix} \vartheta \\ 0 \\ 0 \end{bmatrix}, \begin{bmatrix} -\vartheta \\ 0 \\ 0 \end{bmatrix}, \begin{bmatrix} 0 \\ \vartheta \\ 0 \end{bmatrix}, \begin{bmatrix} 0 \\ -\vartheta \\ 0 \end{bmatrix}, \begin{bmatrix} 0 \\ 0 \\ \vartheta \end{bmatrix}, \begin{bmatrix} 0 \\ 0 \\ -\vartheta \end{bmatrix} \right\}. \tag{11}$$

The generator function, $GF(\pm\vartheta)$, creates the sample set by permuting an element $\theta[i]$ of the original random vector θ one by one by the amount of ϑ. Thus, the scaling parameter, ϑ, controls the spread of the sample points in the p-dimensional parameter space. To calculate the expected value, the discretized integration problem reads as

$$E[g(\theta)] = \int_\Omega g(\theta) pdf_\theta d\theta \approx w_0 g(\theta) + w_1 g(GF(\pm\vartheta)), \tag{12}$$

with $w_0 = 1 - p/\vartheta^2$ and $w_1 = 1/2\vartheta^2$ assuming a standard Gaussian distribution. Moreover, the overall precision of the PEM can be increased gradually by considering higher-order central moments of θ. The more precise approximation scheme [44,45] results in

$$\begin{aligned} E[g(\theta)] = \int_\Omega g(\theta) pdf_\theta d\theta \approx w_0 g(\theta) + \\ w_1 g(GF(\pm\vartheta)) + w_2 g(GF(\pm\vartheta, \pm\vartheta)), \end{aligned} \tag{13}$$

where $GF(\pm\vartheta, \pm\vartheta)$ implies the simultaneous variation of two elements of the given parameter vector θ. Thus, the number of generated sample points, θ_k, scales quadratically with $2p^2 + 1$ for a p-dimensional parameter problem: a good balance of computational load and approximation power.

The four unknown coefficients, ϑ, w_0, w_1, and w_2, can be expressed analytically as

$$\vartheta = \sqrt{\frac{\mu_4}{\mu_2}}, \tag{14}$$

$$w_0 = \frac{-2\mu_2^2\mu_4 p + \mu_{2,2}\mu_2^2(p^2 - p) + 2\mu_0\mu_4^2}{2\mu_4^2}, \tag{15}$$

$$w_1 = \frac{\mu_2^2(\mu_4 + \mu_{2,2}(1-p))}{2\mu_4^2}, \tag{16}$$

$$w_2 = \frac{\mu_2^2\mu_{2,2}}{4\mu_4^2}, \tag{17}$$

where $\mu_0 = \int 1 pdf_\theta d\theta$ is the zeroth central moment, $\mu_2 = \int \theta[i]^2 pdf_\theta d\theta$ the second central moment, $\mu_4 = \int \theta[i]^4 pdf_\theta d\theta$ the fourth central moment, and $\mu_{2,2} = \int \theta[i]^2\theta[j \neq i]^2 pdf_\theta d\theta$ a bivariate central moment. Depending on the PDFs and their central moments, different PEM coefficients are derived. For instance, specifications of different PDFs and their corresponding PEM coefficients are given in Tables 1 and 2, respectively. Here, three different PDF types are considered: (i) normal distribution to represent the uncertainty range of well-known parameter variations; (ii) uniform distribution for cases where almost no information about the parameter variation is available; and (iii) triangle distribution to represent expert knowledge of parameter variations; i.e., the most likely parameter values and plausible upper and lower bounds. Within the individual distribution classes, any desired realization can be derived with a linear transformation step without loss of approximation accuracy, while non-symmetric or non-parametric distributions can be represented by iso-probabilistic but nonlinear transformation rules [42], which might introduce additional approximation errors.

Table 1. Statistical central moments of the standard normal ($\mathcal{N}$), uniform ($\mathcal{U}$), and triangle ($\mathcal{T}$) distribution.

Distribution	μ_0	μ_2	μ_4	$\mu_{2,2}$
$\theta[i] \sim \mathcal{N}(0,1), \quad \forall i = 1,\dots,p$	1	1	3	1
$\theta[i] \sim \mathcal{U}(-1,1), \quad \forall i = 1,\dots,p$	1	1/3	3/5	1/9
$\theta[i] \sim \mathcal{T}(-1,1), \quad \forall i = 1,\dots,p$	1	1/6	1/15	1/36

Table 2. PEM parameters according to the standard normal ($\mathcal{N}$), uniform ($\mathcal{U}$), and triangle ($\mathcal{T}$) distribution.

Distribution	ϑ	w_0	w_1	w_2
$\theta[i] \sim \mathcal{N}(0,1), \quad \forall i = 1,\dots,p$	$\sqrt{3}$	$1 + \frac{p^2 - 7p}{18}$	$\frac{4-p}{18}$	$\frac{1}{36}$
$\theta[i] \sim \mathcal{U}(-1,1), \quad \forall i = 1,\dots,p$	$\frac{3\sqrt{5}}{5}$	$1 + \frac{25p^2 - 295p}{1458}$	$\frac{160 - 25p}{1458}$	$\frac{25}{2916}$
$\theta[i] \sim \mathcal{T}(-1,1), \quad \forall i = 1,\dots,p$	$\frac{\sqrt{10}}{5}$	$1 + \frac{25p^2 - 145p}{288}$	$\frac{85 - 25p}{288}$	$\frac{25}{576}$

Once the deterministic sample points have been derived, the mean and the variance of the model simulations can be approximated as

$$E[y(t,\theta)] \approx \bar{y}(t,\theta) = w_0 y(t,\theta_0) + w_1 \sum_{k=1}^{2p} y(t,\theta_k) + w_2 \sum_{l=2p+1}^{2p^2} y(t,\theta_l), \tag{18}$$

$$
\sigma^2(y(t,\theta)) \approx w_0(y(t,\theta_0) - \bar{y}(t,\theta))(y(t,\theta_0) - \bar{y}(t,\theta))^T +
$$
$$
w_1 \sum_{k=1}^{2p} (y(t,\theta_k) - \bar{y}(t,\theta))(y(t,\theta_k) - \bar{y}(t,\theta))^T + \tag{19}
$$
$$
w_2 \sum_{l=2p+1}^{2p^2} (y(t,\theta_l) - \bar{y}(t,\theta))(y(t,\theta_l) - \bar{y}(t,\theta))^T.
$$

Obviously, Equation (18) can directly be applied to approximate the averaged local sensitivity measure (Equation (7)), whereas the denominator of the Sobol indices (Equation (10)) is approximated with Equation (19). The calculation of the conditional variance terms of the numerator in Equation (10) is given in the following.

The overall set of generated sample points of the PEM provides nested subsets of sample points representing the conditional variance terms. According to the numerator of Equation (10), the expected value is taken over all but the i-th parameter; i.e., the ϑ-value of parameter i is set to zero. The variance, however, is taken exclusively for the i-th parameter; i.e., the corresponding ϑ entry is reset to its original distribution-dependent value. Only the weights, w_0 and w_1, have to be adapted to a one-dimensional problem while leaving the overall PEM samples unchanged, i.e., there is no need for additional costly simulation runs. Please note the weight w_2 is set equal to zero; i.e., sample points generated via $GF(\vartheta, \vartheta)$ are not considered at this point. The described procedure is illustrated in Figure 2 assuming a three-dimensional parameter problem. In Figure 2a, the original sample set generated via Equation (11) is shown. Next, in Figure 2b, three different expected values are determined for $E[y(t)|\theta[1]]$ via the three differently colored sample subsets and different $\theta[1]$ values, respectively. When using these three derived expected values for $\theta[1]$, the conditional variance of $\theta[1]$ can be approximated (Equation (10)). The same procedure is repeated for parameter $\theta[2]$ (Figure 2c) and parameter $\theta[3]$ (Figure 2d) to derive the desired Sobol indices. Please note that the derived samples are not used to quantify the information content [16,46,47], but to quantify the uncertainty of the local sensitivities or to directly derive global parameter sensitivities. In [48,49], the performance of the PEM for calculating global parameter sensitivities and uncertainty analysis is discussed in more detail. Here, the PEM provides appropriate approximations at low computational costs compared to Monte Carlo simulations. In the next step, from these sensitivity measures, the most MB-OED-relevant features have to be extracted by defining proper optimal design measures.

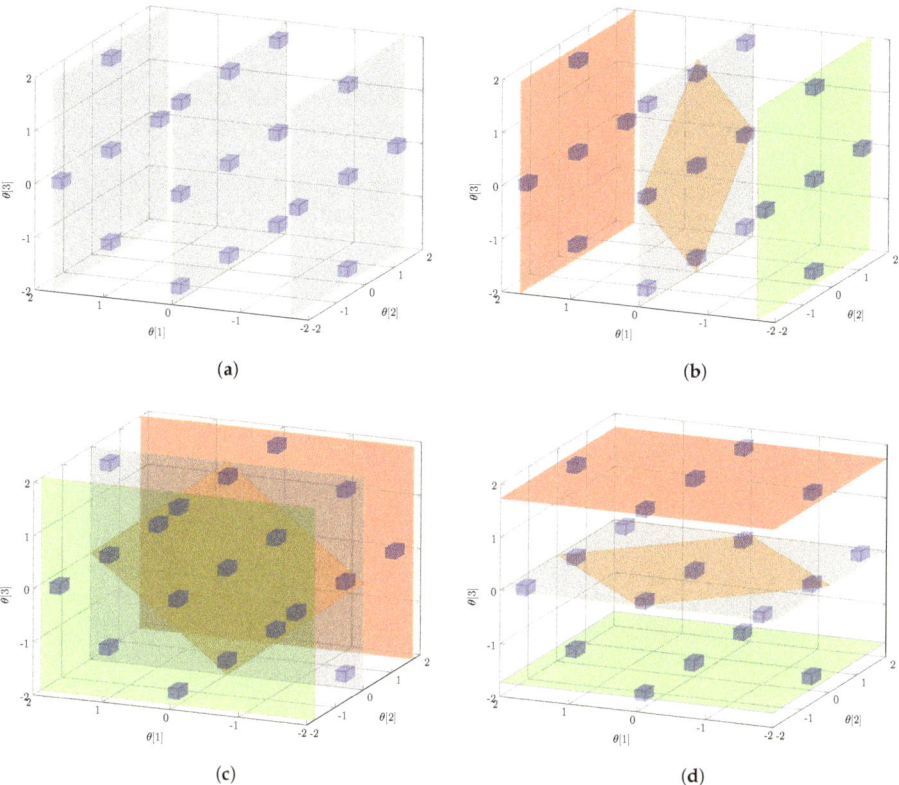

Figure 2. Parameter subset selection: Starting from an unconditioned parameter sample set (**a**), subsets are systematically selected in (**b**–**d**) to determine conditioned expectations and variances, respectively. (**a**) all sample points used to calculate the unconditioned, total uncertainty $\sigma^2(y(t))$; (**b**) sample subsets for ▦ $\theta[1]$, ▦ $\theta[1] + \vartheta$, and ▦ $\theta[1] - \vartheta$; (**c**) sample subsets for ▦ $\theta[2]$, ▦ $\theta[2] + \vartheta$, and ▦ $\theta[2] - \vartheta$; (**d**) sample subsets for ▦ $\theta[3]$, ▦ $\theta[3] + \vartheta$, and ▦ $\theta[3] - \vartheta$.

3. Optimal Design Measures

The actual optimization step of the MB-OED framework calls for a decision criterion that has to express the quality of the analyzed experimental configuration regarding parameter sensitivities. To this end, typically, representative scalar values are chosen for and evaluated with an optimization routine. In the literature, well-known criteria for local sensitivities exist and are summarized below followed by more general global design criteria.

3.1. Local Design Measures

The starting point for most local design criteria is an approximation of the parameter (co)variance matrix, which is defined as

$$C_\theta = E[(\theta - E[\theta])(\theta - E[\theta])^T].$$ (20)

Based on the Cramér–Rao inequality [14] and local sensitivities (Equation (4)), the lower bound of the parameter (co)variance matrix assuming an unbiased estimator reads as

$$C_\theta \geq \text{FIM}^{-1}, \tag{21}$$

where the FIM is given by

$$\text{FIM} = \sum_{t_k}^{K} S^{\text{LSA}}(t_k)^T C_y^{-1}(t_k) S^{\text{LSA}}(t_k). \tag{22}$$

C_y represents the (co)variance matrix of the measurement data, which simplifies to a diagonal matrix assuming uncorrelated data samples. To formulate an optimization problem, a scalar function of the derived parameter (co)variance matrix C_θ is typically evaluated and minimized. Note that the inverse FIM is used as an approximation of the parameter covariance matrix in what follows, $C_\theta \approx \text{FIM}^{-1}$. A common class of these indicators is the so-called alphabetic family of design criteria [14,18] as given in Table 3. Here, λ_{max} and λ_{min} are the maximum and minimum eigenvalues of C_θ, respectively.

Table 3. Local design measures for MB-OED based on the parameter covariance matrix , where E_{LSA}^* refers to the modified E criterion.

Local Design Measures	Cost Functions
A_{LSA}—optimal design	$\Phi_A = trace(C_\theta)$
D_{LSA}—optimal design	$\Phi_D = det(C_\theta)$
E_{LSA}^*—optimal design	$\Phi_{E^*} = \frac{\lambda_{max}(C_\theta)}{\lambda_{min}(C_\theta)}$

Which of these criteria leads to the best optimal experimental design is difficult to predict and depends on the analyzed problem at hand. Moreover, in many practical situations, the involved model parameters are highly correlated; for example, credible parameter estimates are challenging and sometimes impossible to derive for individual parameters. For that reason, dedicated anti-correlation criteria were proposed, aiming to reduce the parameter correlation while increasing the information content of the measurement data at the same time [50]. Here, linear parameter correlation is expressed as

$$Corr[i,j] = \frac{C_\theta[i,j]}{\sqrt{C_\theta[i,i]} \cdot \sqrt{C_\theta[j,j]}}, \tag{23}$$

with $i,j \in \{1,\dots,p\}$. The dominating correlation coefficients are selected directly or added as constraints to parameter variance measures [50], respectively. The basic notation reads as

$$\Phi_{\text{AC}}(Corr) = Corr[i,j]^2 + Corr[k,l]^2, \tag{24}$$

where $Corr[i,j]$ and $Corr[k,l]$ with $i,j,k,l \in \{1,\dots,p\}$ are the two most dominant correlation coefficients. As the (co)variance matrix C_θ is derived by the FIM (Equation (21)), these anti-correlation criteria depend on local parameter sensitivities (Equation (4)), too. As discussed in Section 2.1, local sensitivities may lead to crude approximation errors and sub-optimal experimental designs. In principle, the approximation error can be reduced by the multi-point averaging concept as proposed in Section 2.2 but still ignores nonlinear and multivariate effects as explained below.

While the anti-correlation strategy was successfully demonstrated for chemical and biochemical processes [50,51], it addresses linear parameter correlation solely, which, however, does not imply parameter independence [52]. Thus, nonlinear parameter correlations are neglected, but might be of relevance in MB-OED as well. In Figure 3, parameter dependencies are illustrated showing linear correlation (Figure 3a) and no correlation (Figure 3b) of two model parameters. In Figure 3c, however, the linear correlation coefficient (Equation (23)) is zero, but nonlinear interaction patterns are clearly visible. Ideally, the parameter estimate of one parameter should not impact the estimation outcome of other parameters. The same holds true for multivariate parameter interactions, i.e., the joint effect of more than two parameters that are also not considered by the linear anti-correlation measure introduced in Equation (24). To address nonlinear parameter correlations and multivariate parameter interactions properly, global sensitivities and corresponding optimal design measures have to be analyzed instead.

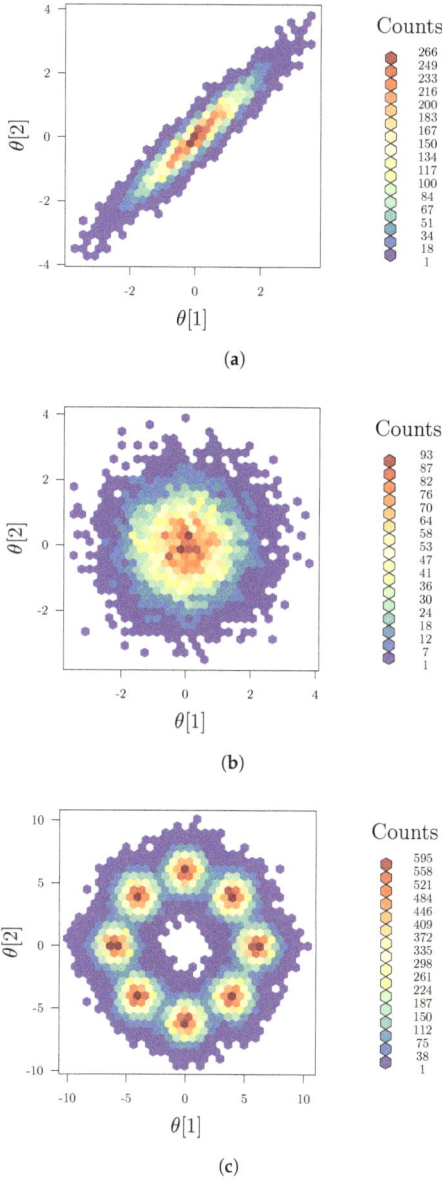

Figure 3. Illustrative example of parameter dependencies: (**a**) linear correlation with $Corr[1,2] = 0.95$; (**b**) no correlation with $Corr[1,2] = 0$; and (**c**) nonlinear correlation but zero linear correlation ($Corr[1,2] = 0$); here, eight standard Gaussian distributions are superimposed in a circular structure and properly rescaled providing a multi-modal density function. (**a**) linear correlated parameters; (**b**) linear uncorrelated parameters; (**c**) nonlinear correlation.

3.2. Global Design Measures

The additional insight given by global sensitivity measures (see Section 2.3) provides a new perspective on MB-OED [19,25,26,53]. As shown in Table 4, local design criteria (Table 3) are typically generalized to hold also for global sensitivities by substituting local sensitivities (Equation (4)) via global sensitivities (Equation (10)) [26,53].

Table 4. Generalized local design measures including global sensitivities.

GSA-Based Design Measures	Cost Functions
A_{GSA}—optimal design	$\Phi_{A_{\text{GSA}}} = trace([S^{\text{GSA}^T} S^{\text{GSA}}]^{-1})$
D_{GSA}—optimal design	$\Phi_{D_{\text{GSA}}} = det([S^{\text{GSA}^T} S^{\text{GSA}}]^{-1})$
E^*_{GSA}—optimal design	$\Phi_{E^*_{\text{GSA}}} = \frac{\lambda_{max}([S^{\text{GSA}^T} S^{\text{GSA}}]^{-1})}{\lambda_{min}([S^{\text{GSA}^T} S^{\text{GSA}}]^{-1})}$

Thus, parameter uncertainties are taken into account naturally. Suboptimal results in MD-OED caused by misspecified reference points for local sensitivities (Equation (4)) can be avoided as much as possible by analyzing the entire PDF of the analyzed model parameters. Moreover, GSA provides additional insight into parameter dependencies, which can be included in the experimental design criteria as well. One attempt in this direction was made recently in [53], where so-called additive sensitivity indices are introduced as modified Sobol indices. Here, the key idea is to account for multivariate parameter interactions and dependencies that impact the overall quality of parameter estimates similar to individual parameter sensitivities. The usage of GSA to express generalized sensitivity matrices for MB-OED may pose some risk as well: (1) Not all GSA-based measures simplify to local measures for linear problems [25], which may lead to sub-optimal experimental designs and counter-intuitive outcomes; (2) including sensitivities of parameter combinations directly in a generalized GSA measure [53] might increase parameter dependencies; i.e., parameter correlations are amplified while limiting individual parameter contributions and, therefore, result in more challenging parameter identification problems; and (3) as the introduced GSA measures are normalized by the total variance factor, the calculated design may lead to an improved parameter sensitivity spectrum, but of lower sensitivities for individual parameters. Thus, the individual Sobol indices are optimized just by decreasing the $\sigma^2(t, y(t))$ term of Equation (10), meaning worse estimates for all model parameters.

To avoid the workaround of transferring global sensitivities into the local framework, the Shannon entropy as a general information measure [54] might be a helpful expression to calculate the desired features for global MB-OED. Intuitively, an optimal experimental design should fulfill the following features: (1) first, after running an experiment, the derived data have to ensure a balanced sensitivity of all parameters; i.e., all parameters should be practically identifiable. Thus, the corresponding Shannon entropy $\Phi_{\text{SE}_{all}}$ of the parameter sensitivities has to be at its maximum and the given criteria at its minimum (Table 5); i.e., there is a uniform distribution of the first-order Sobol indices. Note that $\overline{S^{\text{GSA}}}[i]$ refers to the averaged sensitivities covering all sample time points; (2) at a single time point t_k, however, it is desirable that only a very limited number of parameters contribute at the same time, i.e., to minimize (nonlinear) correlation effects. Thus, the Shannon entropy covering global sensitivities at a single time point $\Phi_{\text{SE}_{t_k}}$ has to be at its minimum; (3) to address multivariate parameter interactions and to minimize them, the gap between the sum of first-order Sobol indices and its theoretical maximum value of one is evaluated Φ_{PD}; see Table 5. Here, a sum of first-order Sobol indices close to one corresponds to low parameter interactions, i.e., a well-posed parameter identification problem; and, (4) finally, the overall output variance Φ_{OU} has to be incorporated as well: the higher the better. A high output variance (Equation (19)) indicates a high parameter sensitivity, which is a prerequisite for credible parameter estimates. Please note that, for consistency, the related cost functions are given as a minimization problems.

Table 5. Global design measures for MB-OED based on Sobol indices and Shannon entropy.

Global Design Measures	Cost Function
(1) Shannon entropy (entire time horizon)	$\Phi_{SE_{all}} = 1/\sum\limits_{i=1}^{p} \overline{S^{GSA}}[i] ln(\overline{S^{GSA}}[i])$
(2) Shannon entropy (at single time point, t_k)	$\Phi_{SE_{t_k}} = \sum\limits_{k=1}^{K} \sum\limits_{i=1}^{p} S^{GSA}{}_{t_k}[i] ln(S^{GSA}{}_{t_k}[i])$
(3) Parameter dependency	$\Phi_{PD} = \sum\limits_{k=1}^{K} (1 - \sum\limits_{i=1}^{p} S^{GSA}{}_{t_k}[i])$
(4) Overall output uncertainty	$\Phi_{OU} = 1/\sum\limits_{k=1}^{K} \sigma^2(y_{t_k})$

In the next step, MB-OED results based on the proposed global design measures are critically compared with the outcome of the traditional local sensitivity measures. To this end, various experimental design conditions are evaluated for two representative case studies in the field of pharmaceutical manufacturing.

4. Case Studies

4.1. Synthesis of an API–Scaffold (DHBD)

During the early stages of API process development, different properties of the unit operations involved are analyzed in order to characterize and to optimize the synthesis as well as the downstream route. Reaction rates are regarded as key descriptors of the synthesis progression as they depend on, temperature, and time. As a first case study for MB-OED, the homogeneous synthesis of 3,4-dihydro-1*H*-1-benzazepine-2,5-dione (DHBD) from the enolized 3-oxocarboxylic ester in wet dimethyl sulfoxide (DMSO) under neutral conditions and at elevated temperatures is presented; see Figure 4.

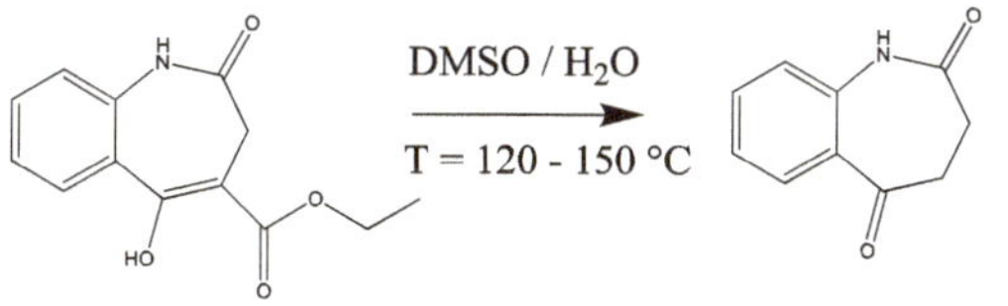

Figure 4. Synthesis of 3,4-dihydro-1*H*-1-benzazepine-2,5-dione (DHBD).

DHBD and its derivatives are pharmaceutically relevant scaffolds utilized for the synthesis of various protein kinase inhibitors and anticancer agents [55–60]. In traditional reaction optimization studies, isothermal syntheses at various temperatures are carried out one by one, and the syntheses are analyzed with offline high-performance liquid chromatography (HPLC) in order to determine reaction kinetics. However, HPLC measurements are tedious in sample preparation and need increased amounts of reactant, which might not be available in the very early stage of API development. Alternatively, we implement the fast-sampling and labor free in situ attenuated total reflectance Fourier transform infrared (ATR-FTIR) spectroscopy in order to quantify the reactant concentration and, subsequently, to calculate reaction rate constants without the need for manual sampling. Combined with non-isothermal temperature profiles during the course of the reaction, temperature-dependent kinetic data, e.g., reaction rate constants at different temperatures and therefore Arrhenius parameters E_A and k_0 of the reaction under study, can be derived from a single experimental run [61]. The quality of the derived reaction rate constants critically depends on the design of the applied non-isothermal temperature profile.

In what follows, we assume an Arrhenius rate expression:

$$k = k_0 \exp\left(-\frac{E_A}{RT}\right),$$ (25)

where k is the rate constant, T is the absolute temperature, R is the ideal gas constant, k_0 is the pre-exponential frequency factor, and E_A the activation energy. Because of the inherent correlation of the two Arrhenius parameters, k_0 and E_A, the parameterization of the Arrhenius equation (Equation (25)) is challenging. That is, independently of the applied experimental setup the correlation cannot be reduced for the Arrhenius rate [62,63] expression. At best, a parameter transformation might be applied to mitigate the correlation effect but changes the meaning of the identified parameters alike [64]. Thus, for this particular MB-OED problem, any anti-correlation criteria, which include the modified E^*-criterion as well, fail. Only the overall uncertainty of the parameter estimates can be reduced in principle. For this very reason, we first derive MB-OED results based on local sensitivities (Equation (4)) and the classical D-criterion (Section 3), which is expected to minimize the volume of the parameter confidence ellipsoid [14]:

$$T^*(t) = \arg\min_{T(t)} \Phi_D,$$ (26)

s.t.

$$\frac{dC_{DMSO}(t)}{dt} = -k(T(t))C_{DMSO}(t)$$ (27)

$$T_{lb} \leq T(t) \leq T_{ub}.$$ (28)

The optimal experimental design problem is described in Equations (26)–(28), where Equation (27) is the dynamic reaction model with the DMSO concentration $C_{DMSO}(t)$, and Equation (28) is the temperature constraint. A temperature profile divided into equidistant and constant subintervals is assumed. For all intervals, upper and lower temperature bounds are given as $T_{lb} = 100\,^{\circ}C$ and $T_{ub} = 150\,^{\circ}C$ while assuming continuous measurements of DHBD. Technically, the Matlab® (R2017a) optimizer fmincon is used to derive an optimal temperature profile; i.e., a profile that provides the most informative data and lowest parameter variations. The performance of the original and optimal temperature profile is validated with 2000 Monte Carlo simulations, where for each simulation and parameter identification, respectively, artificial experiment data are assumed with additive white noise. In Figure 5a, we show a reference temperature profile of five temperature steps of equal step-size, which was chosen by educated guessing. The reference profile might already be a good choice to identify the two Arrhenius parameters as it covers the whole temperature range without any obvious preferences. In Figure 5b, we see that the estimates are of finite variation, but as expected, they are strongly correlated. In Figure 5g,j, the individual parameter distributions represent the parameter uncertainties of E_A and k_0 from a different angle and represent their individual but finite variation. In the next step, the D-optimally designed temperature profile is derived numerically and shown in Figure 5b. Compared to the reference temperature profile, less dedicated temperature steps are visible over the experimental course. However, the range of the temperature values is lower; i.e., it starts at a higher temperature of 135 °C and ends with the highest possible temperature of 150 °C. Assuming the same measurement imperfections, the uncertainty in the identified Arrhenius parameters can be reduced by the optimized temperature profile; i.e., the uncertainty of the individual parameters is lower as indicated by the probability density functions in Figure 5h,k. The parameter correlation, however, as can be seen in Figure 5e, remains at the same level. As the derived optimal temperature profile might be difficult to realize due to the small temperature shifts resulting in operability and control issues, a simplified D^s-optimality temperature profile ignoring temperature shifts that are below

2 K variations (see Figure 5c) could be implemented with almost no performance loss; see Figure 5f,i,l for clarification. Similar lab-relevant constraints might be directly added to the underlying dynamic optimization problem [65] but are beyond the scope of this paper.

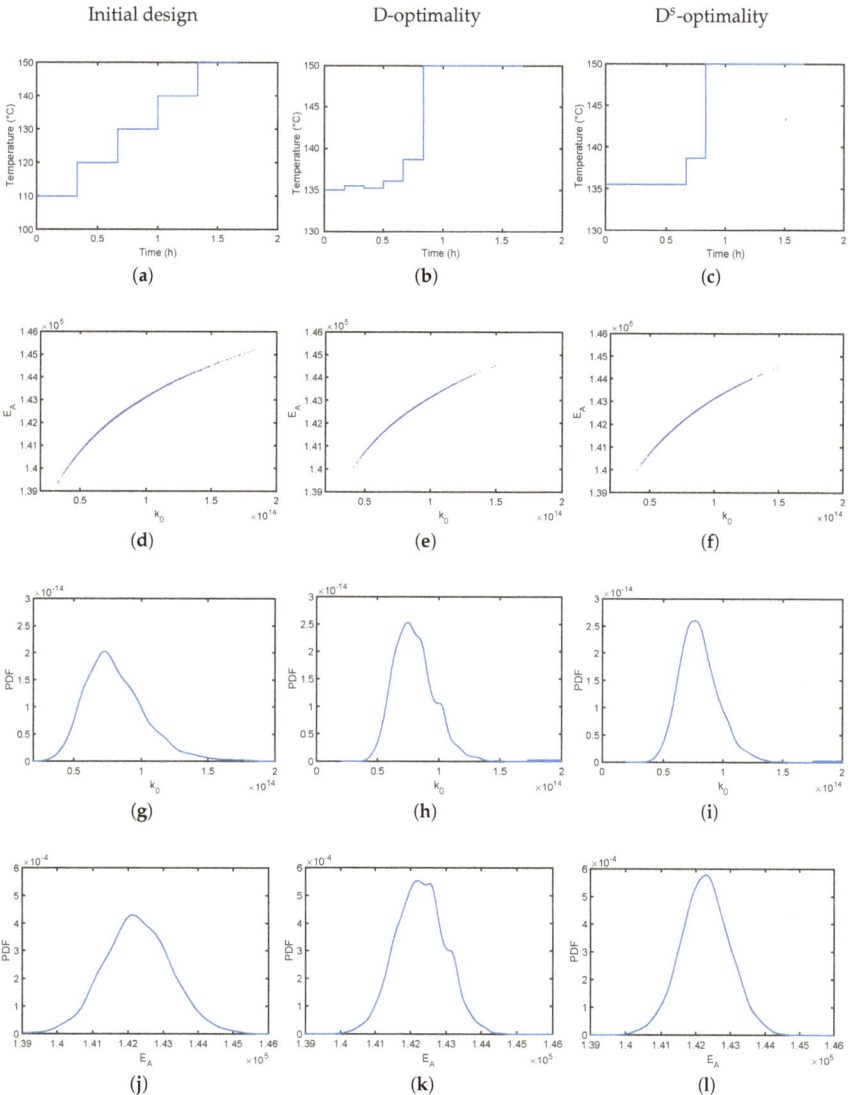

Figure 5. Performance of the initial setting and MB-OED. D^s-optimality means a simplified temperature profile in comparison to D-optimality. (**a**) is the initial; (**b**) the optimally designed; and (**c**) the optimal but simplified temperature profile; (**d–f**) are the scatter plots for the estimated parameters assuming simulated noisy data for 2000 MC simulations; (**g–l**) are the corresponding parameter probability distributions.

Thus far, only local sensitivities have been studied assuming the given reference Arrhenius parameters at which the local sensitivities have to be evaluated. The outcome of this local MB-OED strategy, however, depends critically on the quality of the reference parameters at which the local sensitivities (Equation (4)) are derived. Any deviation of these parameters from their nominal values leads to a change in the local sensitivity values and the D-optimality. In Figure 6, the relative change in the optimized cost function is shown, which is defined as

$$\Delta\Phi_D^{opt}(k_0, E_A) = \frac{\Phi_D^{opt}(k_0, E_A) - \Phi_D^{opt}(k_0^{ref}, E_A^{ref})}{\Phi_D^{opt}(k_0^{ref}, E_A^{ref})}. \tag{29}$$

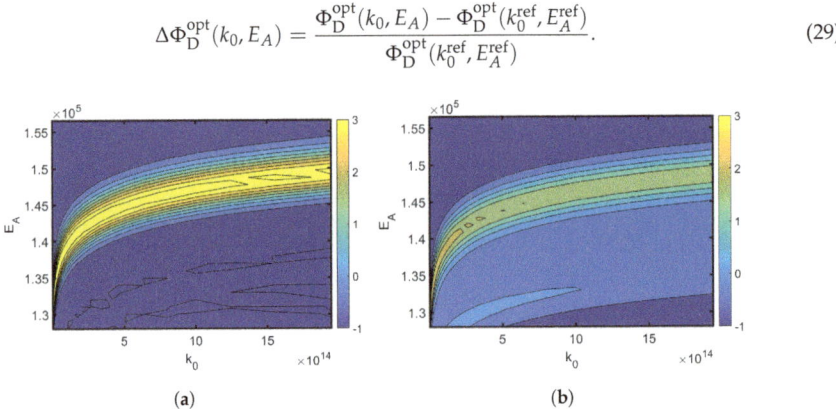

(a)   (b)

Figure 6. Effect of parameter variation. Areas close to zero mean no change in the optimized cost function (Equation (29)) when the Arrhenius parameters are changed. The classical D-optimality shown in (a) degrades drastically with parameter variations. The robust D-optimality shown in (b) is less affected by changes in the parameters. (a) D-optimality; (b) Robust D-optimality.

The results show that a misspecification of the applied reference Arrhenius parameters is likely to result in sub-optimal temperature profiles, less informative measurement data, and higher parameter uncertainties. Alternatively, when the multi-point averaging approach is used (Equation (8)), the calculated design is more robust against reference parameter variations; see Figure 6b. The classical D-optimality shown in Figure 6a degrades drastically with reference variations. The robust D-optimality, however, shown in Figure 6b is less affected by changing the reference parameters. As the reference parameters are typically unknown but are needed to calculate local parameter sensitivities, a robust MB-OED strategy is expected to provide more valuable MB-OED results. Moreover, global parameter sensitivities are evaluated and used for a more credible D-optimality measure (Table 4). The resulting parameter errors for the classical D-optimality, its multi-point averaging realization, and the global sensitivity-based D-design are analyzed. In Figure 7, the multi-point averaging approach and the GSA-based MB-OED lead to more precise parameter estimates in comparison to the classical D-optimality as indicated by the density functions. Here, the GSA-based design results in the most precise estimates; i.e., the probability density functions of Arrhenius parameter have their highest peak close to the true values. Moreover, the multi-point averaging approach and the GSA-based design seem to be less corrupted by the misleading second local minima of the parameter identification problem, which is indicated by the second peak of the probability density functions in Figure 7a,b. Please note, because the proposed PEM sampling strategy is used, only nine sample points for each iteration of the optimizations step are needed to calculate the multi-point averaging or GSA measures in the case of the two Arrhenius parameters.

The applied sensitivity measure has a strong impact on the MB-OED results for the Arrhenius parameters. The classical MB-OED based on local sensitivities is error-prone and is expected to provide sub-optimal experimental designs. For novel APIs, which are available only in very small quantities, each individual experimental run counts. Therefore, MB-OED following a multi-point averaging approach or GSA principles seems to be preferable. Combined with the proposed non-isothermal temperature profile strategy, the optimized temperature profiles ensure the best use of the API-scaffold

DHBD and the most precise estimates of the kinetic rate parameters, k_0 and E_A. In the next step, an MB-OED study of a more complex biochemical synthesis problem is presented where the focus is also on the effect of global design measures.

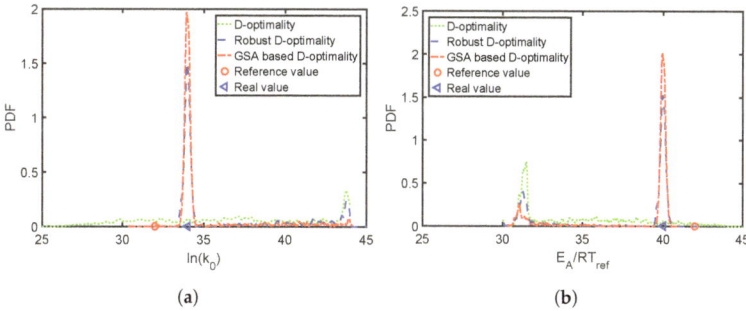

Figure 7. The resulting parameter distributions of the two Arrhenius parameters for the discussed MB-OED strategies. The red circle is the reference value used for MB-OED while the blue triangle is the true parameter value. (**a**) Arrhenius parameter k_0; (**b**) Arrhenius parameter E_A.

4.2. A Fed-Batch Bioreactor

In the second test case study, a fed-batch bioreactor is analyzed. A lumped version of a generic biomass-substrate model reads as follows:

$$\frac{dC_X}{dt} = \mu \cdot C_X - \frac{U}{V} \cdot C_X, \tag{30}$$

$$\frac{dC_S}{dt} = -\sigma \cdot C_X + \frac{U}{V} \cdot (C_{S,in} - C_S), \tag{31}$$

$$\frac{dV}{dt} = U, \tag{32}$$

where C_X is the biomass concentration, C_S is the substrate concentration, V is the liquid volume, and U is the inlet flow rate. The specific growth rate μ and the specific consumption rate σ follow Monod expressions

$$\mu = \mu_{max} \cdot \frac{C_S}{C_S + K_S}, \tag{33}$$

$$\sigma = \frac{\mu}{Y_{X/S} + m}. \tag{34}$$

The nominal parameter values in Equations (30)–(34) and the initial conditions for the dynamic model are listed in Table 6.

Table 6. Parameters as in [66] and initial conditions for the fed–batch bioreactor.

Symbol	Parameter	Unit	Nominal Value
C_{X0}	initial concentration of biomass	$g\,L^{-1}$	0.25
C_{S0}	initial concentration of substrate	$g\,L^{-1}$	3
V_0	initial volume of the liquid phase	L	7
$C_{S,in}$	substrate concentration in the feed	$g\,L^{-1}$	50
U	volumetric feed rate	$L\,h^{-1}$	0–1
μ_{max}	maximum specific growth rate	h^{-1}	0.421
K_S	half velocity constant	$g\,L^{-1}$	0.439
$Y_{X/S}$	yield coefficient of biomass over substrate	-	0.777
m	maintenance factor	h^{-1}	0

For the experimental setups, the following constraints are considered. The duration of the reaction is set to 15 h. The biomass and the substrate concentration are measurable with a sample rate of 0.75 h, which results in a total set of 40 measurement samples. The maximum specific growth rate μ_{max}, the half velocity constant K_S, and the yield coefficient $Y_{X/S}$ should be estimated by minimizing the difference between the measurements and the simulation results. In this simulation study, artificial measurement data [67] are used with additive measurement noise of $\mathcal{N}(0, 6.25 \times 10^{-4})$ and $\mathcal{N}(0, 1 \times 10^{-3})$ for C_x and C_s, respectively. The MB-OED strategy aims at reducing the uncertainty of the parameters by optimizing the feeding policy of the fed-batch bioreactor. Thus, the inlet profile U is parameterized by 20 constant segments of equal size, which are optimized to provide the most informative experimental data. The Monod kinetic parameters, μ_{max}, K_S and $Y_{X/S}$, are treated as uncertain. Their variations are expressed by uniform PDFs of different ranges, i.e., $\mu_{max} \sim U(0.2, 0.8)$, $K_S \sim U(0.2, 0.8)$, and $Y_{X/S} \sim U(0.4, 1.5)$, which might have been derived experimentally with a parameter identification procedure. In addition to μ_{max}, K_S, and $Y_{X/S}$, the model parameters are assumed to be given by the literature without any uncertainty. Thus, please note that only 19 sample points for each iteration of the optimization step are needed to calculate the multi-point averaging or GSA measures when using the proposed PEM sampling strategy.

First, we applied the E^*-design to minimize imbalanced parameter sensitivities and uncertainty. Similar to the previous case study, we compare a default substrate inlet profile (Figure 8c) with the outcome of the classical E^*-design in Figure 8d–f. The resulting parameter scatter plots based on the initial profile (Figure 9a–c) show bimodal behavior indicating some severe nonlinearity and a non-convex optimization problem. For the classical E^*-design, the resulting parameter variations of all three parameters (Figure 9d–f) could be only slightly improved while the parameter correlation for all parameter combinations is clearly visible. The multi-point averaging approach of the E^*-design, in turn, seems to be more appropriate. Applying a low initial substrate concentration (Figure 8g) combined with an optimized impulse-like feeding rate (Figure 8i), the quality of the parameter estimates can be improved as illustrated in Figure 9g–i. In particular, the parameter correlations between K_S and $Y_{X/S}$, and μ_{max} and K_S are reduced. This effect can even be improved by implementing an MB-OED strategy based on GSA. A low initial substrate concentration (Figure 8j) but a more complex feeding rate (Figure 8l) results in more compact parameter scatter plots (Figure 9j–l); i.e., fewer parameter uncertainties and parameter correlations. Finally, in the Shannon 1 study, we also analyze the performance of a Shannon-entropy-based (Table 5) multi-objective design using $w = [1, 1, 0, 1]^T$ as a weighting vector for

$$\Phi_{SE} = w^T [\Phi_{SE_{all}}, \Phi_{SE_{t_k}}, \Phi_{PD}, \Phi_{OU}]. \tag{35}$$

Please note that the third element $w[3]$ was set to zero as the sum of the first-order Sobol indices ($\sum_{i=1}^{p} S^{GSA}_{t_k}[i]$) was always greater or equal to 0.9; i.e., there is no strong multivariate dependency of the analyzed parameters.

Similar to the E^*-design, the derived initial substrate concentration is high, see Figure 8m. For the feeding rate, there are two distinct feeding periods as illustrated in Figure 8o. The resulting parameter estimates, which are summarized in Figure 9m–o, show no significant improvement in the parameter estimates in comparison to the unoptimized initial experimental setting or the E^*-design. The main reason here is the tedious tuning of the multi-objective function, i.e., providing an optimal weighting vector w. For instance, when using $w = [1000, 1, 0, 0.1]^T$ in the Shannon 2 study, the quality of the parameter estimates improves significantly as illustrated in Figure 10. In comparison to the previous MB-OED results, the Shannon-entropy-based design in combination with a proper weight vector w leads to improved parameter estimates, i.e., lower parameter variations and correlations as shown in Figure 10d–f. A low initial substrate concentration (Figure 10a) followed by a gradual increase in the feeding rate (Figure 10c) provide very informative data and precise parameter estimates, respectively. This conclusion can be validated by analyzing Figure 11. The relative parameter errors $|\hat{\theta} - \theta|/\theta$ clearly indicate the second Shannon-entropy-based design as the most suitable one for the parameter identification problem. Please note parameters θ were sampled from their assumed uniform density functions.

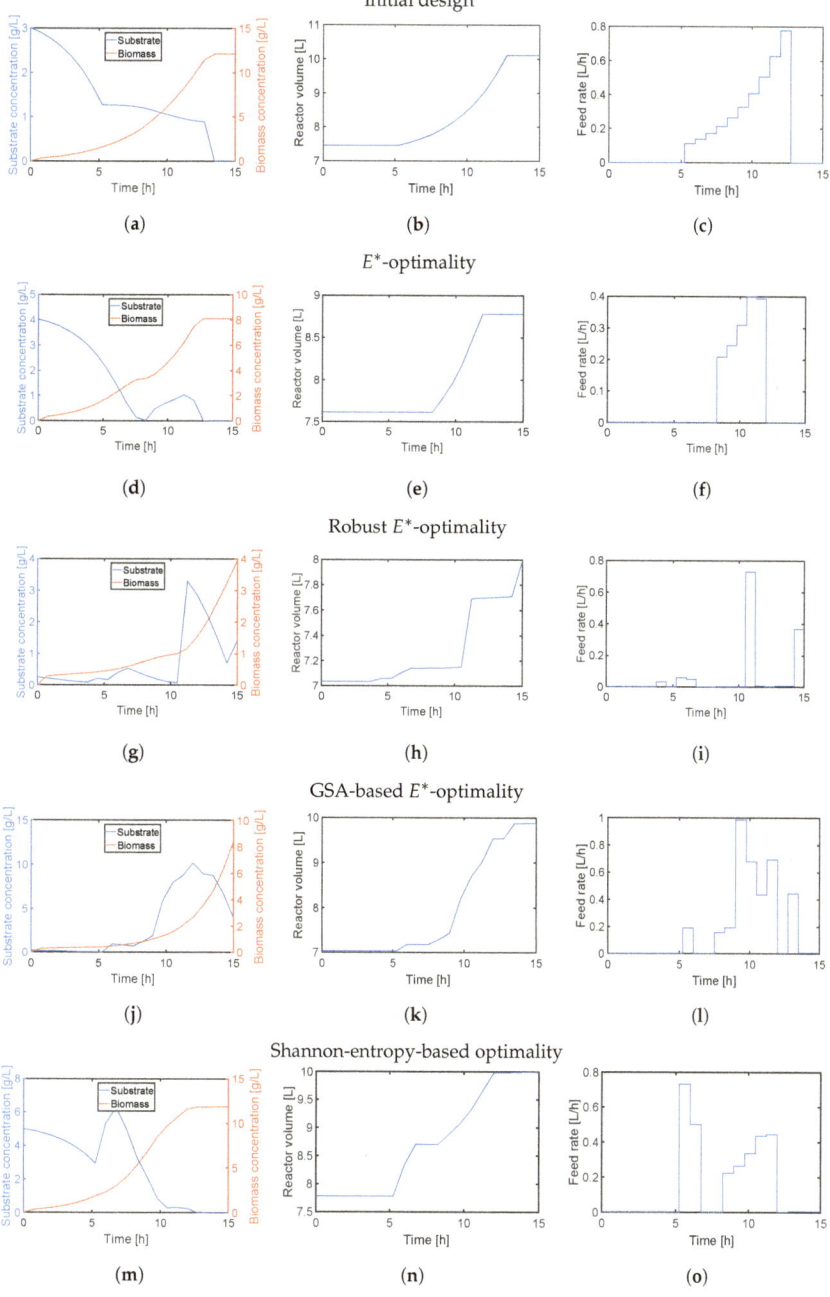

Figure 8. MB-OED results showing the evolution of biomass and substrate concentration (**left**), and volume of the reactor (**middle**) based on the optimized feed rate (**right**).

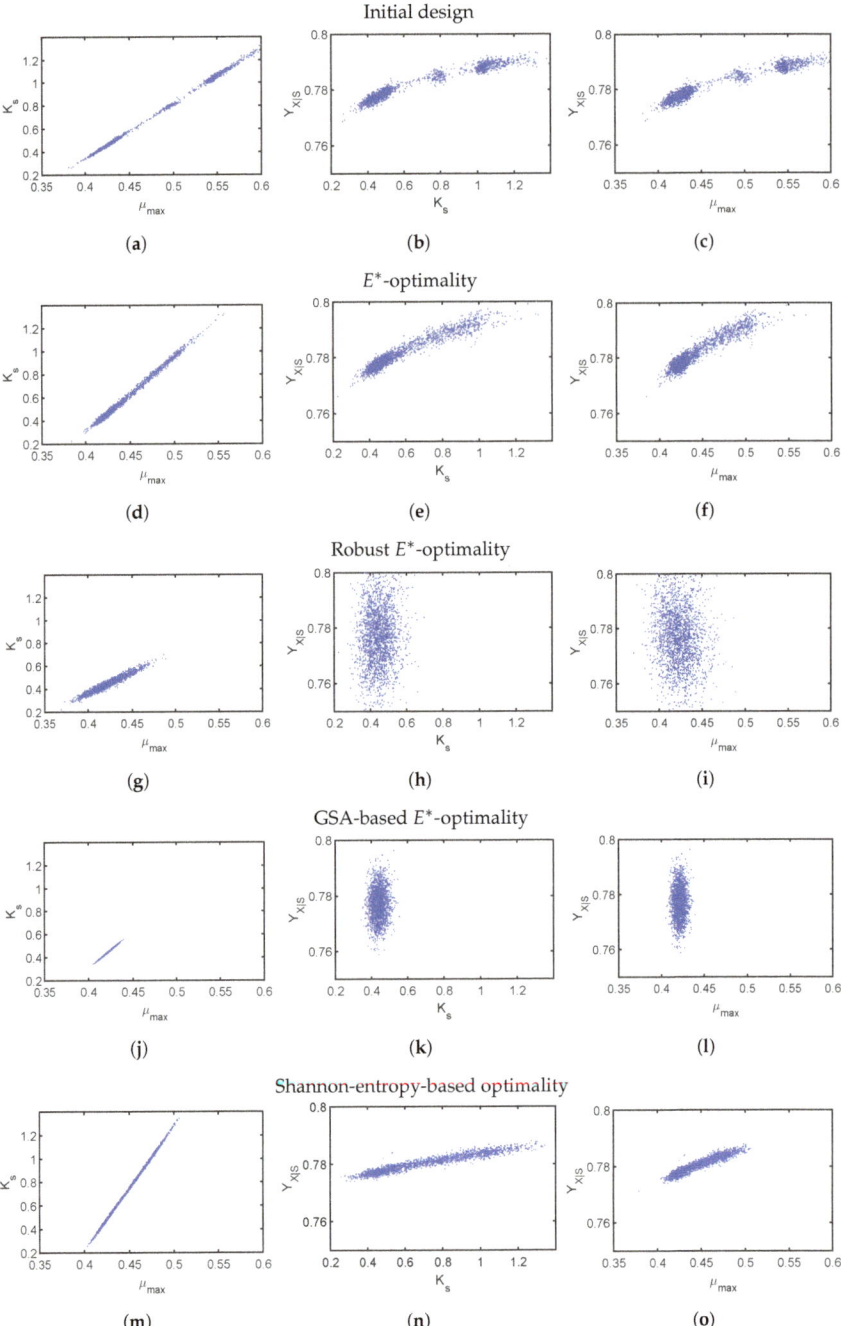

Figure 9. Performance of the experiments from initial design (**a**–**c**), E^*-optimality (**d**–**f**), robust E^*-optimality (**g**–**i**), GSA-based E^*-optimality (**j**–**l**), and Shannon-entropy-based design (**m**–**o**), where scatter plots are used to present the confidence region for the estimated parameters, which result from 2000 MC simulations.

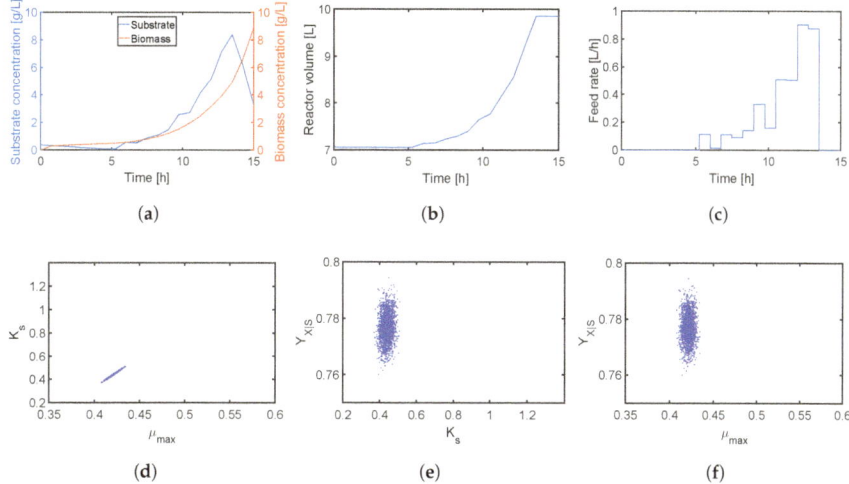

Figure 10. Evolution profiles for three states ((**a**) biomass concentration, substrate concentration; and (**b**) volume of the reactor) and (**c**) the feed rate for the adapted Shannon-entropy-based design with $w = [1000, 0.1, 0, 1]^T$. In (**d–f**), the resulting parameter estimates are shown.

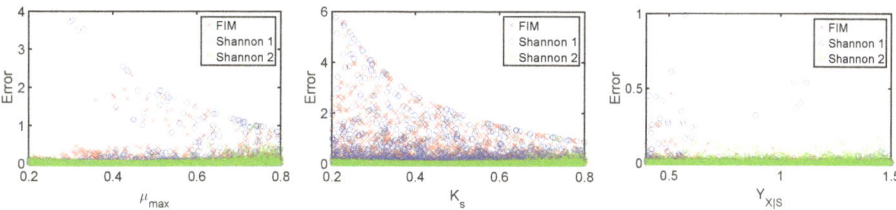

Figure 11. Error of the parameter estimates of μ_{max}, K_S and $Y_{X/S}$. 2000 MC simulations are shown for the classical E^*-design and Shannon-entropy-based design with $w = [1, 1, 0, 1]$, and $w = [1000, 1, 0, 0.1]$.

5. Conclusions

Without a doubt, model-based and computer-aided concepts are valuable tools for translating promising lab findings into efficient processes. Biased model-based results, however, due to model misspecification and model parameter uncertainties, are equally probable. This is particularly true when measurement data are limited in quantity and quality. Thus, in our work, we have successfully demonstrated the benefit of robustification and GSA concepts for MB-OED to gain the most informative data and to improve parameter estimates, respectively. The classical MB-OED, which is based on local sensitivities, is error-prone and is expected to provide sub-optimal experimental designs. For novel APIs, which are available only in very small quantities, each individual experimental run counts. Therefore, MB-OED following a multi-point averaging approach or GSA principles seem to be preferable as demonstrated by the DHBD synthesis problem. Moreover, Shannon-entropy-based design measures, which can account for global parameter sensitivities easily, provide a new angle in MB-OED. Assuming a proper weighting factor, it provides low parameter uncertainties and correlations as illustrated by the biochemical synthesis study.

To ensure the practical relevance, an efficient implementation of the proposed ideas was of particular interest in this study. The point estimate method framework seems to be an attractive alternative compared to state-of-the-art Monte Carlo simulations. On the one hand, it significantly reduces the computational load for advanced sensitivity analyses. On the other hand, its ease in implementation ensures a straightforward adaptation for various problems in MB-OED.

Nevertheless, there are still some open issues and potential research questions that have to be addressed in ongoing research. For instance, there is no systematic analysis under which conditions global sensitivity or robustification concepts perform best and, therefore, need to be explored in greater detail. This could lead to best-practice rules and informed decisions in MB-OED for nonlinear and complex models. An intensive study of tailored design metrics reflecting global sensitivity outcomes may also help to gain improvements in the model building process. In particular, multi-objective design concepts have to be explored and validated in the future. Finally, the proposed ideas have to be systematically extended to incorporate model misspecification as well, i.e., uncertainties related to the model structure itself in addition to the model parameters.

Acknowledgments: We acknowledge support by the German Research Foundation and the Open Access Publication Funds of the Technische Universität Braunschweig. Funding of the "Promotionsprogramm µ-Props" for Xiangzhong Xie and Moritz Rehbein by MWK Niedersachsen, Germany is gratefully acknowledged. Moreover, the anonymous reviewers deserve special thanks for their valuable input.

Author Contributions: R.S., X.X., and M.R. collected information and wrote the manuscript. R.S. conceived the research. X.X. performed the simulations. U.K. and S.S. provided feedback to the content and participated in writing the manuscript.

Conflicts of Interest: The authors declare no conflicts of interest.

References

1. Franz, A.; Song, H.S.; Ramkrishna, D.; Kienle, A. Experimental and theoretical analysis of poly(β-hydroxybutyrate) formation and consumption in Ralstonia eutropha. *Biochem. Eng. J.* **2011**, *55*, 49–58.
2. Grimard, J.; Dewasme, L.; Wouwer, A.V. A review of dynamic models of hot-melt extrusion. *Processes* **2016**, *4*, 19, doi:10.3390/pr4020019.
3. Bück, A.; Neugebauer, C.; Meyer, K.; Palis, S.; Diez, E.; Kienle, A.; Heinrich, S.; Tsotsas, E. Influence of operation parameters on process stability in continuous fluidised bed layering with external product classification. *Powder Technol.* **2016**, *300*, 37–45.
4. Logist, F.; Erdeghem, P.V.; Smets, I.Y.; Impe, J.F.V. Multiple-objective optimisation of a jacketed tubular reactor. In Proceedings of the 2007 European Control Conference (ECC), Kos, Greece, 2–5 July 2007; pp. 963–970.
5. Kunde, C.; Michaels, D.; Micovic, J.; Lutze, P.; Górak, A.; Kienle, A. Deterministic global optimization in conceptual process design of distillation and melt crystallization. *Chem. Eng. Process. Process Intensif.* **2016**, *99*, 132–142.
6. Kaiser, N.M.; Flassig, R.J.; Sundmacher, K. Probabilistic reactor design in the framework of elementary process functions. *Comput. Chem. Eng.* **2016**, *94*, 45–59.
7. Föste, H.; Schöler, M.; Majschak, J.P.; Augustin, W.; Scholl, S. Modeling and validation of the mechanism of pulsed flow cleaning. *Heat Transf. Eng.* **2013**, *34*, 753–760.
8. Schenkendorf, R. Supporting the shift towards continuous pharmaceutical manufacturing by condition monitoring. In Proceedings of the 2016 IEEE 3rd Conference on Control and Fault-Tolerant Systems (SysTol), Barcelona, Spain, 7–9 September 2016.
9. Jelemensk, M.; Klauo, M.; Paulen, R.; Lauwers, J.; Logist, F.; Impe, J.V.; Fikar, M. Time-optimal control and parameter estimation of diafiltration processes in the presence of membrane fouling. *IFAC-PapersOnLine* **2016**, *49*, 242–247.
10. Rogers, A.J.; Inamdar, C.; Ierapetritou, M.G. An integrated approach to simulation of pharmaceutical processes for solid drug manufacture. *Ind. Eng. Chem. Res.* **2014**, *53*, 5128–5147.
11. Rantanen, J.; Khinast, J. The future of pharmaceutical manufacturing sciences. *J. Pharm. Sci.* **2015**, *104*, 3612–3638.
12. Ljung, L. *System Identification: Theory for the User*; Prentice-Hall: Upper Saddle River, NJ, USA, 1987.
13. Balsa-Canto, E.; Alonso, A.A.; Banga, J.R. An iterative identification procedure for dynamic modeling of biochemical networks. *BMC Syst. Biol.* **2010**, *4*, 11, doi:10.1186/1752-0509-4-11.

14. Walter, E.E.; Pronzato, L. *Identification of Parametric Models from Experimental Data*; Springer: New York, NY, USA, 1997; p. 413.
15. Rodríguez-Aragón, L.J.; López-Fidalgo, J. Optimal designs for the Arrhenius equation. *Chemom. Intell. Lab. Syst.* **2005**, *77*, 131–138.
16. Schenkendorf, R.; Kremling, A.; Mangold, M. Optimal experimental design with the sigma point method. *IET Syst. Biol.* **2009**, *3*, 10–23.
17. Galvanin, F.; Marchesini, R.; Barolo, M.; Bezzo, F.; Fidaleo, M. Optimal design of experiments for parameter identification in electrodialysis models. *Chem. Eng. Res. Des.* **2016**, *105*, 107–119.
18. Kiefer, J. Optimum experimental designs. *J. R. Stat. Soc. Ser. B* **1959**, *21*, 272–319.
19. Martínez, E.C.; Cristaldi, M.D.; Grau, R.J. Design of Dynamic Experiments in Modeling for Optimization of Batch Processes. *Ind. Eng. Chem. Res.* **2009**, *48*, 3453–3465.
20. Sinkoe, A.; Hahn, J. Optimal experimental design for parameter estimation of an IL-6 signaling model. *Processes* **2017**, *5*, 49, doi:10.3390/pr5030049.
21. Manesso, E.; Sridharan, S.; Gunawan, R. Multi-objective optimization of experiments using curvature and Fisher information matrix. *Processes* **2017**, *5*, 63, doi:10.3390/pr5040063.
22. Chaloner, K.; Verdinelli, I. Bayesian experimental design: A review. *Stat. Sci.* **1995**, *10*, 273–304.
23. Gotwalt, C.M.; Jones, B.A.; Steinberg, D.M. Fast computation of designs robust to parameter uncertainty for nonlinear settings. *Technometrics* **2009**, *51*, 88–95.
24. Overstall, A.M.; Woods, D.C. Bayesian design of experiments using approximate coordinate exchange. *Technometrics* **2017**, *59*, 458–470.
25. Chu, Y.; Hahn, J. Necessary condition for applying experimental design criteria to global sensitivity analysis results. *Comput. Chem. Eng.* **2013**, *48*, 280–292.
26. Rodriguez-Fernandez, M.; Kucherenko, S.; Pantelides, C.; Shah, N. Optimal experimental design based on global sensitivity analysis. *Comput. Aided Chem. Eng.* **2007**, *24*, 63–68.
27. Bockstal, P.J.V.; Mortier, S.; Corver, J.; Nopens, I.; Gernaey, K.V.; Beer, T.D. Global sensitivity analysis as good modelling practices tool for the identification of the most influential process parameters of the primary drying step during freeze-drying. *Eur. J. Pharm. Biopharm.* **2018**, *123*, 108–116.
28. Scire, J., Jr.; Dryer, F.; Yetter, R. Comparison of global and local sensitivity techniques for rate constants determined using complex reaction mechanisms. *Int. J. Chem. Kinet.* **2001**, *33*, 784–802.
29. Saltelli, A.; Ratto, M.; Tarantola, S.; Campolongo, F. Sensititivity analysis for chemical Models. *Chem. Rev.* **2005**, *105*, 2811–2828.
30. Saltelli, A.; Aleksankina, K.; Becker, W.; Fennell, P.; Ferretti, F.; Holst, N.; Sushan, L.; Wu, Q. Why so many published sensitivity analyses are false. A systematic review of sensitivity analysis practices. *arXiv* **2017**, arXiv:1711.11359.
31. Rao, M.M.; Swift, R.J. *Probability Theory with Applications*; Springer: New York, NY, USA, 2006.
32. Turanyi, T. Sensitivity Analysis of Complex Kinetic Systems. Tools and Applications. *J. Math. Chem.* **1990**, *5*, 203–248.
33. Bauer, I.; Bock, H.G.; Körkel, S.; Schlöder, J.P. Numerical methods for optimum experimental design in DAE systems. *J. Comput. Appl. Math.* **2010**, *120*, 1–25.
34. Ten Broeke, G.; van Voorn, G.; Kooi, B.; Molenaar, J. Detecting tipping points in ecological models with sensitivity analysis. *Math. Model. Nat. Phenom.* **2016**, *11*, 47–72.
35. Kucherenko, S.; Rodriguez-Fernandez, M.; Pantelides, C.; Shah, N. Monte Carlo evaluation of derivative-based global sensitivity measures. *Reliab. Eng. Syst. Saf.* **2009**, *94*, 1135–1148.
36. Rakovec, O.; Hill, M.C.; Clark, M.P.; Weerts, A.H.; Teuling, A.J.; Uijlenhoet, R. Distributed evaluation of local sensitivity analysis (DELSA), with application to hydrologic models. *Water Resour. Res.* **2014**, *50*, 409–426.
37. Zádor, J.; Zsély, I.; Turányi, T. Local and global uncertainty analysis of complex chemical kinetic systems. *Reliab. Eng. Syst. Saf.* **2006**, *91*, 1232–1240.
38. Iooss, B.; Lemaître, P. A review on global sensitivity analysis methods. *arXiv* **2014**, *23*, arXiv:1404.2405.
39. Sobol, I. Global sensitivity indices for nonlinear mathematical models and their Monte Carlo estimates. *Math. Comput. Simul.* **2001**, *55*, 271–280.
40. Buzzard, G.T. Global sensitivity analysis using sparse grid interpolation and polynomial chaos. *Reliab. Eng. Syst. Saf.* **2012**, *107*, 82–89.
41. Caflisch, R.E. Monte Carlo and quasi-Monte Carlo methods. *Acta Numer.* **1998**, *7*, 1–49.

42. Sudret, B. *Uncertainty Propagation and Sensitivity Analysis in Mechanical Models Contributions to Structural Reliability and Stochastic Spectral Methods*; Habilitation a diriger des recherches; Université Blaise Pascal: Clermont-Ferrand, France, 2007.
43. Tyler, G.W. Numerical integration of functions of several variables. *Can. J. Math.* **1953**, *5*, 393–412.
44. Lerner, U.N. Hybrid Bayesian Networks for Reasoning about Complex Systems. Ph.D. Thesis, Stanford University, Stanford, CA, USA, 2002.
45. Evans, D.H. An application of numerical integration techniques to statistical tolerancing. *Technometrics* **1967**, *9*, 441–456.
46. Joshi, M.; Seidel-Morgenstern, A.; Kremling, A. Exploiting the bootstrap method for quantifying parameter confidence intervals in dynamical systems. *Metab. Eng.* **2006**, *8*, 447–455.
47. Heine, T.; Kawohl, M.; King, R. Derivative-free optimal experimental design. *Chem. Eng. Sci.* **2008**, *63*, 4873–4880.
48. Schenkendorf, R.; Mangold, M. Qualitative and quantitative optimal experimental design for parameter identification of a MAP kinase model. *IFAC Proc. Vol.* **2011**, *44*, 11666–11671.
49. Schenkendorf, R.; Mangold, M. Online model selection approach based on unscented Kalman filtering. *J. Process Control* **2013**, *23*, 44–57.
50. Franceschini, G.; Macchietto, S. Novel anticorrelation criteria for design of experiments: Algorithm and application. *AIChE J.* **2008**, *54*, 3221–3238.
51. Ohs, R.; Wendlandt, J.; Spiess, A.C. How graphical analysis helps interpreting optimal experimental designs for nonlinear enzyme kinetic models. *AIChE J.* **2017**, *63*, 4870–4880.
52. Embrechts, P.; Mcneil, A.; Straumann, D. Correlation and dependence in risk management: Properties and pitfalls. In *RISK Management: Value at Risk and Beyond*; Cambridge University Press: Cambridge, UK, 2002; pp. 176–223.
53. Kucerová, A.; Sýkora, J.; Janouchová, E.; Jarušková, D.; Chleboun, J. Acceleration of robust experiment design using Sobol indices and polynomial chaos expansion. In Proceedings of the 7th International Workshop on Reliable Engineering Computing (REC), Bochum, Germany, 15–17 June 2016; pp. 411–426.
54. Lindley, D.V. On a measure of the information provided by an experiment. *Ann. Math. Stat.* **1956**, *27*, 986–1005.
55. Becker, A.; Kohfeld, S.; Lader, A.; Preu, L.; Pies, T.; Wieking, K.; Ferandin, Y.; Knockaert, M.; Meijer, L.; Kunick, C. Development of 5-benzylpaullones and paullone-9-carboxylic acid alkyl esters as selective inhibitors of mitochondrial malate dehydrogenase (mMDH). *Eur. J. Med. Chem.* **2010**, *45*, 335–342.
56. Egert-Schmidt, A.M.; Dreher, J.; Dunkel, U.; Kohfeld, S.; Preu, L.; Weber, H.; Ehlert, J.E.; Mutschler, B.; Totzke, F.; Schͤchtele, C.; et al. Identification of 2-Anilino-9-methoxy-5,7-dihydro-6H-pyrimido[5,4-d][1]benzazepin-6-ones as Dual PLK1/VEGF-R2 Kinase Inhibitor Chemotypes by Structure-Based Lead Generation. *J. Med. Chem.* **2010**, *53*, 2433–2442.
57. Falke, H.; Chaikuad, A.; Becker, A.; Loaëc, N.; Lozach, O.; Jhaisha, S.A.; Becker, W.; Jones, P.G.; Preu, L.; Baumann, K.; et al. 10-Iodo-11H-indolo[3,2-c]quinoline-6-carboxylic Acids Are Selective Inhibitors of DYRK1A. *J. Med. Chem.* **2015**, *58*, 3131–3143.
58. Kunick, C. Fused azepinones with antitumor activity. *Curr. Pharm. Des.* **1999**, *5*, 181–194.
59. Tolle, N.; Kunick, C. Paullones as Inhibitors of Protein Kinases. *Curr. Top. Med. Chem.* **2011**, *11*, 1320–1332.
60. Kunick, C.; Bleeker, C.; Prühs, C.; Totzke, F.; Schächtele, C.; Kubbutat, M.H.; Link, A. Matrix compare analysis discriminates subtle structural differences in a family of novel antiproliferative agents, diaryl-3-hydroxy-2,3,3a,10a-tetrahydrobenzo[b]cycylopenta[e]azepine-4,10(1H,5H)-diones. *Bioorg. Med. Chem. Lett.* **2006**, *16*, 2148–2153.
61. Rehbein, M.C.; Husmann, S.; Lechner, C.; Kunick, C.; Scholl, S. Fast and calibration free determination of first order reaction kinetics in API synthesis using in situ ATR-FTIR. *Eur. J. Pharm. Biopharm.* **2017**, in press.
62. Varga, L.; Szabó, B.; Zsély, I.G.; Zempléni, A.; Turányi, T. Numerical investigation of the uncertainty of Arrhenius parameters. *J. Math. Chem.* **2011**, *49*, 1798–1809.
63. Nagy, T.; Turányi, T. Uncertainty of Arrhenius parameters. *Int. J. Chem. Kinet.* **2011**, *43*, 359–378.
64. Schwaab, M.; Lemos, L.P.; Pinto, J.C. Optimum reference temperature for reparameterization of the Arrhenius equation. Part 1: Problems involving one kinetic constant. *Chem. Eng. Sci.* **2007**, *63*, 2750–2764.
65. Biegler, L.; Grossmann, I.; Westerberg, A. *Systematic Methods for Chemical Process Design*; Prentice Hall: Old Tappan, NJ, USA, 1997.

66. Cappuyns, A.M.; Bernaerts, K.; Smets, I.Y.; Ona, O.; Prinsen, E.; Vanderleyden, J.; Van Impe, J.F. Optimal fed batch experiment design for estimation of monod kinetics of Azospirillum brasilense: From theory to practice. *Biotechnol. Prog.* **2007**, *23*, 1074–1081.
67. Telen, D.; Vercammen, D.; Logist, F.; Van Impe, J. Robustifying optimal experiment design for nonlinear, dynamic (bio)chemical systems. *Comput. Chem. Eng.* **2014**, *71*, 415–425.

 *processes*

Article

Prediction of Metabolite Concentrations, Rate Constants and Post-Translational Regulation Using Maximum Entropy-Based Simulations with Application to Central Metabolism of *Neurospora crassa*

William R. Cannon [1,*], Jeremy D. Zucker [1], Douglas J. Baxter [2], Neeraj Kumar [1], Scott E. Baker [3], Jennifer M. Hurley [4] and Jay C. Dunlap [5]

[1]   Biological Sciences Division, Pacific Northwest National Laboratory, Richland, WA 99352, USA; jeremy.zucker@pnnl.gov (J.D.Z.); Neeraj.kumar@pnnl.gov (N.K.)
[2]   Research Computing Group, Pacific Northwest National Laboratory, Richland, WA 99352, USA; Douglas.Baxter@pnnl.gov
[3]   Environmental Molecular Sciences Laboratory, Pacific Northwest National Laboratory, Richland, WA 99352, USA; Scott.Baker@pnnl.gov
[4]   Department of Biological Sciences, Rensselaer Polytechnic Institute, Troy, NY 12180, USA; hurlej2@rpi.edu
[5]   Department of Molecular and Systems Biology, Geisel School of Medicine at Dartmouth, Hanover, NH 03755, USA; Jay.C.Dunlap@dartmouth.edu
*   Correspondence: William.cannon@pnnl.gov; Tel.: +1-509-375-6732

Received: 21 April 2018; Accepted: 17 May 2018; Published: 28 May 2018

Abstract: We report the application of a recently proposed approach for modeling biological systems using a maximum entropy production rate principle in lieu of having in vivo rate constants. The method is applied in four steps: (1) a new ordinary differential equation (ODE) based optimization approach based on Marcelin's 1910 mass action equation is used to obtain the maximum entropy distribution; (2) the predicted metabolite concentrations are compared to those generally expected from experiments using a loss function from which post-translational regulation of enzymes is inferred; (3) the system is re-optimized with the inferred regulation from which rate constants are determined from the metabolite concentrations and reaction fluxes; and finally (4) a full ODE-based, mass action simulation with rate parameters and allosteric regulation is obtained. From the last step, the power characteristics and resistance of each reaction can be determined. The method is applied to the central metabolism of *Neurospora crassa* and the flow of material through the three competing pathways of upper glycolysis, the non-oxidative pentose phosphate pathway, and the oxidative pentose phosphate pathway are evaluated as a function of the NADP/NADPH ratio. It is predicted that regulation of phosphofructokinase (PFK) and flow through the pentose phosphate pathway are essential for preventing an extreme level of fructose 1,6-bisphophate accumulation. Such an extreme level of fructose 1,6-bisphophate would otherwise result in a glassy cytoplasm with limited diffusion, dramatically decreasing the entropy and energy production rate and, consequently, biological competitiveness.

Keywords: maximum entropy production; mass action kinetics; statistical thermodynamics; metabolism

---

1. Introduction

A grand challenge in biology is to predict the time-dependent behavior of a system. While there have been great successes on the atomistic level that have led to the development of multiscale modeling methods that address phenomena from the femtosecond timescale to the microsecond time scale [1–3], prediction of time-dependent behavior from milliseconds up has been hampered by the lack of rate parameters needed to solve the differential equations governing the behavior.

Many rate parameters have been measured for a few model organisms, but even for those organisms the rate parameters are determined in vitro and do not reflect the in vivo environment. A general solution to this challenge was proposed in 1985 by E. T. Jaynes [4]: "to predict the course of a time-dependent macroscopic process, choose that behavior that can happen in the greatest number of ways while agreeing with whatever information you have—macroscopic or microscopic, equilibrium or nonequilibrium." The approach that Jaynes was advocating was that of maximum path entropy or maximum entropy production. If one applies constraints to the path, then the method is a variational method referred to as maximum caliber [4,5]. Maximum caliber maximizes a path entropy, subject to the imposed constraints. Jaynes, basing his insights on the writings of Gibbs, assured readers that, "in spite of the conceptual simplicity of the approach, its full mathematical expression does prove to be elegant and intricate after all".

Entropy and maximum entropy are often confusing topics, so it is useful to first define the terms used herein and explicitly state what is meant by maximum entropy. Fundamentally, entropy itself is a measure of probability while maximum entropy is simply a characterization or description of a distribution. An increase in entropy is an increase in probability, but to be clear, one should always explicitly specify the probability distribution being considered.

Entropy production should not be confused with entropy change, although both are associated with changes of state [6]. An entropy change for chemical systems is often based on a uniform probability distribution as in the common notion for configurational entropy. Less common in the chemical literature but more frequent in the physics literature is the expression of entropy change between two states, J and K for instance, as a function of the system probability densities, $\Delta S(K, J) = -\log(\Pr(J)/\Pr(K))$ where $\Pr(J)$ and $\Pr(K)$ are probability density functions based on, for instance, the multinomial Boltzmann distribution [6]. In both cases, entropy change is a state function, meaning that the change in entropy due to a change in state does not depend on which process or path was followed from state to state.

Entropy production, on the other hand, is path-dependent. A maximum entropy production path is the thermodynamically optimal or most probable path. The probability density in the case of mass action chemical systems is the multinomial Boltzmann distribution. A change of state due to following a particular path is related to the thermodynamics odds of each respective reaction i, $K_i Q_i^{-1}$, where K_i is the chemical equilibrium constant and Q_i is the reaction quotient. Finding the optimal path for a system is a general problem [7], but is especially important for biology where efficiency and time-to-replication are critical for natural selection. Finally, maximum entropy production and maximum caliber are very similar concepts. A maximum entropy production path is the path that produces a maximal amount of entropy [8,9]. (A minimum entropy production path is a path that dissipates a minimal amount of heat or entropy to the environment. For the purpose of this paper, the two concepts are equivalent and 'least heat' can be used as a working definition of optimal.) A maximum caliber method specifically maximizes the entropy of the path subject to constraints on the system [10]. Either method may be either inferential (applied for the purpose of data analysis) or predictive (employed in a simulation). For the intents and purposes of this paper, we shall not make a distinction between the maximum entropy production and maximum caliber, and will use the term maximum entropy production.

A maximum entropy method is a general term that can refer either to methods that employ an entropy change, or to methods that employ entropy production; it can refer to either prediction through simulation or inference from data. Maximum entropy production specifically refers to a thermodynamically optimal path from one state to another and may be used in the context of either simulations, descriptions of processes, or for the purpose of inference [4]. Jaynes used the term MAXENT to emphasize the use of maximum entropy methods for inference. Regardless of whether maximum entropy is used in the context of inference or prediction, or an entropy change or entropy production, a maximum entropy distribution is the most probable distribution according to the probability density function and boundary conditions.

Maximum entropy production, or more precisely, maximum entropy production rate—the rate at which the maximum amount of entropy is produced—has a long history in physical biology. Lotka first wrote about the concept in 1922, explaining that "in the struggle for existence, the advantage must go to those organisms whose energy-capturing devices are most efficient in directing available energy into channels favorable to the preservation of the species" ([11,12]; reviewed in [13]). At face value, Lotka's statement seems obvious in retrospect, but he went on to advocate that natural selection is based on the physical principles of thermodynamics. Surprisingly, in his report, Lotka states that Boltzmann had been talking about these concepts years earlier. More famously Schrodinger in his 1945 monograph *What is Life?* [14], used the concept of entropy to describe how order, in the form of high-energy compounds in the environment, drives organization within organisms. Prigogine, who received the Nobel prize in 1977 for his work, used the related concept of minimization of entropy production (defined above) to explain the emergence of self-organized systems from non-equilibrium conditions [15]. Recently, Dewar [16] has examined the maximum entropy production concept in detail, noting that when the predictions from maximum entropy approaches fail, it is not the principle that is inadequate but rather the model to which the principle is applied is usually insufficient for the level of prediction needed. While some view this to be a controversial statement, we have observed this to be true in our own work.

Previously, we developed and implemented a maximum entropy production approach to evaluate the dynamics of different versions of the tricarboxylic acid (TCA) cycle found in nature using stochastic kinetics [17] and have more recently generalized the concept to show how simulations can be carried out from knowledge of chemical potentials [18]. In the former study, the rate of the processes were assumed to all occur on the same timescale. In this study, we extend that work to central metabolism including glycolysis and the pentose phosphate pathway using deterministic kinetics. The challenge in both approaches is to find the maximum entropy or most probable distribution, according to a multinomial Boltzmann probability density, consistent with the non-equilibrium boundary conditions. Since rate parameters (and also transition probabilities) have inherent time dependence, a mathematical approach to finding the distribution must be capable of circumventing bottlenecks in phase space that prevent the system from locating the global minima. One solution to this challenge is to widely sample parameter space using ensemble modeling to find the parameters that agree with known behavior or observations [19]. The alternative, as suggested by Jaynes, is to "choose that behavior that can happen in the greatest number of ways", which is to choose the most thermodynamically probable parameters.

In this report, we show that the most thermodynamically probable concentrations can be predicted using a modified version of Marcelin's 1910 equation describing mass action dynamics using reaction affinities [20]. The first step is to obtain the maximum entropy distribution by finding a non-equilibrium steady state that is also a thermodynamic stable state, meaning that the net driving forces on all reactions are equal. When only mass action kinetics are considered in the model and other considerations such as diffusion are ignored, we find that the resulting maximum entropy distribution is appropriate only for the mass action model, but not for a more extensive model that would include diffusion. The discrepancy, as Dewar predicted [16], is not due to the application of the maximum entropy rate principle but rather to the incomplete nature of the model. However, comparison of the predicted metabolite concentrations to expected or observed concentrations allows one to infer points of regulation in the pathways using a loss function similar to that used in machine learning. Once regulation is accounted for, reasonable metabolite concentrations are predicted, as are reaction fluxes. From the concentrations and reaction fluxes, rate parameters are inferred, and full-scale mass action differential equations can be solved, resulting in the time-dependent trajectories of the biological pathways. In doing so, we have implemented Jaynes' proposal of using a simple thermodynamic principle to infer the detailed dynamics of a system without the need to construct the detailed dynamics or parameters from the bottom up. Finally, we evaluate the dynamics of central metabolism of

Neurospora crassa, a model organism for studying the multi-scale dynamics of circadian rhythms. We find that the pentose phosphate pathway can act in a cyclical manner to produce six NADPH for every glucose consumed instead of the generally expected two NADPH per glucose consumed. Together with recent findings that the oxidative pentose phosphate pathway and upper glycolysis are 180 degrees out of phase in the circadian cycle [21], suggests that under nitrogen-limiting conditions when the pentoses are not used for DNA synthesis, the cyclic action of the pentose phosphate pathway may be used to maximize NADPH production for lipid or carbohydrate production.

2. Theory and Methods

Maximum Entropy Optimization Using the Marcelin Equation. Consider a reversible chemical reaction with molecular species $i \in \{A, B, C, D\}$ and unsigned stoichiometric coefficients $v_{i,\alpha}$ for each molecular species i in reaction $\alpha \in \{1, -1\}$ with forward reaction $\alpha = 1$ and the reverse reaction $\alpha = -1$,

$$v_{A,1}n_A + v_{B,1}n_B \overset{rxn\ 1}{\underset{rxn\ -1}{\rightleftharpoons}} v_{C,1}n_C + v_{D,1}n_D \qquad \text{(Scheme 1)}$$

where n_i is the count or concentration of species i. The extent of each of the reactions is given by ξ_1 and ξ_{-1}, such that when $\xi_1 = 0$, the system is in the state where neither of the n_A and n_B reactants have turned into products. When $\xi_1 = 1$, the system is in the state where a stoichiometric amount of the n_A and n_B reactants have turned into products such that there are now $n_C + v_{C,1}$ and $n_D + v_{D,1}$ products and $n_A - v_{A,1}$ and $n_B - v_{B,1}$ reactants. The net extent of each reversible reaction is $\xi_1 - \xi_{-1} \equiv \xi_{1,net}$ and $\xi_{-1} - \xi_1 \equiv \xi_{-1,net}$. Consequently, $d\xi_{1,net}/dt = \dot{\xi}_{1,net}$ is the net flux through reaction 1. The mass action rate law for the reaction in Scheme 1 is,

$$\dot{\xi}_{1,net} = \frac{d\xi_{1,net}}{dt} = k_1 n_A{}^{v_{A,1}} n_B{}^{v_{B,1}} - k_{-1} n_C{}^{v_{C,1}} n_D{}^{v_{D,1}}. \qquad (1)$$

where k_1 and k_{-1} are the rate constants of the forward and the reverse reaction, respectively. Using the signed stoichiometric coefficients $\gamma_{i,1}$ such that $\gamma_{i,1} = -|v_{i,1}|$ for reactants and $\gamma_{i,1} = |v_{i,1}|$ for products, the time-dependence of any molecular species i is,

$$\gamma_{i,1} \frac{dn_i}{dt} = k_1 n_A{}^{v_{A,1}} n_B{}^{v_{B,1}} - k_{-1} n_C{}^{v_{C,1}} n_D{}^{v_{D,1}}. \qquad (2)$$

Although kinetics and thermodynamics are alternate formulations of the law of mass action, Equation (2) is a purely kinetic description in that it does not contain any thermodynamic functions. Thermodynamics can be introduced into Equation (2) by simply factoring out the opposing rate from each term,

$$
\begin{aligned}
\gamma_{i,1} \frac{dn_i}{dt} &= k_1 n_A{}^{v_{A,1}} n_B{}^{v_{B,1}} \left(\frac{k_{-1} n_C{}^{v_{C,1}} n_D{}^{v_{D,1}}}{k_{-1} n_C{}^{v_{C,1}} n_D{}^{v_{D,1}}} \right) - k_{-1} n_C{}^{v_{C,1}} n_D{}^{v_{D,1}} \left(\frac{k_1 n_A{}^{v_{A,1}} n_B{}^{v_{B,1}}}{k_1 n_A{}^{v_{A,1}} n_B{}^{v_{B,1}}} \right) \\
&= k_{-1} n_C{}^{v_{C,1}} n_D{}^{v_{D,1}} \left(\frac{k_1 n_A{}^{v_{A,1}} n_B{}^{v_{B,1}}}{k_{-1} n_C{}^{v_{C,1}} n_D{}^{v_{D,1}}} \right) - k_1 n_A{}^{v_{A,1}} n_B{}^{v_{B,1}} \left(\frac{k_{-1} n_C{}^{v_{C,1}} n_D{}^{v_{D,1}}}{k_1 n_A{}^{v_{A,1}} n_B{}^{v_{B,1}}} \right) \\
&= k_{-1} n_C{}^{v_{C,1}} n_D{}^{v_{D,1}} \left(K_1 \frac{n_A{}^{v_{A,1}} n_B{}^{v_{B,1}}}{n_C{}^{v_{C,1}} n_D{}^{v_{D,1}}} \right) - k_1 n_A{}^{v_{A,1}} n_B{}^{v_{B,1}} \left(K_{-1} \frac{n_C{}^{v_{C,1}} n_D{}^{v_{D,1}}}{n_A{}^{v_{A,1}} n_B{}^{v_{B,1}}} \right),
\end{aligned}
$$

or,

$$= \dot{\xi}_{-1}(t, n_i) e^{A_1(n_i)/RT} - \dot{\xi}_1(t, n_i) e^{A_{-1}(n_i)/RT} \qquad (3)$$

Equation (3) is the equation for the Marcelin–de Donder representation of mass action kinetics [22,23]. The first term describes the time-dependent thermodynamic forces acting on reaction 1 and likewise the second term describes the time-dependent thermodynamic forces acting on the opposing reaction −1. The first term is a product of a purely thermodynamic component, $e^{A_1(n_i)/RT}$, which is the exponential of the thermodynamic driving force $A_1(n_i)$ on reaction 1, and a purely kinetic component, $\dot{\xi}_{-1}(t, n_i)$ which is the time derivative of the extent of the opposing reaction −1,

$\dot{\zeta}_{-1}(t, n_i) = d\zeta_{-1}/dt$. The thermodynamic factor $e^{A_1(n_i)/RT}$ is the odds ratio of the forward reaction to the reverse reaction. The odds are reciprocally related such that $e^{A_1/RT} = e^{-A_{-1}/RT}$. The odds of reaction 1 then change on a time scale determined by $\dot{\zeta}_{-1}(t, n_i)$. That is, the thermodynamic driving force on the forward reaction has a relaxation time that is the time for the reverse reaction to occur. Likewise, the second term describes the odds ratio and time dependence of the odds of the reverse (conjugate) reaction, reaction -1. According to this formulation, positive non-equilibrium forces $(A_1(n_i) > 0)$ will be associated with slower changes in the odds of the forward reaction than the odds of the opposing reaction since the respective relaxation times of the odds are inversely related through the relation,

$$\dot{\zeta}_{-1}(t, n_i)e^{A_1(n_i)/RT} = \dot{\zeta}_1(t, n_i). \tag{4}$$

Equation (4) has the same mathematical form of a fluctuation theorem [24] but is an exact relationship,

$$e^{A_1(n_i)/RT} = \frac{\dot{\zeta}_1(t, n_i)}{\dot{\zeta}_{-1}(t, n_i)}.$$

Generalizing to a large system consisting of Z reactions, the time-dependence of chemical species i is given by,

$$\frac{dn_i}{dt} = \sum_{\alpha}^{Z} \frac{1}{\gamma_{i,\alpha}} \left(\dot{\zeta}_{-\alpha}(t, n_i)e^{A_\alpha(n_i)/RT} - \dot{\zeta}_\alpha(t, n_i)e^{-A_\alpha(n_i)/RT} \right). \tag{5}$$

A convenient thermodynamic optimization procedure can be obtained using the Marcelin formulation of Equation (5). The Marcelin Equation [20] is obtained by setting each of the functions $\dot{\zeta}_\alpha(t, n_i) \neq 0$ to $\dot{\zeta}_\alpha(t, n_i) = c_\alpha$ where c_α is a constant,

$$\frac{dn_i}{dt} = \sum_{\alpha}^{Z} c_\alpha \frac{1}{\gamma_{i,\alpha}} \left(e^{A_\alpha(n_i)/RT} - e^{-A_\alpha(n_i)/RT} \right).$$

That is, the Marcelin Equation sets each of the relaxation rates of the forward and reverse forces to the same rate. Assuming the same relaxation rate for each respective force for all reactions α such that $c_\alpha = c$ removes any kinetic bottlenecks in phase space of the system such that the relative dynamics are governed only by the thermodynamics,

$$\frac{1}{c}\frac{dn_i}{dt} = \sum_{\alpha}^{Z} \frac{1}{\gamma_{i,\alpha}} \left(e^{A_\alpha(n_i)/RT} - e^{-A_\alpha(n_i)/RT} \right). \tag{6}$$

A simulation using Equation (6) will converge to a thermodynamically-optimal steady state, which will be the lowest free energy state given the boundary conditions. The reason for this is that the free energy of chemical systems is the negative log of the multinomial (discrete particle counts) or Dirichlet (continuous particle counts) distribution plus a constant [6]. The multinomial and Dirichlet distributions are members of the exponential family of distributions, which are log-concave when counts are always greater than or equal to zero. Since the free energy is the negative of the log of the distribution, the free energy in this case is a convex surface as a function of the counts. For this thermodynamically-optimal steady state, the net flux through a reaction α is given by,

$$\dot{\zeta}_{\alpha,net} = c\left(e^{A_\alpha(n_i)/RT} - e^{-A_\alpha(n_i)/RT} \right). \tag{7}$$

Accordingly, at the thermodynamically-optimal steady state, the net flux through a reaction is proportional to the net thermodynamic odds. The thermodynamically-optimal steady state is one of many possible kinetic steady states, but its dynamics are such that the state is both kinetically and thermodynamically stable. That is, given a stoichiometric matrix $\mathbf{S}$ of M metabolites $\times$ Z reactions and a vector of the net reaction fluxes $\dot{\xi}$ of length Z, the usual steady state condition applies,

$$\mathbf{S} \cdot \dot{\xi} = 0.$$

The thermodynamic odds of a reaction in Equation (7) are such that $e^{A_\alpha / RT} = K_\alpha Q_\alpha^{-1}$ where K_α is the equilibrium constant and Q_α is the reaction quotient for reaction α. Given Z by Z diagonal matrices of forward and reverse equilibrium constants, $\mathbf{K}^+$ and $\mathbf{K}^-$, and diagonal matrices of the respective forward and reverse reaction quotients, $\mathbf{Q}^+$ and $\mathbf{Q}^-$, the thermodynamically-optimal steady state is the product of the stoichiometric matrix and the net thermodynamic odds of each reaction,

$$\mathbf{S} \cdot ((\mathbf{K}^+ \mathbf{Q}^- - \mathbf{K}^- \mathbf{Q}^+) \cdot \mathbf{1}) = 0.$$

That is, the system is thermodynamically stable as well as kinetically stable. Furthermore, since the parameter c in Equation (7) is arbitrary, the maximum entropy production distribution and the maximum entropy production rate distribution are equivalent. Equation (7) is an important result as relative rate constants can be determined for each reaction α, as will be shown below.

Agreement with Experimental Measurements. It is possible that the thermodynamically-optimal steady state may not have metabolite levels at physiologically realistic values. For example, inference of a steady state model of central metabolism will predict concentrations of fructose 1,6-bisphosphate in excess of 1 M (see below). If the cell were to produce these large concentrations, the cytoplasm would become glassy and practically no diffusion would occur. Hence the entropy production rate would actually approach zero. The issue is that the model assumes that diffusion will remain sufficient regardless of solute concentrations. This is easily remedied without including diffusion in the model by comparing predicted concentrations or counts of metabolite i, $\tilde{n}_i$, to experimentally observed values n_i and reducing the difference between predicted and observed values by optimizing a loss function for each reaction α. As a loss function, we have chosen the log ratio of the observed values to the predicted values of the $M(\alpha)$ reaction products of reaction α,

$$L_\alpha = \log \prod_{i(\alpha)}^{M(\alpha)} \frac{\tilde{n}_{i(\alpha)}}{n_{i(\alpha)}}. \tag{8}$$

When the observed and predicted values agree, $L_\alpha = 0$, and when the predicted values are greater or less than the observed values $L_\alpha > 0$ or $L_\alpha < 0$, respectively. Unfortunately, experimental measurements of metabolite concentrations are hard to obtain; even when available, metabolomics data sets are sparse. Part of the reason is that the chemical properties of small molecules vary widely, so no single experimental design can discriminate and measure each molecular species. Sparsity is also due to the fact that biologically relevant metabolite concentrations may span several orders of magnitude, which is often greater than typical instrument dynamic ranges. Instead, one can simply use rule-of-thumb estimates for metabolite concentrations in place of experimentally observed values. Based on mass spectrometry estimates of absolute concentrations of metabolites [25], a reasonable rule-of-thumb for metabolites is that most metabolites will not exceed millimolar levels. A value of $L_\alpha > 0$ may indicate that the system needs to be modulated or regulated such that the metabolite concentrations stay within a physiological range. The desired regulation can be implemented using any appropriate function, such as a Hill equation, a logistic function or a hyperbolic function. Parameters for the regulation functions are then estimated and the simulations and analysis are repeated, adjusting parameters each time, until a physiological level of metabolite concentrations is achieved. This resulting steady state is thermodynamically optimal conditioned on the required regulation.

Rate Constants. From this steady state with reasonable metabolite concentrations and activity λ_α of reaction α, the rate constants are inferred for each reaction as follows. The net flux $\dot{\zeta}_{\alpha,net}$ is,

$$
\begin{aligned}
\dot{\zeta}_{\alpha,net} &= \lambda_\alpha \cdot k_\alpha \overset{\text{reactants } \alpha}{\underset{i}{\prod}} n_i - \lambda_\alpha \cdot k_{-\alpha} \overset{\text{reactants } -\alpha}{\underset{i}{\prod}} n_i \\
&= \lambda_\alpha \cdot k_\alpha \overset{\text{reactants } \alpha}{\underset{i}{\prod}} n_i \left(1 - \frac{\lambda_\alpha \cdot k_{-\alpha} \overset{\text{reactants } -\alpha}{\underset{i}{\prod}} n_i}{\lambda_\alpha \cdot k_\alpha \overset{\text{reactants } \alpha}{\underset{i}{\prod}} n_i} \right) \\
&= \lambda_\alpha \cdot k_\alpha \overset{\text{reactants } \alpha}{\underset{i}{\prod}} n_i (1 - K_{-\alpha} Q_{-\alpha}^{-1}),
\end{aligned}
$$

where again λ_α is the activity of the enzyme catalyzing reaction α as a function of the regulation. For example, solving for the rate constants of reaction $\alpha = 1$ from Scheme 1 gives,

$$
\begin{aligned}
k_1 &= c \frac{\dot{\zeta}_{1,net}}{\lambda_1 \cdot n_A n_B (1 - K_{-1} Q_{-1}^{-1})}. \\
k_{-1} &= c \frac{K_1}{k_1},
\end{aligned}
\tag{9}
$$

since $\lambda_1 = 1$ (no regulation) in Scheme 1. The usual mass action ODEs using rate constants and regulation are then solved during a simulation. The kinetically accessible energy surface is not necessarily convex because of the introduction of the rate constants—each reaction now has its own time dependence.

Finally, for the readers that are interested in such details, the inference of rate parameters in this manner meets the criteria that Jaynes laid out for MAXENT: *when we make inferences based on incomplete information, we should draw them from that probability distribution that has the maximum entropy permitted by the information we do have* [26]. Here, we have used the maximum entropy distribution from the maximum entropy production simulation to infer parameters that we otherwise know nothing about. However, if one has accurate metabolite measurements or fluxes of the metabolite population that is free in its biological solution (e.g., cytoplasm), then one can also use this information to infer experimentally-based in vivo rate constants as well [18].

Power and Conductance. For complex systems such as biological systems, the power is the change in free energy with respect to time,

$$
P = -\frac{dG}{dt}.
$$

For a reaction, the rate dependence is due to the net change in the extent of a reaction $\dot{\zeta}_{\alpha,net}$ with time such that the rate $\dot{\zeta}_{\alpha,net} = d\zeta_{\alpha,net}/dt$ is the usual mass action rate as exemplified by Equation (1). The power generated by a reaction can then be expressed in terms of the reaction affinity A_α and the extent of the reaction ζ_α,

$$
\begin{aligned}
P_\alpha &= -\frac{dG}{d\zeta_\alpha} \frac{d\zeta_\alpha}{dt} \\
&= A_\alpha \dot{\zeta}_\alpha.
\end{aligned}
$$

This relationship is useful for comparing different chemical processes, as will be seen below. Likewise, the resistance R_α and conductance C_α of reaction α can be calculated at steady state as,

$$
\begin{aligned}
R_\alpha &= \frac{\Delta G_\alpha}{\dot{\zeta}_{\alpha,net}}, \\
C_\alpha &= R_\alpha^{-1}.
\end{aligned}
$$

where ΔG_α is the free energy change across the reaction, which is equal in value to the reaction affinity A_α for a system at steady state with a large number of particles. These latter equations should not be taken to imply that there is a linear relationship between flux and a change in free energy (or flux and resistance). A reaction at steady state has high resistance if the change in free energy is large for the steady state flux relative to other reactions in the pathway. An example of this will be discussed in the results section, in which a regulated reaction has the same flux as other reactions at steady state, but the change in free energy for the reaction is large because the regulation decreases the activity of the corresponding enzyme.

Implementation, Code, Metabolic Model and Parameters. The equations described above for both the optimization and simulation were implemented in the language C in a software program called Boltzmann, which is available as open source code under a Berkeley Software Distribution (BSD) style license [27]. The ODE solver was an implementation of the MATLAB® ode23tb solver, a trapezoidal rule and the backward differentiation solver [28].

Chemical potentials were obtained from component contribution methods [29,30] and adjusted within Boltzmann to the dielectric response ϵ, ionic strength I, and pH of the cell cytoplasm, assumed to be $\epsilon = 0.78$, $I = 0.25$, and pH $= 7.0$ [31].

The compartmentalized *Neurospora crassa* metabolic model originally from Dreyfuss et al. [32] was updated with new information and used for the optimizations and simulations. The updated genome-scale model is publicly available [33]. The model used in this work is a subnetwork of the genome-scale model that includes only central metabolism consisting of upper and lower glycolysis, the TCA cycle, and the pentose phosphate cycle and is available in the supplemental notebooks. In this reduced model, there are 20 variables (metabolite concentrations) and 20 equations (reaction equations).

In order to maintain a non-equilibrium state, boundary concentrations for the initial reactant glucose 6-phosphate and final product CO_2 are set to non-equilibrium values of 2 mM and 0.1 mM, respectively. Likewise, the cofactors CoA, ATP, ADP, orthophosphate, NAD, NADH, NADP and NADPH are also fixed boundary species; these concentrations were taken from a mass spectrometry analysis of absolute metabolite concentrations from Bennett et al. [25] The redox pair employed to shuttle electrons into the mitochondrial respiratory chain were taken to have equal chemical potentials, as done in previous modeling of the TCA cycle [17].

Analysis of simulation data is included in supplementary information/computational notebooks using Python.

3. Results

A metabolic model of *Neurospora crassa* [32] was used along with chemical potentials to maximize the entropy production rate of central metabolism. The dynamics of the each species are governed by the thermodynamic forces acting on each reaction (Equation (6)) and the net flux through each reaction is determined by the thermodynamic odds of the reaction at steady state (Equation (7)).

A map of the flux through the system is shown in Figure 1. As can be seen, the maximum entropy production rate optimization predicts that the fluxes $\dot{\xi}_{\alpha,net}$ in lower glycolysis and the TCA cycle ($\dot{\xi}_{\alpha,net} \approx 6.53$) are twice that of upper glycolysis ($\dot{\xi}_{\alpha,net} = 3.29$), as would be expected since an intermediate in upper glycolysis corresponds to two intermediates in lower glycolysis and the TCA cycle due to the splitting of fructose 1,6-bisphosphate by fructose 1,6-bisphosphate aldolase to effectively two molecules of glyceraldehyde-3-phosphate. The flux values are relative values but can be calibrated to absolute values by knowledge of just one absolute reaction flux.

The maximum entropy production optimization results in metabolite levels tending towards their most probable (Boltzmann) distribution such that,

$$\frac{n_i}{N_T} \propto \frac{e^{-\mu_i{}^0(\epsilon,I,pH)/RT}}{\sum_j^M e^{-\mu_j{}^0(\epsilon,I,pH)/RT}},$$

where $\mu_i^0(\epsilon, I, pH)$ is the standard chemical potential for solute i in an aqueous solution with dielectric constant ϵ, ionic strength I, and constant pH. The values of ϵ, I, pH are chosen to match those in the cell environment, which in this case are assumed to be $\epsilon = 0.15$, $I = 78$ and $pH = 7.0$.

The non-equilibrium boundary conditions for the initial reactant glucose and final product CO_2, and with those for the cofactors CoA, ATP, ADP, orthophosphate, NAD, NADH, NADP and NADPH prevent the intermediates of glycolysis and the TCA cycle from reaching equilibrium.

The resulting maximum entropy production rate concentrations are shown in Figure 2. There are two important aspects of the results shown in Figure 2. First, the consistency of the concentrations of each metabolite indicates that steady state has been reached (the time derivatives are provided in supplementary Notebook S1). Second, the concentrations of metabolites span a range from $10^{-22}M$ for mitochondrial oxaloacetate at the low end to $10^6 M$ for fructose 1,6-bisphosphate and $10^9 M$ for mitochondrial acetyl CoA at the high end. While a concentration of $10^{-22}M$ might be reasonable in a cell (effectively zero concentration in a typical cell with a volume of $10^{-15} - 10^{-12}$ liters), concentrations above $10^{-3}M$ to $10^{-2}M$ are usually not physiologically realistic [25]. For instance, if fructose 1,6-bisphosphate had a concentration much above $10^{-3}M$, the cytoplasm could become so glassy that diffusion of even small molecules would decrease below the level at which metabolism could operate.

However, the issue is not whether the maximum entropy production rate formulation is appropriate for biological systems, but rather that any model is always a reduced representation of a real system [16]. In a complete model of cellular metabolism, diffusion would be included and the maximum entropy production rate solution would balance the rate of diffusion with the tendency for any chemical species to move towards its thermodynamically optimal distribution.

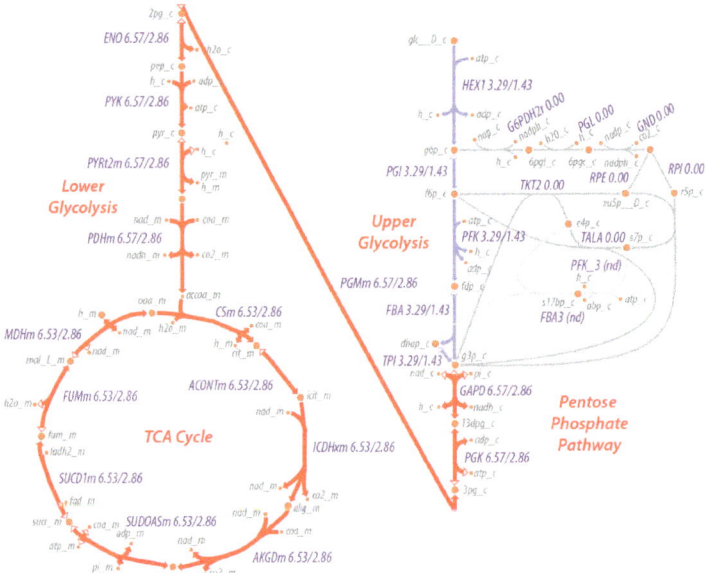

Figure 1. Map of net odds (Equation (7)) for reactions of glycolysis and the tricarboxylic acid (TCA) cycle. Values next to each reaction name indicate the net flux through the reaction, and two flux values are provided for each reaction. The first value (left) is the flux after maximum entropy optimization. The second value (right) is the flux after the same optimization but including regulation at PFK (by ATP) and PDHm (by acetyl-CoA). Reaction and metabolite abbreviations are derived from the BiGG database [34]; full common names are provided as supplementary Tables S1 and S2. The metabolic pathway visualizations here and in Figure 5 were created with Escher [35].

This discrepancy between predicted metabolite concentrations and physiological expectations can be taken advantage of, however, to infer points of post-translational enzyme regulation. To do this, a loss function is formulated (Equation (8)), similar to that used in machine learning approaches, to identify nodes in the system (in this case, reactions or enzymes) that need to be adjusted or regulated in order to match observed results. We applied Equation (8) to the simulation predictions for each reaction using a rule of thumb that physiological concentrations should not exceed $10^{-3} M$. The results are shown in Table 1. Twelve reactions have product concentrations higher than expected. In choosing which reactions to apply regulation to, we take a parsimonious approach based on two principles: (1) regulation of upstream enzymes should have precedence over regulation of downstream enzymes; and (2) the reactions with the highest loss-function value should be evaluated before those with lower loss function values. Using these principles, we chose to apply regulation to only phosphofructokinase (PFK) and mitochondrial pyruvate dehydrogenase (PDHm), which are the two reactions with the highest loss-function values. These are also the two reactions that are most commonly regulated in glycolysis.

The loss function does not help to identify how these reactions/enzymes might be regulated. In experimental studies of *Neurospora crassa*, PFK in *Neurospora crassa* is inhibited by high concentrations of ATP [36]. Early studies on Neurospora PDHm indicated that PDHm is regulated by acetyl-CoA and covalently by phosphorylation [37,38]. We chose to regulate the PFK reaction by ATP concentration and the PDHm reaction by acetyl-CoA concentration using Hill equations.

After regulation of PFK and PDHm are applied, the maximum entropy optimization produces the steady state concentrations shown in Figure 3. Most metabolite concentrations fall within the rule-of-thumb values ($\leq 10^{-3} M$) and reanalysis of the loss function for each reaction finds that only the succinyl-CoA synthetase reaction resulting in the production of succinate has a loss-function value, $L = 3.8$ above the expected value of 0.0. Given that succinate is highly soluble, however, this mild increase of the loss function is reasonable. The predicted concentration of succinate is 3.4 mM. For comparison, the concentration of succinate in exponentially growing *E. coli* under similar conditions was estimated to be 0.57 mM [25].

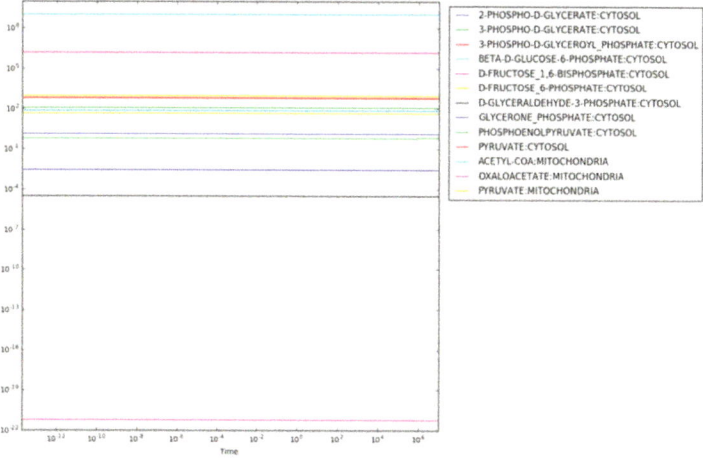

Figure 2. Concentrations as a function of time in maximum entropy optimization without regulation.

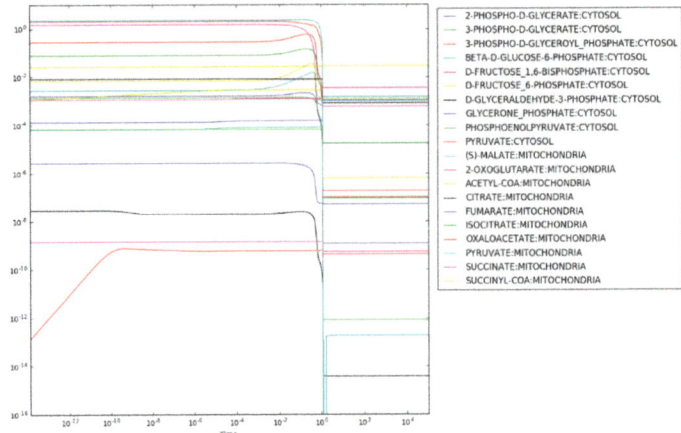

Figure 3. Concentrations as a function of time in maximum entropy optimization with regulation.

Table 1. Product of the reaction product concentrations for the optimization predictions and expected values, and resulting value of the loss function L (Equation (8)). Full common names are provided as supplementary Table S2.

Reaction	Product of Concentrations		
	Predicted	Expected	L
CSm	4.85×10^{-6}	1.00×10^{-6}	1.58
SUCOASm	6.62×10^{-9}	1.00×10^{-9}	1.89
ENO	6.65×10^{-1}	1.00×10^{-3}	6.50
PGM	1.33	1.00×10^{-3}	7.19
HEX1	4.26×10^{-2}	1.00×10^{-6}	10.66
PGI	4.58×10^{1}	1.00×10^{-3}	10.73
GAPD	4.82×10^{-2}	1.00×10^{-6}	10.78
PYRt2m	8.43×10^{2}	1.00×10^{-3}	13.64
PGK	1.09	1.00×10^{-6}	13.90
PYK	5.84	1.00×10^{-6}	15.58
PFK	9.21×10^{2}	1.00×10^{-6}	20.64
PDHm	8.66	1.00×10^{-9}	22.88

In addition to changes in the metabolite concentrations, application of regulation to PFK and PDHm results in reduced flux through both glycolysis and the TCA cycle, as expected. In this case, the flux is reduced by just over 2-fold (6.53/2.86).

With reasonable steady-state metabolite concentrations in hand, rate constants can be inferred from the simulation using Equation (9). The resulting rate constants are provided in supplementary Notebook S2. The inferred rate constants are not generally comparable to rate parameters for the analogous reactions determined from in vitro enzyme kinetic studies, however, since in vitro studies characterize reactions using Michaelis–Menten equations and the simulation studies reported here do not explicitly model enzyme kinetics. However, the flux values from the simulation are consistent with inferred fluxes from metabolic flux analysis studies on the related fungus *Yarrowia lipolytica* [39], which has a very similar central metabolism based on genome analysis and modeling [32,40–42].

From the predicted flux values and reaction-free energies, the power characteristics and resistance of each reaction can be estimated. Shown in Figure 4 are relative values of the (1) reaction free energies; (2) power at each reaction; (3) reaction resistance; and (4) reaction flux for each reaction of glycolysis. Since lower glycolysis and the TCA cycle (supplementary Notebook S3) have identical net fluxes at steady state and similar reaction-free energies, the power characteristics and conductances of these reactions are very similar. The power generated at the PFK reaction is 6-fold higher than the power generated at other reactions of glycolysis. However, while the phosphorylation of fructose 6-phosphate to the highly soluble fructose 1,6-bisphosphate contributes significantly to the overall free energy change of the combined glycolysis-TCA pathway, if the system were allowed to proceed to the maximum entropy distribution (that is, without regulation), all reactions would have very similar power characteristics. It is the applied regulation that keeps the thermodynamic driving force on the PFK reaction much higher than that for the other reactions. Consequently, the chemical resistance at PFK is much higher than for other reactions, as well. Due to the regulation, both the flux through upper glycolysis and the flux through lower glycolysis and the TCA cycle are reduced more than 2-fold. While it has long been stated that PFK is regulated to modulate the flow of material through glycolysis, the simulations suggest that the more nuanced explanation is that regulation acts as a potentiometer to modulate both the flow of material through the reaction and the accumulation of fructose 1,6-bisphosphate in the system. Regardless, flow of material through glycolysis to the TCA cycle is reduced significantly. However, if the causal explanation for the regulation were solely to reduce the flux through the pathway, this would unnecessarily reduce the ability of the organisms using this metabolism to compete since energy production would slowed down significantly, as well, which is counter to Lotka's early conjecture.

Rather than simply not extracting this available energy from glucose at a high rate due to already high levels of ATP, it is reasonable to expect that the energy from glucose is utilized elsewhere. Proteomics studies on Neurospora show that the enzymes of upper glycolysis and the oxidative pentose phosphate pathway oscillate with the circadian cycle but are 180° out of phase with each other [21]. Consequently, simulations of glycolysis, the TCA cycle and the pentose phosphate pathway were carried out following the same steps as described above: (1) maximum entropy production rate optimization without regulation of the pentose phosphate pathway followed by (2) inference of regulation and (3) re-optimization with regulation to obtain steady state metabolite levels. Evaluation of the loss functions for each reaction in the pentose phosphate pathway predicted that glucose-6-phosphate dehydrogenase and phosphogluconolactonase, the first and second steps of the pentose phosphate pathway, should additionally be regulated. Glucose-6-phosphate dehydrogenases are well-known to be regulated by NADP/NADPH, either directly or indirectly, and by phosphorylation [43]. No literature was found regarding post-translational regulation of phosphogluconolactonase. Accordingly, we added regulation to the glucose-6-phosphate dehydrogenase reaction but not to the phosphogluconolactonase reaction. The glucose-6-phosphate dehydrogenase reaction was again regulated by a Hill equation based on the levels of NADPH. We evaluated the kinetics of the system (both oxidative and non-oxidative) under three conditions which differed by the NADP/NADPH ratio. An initial low ratio of NADP/NADPH was taken from an isotope labeling, mass spectrometry analysis of the exponential growth of *E. coli* in which NADP/NADPH $= 2.1 \cdot 10^{-6}/1.2 \cdot 10^{-4} = 0.0175$ [25]. This ratio results in a mild driving force on the NADP/NADPH-dependent reactions of the oxidative branch of the pentose phosphate pathway (enzymes glucose-6-dehydrogenase and phosphogluconate dehydrogenase) of ~-2.7 KJ/mol. A moderate ratio of NADP/NADPH was taken to be NADP/NADPH $= 1$, resulting in driving forces on the latter reactions of ~-3.6 KJ/mol. A high ratio was taken to be the observed values for NAD/NADH in the *E. coli* study such that NADP/NADPH $= 2.6 \cdot 10^{-3}/8.3 \cdot 10^{-5} = 31$, resulting in a driving force of ~-31.2 KJ/mol on these same dehydrogenase reactions.

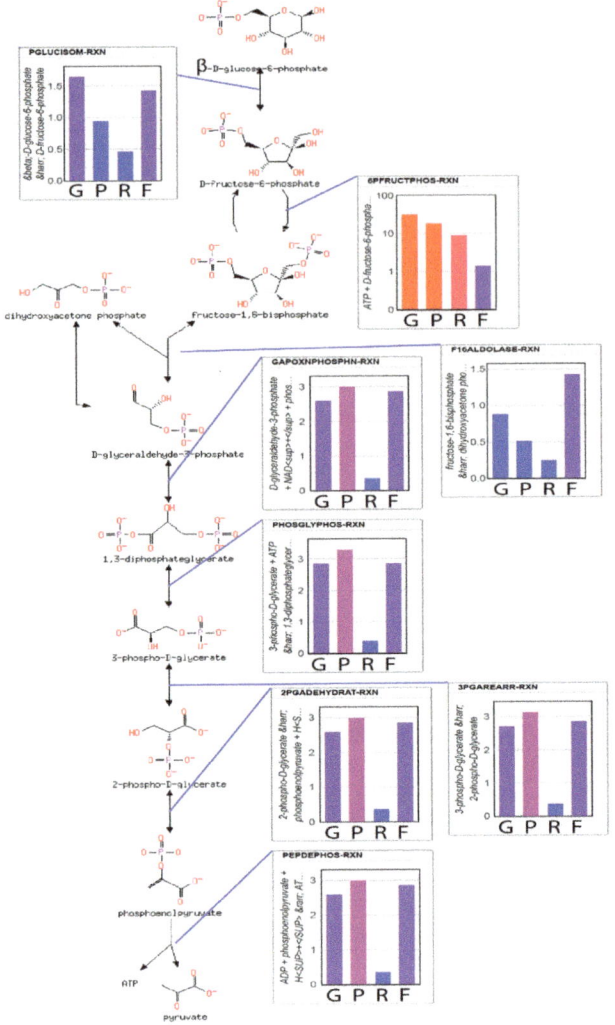

Figure 4. Energetics of glycolysis reactions. Columns from left to right indicate: (G) $-\Delta G_{rxn}$, (P) power, (R) resistance, and (F) flux. Red indicates high values and blue indicates low values. Values are in arbitrary, relative units but the specific values are provided in supplementary Notebook_S3. The phosphofructokinase reaction has dramatically different characteristics than the other reactions because feedback regulation of ATP turns it into a potentiometer. The metabolic pathway visualization was created with Pathway Tools [44].

As expected, the flow of material increases through the pentose phosphate pathway as the NADP/NADPH ratio increases, as shown in Figure 5. At relatively low to moderate ratios of NADP/NADPH, the flow of material is mostly through the non-oxidative branch of the pentose phosphate pathway, while at high ratios of NADP/NADPH, flow through the oxidative branch is maximized, with an average of three cycles through the oxidative pentose phosphate branch for each glucose utilized (the simulation conditions are under non-growth conditions, so pentose phosphates are not drawn off for biosynthetic purposes). That is, rather than acting as a linear pathway from glucose-6-phosphate to glyceraldehyde-3-phosphate, the pentose phosphate pathway acts cyclically to maximize production of NADPH. The overall reaction of the pentose phosphate pathway at high ratios of NADP/NADPH is,

$$\text{Glucose 6} - \text{phosphate} + 6\,\text{NADP}^+ + 3\text{H}_2\text{O} \rightleftharpoons$$
$$\text{glyceraldehyde 3} - \text{phosphate} + 6\,\text{NADPH} + 3\text{CO}_2.$$

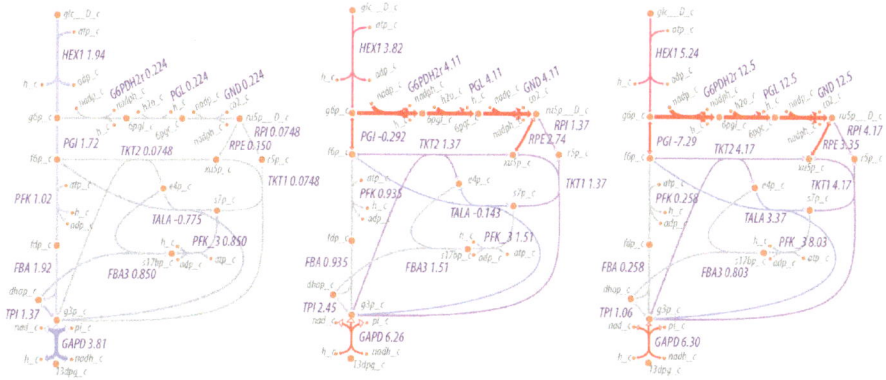

Figure 5. Reaction flux through upper glycolysis and the pentose phosphate pathway as a function of the NADP/NADPH ratio. (**Left**) Low values of the ratio combined with high values of ATP result in approximately equal flow of material through upper glycolysis and the non-oxidative branch of the pentose phosphate pathway, minimizing the production of fructose 1,6-bisphosphate; (**Middle**) a NADP/NADPH ratio of 1 results in flow through each of upper glycolysis, non-oxidative and oxidative pentose phosphate pathways, with the oxidative pentose phosphate pathway containing approximately 65% of the flow of material; (**Right**) a high value of the ratio results in the cycling of flow iteratively through the oxidative pentose phosphate pathway while flow through upper glycolysis is minimal.

4. Discussion

The maximum entropy principle can be used to predict concentrations, obtain optimal rate constants, and calculate the characteristics of biological circuits including energy, power, flux and resistance for individual reactions and entire pathways. In the model of central metabolism used here, enzyme kinetics are represented as the summary reaction of the catalytic process. Dobovišek et al. have shown that the maximum entropy production principle can be used to evaluate enzyme kinetics, as well [45,46]. In this study, for convenience the enzyme catalysts are not explicitly represented in the model except for regulation. It is entirely feasible to do so, however, even without the use of the assumptions built into Michaelis–Menten model that the system be at steady state and far from equilibrium.

The calculated rate constants are optimal for the growth conditions used in the simulation. Similarly, organisms will have rate constants that are optimal for the conditions to which they have adapted. Ideally, to compare predicted metabolite concentrations to those from a wet lab experiment, the organism should be cultured over many generations to ensure adaptation to the laboratory conditions.

One could calculate experimentally based in vivo rate constants for many reactions, as well, following the same procedure outlined above but using experimentally measured concentrations of metabolites free (unbound) in the cytoplasm [18]. But this population is challenging to measure [47] or even estimate as a trend [25]. If such metabolite measurements were available, the hypothesis of maximum entropy production for steady states of evolutionary optimized systems is testable [48].

Park et al. have recently argued that one can calculate free energies of reaction from the respective whole cell concentrations because most cellular metabolites are in free form [49]. The argument is that the total measured metabolite concentration is approximately 300 mM, whereas the protein concentration is approximately 7 mM, and that this suggests that most metabolites are free in solution. However, this argument can be turned around to argue the opposite as well. For example, if three metabolites have concentrations of 100 mM and other species are 1 nM–1 µM, then the pooled concentration of all metabolites may be ~300 mM, depending on the number of other metabolic species. Further suppose that the total pooled protein concentration was 7 mM. In this case, only the three former species may have a significant population free in the cytoplasm. In order to determine the concentration of each free or unbound metabolite experimentally, we need either new experimental isolation technologies or to use predictions such as those discussed herein, but on the genome scale, as prior information in the data analysis. Measured whole-cell population of metabolites can be used to set bounds on reaction-free energies, however.

Using measured whole-cell concentrations of metabolites as the upper bound for metabolites free in solution can also be employed to infer regulation rather than using a rule-of-thumb (e.g., that metabolite concentrations should not exceed 1 mM), as was done here. The experimentally determined concentrations would give more precise estimates of regulation, although the rule-of-thumb values worked well in this study for demonstration purposes.

It is assumed that any predicted concentrations far away from the data are due to the incompleteness of the model and not the maximum entropy production principle. In the cases studied here, it is clear that the predicted high levels of some metabolites would reduce the ability of molecules to diffuse in the cell. In cases such as this, regulation may be inferred by comparison to experimental data or expectations. But it can also be expected that branching reactions for metabolites that are not included in a model may impact the predictions. Branching per se at the point where the metabolite is an intermediate does not necessarily impact metabolite concentrations because the metabolite level is a function of the inward and outward flux producing and consuming the metabolite. The total inward and outward flux may not change when there is a branching reaction if the boundary conditions have not changed; in this case, the output flux splits but the total outward flux may be conserved.

Likewise, changing the boundary conditions would change the overall flux through all pathways. Figure 5 shows the effect on flux through the pentose phosphate pathway as a function of the boundary conditions for NADP and NADPH. The flux through the pentose phosphase pathway changes by 55-fold (12.5/0.224). However, the fluxes in this case are modulated by regulation at the first step in the pathway, the reaction for glucose 6-phosphate dehydrogenase. The corresponding simulations without regulation would change the flux by much more. Small changes in boundary conditions can result in significantly greater flux through pathways because the boundary conditions in this case include internal metabolites as well as external nutrients/waste products. Because the internal metabolites such as NADP/NADPH, NAD/NADH and ATP/(ADP+P$_i$) may be used in many reactions, the impact on flux can be multiplicative.

The ability to completely characterize each individual reaction with regard to flux, energy dissipation, power and resistance can also lead to a more complete characterization of biological circuits, as is done for electrical circuits, and even allow for the use of sophisticated control theory analyses of the operations of a cell. Such developments could be extremely useful for synthetic biology, in that the addition to a cell of an engineered circuit often results in decreased growth and other unintended consequences. Being able to fully characterize the impact of the new circuit before implementing it in the laboratory could dramatically change the design process and success rate.

For example, in an effort to increase the production of fatty acids in *Yarrowia lipolytica*, Wasylenko et al. used 13C-metabolic flux analysis (MFA) in a control strain and an engineered strain to understand where NADPH was primarily being produced, which turned out to be the oxidative pentose phosphate pathway [39]. MFA is a valuable but relatively costly experimental analysis. In comparison, this study used predictive simulations and found that the oxidative branch can act in a cyclical manner to iteratively produce high levels of NADPH. Routinely carrying out modeling studies such as this is promising for bringing more predictive rational design into the development cycle.

Finally, there is a need for faster optimization methods to reach the steady state. While optimization using Equation (6) will find the optimal (least heat dissipation) steady state, the ODEs can become particularly stiff when regulation is not applied. In this case, it is possible for the range of concentrations to vary by many orders of magnitude, leading to very stiff ODEs and long time-to-solution. This may be important as these methods are used to model secondary metabolism as well as to scale up to a genome-scale model, which we plan to do for *Neurospora crassa*.

Scaling up to a genome-scale model will likely bring new challenges. The time scale of metabolism spans the range from milliseconds for individual reactions to on the order of an hour for secondary metabolites to reach steady state. The assumption in Equation (6) that c_α is reaction-independent allows one to find the maximum entropy production distribution at a non-equilibrium steady state. However, one concern is that this would imply that all reactions in the eventual simulation would occur on the same timescale. If all the processes are represented at the same level, for instance that of elementary reactions, then the inferred rates will likely be sufficiently representative of the true processes. However, if one mixes elementary reactions and summary reactions (that represent larger processes) in the model, then it could be the case that there is not a realistic separation of time scales. Summary reactions representing larger processes, for instance, could be rescaled by nc_α instead of c_α, where c_α represents the average transit time through a reaction at the faster scale and n is the number of elementary reactions represented by the summary reaction. Multi-scale simulations can be used in which fast and slow processes are run independently based on the assumption that the fast processes reach steady state quickly on the time scale of the slow processes. If the processes to be represented are scaled appropriately, for instance by nc_α, the processes do not need to be run independently but can be run together using computational singular perturbation methods [50] or other approaches that automatically separate the fast and slow degrees of freedom [51]. Further developments in the methods for inferring regulation, however, will likely be required. Including biomass formation in the form of replication will be challenging, as it will require a chemical potential for biomass to be estimated.

Supplementary Materials: The following are available online at http://www.mdpi.com/2227-9717/6/6/63/s1, Notebook_S1.pdf (Neurospora_no_regulation-supplementary), Notebook_S2.pdf (Neurospora_regulation-supplementary), Notebook_S3.pdf (Neurospora_no_ regulation_rate_constants-supplementary), Table S1. Metabolite Names; Table S2. Reaction Names. The computational notebooks and simulation files are also available at https://github.com/wrcannon/SupplementaryMaterial_Processes2018.

Author Contributions: W.R.C. conceived and designed the analyses and simulations; D.J.B. implemented the C code for the Boltzmann program; J.D.Z. and N.K. contributed to concepts, code, notebooks, graphics and Python code; J.M.H. and J.C.D. provided biological expertise and insight; W.R.C. wrote the paper.

Acknowledgments: This work was jointly funded by the National Institute of Biomedical Imaging and Bioengineering through award U01EB022546 and by the U.S. Department of Energy, Office of Biological and Environmental Research, through project 69513. PNNL is operated by Battelle for the US Department of Energy under Contract DE-AC06-76RLO.

Conflicts of Interest: The authors declare no conflict of interest.

References

1. McCammon, J.A.; Gelin, B.R.; Karplus, M. Dynamics of folded proteins. *Nature* **1977**, *267*, 585–590. [CrossRef] [PubMed]
2. Miao, Y.; McCammon, J.A. Mechanism of the g-protein mimetic nanobody binding to a muscarinic g-protein-coupled receptor. *Proc. Natl. Acad. Sci. USA* **2018**, *115*, 3036–3041. [CrossRef] [PubMed]
3. Warshel, A.; Levitt, M. Theoretical studies of enzymic reactions—Dielectric, electrostatic and steric stabilization of carbonium-ion in reaction of lysozyme. *J. Mol. Biol.* **1976**, *103*, 227–249. [CrossRef]
4. Jaynes, E.T. Macroscopic prediction. In *Complex Systems Operational Approaches in Neurobiology, Physics and Computers*; Haken, H., Ed.; Springer-Verlag: Berlin/Heidelberg, Germany, 1985; Volume 31, pp. 254–269.

5. Presse, S.; Ghosh, K.; Lee, J.; Dill, K.A. Principles of maximum entropy and maximum caliber in statistical physics. *Rev. Mod. Phys.* **2013**, *85*, 1115–1141. [CrossRef]

6. Cannon, W.R. Simulating metabolism with statistical thermodynamics. *PloS ONE* **2014**, *9*, e103582. [CrossRef] [PubMed]

7. Sivak, D.A.; Crooks, G.E. Thermodynamic metrics and optimal paths. *Phys. Rev. Lett.* **2012**, *108*, 190602. [CrossRef] [PubMed]

8. Dyke, J.; Kleidon, A. The maximum entropy production principle: Its theoretical foundations and applications to the earth system. *Entropy* **2010**, *12*, 613–630. [CrossRef]

9. Dewar, R. Information theory explanation of the fluctuation theorem, maximum entropy production and self-organized criticality in non-equilibrium stationary states. *J. Physica A-Math. Gen.* **2003**, *36*, 631–641. [CrossRef]

10. Dixit, P.D.; Wagoner, J.; Weistuch, C.; Presse, S.; Ghosh, K.; Dill, K.A. Perspective: Maximum caliber is a general variational principle for dynamical systems. *J. Chem. Phys.* **2018**, *148*. [CrossRef] [PubMed]

11. Lotka, A.J. Natural selection as a physical principle. *Proc. Natl. Acad. Sci. USA* **1922**, *8*, 151–154. [CrossRef] [PubMed]

12. Lotka, A.J. Contribution to the energetics of evolution. *Proc. Natl. Acad. Sci. USA* **1922**, *8*, 147–151. [CrossRef] [PubMed]

13. Vallino, J.J.; Algar, C.K. The thermodynamics of marine biogeochemical cycles: Lotka revisited. *Annu. Rev. Mar. Sci.* **2016**, *8*, 333–356. [CrossRef] [PubMed]

14. Schrödinger, E. *What is Life? The Physical Aspect of the Living Cell*; Cambridge University Press: Cambridge, UK, 1945; 91p.

15. Prigogine, I. Time, structure, and fluctuations. *Science* **1978**, *201*, 777–785. [CrossRef] [PubMed]

16. Dewar, R.C. Maximum entropy production as an inference algorithm that translates physical assumptions into macroscopic predictions: Don't shoot the messenger. *Entropy* **2009**, *11*, 931–944. [CrossRef]

17. Thomas, D.G.; Jaramillo-Riveri, S.; Baxter, D.J.; Cannon, W.R. Comparison of optimal thermodynamic models of the tricarboxylic acid cycle from heterotrophs, cyanobacteria, and green sulfur bacteria. *J. Phys. Chem. B* **2014**, *118*, 14745–14760. [CrossRef] [PubMed]

18. Cannon, W.R.; Baker, S.E. Non-steady state mass action dynamics without rate constants: Dynamics of coupled reactions using chemical potentials. *Phys. Biol.* **2017**, *14*, 055003. [CrossRef] [PubMed]

19. Tran, L.M.; Rizk, M.L.; Liao, J.C. Ensemble modeling of metabolic networks. *Biophys. J.* **2008**, *95*, 5606–5617. [CrossRef] [PubMed]

20. Marcelin, R. The mechanics of irreversible phenomenon. *Comptes Rendus Hebd. Acad. Sci.* **1910**, *151*, 1052–1055.

21. Hurley, J.M.; Jankowski, M.S.; Crowell, A.; Fordyce, S.; Zucker, J.D.; Kumar, N.; De Los Santos, H.; Purvine, S.; Robinson, E.; Shukla, A.; et al. Circadian proteomic analysis uncovers mechanisms of post-transcriptional regulation in metabolic pathways. 2018; submitted.

22. Gorban, A.N.; Karlin, I.V. Method of invariant manifold for chemical kinetics. *Chem. Eng. Sci.* **2003**, *58*, 4751–4768. [CrossRef]

23. De Donder, T.; van Rysselberghe, P. *Thermodynamic Theory of Affinity: A Book of Principles*; Stanford University Press: Palo Alto, CA, USA, 1936.

24. Seifert, U. Stochastic thermodynamics, fluctuation theorems and molecular machines. *Rep. Prog. Phys.* **2012**, *75*, 126001. [CrossRef] [PubMed]

25. Bennett, B.D.; Kimball, E.H.; Gao, M.; Osterhout, R.; Van Dien, S.J.; Rabinowitz, J.D. Absolute metabolite concentrations and implied enzyme active site occupancy in *Escherichia coli*. *Nat. Chem. Biol.* **2009**, *5*, 593–599. [CrossRef] [PubMed]

26. Jaynes, E.T. On the rationale of maximum-entropy methods. *Proc. IEEE* **1982**, *70*, 939–952. [CrossRef]

27. GitHub. Available online: https://github.com/PNNL-CompBio/Boltzmann (accessed on 20 April 2018).

28. Hosea, M.E.; Shampine, L.F. Analysis and implementation of TR-BDF2. *Appl. Numer. Math.* **1996**, *20*, 21–37. [CrossRef]

29. Fleming, R.M.; Thiele, I. Von Bertalanffy 1.0: A COBRA toolbox extension to thermodynamically constrain metabolic models. *Bioinformatics* **2011**, *27*, 142–143. [CrossRef] [PubMed]

30. Flamholz, A.; Noor, E.; Bar-Even, A.; Milo, R. eQuilibrator–the biochemical thermodynamics calculator. *Nucleic Acids Ress* **2012**, *40*, D770–D775. [CrossRef] [PubMed]

31. Johnson, C.H. Changes in intracellular pH are not correlated with the circadian rhythm of neurospora. *Plant Physiol.* **1983**, *72*, 129–133. [CrossRef] [PubMed]
32. Dreyfuss, J.M.; Zucker, J.D.; Hood, H.M.; Ocasio, L.R.; Sachs, M.S.; Galagan, J.E. Reconstruction and validation of a genome-scale metabolic model for the filamentous fungus neurospora crassa using farm. *PLoS Comput. Biol.* **2013**, *9*, e1003126. [CrossRef] [PubMed]
33. PNNL. Available online: https://cyc.pnnl.gov (accessed on 20 April 2018).
34. Schellenberger, J.; Park, J.O.; Conrad, T.M.; Palsson, B.O. Bigg: A biochemical genetic and genomic knowledgebase of large scale metabolic reconstructions. *BMC Bioinform.* **2010**, *11*, 213. [CrossRef] [PubMed]
35. King, Z.A.; Drager, A.; Ebrahim, A.; Sonnenschein, N.; Lewis, N.E.; Palsson, B.O. Escher: A web application for building, sharing, and embedding data-rich visualizations of biological pathways. *PLoS Comput. Biol.* **2015**, *11*, e1004321. [CrossRef] [PubMed]
36. Tsao, M.U.; Madley, T.I. Kinetic properties of phosphofructokinase of neurospora crassa. *Biochim. BioPhysica Acta* **1972**, *258*, 99–105. [CrossRef]
37. Wieland, O.H.; Hartmann, U.; Siess, E.A. Neurospora crassa pyruvate dehydrogenase: Interconversion by phosphorylation and dephosphorylation. *FEBS Lett.* **1972**, *27*, 240–244. [CrossRef]
38. Harding, R.W.; Caroline, D.F.; Wagner, R.P. The pyruvate dehydrogenase complex from the mitochondrial fraction of neurospora crassa. *Arch. Biochem. Biophys.* **1970**, *138*, 653–661. [CrossRef]
39. Wasylenko, T.M.; Ahn, W.S.; Stephanopoulos, G. The oxidative pentose phosphate pathway is the primary source of nadph for lipid overproduction from glucose in yarrowia lipolytica. *Metab. Eng.* **2015**, *30*, 27–39. [CrossRef] [PubMed]
40. Kerkhoven, E.J.; Pomraning, K.R.; Baker, S.E.; Nielsen, J. Regulation of amino-acid metabolism controls flux to lipid accumulation in yarrowia lipolytica. *NPJ Syst. Biol. Appl.* **2016**, *2*, 16005. [CrossRef] [PubMed]
41. Loira, N.; Dulermo, T.; Nicaud, J.M.; Sherman, D.J. A genome-scale metabolic model of the lipid-accumulating yeast yarrowia lipolytica. *BMC Syst. Biol.* **2012**, *6*, 35. [CrossRef] [PubMed]
42. Pan, P.; Hua, Q. Reconstruction and in silico analysis of metabolic network for an oleaginous yeast, yarrowia lipolytica. *PLoS ONE* **2012**, *7*, e51535. [CrossRef] [PubMed]
43. Stanton, R.C. Glucose-6-phosphate dehydrogenase, NADPH, and cell survival. *IUBMB Life* **2012**, *64*, 362–369. [CrossRef] [PubMed]
44. Karp, P.D.; Latendresse, M.; Paley, S.M.; Krummenacker, M.; Ong, Q.D.; Billington, R.; Kothari, A.; Weaver, D.; Lee, T.; Subhraveti, P.; et al. Pathway tools version 19.0 update: Software for pathway/genome informatics and systems biology. *Brief Bioinform.* **2016**, *17*, 877–890. [CrossRef] [PubMed]
45. Dobovisek, A.; Markovic, R.; Brumen, M.; Fajmut, A. The maximum entropy production and maximum shannon information entropy in enzyme kinetics. *Physica A* **2018**, *496*, 220–232. [CrossRef]
46. Dobovisek, A.; Vitas, M.; Brumen, M.; Fajmut, A. Energy conservation and maximal entropy production in enzyme reactions. *Biosystems* **2017**, *158*, 47–56. [CrossRef] [PubMed]
47. Lu, W.; Su, X.; Klein, M.S.; Lewis, I.A.; Fiehn, O.; Rabinowitz, J.D. Metabolite measurement: Pitfalls to avoid and practices to follow. *Annu. Rev. Biochem.* **2017**, *86*, 277–304. [CrossRef] [PubMed]
48. Unrean, P.; Srienc, F. Metabolic networks evolve towards states of maximum entropy production. *Metab. Eng.* **2011**, *13*, 666–673. [CrossRef] [PubMed]
49. Park, J.O.; Rubin, S.A.; Xu, Y.F.; Amador-Noguez, D.; Fan, J.; Shlomi, T.; Rabinowitz, J.D. Metabolite concentrations, fluxes and free energies imply efficient enzyme usage. *Nat. Chem. Biol.* **2016**, *12*, 482–489. [CrossRef] [PubMed]

50. Lam, S.H.; Goussis, D.A. Understanding complex chemical kinetics with computational singular perturbation. *Sympos. Combust.* **1989**, *22*, 931–941. [CrossRef]
51. Beretta, G.P.; Rivadossi, L.; Janbozorgi, M. Systematic Constraint Selection Strategy for Rate-Controlled Constrained-Equilibrium Modeling of Complex Nonequilibrium Chemical Kinetics: An Automatable and Thermodynamically Consistent, Quasi-Equilibrium Model of Far Nonequilibrium States of Complex Reacting Systems Based on Probing the Fully Detailed Model and Taking a Truncated Singular Value Decomposition of the Resulting Evolution of the Degrees of Disequilibrium. *J. Non-Equilib. Thermodyn.* **2018**, *43*, 121–130.

Article

Minimizing the Effect of Substantial Perturbations in Military Water Systems for Increased Resilience and Efficiency

Corey M. James [1,*], **Michael E. Webber** [2] **and Thomas F. Edgar** [3]

1   Department of Chemistry and Life Science, United States Military Academy, West Point, NY 10996, USA
2   Department of Mechanical Engineering, University of Texas at Austin, Austin, TX 78712, USA;
      webber@mail.utexas.edu
3   Department of Chemical Engineering, University of Texas at Austin, Austin, TX 78712, USA;
      tfedgar@austin.utexas.edu
*   Correspondence: corey.james@usma.edu; Tel.: +1-254-258-0786

Received: 30 August 2017; Accepted: 11 October 2017; Published: 18 October 2017

Abstract: A model predictive control (MPC) framework, exploiting both feedforward and feedback control loops, is employed to minimize large disturbances that occur in military water networks. Military installations' need for resilient and efficient water supplies is often challenged by large disturbances like fires, terrorist activity, troop training rotations, and large scale leaks. This work applies the effectiveness of MPC to provide predictive capability and compensate for vast geographical differences and varying phenomena time scales using computational software and actual system dimensions and parameters. The results show that large disturbances are rapidly minimized while maintaining chlorine concentration within legal limits at the point of demand and overall water usage is minimized. The control framework also ensures pumping is minimized during peak electricity hours, so costs are kept lower than simple proportional control. Thecontrol structure implemented in this work is able to support resiliency and increased efficiency on military bases by minimizing tank holdup, effectively countering large disturbances, and efficiently managing pumping.

Keywords: energy; water; military; control

---

1. Introduction

Large perturbations in municipal water systems are not unique to military bases, but the need to rapidly and accurately correct for disturbances is critical to base resiliency and safety. Resiliency on military bases is vulnerable to deficiencies in water systems created from a variety of disturbances: fires, terrorist activity, large leaks, or loss of chlorination at the inlet. Fires or large breaks in water lines caused by terrorist activity or other natural disasters, create an immediate and unpredicted demand that, without compensation, will quickly deplete water inventory. A large mechanized unit returning from training that requires intensive cleaning and maintenance could also put a similar stress on the water system. Figure 1 shows the physical location of perturbations used in this study at the inlet, or point where water is introduced to the system, and at the demand, which is modeled as a concentration of residential and industrial (military unit) consumers. Figure 2 shows plots of the relative disturbances. This study simulates a disturbance on demand, similar to the ones mentioned above, in the middle of the day when demand is already high. In the event of an extensive demand event, the flow of water required in the system would look more like the disturbance line in the upper plot of Figure 2. The line labeled "normal" depicts a standard day without excessive disturbances and a pattern of use that supports both residential and military operational needs. Typical system inlet chlorine concentration in this study ranges from 0.7–1.0 mg/L. The lower plot of Figure 2 shows a substantial disturbance on inlet concentration that, if not compensated for, would degrade water quality throughout the system and require extensive flushing that creates wasted water and energy. Such an event is a risk to the overall resiliency and security of base operations. Although these perturbations are only a fraction of the possibilities, they were chosen for this study to show the rigor of the proposed model predictive control (MPC) framework because they are large and occur when the system is under the most stress.

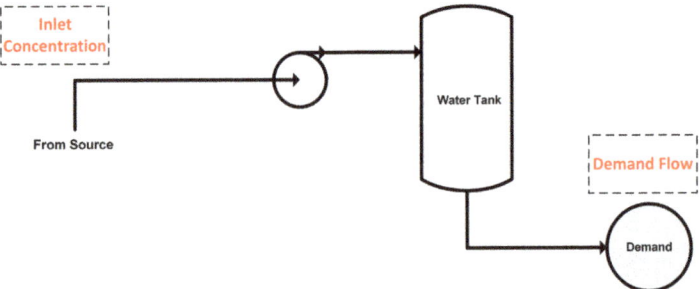

Figure 1. Locations of perturbations used in this study for demand and inlet concentration to demonstrate the effectiveness of model predictive control (MPC).

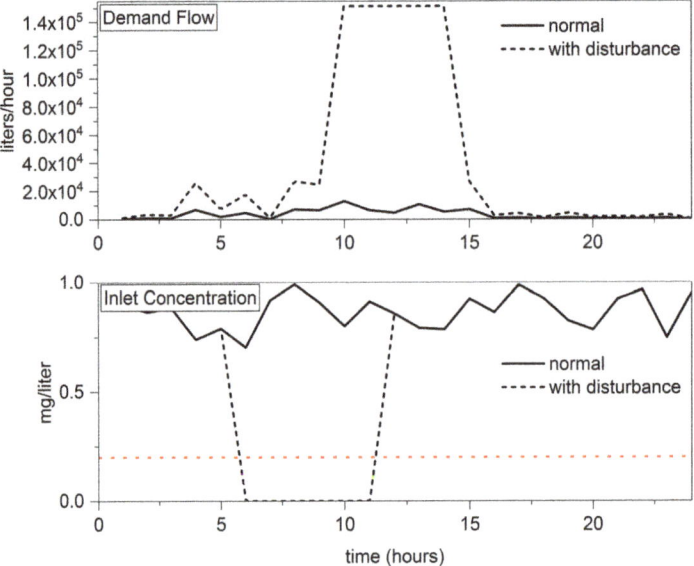

Figure 2. Simulated disturbances used in this study for demand and inlet concentration. The red dotted line at 0.2 mg/L signifies the lower limit of system chlorine concentration allowed to ensure a potable water supply.

1.1. Motivation and Scope

The cost of providing potable water is mostly sensitive to tank water level. Tank water level does drive cost in most, if not all, water systems because pumping water into the system is the only manipulated variable (MV) available. When chlorine concentration decreases below the minimum level because of residence time or a disturbance on inlet concentration, the only MV to manipulate in water systems to increase the controlled variable (CV), or chlorine concentration at demand, c, is pumping flow, q. Especially when disturbances are excessive, a possibility on military bases, manipulating q as the only MV to change c will inevitably lead to excess water introduced into the system. The implication of introducing water in excess of demand is that it will most likely be wasted, or purged, after the chlorine concentration descends below the minimum level allowed by law. An alternate framework that will specifically target the prospect of disturbance variables (DV) of inlet concentration and demand while meeting the goals of the optimization layer is introduced here. The objectives of this work are to: (1) implement local regulatory control loops for tank water level (feedback) and inlet concentration (feedforward), (2) develop a nonlinear multi-input multi-output (MIMO) nonlinear model predictive controller (NMPC) to regulate the CVs with adequate MVs, (3) compare the performance of the MIMO MPC controller to regulatory control alone, (4) integrate chlorine injection as a manipulated variable, and (5) demonstrate the effectiveness of feedforward control on large scale disturbance rejection while minimizing cost and tank water level.

This work proposes a MPC supervisory layer, depicted in Figure 3, to minimize the impact of large, unpredicted disturbances while maintaining the strict constraints and resiliency requirements of military water systems. MPC is a tool used effectively in various process applications throughout the manufacturing and chemical industries [1]. If a reasonably accurate model is available, MPC is a suitable choice to provide supervisory control over processes like water networks where multiple inputs and outputs exist and strict inequality constraints must be met. Previous work by the authors demonstrates accurate nonlinear modeling of military water systems based on the work of Rossman et al. [2,3]. MPC implemented in this work will solve an optimal control problem under constraints with a specified prediction and control horizon. The controller will calculate control moves over the entire control horizon, but only implement the first move before repeating at the next time interval. At each time interval after implementing the first control move, the controller is providing inputs to change the trajectory of the controlled variables to their desired set points [4]. This work's goal is to demonstrate the effective use of MPC on a military water system to reduce cost, water use, and ensure sufficient water inventory is maintained to establish resiliency in the system during times of stress on the system (large, unpredicted disturbances).

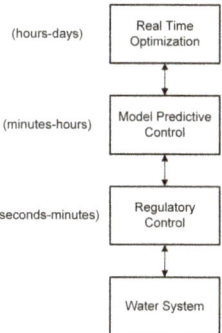

Figure 3. Hierarchy of process control in this work [4].

1.2. Literature Review

Led mainly by Mietek Brdys over the past two decades, a relatively small number of researchers have attempted to model and implement MPC on municipal water systems. MPC was successfully applied originally to minimize the cost of pumping [5–8]. Others took a more general approach using MPC, gaining good optimization and control of water quality and quantity in municipal systems as a whole [9–14]. An excellent feature article was published in 2002 providing an overview of feedback control as it applies to water quality [15]. A couple of key papers expand on the application of

MPC to water systems by making improvements to the controllers or reducing uncertainty in the algorithms [16–18].

Even fewer researchers have used feedforward compensation in efforts to control water networks. The literature shows that the only mention of feedforward control action implemented with model predictive control in water networks is by Sandison [19]. Sandison implemented feedforward compensation on single loop systems with good results, but it is unclear how well the framework would perform under the stress of large inlet disturbances.

A knowledge gap exists in the literature in three areas that will be addressed here: (1) NMPC for the reduction of system volume and storage holdup, (2) feedforward integration to minimize disturbances, and (3) applying NMPC to reduce cost, electricity consumption and increase system resiliency under the unique constraints of military water systems.

2. Research Methods

2.1. System Identification

There are two models employed in this work. The dynamic nonlinear process model used in the real time optimization block of Figure 3 is outlined as [2,3]:

$$\frac{\partial c_{t,k,i}}{\partial t} = \frac{-q_{t,k}}{A_k}\frac{\partial c_{t,k,i}}{\partial x} - \theta c_{t,k,i},$$ (1)

where $c_{t,k,i}$ = chlorine concentration in the pipe bulk flow at time t, in pipe section k, and pipe segment i; $q_{t,k}$ = volumetric flow rate; A_k = cross sectional area of specific pipe sections. The term on the left side of Equation (1) represents the change in chlorine concentration in pipe segments throughout the water distribution network with respect to time. The first term on the right describes the change in concentration axially along the length of each pipe segment, multiplied by the quotient of volumetric flow rate and cross sectional pipe area. It is this term that introduces nonlinear behavior to the model as the flow rate and concentration are changing simultaneously. The second term on the right describes the reaction kinetics multiplied by the chlorine concentration. *Theta* represents the reaction kinetics of chlorine degradation as a function of time and temperature, outlined by Rossman et al. [2]. Further details regarding this model can be found in previous work by James et al. [3]. A series of perturbations were then used with this model and outputs were measured to generate a simpler model for control that would be both efficient and accurate.

A multi-input, multi-output (MIMO) transfer function model was identified to accurately predict tank water level, L, and chlorine concentration at demand, c as a function of four inputs: (1) pumping action, (2) chlorine injection, (3) demand flow, and (4) inlet chlorine concentration and be employed for control purposes in this work. Step changes were made in each input in the aforementioned distributed model and the results of the system identification is depicted in Table 1.

Table 1. Individual step-response models for the identified water system in Figure 5 with four inputs and two outputs. The two blank plots represent no influence from inputs on CVs (CV: controlled variable, MV: manipulated variable, DV: disturbance variables).

	Tank Holdup, CV	Demand Chlorine Concentration, CV
Pump, MV	$G_{11} = \frac{0.5067}{s+0.00022}$	$G_{21} = \frac{4.4\times10^{-7}s+2.7\times10^{-7}}{s^2+0.61s+0.003}$
Chlorine Injection, MV		$G_{22} = \frac{-0.1568s+0.31}{s^2+0.12s+6.6\times10^{-10}}e^{-4s}$
Demand, DV	$G_{13} = \frac{-0.5341s+0.0005}{s^2+0.0009s+4.3\times10^{-9}}$	$G_{23} = \frac{-1.0\times10^{-5}s^2-1.4\times10^{-6}s+4.3\times10^{-6}}{s^3+15.5s^2+10.66s+0.01}$
Initial Concentration, DV		$G_{24} = \frac{-0.07s+0.51}{s^2+4.4s+0.96}$

The System Identification Toolbox™ in the MATLAB® software package (version R2017A, MathWorks, Natick, MA, USA) was used to identify transfer function models for each input/output combination. The relationship between each input and output was captured with at least a 74% fit, so the transfer function models were used in the development of a model predictive controller. The positive result of system identification is evident in Figure 4 where the distributed model and transfer function model describing tank level are very similar. A common practice in MPC development is to use step-response models instead of transfer functions, particularly in highly nonlinear cases [4]. This work did not yet explore the use of step-response modeling to describe the water system and, to the author's knowledge, this technique has not been used in water system modeling.

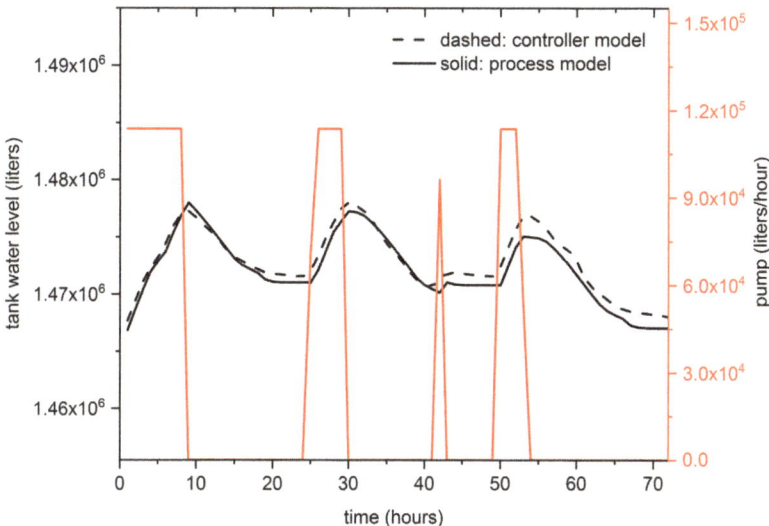

Figure 4. Comparison of tank water volume predicted by the distributed modeling (solid) and the transfer function models shown in Table 1 (dashed). The solid red line, referenced to the right *y*-axis, depicts scheduled and unscheduled pumping by the controller.

2.2. Nonlinear Disturbance Controller Development

If a disturbance can be detected before it enters the process and the process model is sufficiently accurate, feedforward MPC can often provide better disturbance rejection than feedback MPC alone [20]. Most feedforward systems use feedback trim as a means to compensate for errors in modeling and feedforward control discrepancies. However, self-regulating systems that do not require set point tracking, like chlorine concentration in water systems in the presence of large disturbances, can be controlled adequately with feedforward control alone [21].

This section outlines a NMPC based on a dynamic system model, illustrated by the process flowsheet in Figure 5 and the block diagrams in Figures 6–8 that utilize two control loops to reject large process disturbances: (1) a feedback loop that rejects large disturbances on demand, maintains tank holdup above the minimum resiliency constraint, and minimizes the amount of water inventory on hand and (2) a feedforward loop that rejects large disturbances on inlet concentration to ensure that downstream chlorine concentration remains at acceptable levels.

The "control calcuations" block in Figure 6 is made up of the feedforward and feedback loops in Figures 7 and 8. The optimization/scheduling layer provides set points to the control loops based on the model's prediction and inputs on the system such as electricity prices, predicted demand, and inlet concentration. Furthermore, the optimization/scheduling layer provides predictive pump scheduling for the system, which is shown as an input to the feedback control loop in Figure 8.

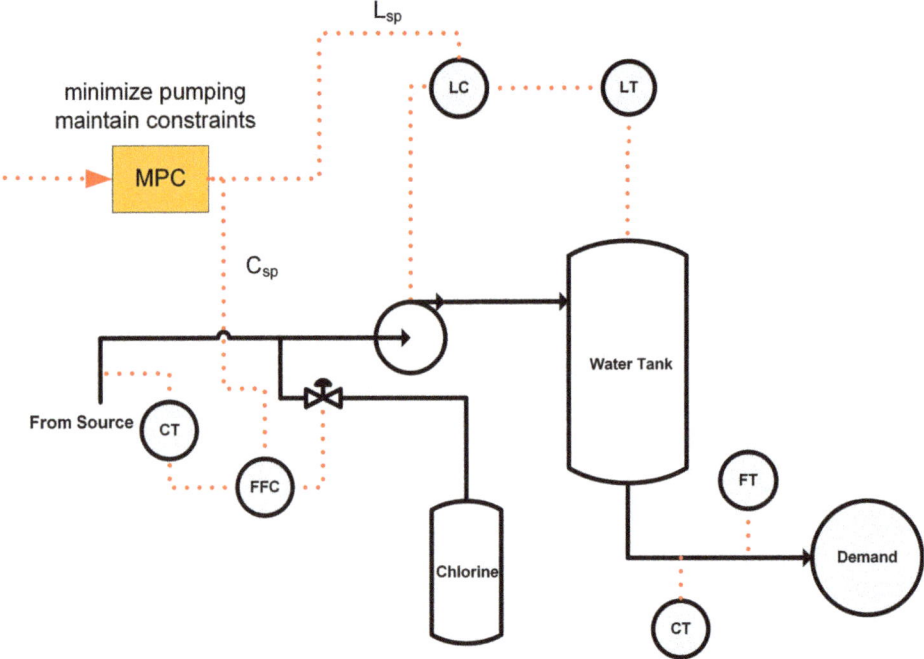

Figure 5. MPC framework with feedforward and feedback structure for disturbance compensation. *LC* and *FFC* are level and feedforward controllers, respectively. *LT*, *FT*, and *CT* are transmitters for level, flow, and concentration, respectively. L_{sp} is the tank level set point and C_{sp} is the inlet chlorine concentration set point.

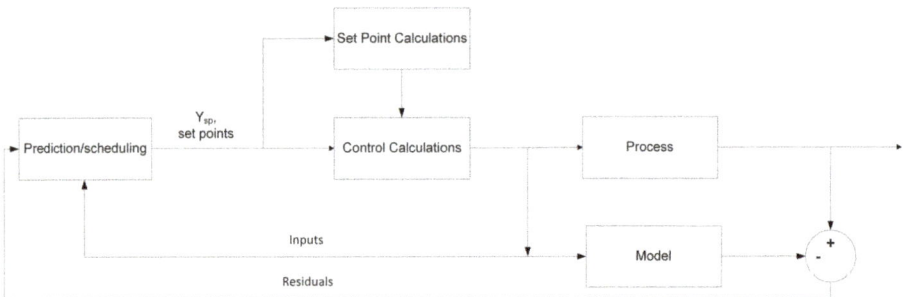

Figure 6. Block diagram for model predictive control with feedforward disturbance compensation, modified from Seborg et al. [4].

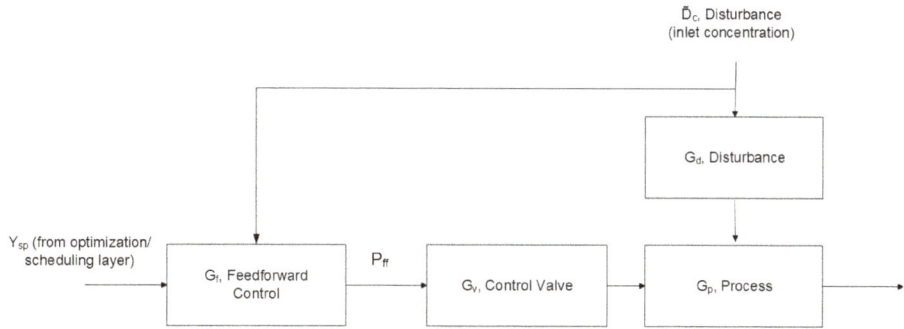

Figure 7. Block diagram for feedforward control. The control action is a portion of the "control calculations" block in Figure 6., modified from Seborg et al. [4].

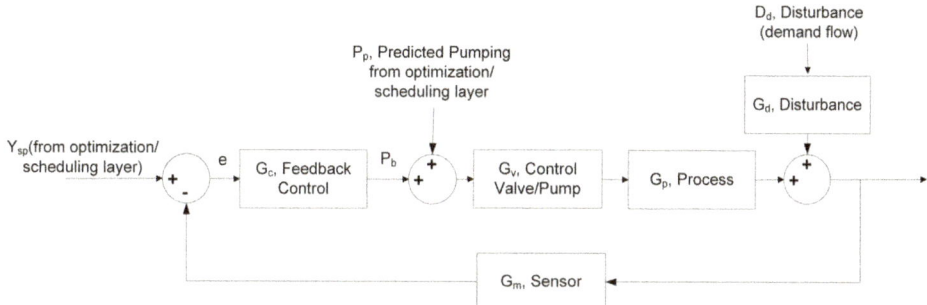

Figure 8. Block diagram for feedback control with included planned pumping from the optimization/scheduling layer. The control action is a portion of the "control calculations" block in Figure 6, modified from Seborg et al. [4].

The nonlinear nature of the model requires that the NMPC problem be formulated using a two-phase approach: estimation phase and control move calculation phase [20]. The estimation phase is stated here as a nonlinear sub-problem:

$$\min_{x_0, p_e, d_e} \sum_{l=0}^{p} \hat{E}_{k+1}^T(k+1) Q \hat{E}_{k+1}(k+1), \tag{2}$$

subject to:

$$
\begin{aligned}
x(t-P) &= x_0, \\
\frac{dx}{dt} &= f(x, u, p_e, d_e, t), \\
y &= g(x, u, p_e, d_e, t), \\
y_{k-P+l}^m &= \text{measured value of } y \text{ when } t = t_{k-P+l}, \\
\hat{E}_{k+1} &= y(t_{k-P+l}) - y_{k-P+l}^m \qquad\qquad \forall t, k,
\end{aligned}
\tag{3}
$$

where x represents the controlled states, namely tank water level and chlorine concentration at demand, p_e and d_e represent estimated system parameters and disturbances, respectively, and $y(t_{k-P+l})$ and

y^m_{k-P+l} represent predicted and measured output, respectively. The prediction horizon, P, is 24 h in this study due to the accessibility of day ahead electricity pricing.

Similar to a linear case, the control move calculation phase is used to calculate the current control action, u_k, plus additional control action and minimize the calculations over the control horizon, M. This work uses two control loops to calculate control moves: a feedback loop to control tank water holdup and a feedforward loop to control chlorine injection in an effort to minimize the effects of large disturbances on inlet chlorine concentration. The feedback loop utilizes a simple proportional control method to allow for averaging level control. Proportional control is also adequate in this case because offset from the set point is allowed:

$$P_b = \bar{p} + P_p + K_b e. \tag{4}$$

In Equation (4), $\bar{p}$ represents the steady state value, P_p is the scheduled pumping action passed from the optimization/scheduling layer, and K_b is the proportional gain for the feedback loop [4]. The error, e is defined as:

$$e = Y_{sp} - Y_m, \tag{5}$$

where Y_{sp} is either the tank level set point, L_{sp} or the inlet chlorine concentration set point, C_{sp} and Y_m is the measured corresponding output values. The feedforward portion of the control phase is treated as "perfect" feedforward control, where the control action is designed to keep the controlled variable exactly at the set point despite dynamic effects from the system [4]:

$$G_f = -\frac{G_d}{G_t G_v G_p}. \tag{6}$$

The dynamic effects of the transmitter and valve, represented by transfer functions G_t and G_v, respectively, are neglected in this study and then G_f is estimated as a lead–lag unit accounting for process and disturbance dynamics, G_p and G_d, respectively. A lead–lag unit is used to estimate the dynamics of the disturbances and process and their effect on the control action. Attempting to use the transfer functions outlined in Table 1 leads to a physically unrealizable controller:

$$G_f = -\frac{K_f(\tau_1 s + 1)}{(\tau_2 s + 1)}. \tag{7}$$

K_f, τ_1, and τ_2 are adjustable parameters in Equation (7). The adjustable parameters were tuned using the steps outlined in Seborg et al. [4]. Due to the dynamics in this system, offset is not a concern and K_f was adjusted until a reasonable control response was achieved. The optimal value used for K_f in this work was determined to be 0.35. τ_1 and τ_2 were set to zero while a trial and error approach was used to establish an appropriate value for K_f. The controlled variable responds faster to the manipulated variable in this system due to its location upstream of the disturbance variable, so the heuristic approach of $\tau_1/\tau_2 = 0.5$ was used to set an initial value for these two parameters. τ_1 and τ_2 were then slightly fine tuned to .01 and .025, respectively, as the disturbance value was adjusted to establish the controller so that it would minimize large disturbances effectively. The controller action, P_{ff}, is then defined as G_f multiplied by the disturbance in inlet chlorine concentration, D_c:

$$P_{ff} = -\frac{K_f(\tau_1 s + 1)}{(\tau_2 s + 1)} D_c. \tag{8}$$

The NMPC control law in Equations (2)–(8) is a multi-variable, proportional control law utilizing a receding horizon approach and a dynamic process model. It is based on predicted error generated by the optimization/scheduling layer shown in Figure 6. The controller tuning parameters are shown in Table 2. To ensure that the slowest dynamics in the system were adequately compensated for, this study used the settling time (t_s) of the demand chlorine concentration control response. It was determined to be 13 h. Heuristics outlined in Seborg et al. were then used to determine values for the control (M) and prediction (P) horizons in Equation (9) [4]. All outputs are weighted equally, so the diagonal elements (q_{ii}) of the output weighting matrix (Q) are assigned a value of unity:

$$\frac{\frac{t_s}{\Delta t}}{3} < M < \frac{\frac{t_s}{\Delta t}}{2},$$

$$P = \frac{t_s}{\Delta t} + M. \tag{9}$$

Table 2. MPC model parameters (MPC: model predictive control).

Parameter	Value	Units
M	5	h
P	24	h
q_{ii}	1	
Δu^c_{lb}	0	L/h
Δu^c_{ub}	3.79	L/h
Δu^v_{lb}	0	L/h
Δu^v_{ub}	113,562	L/h
K^c_c	−0.35	mg/L^2
K^v_c	15	h

3. Results

3.1. Averaging Level Control

The storage tanks in water systems are operated as surge tanks to not only damp out oscillations in the inlet stream, but to provide a constant and predictable pressure to customers. Where downstream flow rates change gradually, water levels can be maintained within specified upper and lower limits, and steady-state mass balances can be satisfied at all times, averaging level control is appropriate and is often employed successfully when conditions warrant [4,21].

Using averaging level control in this system contributes greatly to the systems' ability, guided by robust control and optimization, to minimize large disturbances. Figure 9 shows the results of averaging level control on the model system, giving the controller flexibility to hold inventory when predictions require it.

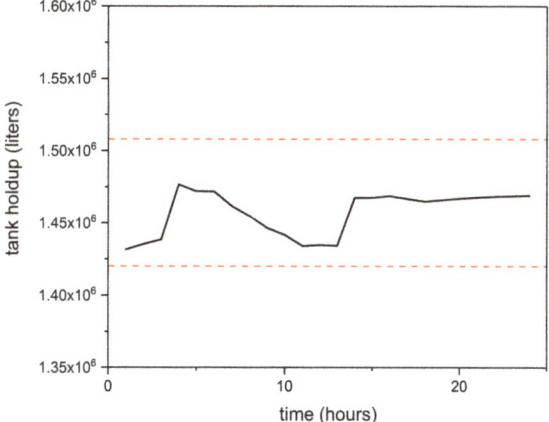

Figure 9. Average level control of tank holdup. Red dotted lines represent upper and lower constraints on tank holdup.

The NMPC controller described in the previous section was implemented using the General Algebraic Modeling System (GAMS) and the MATLAB Simulink environment. GAMS performed the optimization and scheduling of pumping action based on the system model. The remainder of this section is dedicated to presenting the results of this work.

3.2. Closed Loop Controller Response

Figure 10 shows the control action affecting tank water holdup under large disturbances. The controller provides an immediate and accurate response to a large demand load on the system beginning at hour eight and continuing to hour fifteen. Although there is approximately one hour of delay in the system's reaction to the demand, the effect of the large demand on the CV is minimal and the critical tank holdup is maintained. The input curve at the top of Figure 10 shows two large control actions: (1) scheduled pumping during non-peak hours (night) that was completed to avoid pumping during the peak hours of the day and (2) un-scheduled pumping during peak hours due to closed-loop control attempting to maintain the system within constraints during a large disturbance event. When compared to closed-loop proportional control under the same conditions, MPC is the clear choice to ensure the effect of disturbances is minimized. Figure 10 shows the proportional controller recovering the system to the set point, but the response is delayed for approximately eight hours and it allows the tank holdup level to decrease well below the lower constraint of 1.42×10^6 L. The military's need for aggressive adherence to the lower constraint on tank holdup for resiliency purposes means that regulatory control alone is insufficient.

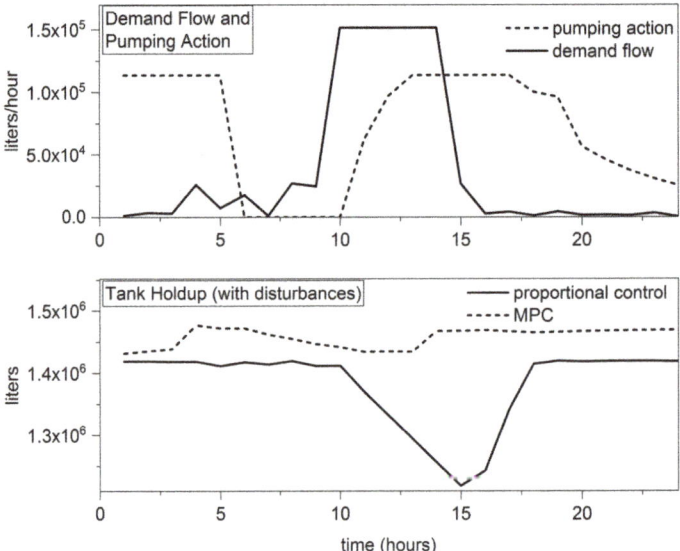

Figure 10. Controller action to minimize disturbance on tank holdup. In the bottom plot, MPC is compared to proportional control on tank holdup under the same conditions.

Large disturbances on inlet concentration pose a unique challenge to military water systems. Due to the size of water systems, the distance between the inlet and the CV of concern is usually extensive. Distance and chlorine reaction kinetics combine to create a time delay on the demand chlorine concentration. If a large disturbance on inlet concentration were to occur currently, the majority of the water system would be contaminated before the disturbance was detected. Feedforward control shown in Figure 11 demonstrates an effective solution to large disturbances. Because the disturbance is recognized immediately by the controller as well as the plant, control action begins to compensate immediately after the disturbance occurs. The CV continues to decrease for at least two hours after compensation due to the system time delay, but the CV remains well within constraints. Conversely, the CV decreases below the lower chlorine concentration constraint when model based control with feedforward action is not employed. Without feedforward control, the system time delay dominates, ensures that the concentration descends below the lower constraint, and spends at least twelve hours re-establishing the steady state.

While conserving water and electricity are motivations for this work, reducing costs to the military while maintaining resiliency receive priority when inefficiencies are concerned.

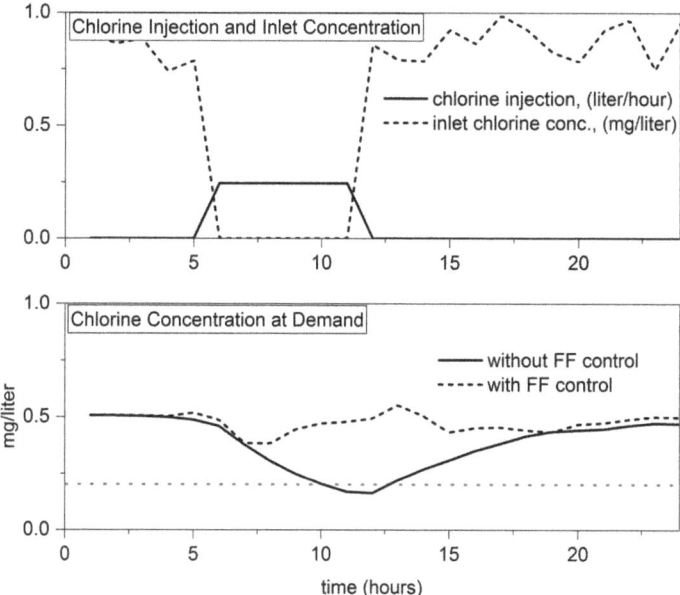

Figure 11. Feedforward control response to large disturbances on inlet chlorine concentration and the effect on demand chlorine concentration.

3.3. Robustness Analysis

Effective control systems are required to adequately counter process disturbances while also providing satisfactory performance for a wide range of process conditions and a reasonable degree of model inaccuracy [4]. The ability to control the system under a variety of conditions is known as robustness. Figures 10–13 shown earlier in this section demonstrate the performance of the control hierarchy in this work. Analysis was conducted on the control hierarchy to determine if it was effective beyond the scope of the figures earlier in this section. Figure 14 shows great control system response in terms of its ability to rapidly and smoothly compensate for a range of unpredicted demand events. The demand events are arrayed between a normal operation status with no large demand events and a disturbance to the system of 151,000 L. In all cases, resiliency within the system is shown by a rapid response and the ability to maintain the tank water level well above the lower safety threshold of 1,400,000 L.

Figure 15 demonstrates, with respect to the controlled variable of demand chlorine concentration, that the control hierarchy is capable of efficiently handling a wide range of disturbances. Under normal conditions in this model system, an inlet concentration that ranges from 0.7 mg/L to 1.0 mg/L will lead to a demand chlorine concentration of approximately 0.5 mg/L. If large disturbances on inlet chlorine concentration were allowed to persist without compensation, contamination and water waste will result. Figure 15 shows that the control system will inject chlorine to adequately compensate for substantial losses in inlet chlorine ranging from 0 mg/liter to 0.4 mg/L. Each disturbance is smoothly compensated for and the overall result is the entire model water system maintaining a chlorine concentration above the minimum safe level of 0.2 mg/L.

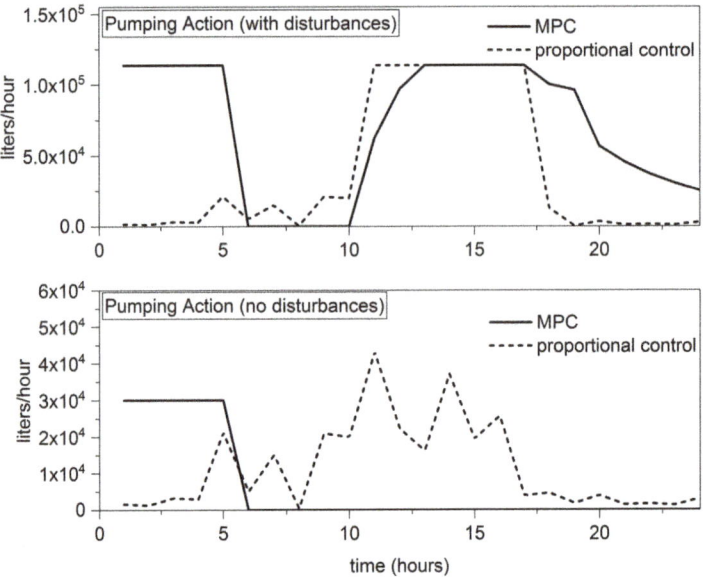

Figure 12. Comparison of pumping controlled by proportional control alone and model predictive control and how those strategies react to large disturbances.

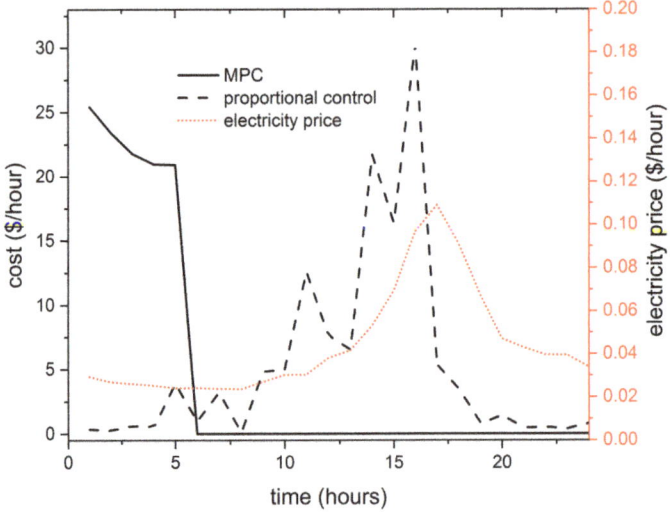

Figure 13. Cost of pumping over a 24 h period when proportional control and model predictive control are employed on the water system.

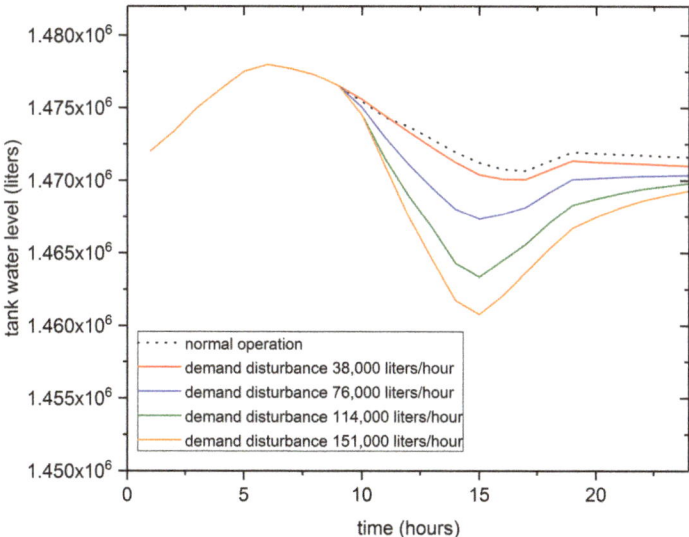

Figure 14. Controller response to a range of demand disturbances in order to maintain tank water level.

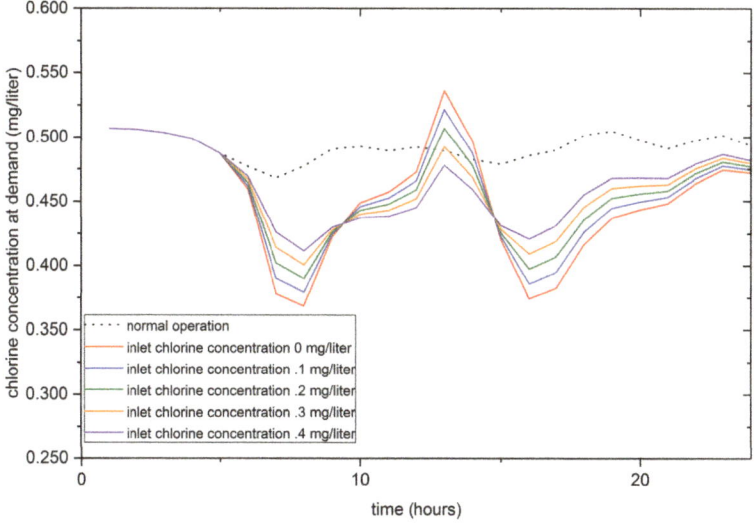

Figure 15. Controller response to a range of inlet chlorine concentration disturbances to ensure downstream demand chlorine concentration remains above the safe limit of 0.2 mg/L.

4. Discussion

Since the majority of the cost of providing potable water to installations is related to pumping water to fill tanks, it follows that any efforts to minimize cost should begin there. As shown in Figure 12, the MPC framework outlined earlier in this paper attempts to pump in a pattern that avoids the more expensive times of the day based on day ahead electricity pricing. Figure 12 shows the action of MPC with and without large disturbances present. Without disturbances, the controller will ensure that pumping is accomplished in the most inexpensive manner possible and will only pump during the unfavorable hours of the day if it needs to in order to maintain critical system constraints. Safety and resiliency are prioritized by the controller. Regulatory control alone will also effectively manage tank holdup in the absence of disturbances, but does not discriminate against higher prices in the afternoon, leading to excessive cost. Because of its predictive nature, MPC has the effect of having water inventory on hand when large disturbances occur that allow it to respond in a more efficient and inexpensive manner. Regulatory control alone with no predictive capability is at a disadvantage and does not perform as well because it did not store an additional amount of inventory during the more inexpensive times of the day. Figure 13 shows the cost comparison of regulatory control and MPC with no disturbances present. A 12% decrease in cost is realized when MPC is implemented. This decrease is realized because the controller manipulates the pumping action based on the cost of electricity and regulatory control does not.

5. Conclusions

In conclusion, large disturbances within military water systems can be adequately controlled using nonlinear model predictive control. Due to distance, time delays, time scales, and reaction kinetics, multiple types of control (feedback, feedforward, etc.) should be employed in the regulatory layer to compensate for disturbances effectively while maintaining constraints for safety and resiliency. Water system modeling is hindered by the existence of different time scales of phenomena within the system. The solution of this controller follows the process control hierarchical structure in Figure 3, employing the regulatory control layer to manage the fast dynamics of the system while the supervisory controller developed in this paper effectively managed the slow dynamics [22]. The NMPC framework outlined lowers costs and reduces waste, while improving resiliency and safety.

Acknowledgments: This work was funded by the United States Army as part of the Advanced Civil Schooling (ACS) program.

Author Contributions: Corey M. James, Michael E. Webber, and Thomas F. Edgar conceived the idea to place rigorous controls on military water systems, while Corey M. James designed and wrote the code to implement control on the system. Corey M. James, Michael E. Webber, and Thomas F. Edgar analyzed the data and Corey M. James authored the paper. Michael E. Webber and Thomas F. Edgar helped review the paper.

Conflicts of Interest: The authors declare no conflict of interest. The founding sponsors had no role in the design of the study; in the collection, analyses, or interpretation of data; in the writing of the manuscript, and in the decision to publish the results.

References

1. Qin, S.J.; Badgwell, T.A. A survey of industrial model predictive control technology. *Control Eng. Pract.* **2003**, *11*, 733–764.
2. Rossman, L.A.; Clark, R.M.; Grayman, W.M. Modeling chlorine residuals in drinking-water distribution systems. *J. Environ. Eng.* **1994**, *120*, 803–820.
3. James, C. Reducing the Cost of Operational Water on Military Bases Through Modeling, Optimization, and Control. Ph.D. Thesis, The University of Texas at Austin, Austin, TX, USA, 2017.
4. Seborg, D.E.; Mellichamp, D.A.; Edgar, T.F.; Doyle, F.J., III. *Process Dynamics and Control*; John Wiley & Sons: Hoboken, NJ, USA, 2016.
5. Ormsbee, L.; Reddy, L.; Chase, D. *Comparison of Three Nonlinear Control Algorithms for the Optimal Operation of Water Supply Pumping Systems*; Integrated Computer Applications in Water Supply (Vol. 1); John Wiley & Sons, Inc.: Hoboken, NJ, USA, 1994; pp. 259–271.
6. Ormsbee, L.E.; Lansey, K.E. Optimal control of water supply pumping systems. *J. Water Resour. Plan. Manag.* **1994**, *120*. doi:10.1061/(ASCE)0733-9496(1994)120:2(237).
7. Van Staden, A.J.; Zhang, J.; Xia, X. A model predictive control strategy for load shifting in a water pumping scheme with maximum demand charges. *Appl. Energy* **2011**, *88*, 4785–4794.
8. Yu, G.; Powell, R.; Sterling, M. Optimized pump scheduling in water distribution systems. *J. Optim. Theory Appl.* **1994**, *83*, 463–488.

9. Duzinkiewicz, K.; Brdys, M.; Chang, T. Hierarchical model predictive control of integrated quality and quantity in drinking water distribution systems. *Urban Water J.* **2005**, *2*, 125–137.

10. Brdys, M.A.; Chang, T.; Duzinkiewicz, K.; Chotkowski, W. Hierarchical control of integrated quality and quantity in water distribution systems. In Proceedings of the ASCE 2000 Joint Conference on Water Resources Engineering and Water Resources Planning and Management, Minneapolis, MN, USA, 30 July–2 August 2000.

11. Chang, T.; Brdys, M.; Duzinkiewicz, K. Decentralized robust model predictive control of chlorine residuals in drinking water distribution systems. In Proceedings of the World Water and Environmental Resources Congress, World Water and Environmental Resources Congress and Related Symposia, Philadelphi, PA, USA, 23–26 June 2004; pp. 23–26.

12. Wang, J.; Brdys, M.A. Optimal operation of water distribution systems under full range of operating scenarios. In Proceedings of the 6th UKACC International Control Conference, Glasgow, UK, 30 August–1 September 2006.

13. Drewa, M.; Brdys, M.; Ciminski, A. Model predictive control of integrated quantity and quality in drinking water distribution systems. In Proceedings of the 8th International IFAC Symposium on Dynamics and Control of Process Systems, Cancún, Mexico, 6–8 June 2007.

14. Brdys, M.; Chang, T.; Duzinkiewicz, K. Intelligent model predictive control of chlorine residuals in water distribution systems. In Proceedings of the ASCE Water Resource Engineering and Water Resources Planning and Management, Orlando, FL, USA, 20–24 May 2001; pp. 391–401.

15. Polycarpou, M.M.; Uber, J.G.; Wang, Z.; Shang, F.; Brdys, M. Feedback control of water quality. *IEEE Control Syst.* **2002**, *22*, 68–87.

16. Trawicki, D.; Duzinkiewicz, K.; Brdys, M. Optimising model predictive controller for hierarchical control of integrated quality and quantity in drinking water distribution systems. In Proceedings of the IFAC I International Conference on Technology, Automation and Control of Wastewater Systems-TiASWiK'02, Gdansk-Sobieszewo, Poland, 19–21 June 2002; pp. 19–21.

17. Chang, T.; Brdys, M.; Duzinkiewicz, K. Quantifying uncertainties for chlorine residual control in drinking water distribution systems. *Proc. ASCE World Water Environm. Resour.* **2003**, *64*, 28–37.

18. Wang, Y.; Puig, V.; Cembrano, G. Non-linear economic model predictive control of water distribution networks. *J. Process Control* **2017**, *56*, 23–34.

19. Sandison, J. Controlling chlorination in drinking water: Water and waste water. *IMIESA* **2006**, *31*, 32–37.

20. Brosilow, C.; Joseph, B. *Techniques of Model-Based Control*; Prentice Hall Professional: Upper Saddle River, NJ, USA, 2002.

21. Shinskey, F.G. *Process Control Systems: Application, Design and Tuning*; McGraw-Hill, Inc.: London, UK, 1990.

22. Touretzky, C.R.; Baldea, M. Nonlinear model predictive control of energy-integrated process systems. *Syst. Control Lett.* **2013**, *62*, 723–731.

Article

Optimal Multiscale Capacity Planning in Seawater Desalination Systems

Hassan Baaqeel [1,2] and Mahmoud M. El-Halwagi [1,*]

1   Chemical Engineering Department, Texas A&M University, College Station, TX 77843-3122, USA;
    hassanmb@tamu.edu
2   Chemical Engineering Department, King Fahad University of Petroleum & Minerals,
    Dhahran 31261, Saudi Arabia
*   Correspondence: el-halwagi@tamu.edu; Tel.: +1-979-845-3484

Received: 13 May 2018; Accepted: 25 May 2018; Published: 1 June 2018

Abstract: The increasing demands for water and the dwindling resources of fresh water create a critical need for continually enhancing desalination capacities. This poses a challenge in distressed desalination network, with incessant water demand growth as the conventional approach of undertaking large expansion projects can lead to low utilization and, hence, low capital productivity. In addition to the option of retrofitting existing desalination units or installing additional grassroots units, there is an opportunity to include emerging modular desalination technologies. This paper develops the optimization framework for the capacity planning in distressed desalination networks considering the integration of conventional plants and emerging modular technologies, such as membrane distillation (MD), as a viable option for capacity expansion. The developed framework addresses the multiscale nature of the synthesis problem, as unit-specific decision variables are subject to optimization, as well as the multiperiod capacity planning of the system. A superstructure representation and optimization formulation are introduced to simultaneously optimize the staging and sizing of desalination units, as well as design and operating variables in the desalination network over a planning horizon. Additionally, a special case for multiperiod capacity planning in multiple effect distillation (MED) desalination systems is presented. An optimization approach is proposed to solve the mixed-integer nonlinear programming (MINLP) optimization problem, starting with the construction of a project-window interval, pre-optimization screening, modeling of screened configurations, intra-process design variables optimization, and finally, multiperiod flowsheet synthesis. A case study is solved to illustrate the usefulness of the proposed approach.

Keywords: desalination; multi-effect distillation; membrane distillation; process integration; optimization; scheduling

---

1. Introduction

In arid regions of the world, thermal desalination technologies, such as multiple effect distillation (MED), are mainstream for producing desalinated water for both residential and industrial sectors. Desalination technologies, in general, and thermal desalination technologies, specifically, are generally characterized by their high capital intensity. For example, fixed cost charges in MED typically account for 40–50% of the unit cost of production, while it is 30% in RO systems. When examining the capacities of desalination projects in arid areas, such as the Gulf countries, one cannot help but notice the widespread use of large capacity desalination projects. Large desalination plants were justified in the past to cope with the booming population in the area. For example, the population growth rate in Saudi Arabia has increased incessantly from 3% in 1960 to over 6% in 1982 [1]. However, it has plateaued since then, at around 2%. Despite that, the trend of installing large desalination projects has continued in recent years. In 2014, Saudi Arabia built one of the world's largest desalination plants, with a design capacity of 226 million imperial gallons per day (MIGD), using multistage flash (MSF) and reverse osmosis (RO) [2]. The country is expected to spend $27 billion in the next 20 years towards desalination projects. Hence, planning for capacity expansion of desalination systems in such situations poses a great multiperiod multiscale optimization opportunity.

Large investments are typically justified by the economies of scale associated with large projects and, in some cases, additional technological and operational limitations. However, as many technological and operational limitations diminish with the maturity of desalination technologies, the design capacity of future desalination projects is primarily an economic optimization problem. The downside of large investments lies in the higher fixed operating cost associated with larger underutilized systems and lower capital productivity.

The optimization of the capacity planning problem has been widely studied from different vistas. In the field of operational research, it is studied under the problem formulation of "Time-Capacity Optimization". The essence of this field is to optimize the size of future investment, taking advantages of the economies of scale exhibited by larger investments and at the same time, minimizing cost associated with money value of time. In a temporal order, Manne [3] was among the first to develop an analytical solution for the case of constant linear demand growth with an infinite horizon in his book "Investment for Capacity Expansion". Scarato [4] and Shuhaibar [5] explored the time-capacity expansion problem using Manne's framework in urban water systems and MSF desalination systems, respectively. Both studies have contended that the cost function is flat near the optimum point. Other papers have examined the problem in other applications, such as in the planning of hydroelectric projects [6], waste treatment systems [7], and power systems [8,9]. The problem was reconstructed by Neebe and Rao [10] for discrete technology selection with fixed capacities. Several studies [11–14] in the field of PSE (e.g., process systems engineering) address the capacity planning optimization for both deterministic and stochastic problems, and at various applications and solution techniques. In water desalination network design, process synthesis techniques have been employed for the design of desalination units of a specific technology. Example research in the synthesis of reverse osmosis networks includes the work by El-Halwagi [15] and subsequent research contributions, e.g., [16,17]. Design and optimization techniques have been developed to assess several configurations of MSF systems for various criteria [18,19]. Druetta et al. [20] evaluated the detailed design of MED seawater desalination systems for the minimization of total cost, where mixed-integer nonlinear programming (MINLP) model is employed to determine the nominal optimal sizing of system's equipment. Gabriel et al. [21] used linearization techniques to achieve global solutions of the design of MED systems. Several research contributions have been made in the area of optimizing the synthesis of MD networks for various applications [22–25]. Other research has focused on the synthesis of hybrid desalination systems. Bamufleh et al. [26] developed the framework for synthesis of a MED–MD desalination system that is thermally coupled with industrial process. Al-Aboosi and El-Halwagi [27] developed an approach for the optimization of the design of RO–MED hybrid systems using a water-energy nexus approach. Huang et al. integrated multiple desalination technologies with combined heat and power in industrial and power plants [28]. Kermani et al. [29] provided a review of water-heat nexus with a meta-analysis of network features.

Notwithstanding previous research in the field, to the extent of the authors' knowledge, no optimization framework has been established for the multiscale optimization of capacity planning in water desalination systems taking into consideration mass and heat integration opportunities with emerging desalination technologies. This paper aims at developing an optimization formulation for the capacity expansion planning that systematically extracts the optimal process design over time from numerous alternatives while considering retrofit options of existing desalination units and heat and mass integration between desalination systems. It will also simultaneously optimize the intra-process design and operating variables. This work seeks to answer the following questions in the context of capacity planning of desalination systems:

- What is the optimum staging/sizing of the new desalination units that optimize the selected objective function? Which technologies should be selected for water demand satisfaction?
- What are the optimum design and operating variables (i.e., evaporator's area, top brine temperature, etc.) for the existing and newly installed desalination units over the planning horizon?
- How shall existing and new desalination units in each planning interval be integrated (i.e., mass and heat integration) for the optimization of the objective function?

2. Problem Statement

The multiscale capacity planning problem in desalination network may be stated as follows: Given is a water desalination system subject to capacity expansions within a planning horizon of N_t years. The horizon is discretized to annual counterparts, $INTERVAL = (t|t = 1, 2, \ldots, N_t)$, where $t = 1$ represents the initial time of the planning horizon, satisfied by the initial desalination system. Expansion projects commence at $t = 2$. In each interval, the total water desalination capacity of the system is denoted D_t, while the water demand at a given period is denoted d_t. Desalinated water price, Pr_t, may vary in each interval.

The set $CONFIG = (i|i = 1, 2, \ldots, N_i)$ of desalination configurations are considered to meet the water demand increase. Two subsets for each configuration exist. The set $INLET_i = (m|m = 1, 2, \ldots, N_{m_i})$ represents the inlet nodes for ith configuration. The set $OUTLET_i = (n|n = 1, 2, \ldots, N_{n_i})$ represents the outlet node for ith configuration. For example, the configuration of MED–RO desalination depicted in Figure 1a has two inlet nodes and four outlet nodes, while the one depicted in Figure 1b has one inlet node and three outlet nodes.

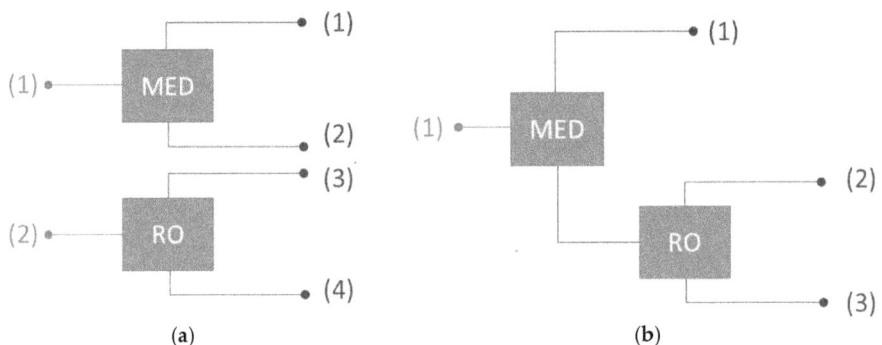

Figure 1. Example configurations for water desalination (**a**) with two inlet nodes and four outlet nodes, (**b**) with one inlet node and three outlet nodes.

The initial design of the system is fixed with a known distillate capacity of $D_t = D_1$. Due to distillate demand growth in the horizon, expansion of the desalination system is required to meet the planning horizon water demand. Saline water feed, F_t^{sw}, with a fixed salinity, x^{sw}, is available as a feedstock to new desalination units. On the overall system's level, a constraint exists on the total brine reject flowrate B_{max}^{reject} from the system, while the salinity of the system's brine and distillate are constrained by x_{max}^b and x_{max}^d, respectively.

In the context of this study, the objective is to maximize the net present value (NPV) of capital investment portfolio in the system accounting for annual revenue, fixed and operating cost, and book values. However, the formulation may be adjusted to target other objectives such as other economic, environmental, and reliability objectives. At a given minimum rate on investment (r), the objective is to determine the optimal planning for the desalination system capacity that maximizes the total net present value (NPV) while fulfilling water-demand forecast.

Figure 1 is a schematic representation of the multiperiod capacity planning problem in a desalination system. The superstructure shows the multiperiod interactions between the potential configurations in each interval.

3. Synthesis Approach

The representation in Figure 2 typifies the synthesis approach for the system. The change in the system's distillate capacity is measured by the added capacity at each interval, ΔD_t. It is assumed that the inherited system design from a preceding interval is fixed, and changes in the system are limited to the selected configuration added at the current interval and its design variables. Nonetheless, all intervals' designs will be solved simultaneously. In each interval, all possible configurations are evaluated, each named a $SUBSYSTEM_{i,t}$. For example, the first configuration in the second interval holds the notation $SUBSYSTEM_{1,2}$.

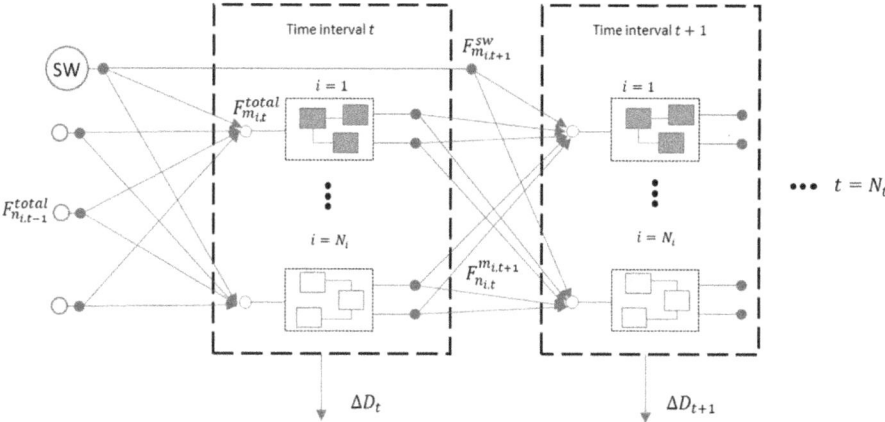

Figure 2. Multiperiod superstructure of the desalination capacity planning problem.

The multiperiod superstructure is rich enough to embed many potential designs of interest in the desalination system. For example, hybridization of desalination configurations can be done across intervals. However, the complexity of the superstructure can be prohibitive for a feasible mathematical optimization. For example, there is a total of 9,765,625 possible designs for a 10-interval horizon, and 5 five considered desalination configurations.

Our approach for a feasible optimization is shown in Figure 3. A pre-optimization screening of configurations is carried out on the system. The extensive list of desalination configurations is screened based on characteristic data of the system and knowledge on the commerciality, maturity, and economic efficacy of each configuration. In this step, unfeasible configurations, either economically or technically, based on parameters such as feed water salinity, distillate quality, range of distillate capacity, and minimum required recovery are, first, identified. The pre-optimization screening can be done in one of two ways:

- Complete elimination of configurations as a possible element of the optimal policy. This is applied on configurations with no hope of making it in the optimal flowsheet of the water system. For example, previous research and experience indicates the efficacy of RO in desalinating low- and medium-salinity water feed (i.e., brackish water) compared to MED. However, reliability and performance issues hinder its application for high-salinity water desalination. Hence, knowledge of the feed-water quality enables the elimination of some desalination technologies and configurations. Other factors for the screening of candidate configurations are listed in Figure 3 that include, but are not limited to, their ability to achieve product quality (e.g., boron separation), meet a system constraint (e.g., brine salinity), and achieve an acceptable level of commerciality.

- Disjunction of configuration's selection based on the problem's parameters. For example, the selection between simple MED and MD desalination configurations can be modelled by a disjunctive inequality based on the targeted design capacity, Equation (1). Assuming previous knowledge of the technical and economical feasible capacity range for each configuration, the disjunction can be reformulated using common disjunctive inequality solution techniques, such as convex hull or big-M reformulation.

$$\begin{bmatrix} y_{MD} \\ D_{MD}^{min} \leq D_{subsystem} \leq D_{MD}^{max} \\ \vdots \end{bmatrix} \vee \begin{bmatrix} y_{MED} \\ D_{MED}^{min} \leq D_{subsystem} \leq D_{MED}^{max} \\ \vdots \end{bmatrix} \tag{1}$$

Next, key intra-process design variables are screened. Candidate design variables for the application of Bellman's principle of optimality are locally optimized within the configuration. In the cases where the design variable's optimality depends on the design capacity of the potential desalination configuration, a profile of the design variable's optimal policy with the design capacity is developed. The outputs from the disjunction, configurations' modelling, and intra-process design variables optimization are entered into the overall system optimization model. In the next section, the general formulation of the problem is presented, followed by a discussion on the special case of optimizing multiple effect distillation (MED) desalination systems.

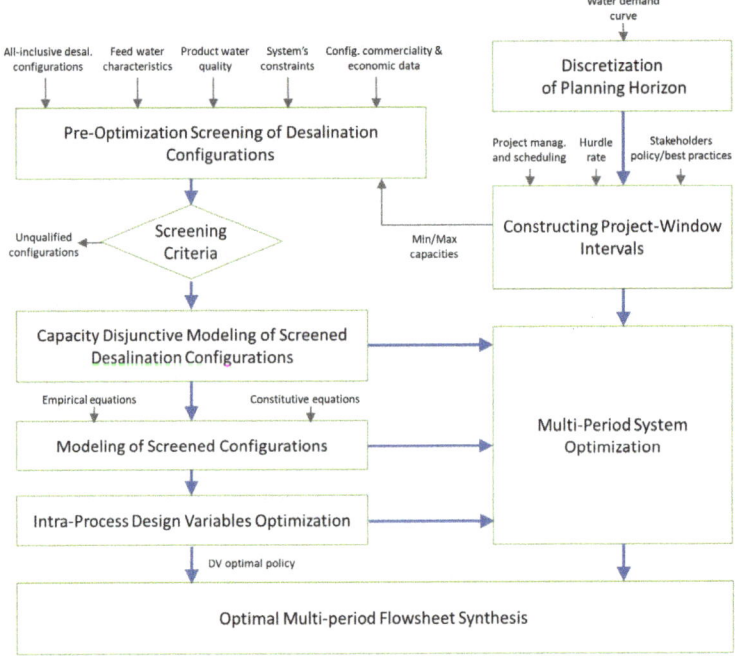

Figure 3. Optimization approach for the multiperiod capacity planning problem.

4. General Formulation

The objective function, presented later in the formulation, is subject to the following constraints:

4.1. System's Distillate Capacity

The capacity of the system shall meet or exceed water demand at any tth interval, as expressed in Equation (2), where D_t is the interval system total capacity, and d_t is the interval's water demand.

$$D_t \geq d_t \ \forall t \tag{2}$$

The total distillate capacity can change across the multiperiod horizon. The total system's capacity at a given interval is the summation of the total distillate capacity from the previous interval and the added distillate capacity, ΔD_t, at the interval, as given by:

$$D_t = D_{t-1} + \Delta D_t \ \forall t \tag{3}$$

The added capacity at any interval consists of the distillate capacity of the subsystems, $D_{i,t}$, installed in the interval.

$$\Delta D_t = \sum_i D_{i,t} \ \forall t \tag{4}$$

4.2. Subsystem's Mass Balance

The mass balance on each subsystem (i.e., configuration) is given by:

$$\sum_m F_{m_{i,t}}^{total} - \sum_n F_{n_{i,t}}^{total} - D_{i,t} = 0 \ \forall t \ \forall i, \tag{5}$$

where $F_{m_{i,t}}^{total}$ is the mass flowrate to the mth inlet node of a given subsystem. The mass flowrates in all inlet nodes constitute the total inlet feed to the subsystem. Conversely, $F_{n_{i,t}}^{total}$ is the mass flowrate for the mth outlet node of the subsystem. The mass flowrates in all outlet nodes from all subsystems constitute the interval's brine reject, as in Equation (6), where $B_{i,t}$ and B_t are the brine flowrate of a subsystem and the interval, respectively.

$$B_{i,t} = \sum_n F_{n_{i,t}}^{total} \ \forall t \ \forall i \tag{6}$$

$$B_t = \sum_i B_{i,t} \ \forall t \ \forall i \tag{7}$$

4.3. Subsystem's Inlet and Outlet Nodes

Given $F_{n_{i,t}}^{m_{i,t+1}}$ denotes the flow from nth node in $SUBSYSTEM_{i,t}$ to the mth inlet node in $SUBSYSTEM_{i,t+1}$, the split of nth outlet node is modelled as follows:

$$F_{n_{i,t}}^{total} = \sum_m F_{n_{i,t}}^{m_{i,t+1}} \ \forall t \ \forall i. \tag{8}$$

The mixing in mth inlet node in any subsystem is given by:

$$F_{m_{i,t}}^{total} = F_{m_{i,t}}^{sw} + \sum_n F_{n_{i,t-1}}^{m_{i,t}} \ \forall t \ \forall i, \tag{9}$$

where $F_{m_{i,t}}^{sw}$ is the fresh water (i.e., seawater) mass flowrate to the inlet node. Similar to Equation (3) for distillate capacity, the total seawater flowrate at any interval increases over intervals as in Equation (10).

$$F_t^{sw} = F_{t-1}^{sw} + \sum_m F_{m_{i,t}}^{sw} \ \forall t \ \forall i \tag{10}$$

4.4. Subsystem's Modeling Equations and Constraints

Each desalination configuration is described by a distinct vector of modelling equations and constraints that characterize the performance and limitations of the subsystems employing the configuration. In addition to the design capacity and compositions, a configuration is characterized by the vectors $DV_{i,t}$ for design variables, $OV_{i,t}$ for operational variables, and $SV_{i,t}$ for state variables.

$$\phi_i(D_{i,t}, x_{i,t}^d, x_{i,t}^b, DV_{i,t}, OV_{i,t}, SV_{i,t}) = 0 \ \forall t \ \forall i \tag{11}$$

$$\phi_i(D_{i,t}, x_{i,t}^d, x_{i,t}^b, DV_{i,t}, OV_{i,t}, SV_{i,t}) \geq 0 \ \forall t \ \forall i \tag{12}$$

Key constraints for desalination configurations include the design capacity as in Equation (13), and limits on some design variables (i.e., membrane area), as in Equation (14).

$$D_i^{min} \leq D_{i,t} \leq D_i^{max} \ \forall t \ \forall i \tag{13}$$

$$DV_i^{min} \leq DV_{i,t} \leq DV_i^{max} \ \forall t \ \forall i \tag{14}$$

In some cases, the limitation on the design variable extends across intervals, for example, the maximum RO modules in series, or a constraint on the maximum number of evaporative effects in series across all intervals. Such constraints may be captured by the following:

$$\sum_t DV_{i,t} \leq DV_i^{max} \ \forall t \ \forall i. \tag{15}$$

One key constraint for the system, to be met in all intervals, is the salinity constraint in both the brine and distillate. The following provide the component mass balances for the system's distillate and brine, respectively:

$$x_t^d \, D_t = x_{t-1}^d \, D_{t-1} + \sum_i x_i^d \Delta D_t \ \forall t \ \forall i, \tag{16}$$

$$x_t^b \, B_t = \sum_i x_i^b B_{i,t} \ \forall t \ \forall i. \tag{17}$$

Given a fixed maximum salinity on the brine, x_{max}^b and the distillate x_{max}^d, the respective constraints are given by

$$x_t^b \leq x_{max}^b \ \forall t, \tag{18}$$

$$x_t^d \leq x_{max}^d \ \forall t. \tag{19}$$

4.5. System's Costing and Objective Function

The total capital investment of each subsystem is correlated with the design variables and design capacity. Total operating cost correlates with the actual distillate production at the interval, as well as, all design and operating variables of the constituent subsystems.

$$TCI_{i,t} = f_i^c(D_{i,t}, DV_{i,t}) \ \forall i \ \forall t \tag{20}$$

$$TCI_t = \sum_i TCI_{i,t} \ \forall i \ \forall t \tag{21}$$

$$AOC_t = f_i^o(d_t, DV_{i,t}, OV_{i,t}) \ \forall i \ \forall t \tag{22}$$

The proposed objective function is the maximization of the net present value (NPV) as an economic metric of the desalination system as given by Equation (23). Thus, other economic metrics, such as internal rate of return (IRR), may easily be used instead. The terms V_t, AOC_t, and TCI_t represent the revenue, annual operating cost, and total capital investment at tth interval, respectively. All cash flows are properly discounted with the underlying assumption of the cash flow's realization at the beginning of the year. A linear depreciation model with no salvage value is assumed to estimate the system's book value, BV_t, at the end of the planning horizon. The service life, SL, is assumed constant for all units in the desalination system.

$$Maximize\ NPV = \sum_t \frac{REV_t - AOC_t - TCI_t}{(1+r)^{(t-2)}} - \frac{BV_t}{(1+r)^{(H-1)}}\ \forall t \tag{23}$$

$$BV_t = TCI_t \times argmax\left(0, 1 - \frac{H - (t-1)}{SL}\right)\ \forall t \tag{24}$$

$$REV_t = D_t \times Pr_t\ \forall t \tag{25}$$

To model the project-window intervals stipulated in synthesis approach (e.g., Figure 2), a new constraint is introduced on the allowable intervals for plant's installation. It is unlikely for capacity expansion projects to sequence in annual or biannual basis for economic and other considerations (i.e., safety, reliability, project management, etc.). Assuming a fixed period between project windows, τ, the constraint is enforced by assuming zero added desalination capacity for potential desalination plants in between project-permissible intervals, as given by

$$\Delta D_t = 0\ \forall t \in INTERVAL : t \neq \tau, 2\tau, \ldots, n\tau. \tag{26}$$

5. MED Special Case Formulation

Characterized by large design capacity and its capacity for integration with other desalination technologies, seawater multiple effect distillation (MED) desalination systems are good candidates for the presented optimization formulation. In this section, a shortcut method is proposed for the modelling and optimization of capacity expansion planning in MED desalination system. A set of three technologies are considered as modifications in the network to meet water demand. The conventional option is installing a new grassroots MED unit. Alternatively, existing MED units may be retrofitted with additional evaporative effects for additional water recovery or integrated with MD for brine treatment.

Next, a step-by-step application of the synthesis approach in Figure 2 on MED desalination systems for capacity expansion is carried out to develop a shortcut method for the special case.

5.1. Desalination Configuration Screening

A strategy of screening unfeasible desalination technologies and technologies that do not integrate with the existing system is adopted. MED and MD are the two technologies considered, forming three distinct configurations: new standalone MED unit, new evaporative effects to existing MED units, and MD unit for brine treatment. MD desalination of fresh seawater is eliminated based on previous techno-economic analysis and research on MD [26].

5.2. Capacity Disjunctive Modelling

In this step, the search space for the optimal flowsheet is reduced by applying the predetermined knowledge on the optimality of each screened configuration. For the retrofit configuration (EE), a capacity range, D_{EE}^{low} and D_{EE}^{high} is determined in which retrofit is part of the optimality policy. The decision is based on three distinct features of this configuration: limited distillate production, higher energy efficiency, and modest capital investment. The disjunction is expressed mathematically as follows:

$$I_{EE,t}\ D_{EE}^{low} \leq D_t \leq I_{EE,t}\ D_{EE}^{high}\ \forall t, \tag{27}$$

$$I_{EE,t}\ D_{EE}^{high} \leq D_t \leq I_{EE,t}\ D_{MED}^{max}\ \forall t, \tag{28}$$

where $I_{i,t}$ is a binary variable for each desalination subsystem. In the case where one configuration is allowed in each interval, the sum of the binary variables at any given interval must not exceed one.

$$\sum_i I_{i,t} \leq 1 \; \forall t \; \forall i \tag{29}$$

5.3. Configuration's Modeling

Various models were evaluated for use in the special formulation for the MED configuration [21,30–32]. A modified version of the MED model presented by El-Halwagi [32] is used here. The modification intends to make the model suitable for capacity expansion optimization applications, in which retrofitting the system with additional effects is considered. All the above mathematical models target a grassroots design. Hence, the implication of adding additional effects on an existing MED unit on water production, steam consumption, and capital and operating cost are not easily inferred.

For a given MED system with a variable number of effects, N_{eff}, the total MED distillate production is given by

$$D_{MED} = \sum_{n=1}^{N_{eff}} D_n, \tag{30}$$

where n is the evaporative effect number. D is the distillate water mass flowrate. The heat load of each evaporator, $Q_{evp,n}$ is estimated by the heat of vaporization, λ_n, at the temperature of the evaporator:

$$Q_{evp,n} = \lambda_n D_n. \tag{31}$$

Several types of evaporators may be used, including falling film, rising film, and forced circulation. Assuming a horizontal-tube falling film evaporator (HTFFE), the evaporator's design (i.e., area) is given by

$$Q_{evp,n} = U_{HTFFE,n} \, A_{HTFFE,n} \Delta T_{LM,n}. \tag{32}$$

For a conceptual design of a water system, like the one treated in this paper, the following simplifying assumptions are deemed acceptable, reducing the MED distillate capacity equation to Equation (35):

- The log-mean temperature difference may be assumed equal to the temperature difference between the vapor temperature in the tubes and the evaporator's temperature, Equation (33).
- The temperature difference between all effects are equal. Therefore, the MED temperature profile is estimated by Equation (34).
- All evaporators are identical in size.

$$\Delta T_{LM,n} = T_{n-1} - T_n \tag{33}$$

$$\Delta T_{LM,n} = \frac{(T_s - T_c)}{N_{eff} + 1} \tag{34}$$

$$D_{MED} = \frac{A_{HTFFE} \, (T_s - T_c)}{\left(N_{eff} + 1\right)} \sum_n^{N_{eff}} \frac{U_{HTFFE,n}}{\lambda_n} \tag{35}$$

Several correlations exist for U_{HTFFE} and λ with temperature, examples of which are presented in Equations (36) and (37). Assuming a linear correlation of both parameters with temperature, the summation term of the U/λ ratio may be correlated to three design variables: T_s, T_c, and Neff. Numerical analysis of the term shows a linear correlation of the term with N_{eff} at a fixed T_s and T_c values. Therefore, Equation (35) can be rewritten as

$$U_{HTFFE,n} = 0.8552 + 0.0047 \, T_n, \tag{36}$$

$$\lambda_n = -2.7532\,T_n + 3278.8, \tag{37}$$

$$D_{MED} = \frac{A_{HTFFE}(T_s - T_c)}{(N_{eff} + 1)}\left(\alpha_{MED}\,N_{eff}\right). \tag{38}$$

α_{MED} is a design parameter, estimated from the steam and cooling water temperature available at the facility. It is linearly estimated by Equation (39), where a_s and a_c are two scalar values.

$$\alpha_{MED} = \alpha_s T_s + \alpha_c T_c \tag{39}$$

Hence, Equation (35) may be rewritten as follows:

$$D_{MED} = \alpha_{MED}\,(T_s - T_c)\left(\frac{N_{eff}}{N_{eff} + 1}\right)A_{HTTFE}. \tag{40}$$

The new equation correlates, conveniently, the unit's total water production with the area and number of evaporative effects. Gained output ratio, GOR, is a useful estimate of the unit's thermal efficiency. It is defined by Equation (40) and empirically estimated by Equation (42).

$$M^s = \frac{D}{GOR} \tag{41}$$

$$GOR = N_{eff}(0.98)^{N_{eff}}{}^a \tag{42}$$

Combining Equations (39)–(41), the total steam consumption is given by

$$M^s = \frac{\alpha_{MED}(T_s - T_c)A_{HTTFE}}{(N_{eff} + 1)(0.98)^{N_{eff}}}. \tag{43}$$

It is reckoned that Equations (40) and (43) are very useful in modeling the second configuration, the evaporative effects retrofit (EE). Figure 4 shows the distillate capacity and steam consumption incremental change with the increase or decrease of an evaporative effect, based on the above model.

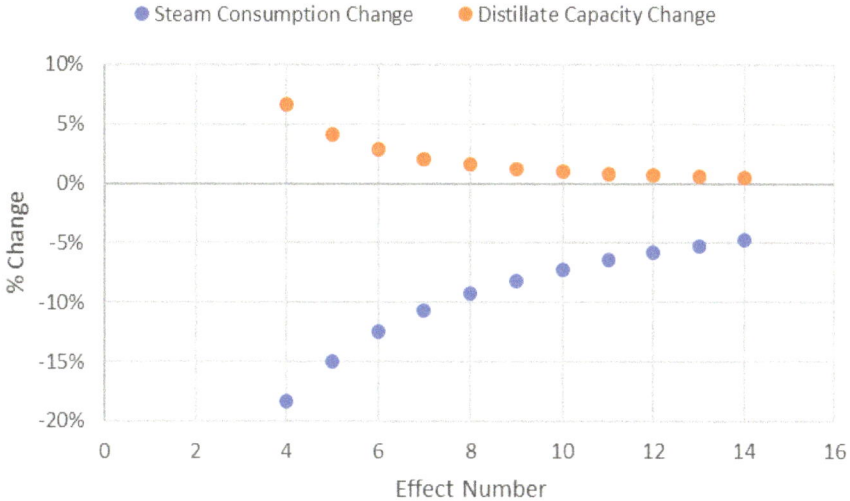

Figure 4. Net change in distillate production and steam consumption with e increase in number of evaporation effects.

5.4. Intra-Process Design Variables Optimization

In the special case, three design variables within the considered configurations are investigated: top brine temperature (TBT) in MED, number of effects in MED, and bulk feed temperature (TBF) in MD. The optimization TBT and TBF variables are associated with a tradeoff between operating and capital cost in their respective units, and have no association with the inter-process design of the system (i.e., subsystem capacity). Therefore, they are locally optimized, and inferred as a constituent of the global optimization solution.

The optimization of MED number of effects must be solved simultaneously with the multiperiod planning optimization. Alternatively, an optimal policy of the effect's number and the subsystem capacity is generated, and fed to the system multiperiod optimization.

6. Case Study

6.1. Case Study Description

The case study considers the capacity planning of an industrial water desalination system with five identical MED units, each consisting of 6 evaporative effects capable of producing 300 kg/s of distilled water. All the units are currently fully exhausted by the water demand. Due to a planned expansion in the industrial facility, water demand in a horizon of 30 years is expected to drastically increase in the next 20 years, followed by slim increases in the remaining 10 years. The demand curve is shown in Figure 5.

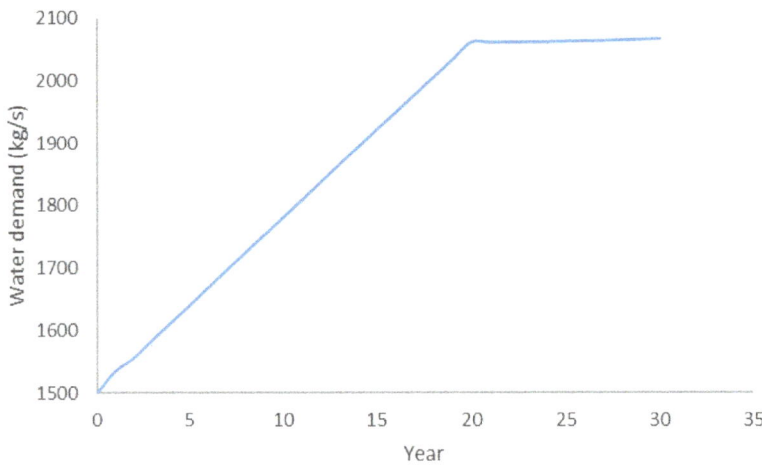

Figure 5. Water demand curve for the case study.

Seawater feed and brine parameters, as well as cost parameters, are shown in Table 2. The price of water product is fixed throughout the planning period at $1.5 per m^3. Project windows are assigned every 5 years (i.e., $\tau = 5$). Additional parameters are listed in Table 1.

For MD, a polypropylene hollow-fiber membrane MD020CP2N manufactured by Microdyn is used. The hollow fibers have a length, inner diameter, and outer diameter of 0.45 m, 1.5 mm, and 2.8 mm. Remaining details on the membrane can be found in [33]. MD design and costing model was adopted from Elsayed et al. [34]. For this case study, membrane permeability, B_w, is assumed constant at $1.92 \times 10^{-7} \frac{kg}{m^2 \cdot s \cdot Pa}$. Annualized fixed cost and annual operating cost for MD system is given by [33]

$$TCI_{MD} = 459\, A_m + 13,117(1 + \gamma)F_{MD} \qquad (44)$$

$$AOC_{MD} = c_{HU}\left(\frac{J_w H_{vw}}{\eta_m}\right) + (1411 + 43(1 - \xi) + 1613(1 + \gamma))F_{MD}, \tag{45}$$

where c_{HU} is cost of heating utility, ξ is water recovery, γ is ratio of MD recycle flowrate to feed flowrate. Design and costing equations for MED are adopted from El-Halwagi [32]. Assuming a Lang-factor of 3.5, the total capital cost is given by

$$TCI_{MED} = 24,600 \, N_{eff} A_{HTFFE}^{0.6}. \tag{46}$$

It is desirable to synthesize the expansion of the system for the planning horizon of 30 years, considering the three desalination configurations listed for the MED special case. The objective is to develop an optimal investment strategy to maximize net present value of the system. A minimum rate on investment for stakeholders is 15% (e.g., hurdle rate). In scenario #2, an environmentally-driven limitation of 3600 $\frac{m^3}{hr}$ on the seawater mass flowrate is considered.

Table 1. Design basis for the case study.

Parameter	Symbol	Value	Unit
Seawater feed temperature	T_{sw}	298	K
Seawater salinity	x^{sw}	30,000	ppm
MED brine maximum salinity	x_{MED}^b	75,000	ppm
Steam temperature	T_s	373	K
Cooling water temperature	T_c	298	K
Steam price	c_{HU}	2.5	$/MJ
MED minimum capacity	D_{MED}^{min}	50	kg/s
MED maximum capacity	D_{MED}^{max}	350	kg/s

6.2. Solution

The horizon was discretized to annual intervals, $N_t = 30$, with allowable expansion windows every 5 years. The optimization formulations are solved using the software LINGO® [35]. The problem is formulated as MINLP with 1337 variables, and solved on Intel Core i7-6700 CPU with 16 GB RAM in 274 s. A summary of the optimization results for all the scenarios are shown in Table 2.

Table 2. Summary of results for the case study.

	Units	Base Case	Scenario #1	Scenario #2
NPV	10^6 $/year	−7.1	10.5	6.9
New MED Units (MED)		1	3	3
Retrofitted effects (EE)		0	1	1
Total MD area (MD)	m^2	0	0	10,800
MED top brine temperature	K	358	358	358
MD feed temperature	K	363	363	363

As a benchmark for the case study results, the results of a base case scenario of installing one MED system capable of handling the full capacity required for the planning horizon is presented.

First, two intra-process design variables are optimized locally: the TBT and TBF temperature. Within specific design limits, both involve a tradeoff between capital and operating cost. Higher top brine temperature in MED, for example, yields higher thermal efficiency, (i.e., lower specific latent heat of vaporization) and lower specific evaporative area (i.e., higher heat transfer coefficient). On the other hand, higher unit cost of steam as well as reliability and operability issues (i.e., scaling) may be incurred. The optimum TBT and TBF temperatures for all the scenarios are 358 K and 363 K, respectively. Additionally, the optimal policy for MED number of effects vs MED capacity is developed and presented in Table 3.

Table 3. Number of effects optimal policy for various MED capacities.

Number of Effects	Min. Capacity Limit (kg/s)	Max. Capacity Limit (kg/s)
6	30	55
7	55	94
8	95	150
9	151	270

Scenario #1. The optimization formulation is solved without any constraints on the seawater mass flowrate or water recovery. The solution is shown by Figure 6. Three MED units are installed in interval 1, 10, and 15. In the period of sluggish demand increase (e.g., year 20–30), retrofit of the largest MED unit with two additional effects was included in the solution. This exploits the advantage of the retrofit option: modest increase in production with positive gain in thermal efficiency.

Scenario #2: MED units are limited in water recovery by the maximum salinity in brine. Hence, with the introduction of a constraint on seawater feed, MD became a constituent of the optimal flowsheet to satisfy the required water demand with the limited fresh feed through brine treatment. MD was introduced to the flowsheet at year 15 with a total area of 10,800 m^2 producing a total of 54 kg/s. The total MD capital investment is estimated to be \$9.87 millions. The solution is shown by Figure 7.

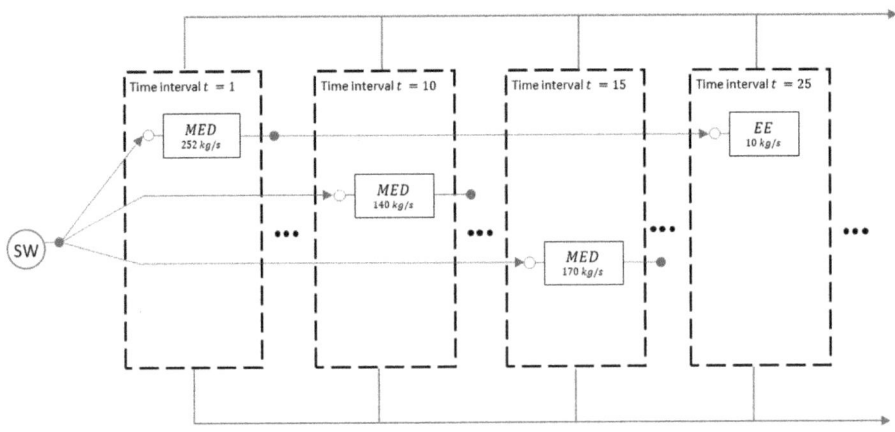

Figure 6. Results for Scenario #1.

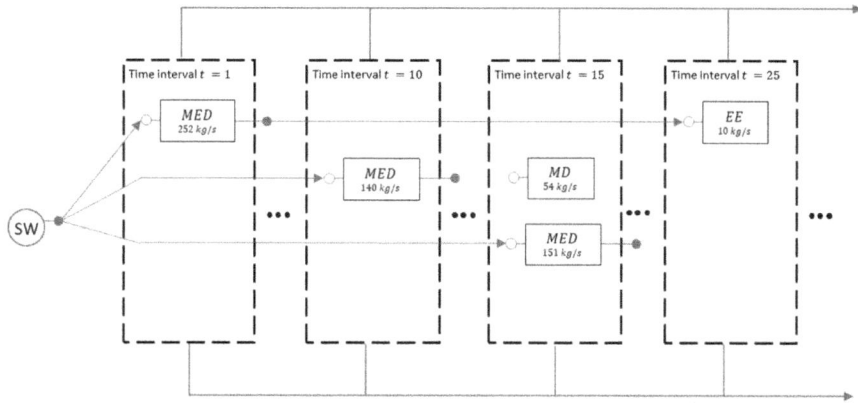

Figure 7. Results for Scenario #2.

7. Conclusions

This paper has introduced a multiperiod optimization approach for the capacity-expansion planning of water desalination systems to satisfy a forecasted demand growth over a given time horizon. The approach was illustrated in a general formulation, and in a special case for the capacity expansion in MED desalination systems, where three options were considered for capacity expansion: grassroots MED, existing MED retrofits, and MD desalination. The presented formulation simultaneously optimizes design capacities, period of installation, as well as technology-specific design variables, such as the number of MED effect, TBT, and MD feed temperature. A case study has been solved for three different scenarios to illustrate the merits of the presented approach. The results have illustrated the impact of considering alternative options for capacity expansion on the optimal design. Notwithstanding the economic challenges facing emerging technologies, alternative options to meeting demand growth, such as MD, provide advantages to the system by providing valuable flexibility and modularity in the design stage to maximize economic return.

Author Contributions: This paper is a joint collaboration between the two authors. Both authors contributed to the conception of the problem statement; solution approach; and optimization framework. H.B. carried out the computational aspects of the case study under the supervision of M.E.-H. Both authors contributed to writing and editing of the manuscript.

Conflicts of Interest: The authors declare no conflict of interest.

Nomenclature

A_{HTFFE}	area of MED evaporators, m^2
A_m	area of MD module, m^2
AOC_t	annual operating cost at tth interval
$B_{i,t}$	brine flowrate of a subsystem, kg/s
B^t	system's brine flowrate, kg/s
BV_t	end-of-horizon book value of investments at tth interval
c_{HU}	unit price of heating utility
D_t	system total distillate capacity at an interval, kg/s
$D_{i,t}$	subsystem's water capacity, kg/s
d_t	water demand at an interval, kg/s
ΔD_t	added distillate capacity at an interval, kg/s

$DV_{i,t}$	design variables of a subsystem
$F_{m_{i,t}}^{total}$	mass flowrate to the mth inlet node, kg/s
$F_{n_{i,t}}^{total}$	mass flowrate for the nth outlet node, kg/s
$F_{n_{i,t}}^{m_{i,t}}$	mass flowrate from nth outlet node to mth inlet node, kg/s
F^{sw}	seawater mass flowrate, kg/s
GOR	gained output ratio
H_{vw}	water heat of vaporization
$I_{i,t}$	binary variables for the existence of ith configuration at tth interval
J_w	water mass flux in MD, kg/(s m^2)
M^s	MED steam mass flowrate, kg/s
NPV	net present value
N_{eff}	number of evaporative effects
$OV_{i,t}$	operating variables of a subsystem
Q_{evp}	heat duty of MED evaporative effect, W
REV_t	revenue at tth interval
r	minimum rate of return
$SV_{i,t}$	state variables of a subsystem
SL	average subsystem service life
TCI_t	total capital cost of tth interval
T_s	temperature of heating steam in MED, K
T_c	temperature of cooling medium in MED, K
U_{HTFFE}	overall heat transfer coefficient of horizontal-tube falling film evaporators, W/(m^2 K)
$x_{i,t}^d$	distillate salinity from a subsystem
$x_{i,t}^b$	brine salinity from a subsystem
x_t^d	overall distillate salinity of the system
x_t^b	overall brine salinity of the system

Indices

i	index for desalination configuration
m	index for configuration's inlet node
n	index for configuration's outlet notes or number of effects
t	index for hourly time interval

Greek

η	thermal efficiency
α	MED temperature-dependent design parameter
γ	ratio of MD recycle flowrate to fresh feed
ξ	MD water recovery
λ_n	heat of evaporation at a given MED effect, W

References

1. *Population Growth (Annual %)*; The World Bank. Available online: https://data.worldbank.org/indicator/SP.POP.GROW?locations=SA (accessed on 1 November 2017).
2. *Ras Al Khair Desalination Plant*; Water-Technology. Available online: https://www.water-technology.net/projects/ras-al-khair-desalination-plant. (accessed on 1 May 2018).
3. Manne, A.S. *Investment for Capacity Expansion*; The MIT Press: Cambridge, MA, USA, 1967.
4. Scarato, R.F. Time-capacity expansion of urban water systems. *Water Resour. Res.* **1969**, *5*, 929–936. [CrossRef]
5. Shuhaibar, Y.K. *Staging of Investment in Desalination Facilities and Associated Storage Facilities*; The University of Arizona: Tucson, AZ, USA, 1972.
6. Hreinsson, E.B. Economies of scale and optimal selection of hydroelectric projects. In Proceedings of the International Conference on Electric Utility Deregulation and Restructuring and Power Technologies (DRPT 2000), London, UK, 4–7 April 2000; pp. 284–289.
7. Rachford, T.M.; Scarato, R.F.; Tchobanoglous, G. Time-capacity expansion of waste treatment systems. *J. Sanit. Eng. Div.* **1969**, *95*, 1063–1078.

8. Billinton, R.; Karki, R. Capacity expansion of small isolated power systems using PV and wind energy. *IEEE Trans. Power Syst.* **2001**, *16*, 892–897. [CrossRef]
9. Malcolm, S.A.; Zenios, S.A. Robust optimization for power systems capacity expansion under uncertainty. *J. Oper. Res. Soc.* **1994**, *45*, 1040–1049. [CrossRef]
10. Neebe, A.M.; Rao, M.R. Sequencing capacity expansion projects in continuous time. *Manag. Sci.* **1986**, *32*, 1467–1479. [CrossRef]
11. Sahinidis, N.V.; Grossmann, I.E.; Fornari, R.E.; Chathrathi, M. Optimization model for long range planning in the chemical industry. *Comput. Chem. Eng.* **1989**, *13*, 1049–1063. [CrossRef]
12. Maravelias, C.T.; Sung, C. A projection-based method for production planning of multiproduct facilities. *AIChE J.* **2009**, *55*, 2614–2630.
13. Iyer, R.; Grossmann, I.E. Synthesis and operational planning of utility systems for multiperiod operation. *Comput. Chem. Eng.* **1998**, *22*, 979–993. [CrossRef]
14. Mitra, S.; Pinto, J.M.; Grossmann, I.E. Optimal multi-scale capacity planning for power-intensive continuous processes under time-sensitive electricity prices and demand uncertainty. Part I: Modeling. *Comput. Chem. Eng.* **2014**, *65*, 89–101. [CrossRef]
15. El-Halwagi, M.M. Synthesis of reverse-osmosis networks for waste reduction. *AIChE J.* **1992**, *38*, 1185–1198. [CrossRef]
16. Khor, C.S.; Foo, D.C.Y.; El-Halwagi, M.M.; Tan, R.R.; Shah, N. A superstructure optimization approach for membrane separation-based water regeneration network synthesis with detailed nonlinear mechanistic reverse osmosis model. *Ind. Eng. Chem. Res.* **2011**, *50*, 13444–13456. [CrossRef]
17. Vince, F.; Marechal, F.; Aoustin, E.; Bréant, P. Multi-objective optimization of RO desalination plants. *Desalination* **2008**, *222*, 96–118. [CrossRef]
18. El-Dessouky, H.; Alatiqi, I.; Ettouney, H. Process synthesis: The multi-stage flash desalination system. *Desalination* **1998**, *115*, 155–179. [CrossRef]
19. El-Halwagi, M.M. A Shortcut Approach to the Design of Once-Through Multi-Stage Flash Desalination Systems. *Desalin. Water Treat.* **2017**, *62*, 43–56. [CrossRef]
20. Druetta, P.; Aguirre, P.; Mussati, S. Minimizing the total cost of multi effect evaporation systems for seawater desalination. *Desalination* **2014**, *344*, 431–445. [CrossRef]
21. Gabriel, K.J.; Linke, P.; El-Halwagi, M.M. Optimization of multi-effect distillation process using a linear enthalpy model. *Desalination* **2015**, *365*, 261–276. [CrossRef]
22. Elsayed, N.A.; Barrufet, M.A.; El-Halwagi, M.M. An Integrated Approach for Incorporating Thermal Membrane Distillation in Treating Water in Heavy Oil Recovery using SAGD. *J. Unconv. Oil Gas Resour.* **2015**, *12*, 6–14. [CrossRef]
23. Elsayed, N.A.; Barrufet, M.A.; Eljack, F.T.; El-Halwagi, M.M. Optimal Design of Thermal Membrane Distillation Systems for the Treatment of Shale Gas Flowback Water. *Int. J. Membr. Sci. Technol.* **2015**, *2*, 1–9.
24. González-Bravo, R.; Nápoles-Rivera, F.; Ponce-Ortega, J.M.; Nyapathi, M.; Elsayed, N.A.; El-Halwagi, M.M. Synthesis of Optimal Thermal Membrane Distillation Networks. *AIChE J.* **2015**, *61*, 448–463. [CrossRef]
25. González-Bravo, R.; Elsayed, N.A.; Ponce-Ortega, J.M.; Nápoles-Rivera, F.; Serna-González, M.; El-Halwagi, M.M. Optimal Design of Thermal Membrane Distillation Systems with Heat Integration with Process Plants. *Appl. Therm. Eng.* **2014**, *75*, 154–166. [CrossRef]
26. Bamufleh, H.; Abdelhady, F.; Baaqeel, H.M.; El-Halwagi, M.M. Optimization of multi-effect distillation with brine treatment via membrane distillation and process heat integration. *Desalination* **2017**, *408*, 110–118. [CrossRef]
27. Al-Aboosi, F.Y.; El-Halwagi, M.M. An Integrated Approach to Water-Energy Nexus in Shale Gas Production. *Processes* **2018**, *6*, 52. [CrossRef]
28. Huang, X.; Luo, X.; Chen, J.; Yang, Z.; Chen, Y.; Ponce-Ortega, J.M.; El-Halwagi, M.M. Synthesis and dual-objective optimization of industrial combined heat and power plants compromising the water–energy nexus. *Appl. Energy* **2018**, *224*, 448–468. [CrossRef]
29. Kermani, M.; Kantor, I.D.; Maréchal, F. Synthesis of heat-integrated water allocation networks: A meta-analysis of solution strategies and network features. *Energies* **2018**, *11*, 1158. [CrossRef]
30. El-Dessouky, H.; Alatiqi, I.; Bingulac, S.; Ettouney, H. Steady-state analysis of the multiple effect evaporation desalination process. *Chem. Eng. Technol.* **1998**, *21*, 437–451. [CrossRef]

31.   Mistry, K.H.; Antar, M.A.; Lienhard, V.J.H. An improved model for multiple effect distillation. *Desalin. Water Treat.* **2013**, *51*, 807–821. [CrossRef]

32.   El-Halwagi, M.M. Water-energy nexus for thermal desalination processes. In *Sustainable Design through Process Integration: Fundamentals and Applications to Industrial Pollution Prevention, Resource Conservation, and Profitability Enhancement*; Butterworth-Heinemann: Oxford, UK, 2017; Chapter 17, pp. 441–506.

33.   Al-Obaidani, S.; Curcio, E.; Macedonio, F.; di Profio, G.; Al-Hinai, H.; Drioli, E. Potential of membrane distillation in seawater desalination: Thermal efficiency, sensitivity study and cost estimation. *J. Membr. Sci.* **2008**, *323*, 85–98. [CrossRef]

34.   Elsayed, N.A.; Barrufet, M.A.; El-Halwagi, M.M. Integration of thermal membrane distillation networks with processing facilities. *Ind. Eng. Chem. Res.* **2013**, *53*, 5284–5298. [CrossRef]

35.   Schrage, L. *Optimization Modeling with LINGO*, 6th ed.; LINDO Systems: Chicago, UL, USA, 2006.

MDPI

St. Alban-Anlage 66

4052 Basel

Switzerland

Tel. +41 61 683 77 34

Fax +41 61 302 89 18

www.mdpi.com

Processes Editorial Office

E-mail: processes@mdpi.com

www.mdpi.com/journal/processes

Lightning Source UK Ltd.
Milton Keynes UK
UKHW052242020219
336570UK00003B/41/P

9 783038 975250